# Methods in Molecular Biology

*Series Editor*
John M. Walker
School of Life and Medical Sciences
University of Hertfordshire
Hatfield, Hertfordshire, UK

For further volumes:
http://www.springer.com/series/7651

For over 35 years, biological scientists have come to rely on the research protocols and methodologies in the critically acclaimed Methods in Molecular Biology series. The series was the first to introduce the step-by-step protocols approach that has become the standard in all biomedical protocol publishing. Each protocol is provided in readily-reproducible step-by step fashion, opening with an introductory overview, a list of the materials and reagents needed to complete the experiment, and followed by a detailed procedure that is supported with a helpful notes section offering tips and tricks of the trade as well as troubleshooting advice. These hallmark features were introduced by series editor Dr. John Walker and constitute the key ingredient in each and every volume of the Methods in Molecular Biology series. Tested and trusted, comprehensive and reliable, all protocols from the series are indexed in PubMed.

Manuscript Instructions https://drive.google.com/file/d/1K4Az5vpAFVzYadJ3jKx2-QSo7Dqa_SNA/view?usp=sharing

# The Cytoskeleton

## Methods and Protocols

Edited by

**Noa B. Martín-Cófreces**

*Immunology Service, Instituto de Investigación Sanitaria del Hospital Universitario La Princesa (IIS Princesa), Madrid, Spain*

**Francisco Sánchez-Madrid**

*Immunology Service, Hospital Universitario de la Princesa, Universidad Autónoma de Madrid,, Madrid, Spain*

**José María Valpuesta**

*Centro Nacional de Biotecnologia, Madrid, Madrid, Spain*

Humana Press

*Editors*
Noa B. Martín-Cófreces
Immunology Service
Instituto de Investigación Sanitaria del Hospital
Universitario La Princesa (IIS Princesa)
Madrid, Spain

Francisco Sánchez-Madrid
Immunology Service
Hospital Universitario de la Princesa
Universidad Autónoma de Madrid
Madrid, Spain

José María Valpuesta
Centro Nacional de Biotecnologia
Madrid, Madrid, Spain

ISSN 1064-3745 ISSN 1940-6029 (electronic)
Methods in Molecular Biology
ISBN 978-1-0716-5514-6 ISBN 978-1-0716-5515-3 (eBook)
https://doi.org/10.1007/978-1-0716-5515-3



This Humana imprint is published by the registered company Springer Science+Business Media, LLC, part of Springer Nature.
The registered company address is: 1 New York Plaza, New York, NY 10004, U.S.A.

# Preface

The cytoskeleton is a complex and dynamic network of filaments that facilitates intracellular organization and transport, helping communication between organelles. It also provides the cell with mechanical abilities for adhesion and movement through regulated interactions with plasma membrane receptors, controlling connections with adjacent cells and the extracellular matrix.

Our current and expanding knowledge encompasses not only the fundamental building blocks of microfilaments (actins), intermediate filaments, and microtubules (tubulins) but also the proteins that interact with and regulate the dynamics of polymerization and depolymerization, stabilize filaments, and facilitate the transport and docking of cargoes along the filaments through specific molecular motors. The collection of chapters that form the present book, part of the *Methods in Molecular Biology* series, aims to provide a summary of current knowledge on cytoskeletal structural and dynamic aspects that regulate cell organization and affect cell–cell interactions.

The discovery of actin in 1942 in the Albert Szent-Györgyi laboratory represented a cornerstone for biomedical sciences, allowing initial molecular studies on the contractility of muscles. Tubulin was discovered more recently, in 1967–1968, by Gary Borisy and Edwin Taylor as a colchicine-binding protein. Since then, extensive research has been conducted to describe the polymers formed by these building blocks of the cytoskeleton and to understand the dynamics and regulation of their polymerization and proteostasis in cells. A long-standing question has been how the interplay between the actin and tubulin cytoskeletons is regulated in cells. It is noteworthy that actin and tubulin cytoskeletons have evolved in parallel and even use the same chaperone in eukaryotes, namely the CCT chaperonin (Chaps. 1, 2, 15–17). However, the process of folding tubulin is more complex than that of actin due to the involvement of several tubulin-binding cofactors and the requirement for heterodimer formation. To study the dynamics of the cytoskeleton, it is therefore crucial to understand the structural features of the proteins.

Several methodological strategies for the production and isolation of relevant proteins and organelles have been developed and are herein described to help researchers in this field (Chaps. 3–5). Assays describing methods to study in vitro, in vivo, and ex vivo actin and tubulin polymerization and motor transport complement these methodologies (Chaps. 6–10), together with methods for direct visualization (Chaps. 11–14). Cryoelectron microscopy (CryoEM) is a powerful technique to characterize the structures of macromolecular complexes generated during cytoskeleton formation and disassembly (Chaps. 1–3, 16, and 17). Posttranslational modifications and protein localization are also relevant to understanding the function and structure of cytoskeletal components, and biochemical methodologies such as advanced proteomics are explored herein (Chaps. 18 and 19). Finally, bioinformatic methodologies are described to help analyze the data obtained by the authors (Chaps. 19

and 20). We thank all the contributors to this book for helping to spread good practices in science and providing other researchers with these valuable tools.

Madrid, Spain — Noa B. Martín-Cófreces
Madrid, Spain — Francisco Sánchez-Madrid
Madrid, Spain — José María Valpuesta

# Contents

*Preface* . . . . . . . . . . . . . . . . . . . . . . . . . . . . . . . . . . . . . . . . . . . . . . . . . . . . . . . . . *v*
*Contributors* . . . . . . . . . . . . . . . . . . . . . . . . . . . . . . . . . . . . . . . . . . . . . . . . . . . . . *ix*

1 Purification of the Eukaryotic Chaperonin CCT from Native Tissues . . . . . . . . 1
***Virginia Rodríguez-Martín, Sergio Pipaón, Jorge Gutiérrez-Seijo, José María Valpuesta and Jorge Cuéllar***

2 Single-Step Purification of Endogenous Human Chaperonin CCT . . . . . . . . . . 17
***Javier M. Rodríguez, Esther Martín-Forero, Jorge Gutiérrez-Seijo, Jimena Muntaner, Carmen Majano, José María Valpuesta and Jorge Cuéllar***

3 Integrated Protocol for γTuRC Expression and Purification for Microtubule Nucleation and Cryo-EM Analysis . . . . . . . . . . . . . . . . . . . . . . . . . . . . . . . . 39
***Fabian Zimmermann and Marina Serna***

4 Isolation of Centrosomal Fractions for Proteomic Studies in Lymphocytes . . . . 55
***Marta Lozano-Prieto, Francisco Javier Chichón and Francisco Sánchez-Madrid***

5 Method for Analyzing Nucleoskeletal Components from the Isolated Nucleus . . . . . . . . . . . . . . . . . . . . . . . . . . . . . . . . . . . . . . . . . . . . . . . . . . 67
***Marina Armesto, Christopher P. Toseland and Javier Redondo-Muñoz***

6 In Vitro Analysis of Profilin 1 Binding to Recombinant Actin . . . . . . . . . . . . . . 77
***Leticia Villadangos, David Escot and Juan M. Serrador***

7 In Vitro Production of Full-Length Myosin 2 and Visualization of Actin Filament Motility by Total Internal Reflection Microscopy . . . . . . . . . . . . . . . . 87
***Rafael Pérez-Díaz, Hugo Ramos-Solano, Marina Garrido-Casado, Vanessa C. Talayero, Johannes N. Greve and Miguel Vicente-Manzanares***

8 Reconstituting the Motility of Intraflagellar Transport Trains Ex Vivo . . . . . . . 111
***Aditya Chhatre, Ludek Stepanek, Stefan Diez and Gaia Pigino***

9 Planar Immunological Synapses Formed by Human CD4 T Cells as In Vitro System to Study Actin and Tubulin Dynamics Through TIRF Microscopy . . . . 129
***Álvaro Gómez-Morón, Marta Lozano-Prieto and Noa B. Martín-Cófreces***

10 Experimental Tools and Techniques for Studying Microtubule Nucleation at the Centrosome and the Golgi Apparatus . . . . . . . . . . . . . . . . . . . . . . . . . . . 147
***Chiara Marcozzi, Maria P. Gavilán and Rosa M. Ríos***

11 Study of Unconventional Cytoskeleton Components Regulating Actin Contraction . . . . . . . . . . . . . . . . . . . . . . . . . . . . . . . . . . . . . . . . . . . . . . . . 173
***Jing Zhang, Weiwei Qin, Che Hu, Wei Miao and Jie Xiong***

12 Methods to Study the Role of Cytoskeletal Motor Proteins in the Organization of the Postsynaptic Membrane . . . . . . . . . . . . . . . . . . . . . . . . . . . 205
*Amanda L. Neisch, Thomas Pengo and Thomas S. Hays*

13 STED Super-Resolution Microscopy for Studying Actin Cytoskeleton Organization . . . . . . . . . . . . . . . . . . . . . . . . . . . . . . . . . . . . . . . . . . . . . . . . . . . . . 223
*Marta Sampietro, Elvira Arza, Cristina Scielzo and Valeria R. Caiolfa*

14 Immunofluorescence-Based Study of Cargo Transport Mediated by Cytoskeleton-Associated Molecular Complexes. . . . . . . . . . . . . . . . . . . . . . . . . . . 243
*Francesca Finetti and Cosima T. Baldari*

15 CryoEM Strategies for Elucidating the Closed State Architecture of CCT. . . . . 263
*Jorge Gutiérrez-Seijo, Ana Cuervo, Sergio Pipaón, David Gil-Cantero, Carmen Majano, César Santiago, José María Valpuesta and Jorge Cuéllar*

16 The Open Conformation of CCT: How to Achieve High Resolution Using CryoEM . . . . . . . . . . . . . . . . . . . . . . . . . . . . . . . . . . . . . . . . . . . . . . . . . . . . . . . . 283
*Jorge Gutiérrez-Seijo, César Santiago, Sergio Pipaón, David Gil-Cantero, Ana Cuervo, Jorge Cuéllar and José María Valpuesta*

17 Methods to Study CCT Expression and Impact on Cytoskeleton . . . . . . . . . . . 301
*Camila Scagnetti, Hugo Yébenes, Silvia Requena and Noa B. Martín-Cófreces*

18 Methods to Study Arginylation of Tubulin . . . . . . . . . . . . . . . . . . . . . . . . . . . . 323
*Brittany MacTaggart and Anna Kashina*

19 Posttranslational Dynamics of the Subcellular Proteome by Mass Spectrometry-Based Proteomics . . . . . . . . . . . . . . . . . . . . . . . . . . . . . . . . . . . . 331
*Ana Martínez-Val, Jose Manuel Rodríguez, Jesper V. Olsen and Jesús Vázquez*

20 Computational and Statistical Framework for Quantitative Proteomics Analysis . . . . . . . . . . . . . . . . . . . . . . . . . . . . . . . . . . . . . . . . . . . . . . . . . 349
*Ismail Kirrout and Nuria Montes*

*Index*. . . . . . . . . . . . . . . . . . . . . . . . . . . . . . . . . . . . . . . . . . . . . . . . . . . . . . . . . . . . . . *397*

## Contributors

MARINA ARMESTO • *Department of Biomedicine, Centro de Investigaciones Biológicas Margarita Salas, Madrid, Spain*

ELVIRA ARZA • *Unit of Microscopy and Dynamic Imaging, Centro Nacional de Investigaciones Cardiovasculares Carlos III (F.S.P.), Madrid, Spain*

COSIMA T. BALDARI • *Department of Life Sciences, University of Siena, Siena, Italy*

VALERIA R. CAIOLFA • *Unit of Microscopy and Dynamic Imaging, Centro Nacional de Investigaciones Cardiovasculares Carlos III (F.S.P.), Madrid, Spain; Experimental Imaging Centre, IRCCS-Ospedale San Raffaele, Milano, Italy*

ADITYA CHHATRE • *Biotechnology Center, TUD Dresden University of Technology, Dresden, Germany*

FRANCISCO JAVIER CHICHÓN • *Cryo-Electron Microscopy Facility (cryoEM-CSIC) and Department of Macromolecular Structures, Centro Nacional de Biotecnología, CSIC, Madrid, Spain*

JORGE CUÉLLAR • *Centro Nacional de Biotecnología, Consejo Superior de Investigaciones Científicas, Madrid, Spain*

ANA CUERVO • *Centro Nacional de Biotecnología, Consejo Superior de Investigaciones Científicas, Madrid, Spain*

STEFAN DIEZ • *Cluster of Excellence Physics of Life, TUD Dresden University of Technology, Dresden, Germany; B CUBE—Center for Molecular Bioengineering, TUD Dresden University of Technology, Dresden, Germany; Max Planck Institute of Molecular Cell Biology and Genetics, Dresden, Germany*

DAVID ESCOT • *Interactions with the Environment Program, Development and Function of the Immune System Unit, Centro de Biología Molecular Severo Ochoa (CBMSO), Consejo Superior de Investigaciones Científicas (CSIC)–Universidad Autónoma de Madrid (UAM), Madrid, Spain*

FRANCESCA FINETTI • *Department of Life Sciences, University of Siena, Siena, Italy*

MARINA GARRIDO-CASADO • *Molecular Mechanisms Program, Centro de Investigación del Cáncer, Consejo Superior de Investigaciones Científicas (CSIC)-University of Salamanca, Salamanca, Spain*

MARIA P. GAVILÁN • *Centro Andaluz de Biología Molecular y Medicina Regenerativa (CABIMER), Universidad de Sevilla-CSIC-Universidad Pablo de Olavide, Seville, Spain; Departamento de Citología e Histología Normal y Patológica, Facultad de Medicina, Universidad de Sevilla, Seville, Spain*

DAVID GIL-CANTERO • *Centro Nacional de Biotecnología, Consejo Superior de Investigaciones Científicas, Madrid, Spain*

ÁLVARO GÓMEZ-MORÓN • *Immunology Service, Instituto de Investigación Sanitaria del Hospital Universitario La Princesa (IIS Princesa), Madrid, Spain; Department of Immunology, Ophthalmology and ENT, School of Medicine, Universidad Complutense de Madrid, Madrid, Spain*

JOHANNES N. GREVE • *Institute for Biophysical Chemistry, Hannover Medical School, Fritz-Hartmann Centre for Medical Research, Hannover, Germany*

JORGE GUTIÉRREZ-SEIJO • *Centro Nacional de Biotecnología, Consejo Superior de Investigaciones Científicas, Madrid, Spain*
THOMAS S. HAYS • *Department of Genetics, Cell Biology and Development, University of Minnesota, Minneapolis, MN, USA*
CHE HU • *Institute of Hydrobiology, Chinese Academy of Sciences, Hubei Province, Wuhan, China*
ANNA KASHINA • *Department of Biomedical Sciences, School of Veterinary Medicine, University of Pennsylvania, Philadelphia, PA, USA*
ISMAIL KIRROUT • *Servicio de Inmunología, Instituto de Investigación Sanitaria del Hospital Universitario la Princesa (IIS Princesa), Madrid, Spain; Centro de Biotecnología y Genómica de Plantas (CBGP) UPM-INIA/CSIC and E.T.S. Ingeniería Agronómica, Alimentaria y de Biosistemas., Universidad Politécnica de Madrid, Pozuelo de Alarcón, Spain*
MARTA LOZANO-PRIETO • *Immunology Service, Instituto de Investigación Sanitaria del Hospital Universitario La Princesa (IIS Princesa), Madrid, Spain; Centro de Investigación en Red Cardiovascular (CIBERCV), Instituto de Salud Carlos III, Madrid, Spain*
BRITTANY MACTAGGART • *Department of Biomedical Sciences, School of Veterinary Medicine, University of Pennsylvania, Philadelphia, PA, USA*
CARMEN MAJANO • *Centro Nacional de Biotecnología, Consejo Superior de Investigaciones Científicas, Madrid, Spain*
CHIARA MARCOZZI • *Centro Andaluz de Biología Molecular y Medicina Regenerativa (CABIMER), Universidad de Sevilla-CSIC-Universidad Pablo de Olavide, Seville, Spain*
NOA B. MARTÍN-CÓFRECES • *Immunology Service, Instituto de Investigación Sanitaria del Hospital Universitario La Princesa (IIS Princesa), Madrid, Spain; Centro de Investigación en Red Cardiovascular (CIBERCV), Instituto de Salud Carlos III, Madrid, Spain*
ANA MARTÍNEZ-VAL • *Cardiovascular Proteomics Laboratory, Centro Nacional de Investigaciones Cardiovasculares Carlos III (CNIC), Madrid, Spain; Centro de Investigación Biomédica en Red de Enfermedades Cardiovasculares (CIBERCV), Madrid, Spain*
ESTHER MARTÍN-FORERO • *Centro Nacional de Biotecnología, Consejo Superior de Investigaciones Científicas, Madrid, Spain*
WEI MIAO • *Institute of Hydrobiology, Chinese Academy of Sciences, Hubei Province, Wuhan, China*
NURIA MONTES • *Unidad de Metodología, Instituto de Investigación Sanitaria Princesa, Madrid, Spain*
JIMENA MUNTANER • *Centro Nacional de Biotecnología, Consejo Superior de Investigaciones Científicas, Madrid, Spain*
AMANDA L. NEISCH • *Department of Biological Sciences, Rutgers University, Newark, NJ, USA*
JESPER V. OLSEN • *Novo Nordisk Foundation Center for Protein Research, Department of Cellular and Molecular Medicine, Faculty of Health and Medical Sciences, University of Copenhagen, Copenhagen, Denmark*
THOMAS PENGO • *Minnesota Supercomputing Institute, Minneapolis, MN, USA*

RAFAEL PÉREZ-DÍAZ • *Molecular Mechanisms Program, Centro de Investigación del Cáncer, Consejo Superior de Investigaciones Científicas (CSIC)-University of Salamanca, Salamanca, Spain*
GAIA PIGINO • *Human Technopole, Milan, Italy*
SERGIO PIPAÓN • *Centro Nacional de Biotecnología, Consejo Superior de Investigaciones Científicas, Madrid, Spain*
WEIWEI QIN • *Institute of Hydrobiology, Chinese Academy of Sciences, Hubei Province, Wuhan, China*
HUGO RAMOS-SOLANO • *Molecular Mechanisms Program, Centro de Investigación del Cáncer, Consejo Superior de Investigaciones Científicas (CSIC)-University of Salamanca, Salamanca, Spain*
JAVIER REDONDO-MUÑOZ • *Department of Biomedicine, Centro de Investigaciones Biológicas Margarita Salas, Madrid, Spain*
SILVIA REQUENA • *Servicio de Inmunología, Instituto Investigación Sanitaria Princesa (IIS Princesa), Hospital Universitario de la Princesa, Universidad Autónoma de Madrid, Madrid, Spain*
ROSA M. RÍOS • *Centro Andaluz de Biología Molecular y Medicina Regenerativa (CABIMER), Universidad de Sevilla-CSIC-Universidad Pablo de Olavide, Seville, Spain*
JAVIER M. RODRÍGUEZ • *Centro Nacional de Biotecnología, Consejo Superior de Investigaciones Científicas, Madrid, Spain*
JOSE MANUEL RODRÍGUEZ • *Cardiovascular Proteomics Laboratory, Centro Nacional de Investigaciones Cardiovasculares Carlos III (CNIC), Madrid, Spain; Centro de Investigación Biomédica en Red de Enfermedades Cardiovasculares (CIBERCV), Madrid, Spain*
VIRGINIA RODRÍGUEZ-MARTÍN • *Centro Nacional de Biotecnología, Consejo Superior de Investigaciones Científicas, Madrid, Spain*
MARTA SAMPIETRO • *Division of Experimental Oncology, Malignant B Cells Biology and 3D Modeling Unit, IRCCS Ospedale San Raffaele, Comprehensive Cancer Centre, Milano, Italy; Unit of Microscopy and Dynamic Imaging, Centro Nacional de Investigaciones Cardiovasculares Carlos III (F.S.P.), Madrid, Spain*
FRANCISCO SÁNCHEZ-MADRID • *Immunology Service, Hospital Universitario de la Princesa, Instituto Investigación Sanitaria Princesa (IIS Princesa), Madrid, Spain; Centro de Investigación Biomédica en Red, Enfermedades Cardiovasculares (CIBERCV), Madrid, Spain; Universidad Autónoma de Madrid (UAM), Madrid, Spain*
CÉSAR SANTIAGO • *Centro Nacional de Biotecnología, Consejo Superior de Investigaciones Científicas, Madrid, Spain*
CAMILA SCAGNETTI • *Servicio de Inmunología, Instituto Investigación Sanitaria Princesa (IIS Princesa), Hospital Universitario de la Princesa, Universidad Autónoma de Madrid, Madrid, Spain*
CRISTINA SCIELZO • *Division of Experimental Oncology, Malignant B Cells Biology and 3D Modeling Unit, IRCCS Ospedale San Raffaele, Comprehensive Cancer Centre, Milano, Italy*
MARINA SERNA • *Macromolecular Structure Department, National Center for Biotechnology, Spanish National Research Council (CNB-CSIC), Madrid, Spain*
JUAN M. SERRADOR • *Interactions with the Environment Program, Development and Function of the Immune System Unit, Centro de Biología Molecular Severo Ochoa (CBMSO), Consejo Superior de Investigaciones Científicas (CSIC)–Universidad Autónoma de Madrid (UAM), Madrid, Spain*

LUDEK STEPANEK • *Institute of Molecular Genetics, Czech Academy of Sciences, Prague, Czech Republic*

VANESSA C. TALAYERO • *Molecular Mechanisms Program, Centro de Investigación del Cáncer, Consejo Superior de Investigaciones Científicas (CSIC)-University of Salamanca, Salamanca, Spain*

CHRISTOPHER P. TOSELAND • *School of Medicine and Population Health, University of Sheffield, Sheffield, UK*

JOSÉ MARÍA VALPUESTA • *Centro Nacional de Biotecnología, Consejo Superior de Investigaciones Científicas, Madrid, Spain*

JESÚS VÁZQUEZ • *Cardiovascular Proteomics Laboratory, Centro Nacional de Investigaciones Cardiovasculares Carlos III (CNIC), Madrid, Spain; Centro de Investigación Biomédica en Red de Enfermedades Cardiovasculares (CIBERCV), Madrid, Spain*

MIGUEL VICENTE-MANZANARES • *Molecular Mechanisms Program, Centro de Investigación del Cáncer, Consejo Superior de Investigaciones Científicas (CSIC)-University of Salamanca, Salamanca, Spain*

LETICIA VILLADANGOS • *Interactions with the Environment Program, Development and Function of the Immune System Unit, Centro de Biología Molecular Severo Ochoa (CBMSO), Consejo Superior de Investigaciones Científicas (CSIC)–Universidad Autónoma de Madrid (UAM), Madrid, Spain*

JIE XIONG • *Institute of Hydrobiology, Chinese Academy of Sciences, Hubei Province, Wuhan, China*

HUGO YÉBENES • *Facultad de Ciencias de la Salud, Universidad Francisco de Vitoria, Pozuelo de Alarcón, Spain*

JING ZHANG • *Institute of Hydrobiology, Chinese Academy of Sciences, Hubei Province, Wuhan, China*

FABIAN ZIMMERMANN • *Mechanisms of Disease Programme, Institute for Research in Biomedicine (IRB Barcelona), The Barcelona Institute of Science and Technology (BIST), Barcelona, Spain; Max Planck Institute of Molecular Physiology, Department of Mechanistic Cell Biology, Dortmund, Germany*

Chapter 1

# Purification of the Eukaryotic Chaperonin CCT from Native Tissues

Virginia Rodríguez-Martín, Sergio Pipaón, Jorge Gutiérrez-Seijo, José María Valpuesta and Jorge Cuéllar

**Abstract**

The eukaryotic CCT (chaperonin containing TCP1; also known as TRiC) is an essential molecular chaperone involved in the folding of nearly 10 % of newly synthesized proteins, including key cytoskeletal components such as actin and tubulin. It belongs to the Hsp60 family (60 kDa Heat Shock Proteins) or chaperonins, and it is composed of two stacked rings formed by eight different subunits. The CCT structural complexity makes its heterologous expression and purification highly challenging. To circumvent these difficulties, native purification from tissue remains a reliable approach. This protocol details the purification of bovine CCT from bull testis, a tissue naturally enriched in the complex due to its role in spermatogenesis. Originally developed by Frydman et al. in 1992 and extensively optimized by Prof. José María Valpuesta's group at CNB-CSIC over the past 25 years, the protocol has enabled key advances, including the crystallization and structural determination of bovine CCT. It continues to serve as a foundational tool for biochemical and structural studies of this essential chaperonin.

**Keywords** CCT, Molecular chaperone, Chaperonin, Protein folding, Purification from native tissue
Virginia Rodríguez-Martín and Sergio Pipaón are the first authors.

## 1 Introduction

Chaperonins are a family of chaperones whose structure consists of two juxtaposed oligomeric rings with an internal cavity, where the folding process takes place [1]. CCT (chaperonin containing TCP-1; also known as TRiC, the most complex of all chaperonins, is an oligomeric protein complex of 1 MDa, where each of its rings is composed of eight homologous subunits of 55–60 kDa (CCT1-CCT8), encoded in different paralogous genes and arranged in a specific order [2–4]. In eukaryotic cells, the chaperonin CCT plays a crucial role in guiding the proper folding of a significant portion of proteins, among them essential cytoskeletal components such as actin and tubulin [5].

Noa B. Martín-Cófreces et al. (eds.), *The Cytoskeleton*, Methods in Molecular Biology, vol. 3070,
https://doi.org/10.1007/978-1-0716-5515-3_1,

The purification of CCT is essential for studying its biological function and its implications in disease, as it enables a wide range of analyses, from in vitro activity assays such as ATPase and client folding assays (i.e., actin or tubulin, its most important substrates) [6–8] or cross-linking and mass spectrometry analysis [3], to structural studies including cryo-electron microscopy [9]. However, the structural complexity of CCT presents a significant challenge for heterologous expression systems, as the proper assembly of its eight distinct subunits is crucial and yet not fully understood [10].

Due to these challenges, purification from biological tissue provides a simpler alternative for isolating the complex, albeit with low recovery yields. This protocol focuses on obtaining bovine CCT from bovine testicles, a tissue highly enriched in this protein due to its role in spermatogenesis [11].

This protocol is based on the original method described by J. Frydman et al. in 1992 [12]. Since then, the procedure has undergone extensive modifications until it was formally established in 2000 [13]. For over 25 years, the laboratory of Prof. José María Valpuesta at the National Center for Biotechnology (CNB-CSIC) in Madrid has been purifying bovine CCT from bovine testis. Throughout this time, the group has introduced significant improvements to the protocol (see Fig. 1). These refinements ultimately led to the successful crystallization of pure CCT, as reported by Muñoz I.G. and Yébenes H. in 2011 [8], enabling the determination of the bovine CCT structure by X-ray crystallography. This protocol remains highly relevant, and it is widely used in laboratories worldwide for a variety of purposes.

## 2 Materials

### 2.1 General Inventory

1. High-speed blender (minimum 450 watts) for protein homogenization.
2. Potter-Elvehjem PTFE pestle and glass tube coupled with a vertical drill (230 V) to improve disruption of cell membranes.
3. Centrifuge and ultracentrifuge; rotor with swinging buckets.
4. Ammonium sulfate, molecular biology grade for protein precipitation.
5. Molecular weight standards for SDS-PAGE gels.
6. Fast Protein Liquid Chromatography (FPLC) system for the affinity and size exclusion chromatographies.
7. Dialysis membranes (16 mm inner diameter).
8. Bovine Serum Albumin (BSA) for reference standard in protein quantification assays.
9. Carbon Film 300 Mesh Copper Grids.
10. Glow discharge unit.

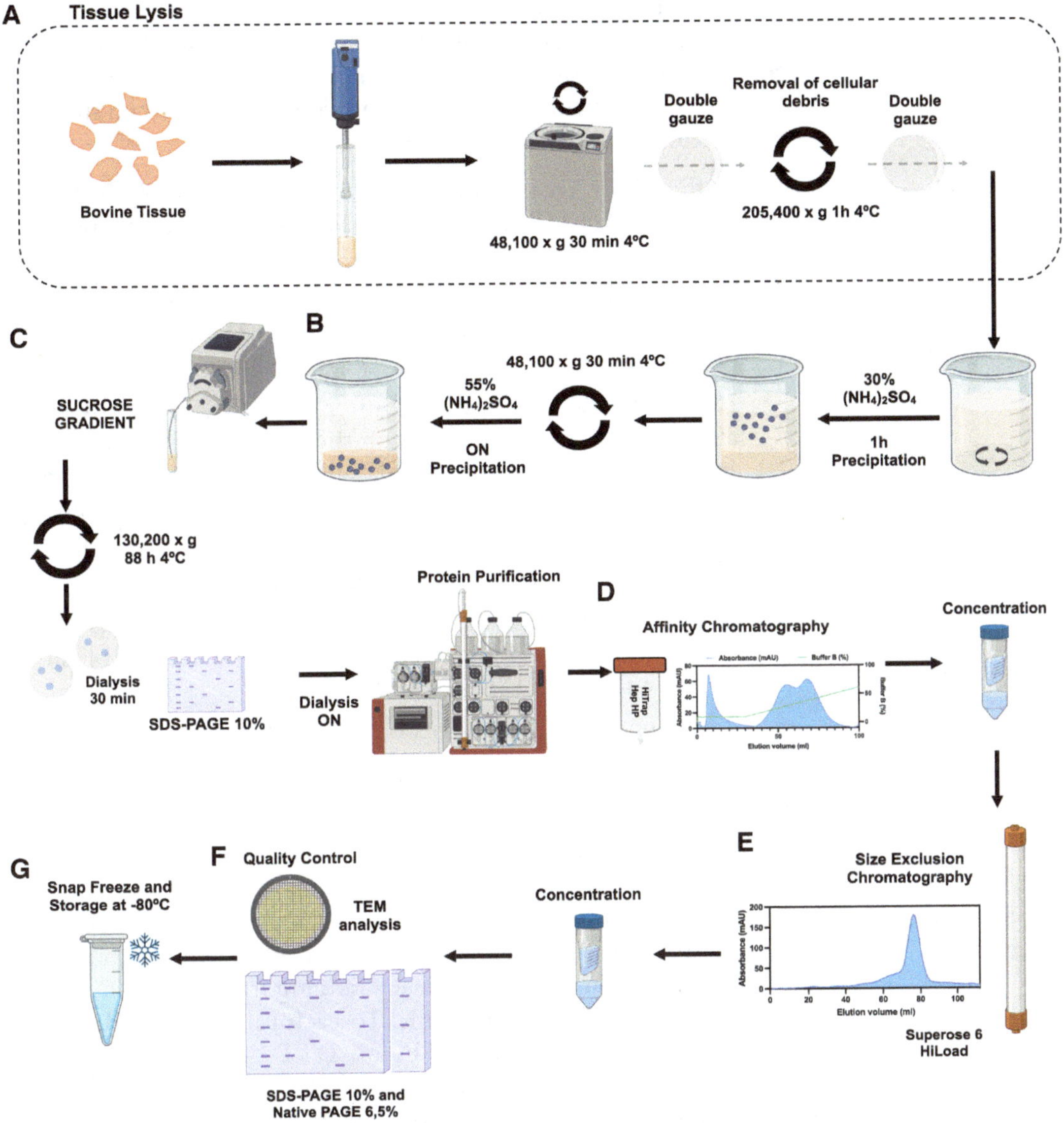


**Fig. 1** CCT Purification Workflow from Bovine Tissue. (**A**) The process begins with tissue lysis, where homogenized bovine tissue undergoes centrifugation to remove the cellular debris. (**B**) CCT is selectively precipitated using ammonium sulfate in two sequential (30% and 55%) precipitations followed by centrifugation. (**C**) The sample is further enriched based on its density using a sucrose gradient step, followed by dialysis and SDS-PAGE analysis for quality assessment. (**D**) Next, an affinity chromatography is performed, which isolates target proteins based on specific binding interactions to the column matrix. (**E**) The resulting CCT-enriched fractions are pooled, concentrated, and subjected to size exclusion chromatography to separate them from contaminants with different molecular sizes. (**F**) After a final concentration step, the purified CCT sample undergoes quality control analysis via SDS-PAGE 10%, native PAGE 6.5 %, and transmission electron microscopy (TEM) imaging. (**G**) The final sample is aliquoted, snap-frozen, and stored at −80°C for long-term preservation

### 2.2 Bovine Tissue Extracts

A meat supplier provides bovine testicular tissue following ethical guidelines. In the laboratory, the tissue is processed by removing the outer connective tissue layers, including the *tunica albuginea*, to retain the testicular matrix. The processed tissue is then stored in 50 mL tubes at −80°C (*see* **Note 1**).

### 2.3 Buffers

1. Buffer H for Tissue Lysis: 20 mM Tris buffer pH 7.35, 10 mM KCl, 1 mM EDTA, 1 mM PMSF, 1 mM DTT, and protease inhibitors (*see* **Note 2**).
2. Buffer A, for Gradients and Affinity Chromatography: 50 mM Tris buffer pH 7.35, 150 mM NaCl, 5 mM $MgCl_2$, 10% glycerol, 1 mM PMSF, and 1 mM DTT (*see* **Note 3**).
3. 10× Buffer A: 500 mM Tris buffer pH 7.35, 1500 mM NaCl, 50 mM $MgCl_2$. When preparing the 1× Buffer, just before use, add 10% glycerol, 1 mM PMSF, and 1 mM DTT (*see* **Note 3**).
4. Buffer SE.A, for Size Exclusion Chromatography: 50 mM HEPES buffer pH 7.35, 150 mM NaCl, 5 mM $MgCl_2$, 10% glycerol, and 1 mM PMSF (*see* **Note 4**).
5. Buffer B, for Affinity Chromatography: 50 mM Tris buffer pH 7.35, 1000 mM NaCl, 5 mM $MgCl_2$, 10 % glycerol, 1 mM PMSF, and 1 mM DTT (*see* **Note 5**).
6. Resolving buffer: 1 M Tris-HCl, pH 8.8.
7. Stacking buffer: 1 M Tris-HCl, pH 6.8.
8. 10% SDS in water.
9. 2% uranyl acetate solution.

### 2.4 Sucrose Gradients

1. Sucrose gradient solutions: 20%–50% (w/w) in 2.5 % increments in buffer A. Prepare thirteen 50 mL tubes and weigh sucrose amounts to make layers ranging from 20%–50% (w/w) in 2.5% increments in 50 mL. To each tube, add 5 mL of 10× Buffer A (containing PMSF and DTT but without glycerol) and adjust the final volume to 50 mL with deionized water. Mix well until fully dissolved. Store the sucrose solutions at −20°C until you are ready to prepare the gradients or proceed to the following step (*see* **Note 6**).
2. Take six ultracentrifuge-compatible tubes (rated for at least 130, 200 x*g*) and place them upright on dry ice to chill.
3. Layer the gradients starting with the sucrose gradient solutions containing the highest sucrose concentration (50%); adjust equal volumes for each layer. The volume used depends on the total volume of the ultracentrifuge tubes. At the top of each gradient, leave approximately 3 mL of space for sample loading.
4. Carefully layer each sucrose solution sequentially from highest to lowest concentration (down to 20%).

5. After adding each layer, allow it to freeze completely before adding the next one.
6. Once all thirteen layers are frozen and stacked, store the prepared gradients at −20°C until use (*see* **Note 7**).

### 2.5 Gels

1. A minimum of ten 10% denaturing SDS-PAGE gels (*see* **Note 8**) are required for the entire purification protocol: six for analyzing the fractionated gradients (three for each round of purification), two for analyzing the fractions from affinity chromatography (one per round), one for analyzing the fractions from size exclusion chromatography, and one for assessing the effectiveness of the final concentration step. Additionally, a 6.5 % native gel is required for the quality control of the purification. To prepare denaturing acrylamide-bisacrylamide gels, assemble the gel casting mold, ensuring a tight seal to prevent leakage.
2. Prepare the resolving gel solution: 0.375 M Resolving buffer containing 10% acrylamide:bisacrylamide (29:1 stock), 0.1% SDS (Sodium dodecyl sulfate), followed by 0.1% APS (ammonium persulfate) and 0.05% TEMED (N,N,N′,N′-Tetramethylethylenediamine) to initiate polymerization.
3. Pour the resolving gel in the casting module of your electrophoretic system.
4. Overlay with isopropanol or water to ensure a flat surface, and allow it to polymerize.
5. Prepare the stacking gel solution: 0.375 M stacking buffer containing 4% acrylamide:bisacrylamide (29:1 stock), 0.1% SDS (Sodium dodecyl sulfate). Add 0.1% APS (ammonium persulfate) and 0.05% TEMED (N,N,N′,N′-Tetramethylethylenediamine) to initiate polymerization once the resolving gel has become solid.
6. Remove the water or isopropanol overlay on top of the resolving gel.
7. Wash with water in case of isopropanol.
8. Pour the stacking gel, insert the comb, and let it polymerize completely. Protect from light. Gels can be stored at 4°C for a month, provided they do not dry.
9. To prepare native acrylamide-bisacrylamide gels, assemble a dedicated gel casting mold, ensuring a tight seal to prevent leakage, and that there is no SDS present.
10. Prepare the resolving gel solution: 0.375 M Resolving buffer containing 6.5% acrylamide:bisacrylamide (29:1 stock), followed by 0.1% APS (ammonium persulfate) and 0.05% TEMED (N,N,N′,N′-Tetramethylethylenediamine) to initiate polymerization.

11. Pour the resolving gel in the casting module of your electrophoretic system.
12. Overlay with isopropanol or water to ensure a flat surface, and allow it to polymerize.
13. Prepare the stacking gel solution: 0.375 M Stacking buffer containing 4% acrylamide:bisacrylamide (29:1 stock). Add 0.1% APS (ammonium persulfate) and 0.05% TEMED (N,N,N′,N′-Tetramethylethylenediamine) to initiate polymerization once the resolving gel has become solid.
14. Remove the water or isopropanol overlay on top of the resolving gel.
15. Wash with water in case of isopropanol.
16. Pour the stacking gel, insert the comb, and let it polymerize completely. Protect from light.

### 2.6 Gel Staining Solutions

1. Fixation solution: 40% ethanol, 10% acetic acid in water.
2. Staining solution: 50% ethanol in water.
3. Coomassie solution: 0.25% Coomassie Brilliant Blue G250 and 50% ethanol in water.

Using this method, a destaining solution is not required. However, gels can be washed with water after staining to increase the contrast of the protein bands.

## 3 Methods

### 3.1 Tissue Lysis

1. Thaw tissue overnight at 4°C.
2. Weigh the content of four 50 mL tubes filled with tissue, which will usually be around 200 g.
3. Mix the contents of the tubes with 200 mL of buffer H.
4. Homogenize the tissue in a blender at 4°C. Blend at maximum speed in 2 second pulses to avoid overheating the sample.
5. Collect the mixture in a cooled beaker (*see* **Note 9**).
6. Secure a Potter-Elvehjem homogenizer (of approximately 15 mL) in the chuck of a vertical drill.
7. Transfer the liquefied tissue into the glass tube.
8. Disrupt cell membranes through twenty controlled up-and-down strokes. Use a constant low-speed setting for the vertical drill to reduce the heating of the sample (*see* **Note 10**).
9. Process the total sample volume in batches until fully completed (*see* **Note 11**).

### 3.2 Removal of Tissue Debris

1. Centrifuge the homogenate at 48,000 x*g* for 30 min at 4°C.
2. Collect the supernatants in cooled-ice tubes
3. Filter through a double-layered gauze (*see* **Note 12**).
4. Ultracentrifuge the filtered supernatant at 205,000 x*g* for 1 h at 4°C.
5. Filter the supernatants through a double-layered gauze, maintaining a temperature of 4°C throughout the process (*see* **Note 12**).

### 3.3 Ammonium Sulfate Precipitation

1. Measure the filtered volume in a graduated cylinder and keep it at 4°C.
2. Carry out the precipitation by slowly adding ammonium sulfate to 30% saturation (16.4 g per 100 mL of sample) under constant stirring in a beaker (*see* **Note 13**).
3. Allow the mixture to precipitate for 1 h.
4. Centrifuge the sample at 48,000 x*g* for 30 min at 4°C
5. Filter the supernatant through a double-layered gauze.
6. Precipitate the filtered supernatant to 55% saturation by slowly adding ammonium sulfate (14.8 g per 100 mL of sample) under constant stirring in a beaker (*see* **Note 13**).
7. Allow the mixture to precipitate overnight at 4°C.

### 3.4 Gradient Ultracentrifugation

1. Thaw enough frozen gradients to hold the totality of the sample in a 4°C cold room, positioned as vertically as possible in a rack. Each gradient can hold a maximum of 3 mL of sample (*see* **Note 14**).
2. Centrifuge the sample from Sect.3.3. at 48,000 x*g* for 30 min at 4°C.
3. Discard the supernatant (*see* **Note 15**).
4. Detach each pellet carefully from the bottom of the tube using a borosilicate glass rod.
5. Leave each pellet to soak in 1 mL of buffer A on ice. Allow sufficient soaking (around 5 min) (*see* **Note 16**).
6. Resuspend carefully by pipetting until fully homogenized.
7. Once the pellets are resuspended, pool them together in a tube.
8. Dialyze the pooled pellets in a dialysis membrane with 16 mm of inner diameter against 1 L of buffer A at 4°C for 1 h (*see* **Note 17**).
9. Transfer the dialysis membrane with the sample to a new container and further dialyze it against 2 L of buffer A at 4°C for 2 h.
10. Introduce the 6 thawed gradients inside compatible centrifugal buckets.

11. Load the dialyzed sample carefully onto the six gradients (*see* **Note 18**).
12. Weight and balance the buckets with the gradients in pairs.
13. Ultracentrifuge the samples at 130,000 x*g* for 88 h at 4°C.

### 3.5 Gradient Collection and Fractionation

1. Use a peristaltic pump connected to a fraction collector to recover the gradient from bottom to top. To do this, connect the inlet tubing of the peristaltic pump to a hollow metal rod (similar to a syringe needle), which is carefully inserted into the bottom of the gradient tube to avoid disturbing the sucrose layers.
2. Once the tip is positioned at the base, initiate the fractionation at medium-slow speed.
3. Collect fractions at a volume of 2.2 mL each, which should result in 14 fractions (*see* **Note 19**).
4. Rinse the pump circuit with distilled water and purge it with air after each gradient to clearly define the starting point of the next gradient's volume.
5. To remove the excess sucrose in the samples, 30 μL of each of the fractions is dialyzed against buffer A (without PMSF, DTT, and glycerol) for 30 min, using 0.025 μm dialysis membranes (*see* **Note 20**).
6. Load the dialyzed samples onto 10% SDS-PAGE gels; optimally three (see Fig. 2A for sample organization).
7. Fix the proteins in the gel with fixation solution (approximately 10 mL per gel) for 5 min.
8. Discard the solution.
9. Incubate the gel in staining solution (approximately 10 mL per gel) supplemented with 100 μL of Coomassie Blue solution. Stain for 1 h or until the bands are clearly visible (*see* **Note 21**).
10. Destain the gels in water for 30 min.
11. CCT subunits are recognized as a cluster of bands around 60–70 kDa of molecular weight.
12. Pool the CCT-enriched fractions and dialyze them in a membrane with 16 mm of inner diameter against 2 L of buffer A overnight at 4°C (*see* **Note 17**).

### 3.6 Affinity Chromatography

1. Filter the dialyzed sample sequentially by using a syringe with 0.45 μm and 0.22 μm pore size filters (*see* **Note 22**).
2. Filter buffer A and buffer B using a 0.22 μm pore size filter.
3. Subject the sample to affinity chromatography by using a HiTrap Heparin HP 5 mL column (*see* **Note 23**).

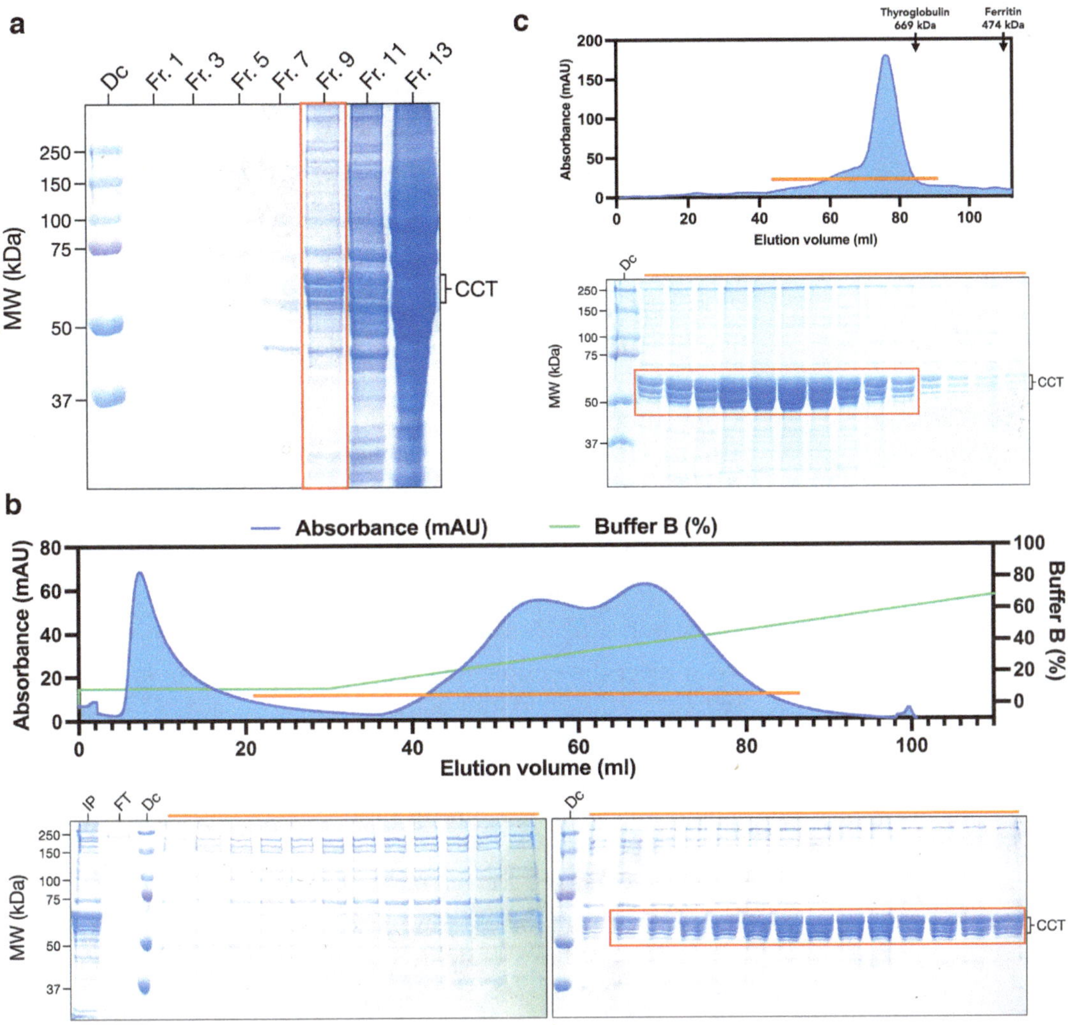


**Fig. 2** Sucrose gradients, affinity chromatography, and size exclusion chromatography. (**A**) 10 % SDS-PAGE analysis of the fractions resulting from a single sucrose gradient. This layout displays a detailed pattern of the CCT distribution along the gradient volume. The lane highlighted in red corresponds to the fraction selected for further purification. Dc, Dual Color molecular weight marker; Fr. 1–13, Gradient fractions loaded onto the gel. (**B**) 10% SDS-PAGE analysis of the fractions resulting from the affinity chromatography step and its corresponding chromatogram. In this step, contaminants from the sample that elute at Buffer B concentrations different from CCT can be discarded. A prominent contaminant peak eluting with 20% buffer B can be observed at around 10 mL. The orange bar indicates the fractions from the chromatogram that were loaded onto the gel, and the lanes highlighted in red correspond to the fractions selected for further purification. IP, Input; FT, Flow Through; Dc, Dual Color molecular weight marker. (**C**) 10% SDS-PAGE analysis of the fractions resulting from the size exclusion chromatography step and its corresponding chromatogram. This step allows for the removal of contaminants with a molecular weight different from CCT. The orange bar indicates the fractions from the chromatogram that were loaded onto the gel, and the lanes highlighted in red correspond to the fractions selected for further purification. Calibration references for the used column are shown

4. Follow the stepwise elution protocol:

A. An isocratic step to 10% buffer B
B. A gradient from 10%–50% buffer B in 100 mL
C. A final isocratic step to 100% buffer B (Fig. 2B).

5. Analyze the fractions corresponding to the elution peak of protein (measured by a coupled U.V detector) by 10% SDS-PAGE as in Sect. 3.5.6. and subsequent steps (see Fig. 2B).
6. Pool the fractions containing CCT subunits.

### 3.7 Size Exclusion Chromatography

1. Concentrate the pooled fractions from one or two steps of affinity chromatography (*see* **Note 24**) to a volume <5 mL. A device based on centrifugation combined with a low-binding, filtering membrane of 100 kDa pore size is recommended (*see* **Note 25**). Centrifuge the sample volume to concentrate it up to 15,000 x*g* for 10 min at 4°C (*see* **Note 26**).
2. Filter buffer SE.A using a 0.22 μm pore size filter.
3. To subject the sample to size exclusion chromatography (SEC; see Fig. 2C), load the supernatant onto a SEC column for high-resolution purification of large biomolecules such as a Superose 6 HiLoad column using a 5 mL loop (*see* **Note 27**).
4. The buffer is passed through the loop to ensure the sample is fully carried into the column. As the separation proceeds, fractions are collected for further analysis.
5. Analyze the fractions corresponding to the elution peak of protein (measured by a coupled U.V detector) by 10% SDS-PAGE as in Sect. 3.5.5. (see Fig. 2c).
6. Pool the fractions containing CCT (see Fig. 2c).

### 3.8 Quality Control and Concentration Estimation

1. Measure the protein concentration using a spectrophotometer at 280 nm ($\varepsilon_{CCT}$= 364,660 L mol$^{-1}$ cm$^{-1}$; $MW_{CCT}$=960,000 Da).
2. Alternatively, estimate the protein concentration by quantifying serial dilutions of the protein and comparing them to a BSA protein standard (see Fig. 3A).
3. If concentration is low, use a centrifuge-dependent device with 100 kDa pore size membrane to concentrate it (*see* **Notes 25** and **28**).
4. Run a 6.5% native gel with purified CCT samples to confirm that the protein remains in its oligomeric state. The native gel is run at room temperature, starting with a current limit of 10 mA for the first 15 min, followed by 20 mA for approximately 2 h.

A reference CCT sample with a previously validated oligomeric state should be included as a reference (see Fig. 3B for sample organization).

5. Reserve an aliquot for Transmission Electron Microscopy (TEM).
6. Aliquot the sample and snap-froze in liquid nitrogen to store at −80°C (*see* **Note 29**).

### *3.9 Visualization Through TEM*

A negative staining grid with 0.075 μM–0.1 μM of CCT can be prepared to verify visually the integrity of the CCT, using TEM (see Fig. 3C).

1. Glow-discharge carbon-coated 300 mesh copper grids at 25 mA for 15 s.
2. Place the carbon side of the grid onto a drop of 5 μL of sample for 1 min at room temperature on a piece of hydrophobic material.
3. Blot gently the grids with high-quality filter paper.
4. Place the carbon side of the grid onto a drop of 5 μL of 2% uranyl acetate for 1 min at room temperature on a piece of hydrophobic material.
5. Blot the grids gently with high-quality filter paper.
6. Left the grids to air dry. Grids are then ready for imaging.

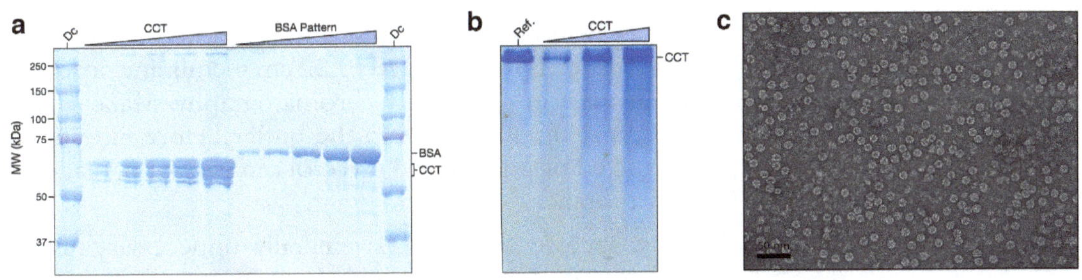


**Fig. 3** Quality control and concentration estimation of the purification. (**A**) 10% SDS-PAGEanalysis of the final concentration step. Increasing known volumes of the purified CCT are loaded onto a gel (0.125 μl, 0.25 μl, 0.5 μ l, 1 μl and 2 μl are shown) alongside increasing known amounts of BSA (0.25 μg, 0.5 μg, 1 μg, 2 μg, and 4 μg are shown). By quantifying the band intensities in the gel, the amount of the CCT complex can be interpolated with the calibration curve resulting from the BSA. This allows the calculation of the concentration obtained in the purified volume, while also being useful to evaluate the presence of contaminants. Dc, Dual Color molecular weight marker. (**B**) 6.5% native electrophoresis analysis of the final concentration step. This is used to assess the oligomerization state of the purified CCT. Increasing known volumes of the purified CCT are loaded onto a gel (1 μl, 2 μl, and 4 μl are shown). Ref, 15 μg of reference bovine CCT. (**C**) Negatively stained sample (0.1 μM) observed by transmission electron microscopy to further confirm that the purified CCT assemblies into oligomers

## 4 Notes

1. Bovine testicular tissue must be collected from freshly slaughtered animals and processed and frozen on the same day. Other sources can be used as starting material, like human cell pellets, from which human CCT could be purified.
2. For buffer H, add PMSF and DTT on the day of use, along with an appropriate number of Protease Inhibitor tablets (we use the cOmplete tablets 04693116001 from Roche), one for each 30 mL of buffer. It is advised to prepare a 2× stock (without PMSF, DTT, and protease inhibitors) of this buffer for easy preparation on the day of use.
3. For buffer A, add glycerol, PMSF, and DTT on the day of use. This buffer needs to be filtered (0.22 μm membrane pore size) before using it for affinity chromatography. It is advisable to prepare a 10× stock (without glycerol, PMSF, and DTT) of this buffer for easy preparation on the day of use (as described on 2.3.3). Make sure to properly mix the glycerol into the buffer before filtering, as poorly mixed or undissolved glycerol can clog or damage the filter membrane.
4. For buffer SE.A, add glycerol and PMSF on the day of use. This buffer needs to be filtered (0.22 μm membrane pore size) before using it for size exclusion chromatography. Make sure to properly mix the glycerol into the buffer before filtering, as poorly mixed or undissolved glycerol can clog or damage the filter membrane.
5. For buffer B add glycerol, PMSF and DTT on the day of use. This buffer needs to be filtered (0.22 μm membrane pore size) before using it in the affinity chromatography. Make sure to properly mix the glycerol into the buffer before filtering, as poorly mixed or undissolved glycerol can clog or damage the filter membrane.
6. Filtering sucrose solutions is generally unnecessary, as the risk of contamination is minimal when stored at −20°C. Additionally, their high viscosity can make filtration difficult.
7. Sucrose gradients can also be prepared using a gradient maker.
8. 10% denaturing gels are optimal for resolving CCT subunits, which form a recognizable cluster of bands around 55–60 kDa.
9. It is crucial to maintain the sample at 4°C throughout the entire protocol. All equipment and materials should be precooled to 4°C before use. Additionally, the use of protease inhibitors throughout the purification, such as PMSF or inhibitor cocktail tablets, is crucial to prevent the degradation of enriched CCT by cellular proteases.

10. Since the entire volume of the tissue suspension cannot be homogenized at once, process it in small batches using the Potter-Elvehjem homogenizer. Take care to avoid overflow when inserting the pestle into the container. For improved cell disruption, the homogenization step can be repeated a second time.

11. Expect 350 mL of final volume for 200 g of tissue blended in 200 mL of buffer H.

12. The homogenized tissue is first centrifuged to remove cellular debris. The subsequent ultracentrifugation step is used to remove mitochondria, lysosomes, and other large organelles. Since CCT is soluble and has a lower molecular weight, it remains in the supernatant. Importantly, centrifugation does not remove lipid contaminants, which float on top of the sample. Rather, these are eliminated by filtering the supernatants through double-layered gauze, as described.

13. Ammonium sulfate should be added gradually, without allowing it to foam up, at a rate of a few grams every five min. Alternatively, a saturated solution (10 0%) of ammonium sulfate can be added in these steps, at the cost of greatly increasing the sample volume. This precipitation step consists of two phases: first, proteins that precipitate below 30% ammonium sulfate saturation are removed, followed by the selective precipitation of CCT, which occurs between 30%–45% saturation.

14. When starting the purification with 200 g of tissue, the expected volume of sample to be loaded onto the sucrose gradients is around 18 mL. Therefore, you should expect to use approximately six sucrose gradients to process the entire volume.

15. A small amount of the precipitated sample can be stored at this point to check the efficiency of the precipitation by SDS-PAGE.

16. Ensure that all of the pellets are covered by buffer A for proper soaking. Since the pellet is usually slanted towards one side of the bottom of the tube, tilting the tubes over a bed of ice is recommended.

17. To ensure optimal dialysis, it is advised to use a 1 L graduated cylinder pre-filled with buffer and maintained at 4°C. Cut a piece of dialysis tubing slightly longer than the required length, and fully submerge it in buffer to soften the membrane. Avoid touching the membrane itself—handle only the ends that will not be in contact with the sample. Tie a secure knot at one end, gently pushing the knot inward to avoid stressing the portion of the membrane that will contain the sample. Then, invert the tubing, carefully load the

sample, remove any air bubbles, and tie a second knot to seal the other end. Suspend the dialysis bag in the cylinder using either a clamp or a float, ensuring it does not touch the bottom of the cylinder, where a magnetic stir bar is placed to maintain gentle agitation. Contact between the membrane and the stir bar can damage the tubing and compromise the sample. Perform the dialysis at 4°C for 1 h as described.

18. When loading the sample onto the sucrose gradients, it should be carefully dispensed along the inner wall of the tube to ensure it gently settles on the upper layers without disturbing the gradient distribution. If flexible tubes are used for the sucrose gradients, it is crucial to fill them up completely—even if they are just counterbalances—to prevent collapse.

19. Gradient fractionation can be performed either manually or automatically, depending on the available resources. An automatic fraction collector can be set to collect 2.2 mL per tube, but manual collection is also feasible by determining how many drops from the outlet correspond to this volume and then counting the same number of drops (around 40) for each fraction. As long as the method is applied consistently across all tubes, it yields reliable results.

20. A recommended dialysis membrane is (VSWP02500, Merck Group)

21. After adding the fixation and staining solution, the gel in the solution can be optionally heated in a microwave for 20 s, or just until the solution starts to boil. This improves the staining.

22. Non-sterile, low-binding polypropylene (PP) filters with cellulose membranes are used. Recommended filters are commercially available (10463050 and 10463060, Cytiva).

23. In this protocol, we have described the purification process using a series of FPLC columns from a particular company (HiTrap™ Heparin HP affinity column, 17040701 (Cytiva)). However, equivalent columns with similar properties can be used instead. The key aspects to consider when selecting alternative columns are their binding capacity, resolution, and compatibility with the buffers and conditions described in this method.

24. In order to increase the purification yields, Steps 3.1–3.6 can be repeated. Then, the pools from both affinity chromatographies (the first of which should have been properly frozen) are combined and used in the final size exclusion chromatography step.

25. Amicon® filters with Ultracel® membranes, made of regenerated cellulose, are used for this step. Recommended units are available commercially (UFC9100; Millipore, Merck Group).

26. The sample is centrifuged instead of filtered because its volume after concentration is too low, and filtration would lead to significant sample loss. A high-speed centrifugation allows for the removal of impurities that might otherwise clog the FPLC filters. The supernatant is carefully collected without disturbing the bottom of the tube.
27. In this protocol, we have described this step of the purification process using a prepacked FPLC column from a particular company (Superose 6 HiLoad column, 29323952 (Cytiva)). Equivalent columns with similar properties can be used instead. The key aspects to consider when selecting alternative columns are their binding capacity, resolution, and compatibility with the buffers and conditions described in this method.
28. The expected yield obtained using this protocol is around 15 mg of CCT for each 200 g of initial bovine tissue.
29. It is crucial that the final buffer, in which the protein is snap-frozen, contains 10% glycerol to enhance sample preservation. CCT can be stored at −80°C or lower for up to 1 year without loss of activity.

## Acknowledgments

This research was supported by the Spanish Ministry of Science and Innovation and Agencia Estatal de Investigación through the grant PID2022-137175NB-I00 (AEI/FEDER, UE) and AEI/10.13039/501100011033, through the "Severo Ochoa" Programme for Centres of Excellence in R&D (CEX2023-001386-S) to J.M.V. and J.C. We also acknowledge the support of the CaixaResearch Health 2023 grant (HR23-00837) to J.M.V. In addition, J.C. acknowledges the support of a grant from Fundación Ramón Areces. The authors acknowledge the technical support of the cryoEM CNB-CSIC facility.

### References

1. Bueno-Carrasco MT, Cuéllar J (2018) Mechanism and function of the Eukaryotic Chaperonin CCT. In: eLS, John Wiley & Sons, Ltd., Chichester, UK, pp 1–9. https://doi.org/10.1002/9780470015902.a0028208
2. Pipaón S, Gragera M, Teresa Bueno-Carrasco M et al (2021) Chaperonins: nanocarriers with biotechnological applications. Nanomaterials 11: 1–12. https://doi.org/10.3390/nano11020503
3. Kalisman N, Adams CM, Levitt M (2012) Subunit order of eukaryotic TRiC/CCT chaperonin by cross-linking, mass spectrometry, and combinatorial homology modeling. Proc Natl Acad Sci USA 109:2884–2889. https://doi.org/10.1073/pnas.1119472109
4. Leitner A, Joachimiak LA, Bracher A et al (2012) The molecular architecture of the eukaryotic chaperonin TRiC/CCT. Structure 20:814–825. https://doi.org/10.1016/j.str.2012.03.007
5. Grantham J (2020) The molecular chaperone CCT/TRiC: an essential component of

proteostasis and a potential modulator of protein aggregation. Front Genet 11:1–7. https://doi.org/10.3389/fgene.2020.00172
6. Reissmann S, Joachimiak LA, Chen B et al (2012) A gradient of ATP affinities generates an asymmetric power stroke driving the chaperonin TRIC/CCT folding cycle. Cell Rep 2:866–877. https://doi.org/10.1016/j.celrep.2012.08.036
7. Balchin D, Miličić G, Strauss M et al (2018) Pathway of actin folding directed by the eukaryotic chaperonin TRiC. Cell 174:1507–1521. https://doi.org/10.1016/j.cell.2018.07.006.e16
8. Muñoz IG, Yébenes H, Zhou M et al (2011) Crystal structure of the open conformation of the mammalian chaperonin CCT in complex with tubulin. Nat Struct Mol Biol 18:14–19. https://doi.org/10.1038/nsmb.1971
9. Cuéllar J, Martín-Benito J, Scheres SHW et al (2008) The structure of CCT–Hsc70NBD suggests a mechanism for Hsp70 delivery of substrates to the chaperonin. Nat Struct Mol Biol 15:858–864. https://doi.org/10.1038/nsmb.1464
10. Betancourt Moreira K, Collier MP, Leitner A et al (2023) A hierarchical assembly pathway directs the unique subunit arrangement of TRiC/CCT. Mol Cell 83:3123–3139. https://doi.org/10.1016/j.molcel.2023.07.031.e8
11. Souès S, Kann ML, Fouquet JP, Melki R (2003) The cytosolic chaperonin CCT associates to cytoplasmic microtubular structures during mammalian spermiogenesis and to heterochromatin in germline and somatic cells. Exp Cell Res 288:363–373. https://doi.org/10.1016/S0014-4827(03)00248-9
12. Frydman J, Nimmesgern E, Erdjument-Bromage H et al (1992) Function in protein folding of TRiC, a cytosolic ring complex containing TCP-1 and structurally related subunits. EMBO J 11:4767–4778. https://doi.org/10.1002/j.1460-2075.1992.tb05582.x
13. Ferreyra RG, Frydman J (2015) Purification of the cytosolic chaperonin TRiC from Bovine Testis. In: Chaperonin protocols. Humana Press, New Jersey, pp 153–160

Check for updates

# Chapter 2

# Single-Step Purification of Endogenous Human Chaperonin CCT

**Javier M. Rodríguez, Esther Martín-Forero, Jorge Gutiérrez-Seijo, Jimena Muntaner, Carmen Majano, José María Valpuesta and Jorge Cuéllar**

## Abstract

The eukaryotic chaperonin CCT (chaperonin-containing TCP-1) is essential for protein homeostasis in eukaryotic cells, and its dysfunction has been linked to an increasing number of pathologies, including cancer, neurodegeneration, and infectious diseases. Traditional CCT purification methods are complex, multi-step processes, while newer affinity chromatography approaches rely on tagged, non-native CCT. Here, we describe a single-step method to purify endogenous CCT from a custom Expi293F cell line generated using the Sleeping Beauty transposon system. This cell line inducibly expresses a phosducin-like protein (PhLP1) with a C-terminal Twin-Strep-tag, which binds CCT with high affinity. Using Strep-Tactin XT resin, we isolate CCT directly from cell lysates via the PhLP1–CCT interaction, achieving high-purity CCT with yields comparable to traditional methods. This rapid, specific approach simplifies CCT purification while maintaining scalability and efficiency.

**Keyword** CCT, Molecular chaperone, Chaperonin, PhLP1, Single-step purification, Expi293F cells, Twin-Strep-tag, Affinity chromatography

## 1 Introduction

The eukaryotic chaperonin CCT, also known as TRiC, is a highly abundant [1] and essential molecular chaperone found in the cytoplasm of all eukaryotic cells. Its main function is to assist a discrete set of client proteins in reaching their native conformations [2, 3]. CCT is a roughly cylindrical complex of approximately 1 MDa in size, composed of two back-to-back rings of eight distinct but similar (~30%–35% sequence identity) 55–60 kDa protein subunits (CCT1–8 in yeast, CCTα–CCTθ in mammals), encoded by paralogous genes [4–6]. Each ring encloses a central cavity in which substrate proteins fold in an ATP-dependent conformational cycle [7].

Noa B. Martín-Cófreces et al. (eds.), *The Cytoskeleton*, Methods in Molecular Biology, vol. 3070,
https://doi.org/10.1007/978-1-0716-5515-3_2,

Although CCT can interact directly with some of its substrates, its function is better understood as part of a dynamic network in which it transiently associates with other chaperones like HSP70 [8, 9], and with a diverse set of co-chaperones, like the prefoldin complex [10, 11] and the PhLP1 [12–15]. Those molecules participate in the recruitment of their substrates, modulate the folding process, and regulate the release of the folded proteins, among other auxiliary functions. CCT clients tend to be multidomain proteins with intricate topologies, β-sheet-rich proteins with complex folding trajectories, and aggregation-prone proteins [16–19]. The best characterized CCT substrates are the essential cytoskeletal proteins actin and tubulin [2, 20], which are entirely dependent on CCT for proper folding. Relevant CCT substrates include proteins implicated in the regulation of cell cycle, T-cell immunity, autophagy, signal transduction, cancer, and neurological disorders [21, 22]. Due to the physiological relevance of its clients, CCT dysfunction is associated with an increasing number of diseases, making it an attractive target for biomedical interventions, which makes the availability of robust methods for the preparation of CCT of considerable relevance.

CCT has historically been purified using classical biochemical methods followed by multiple chromatographic steps, from bovine testis [23, 24] [see Purification of the eukaryotic chaperonin CCT from native tissues], rabbit reticulocyte lysates [25], cultured cell lines [26] or *Trichoplusia ni* insect cells infected with baculovirus expressing the complete set of human subunits [10]. Although effective, these methods are labor-intensive, technically demanding, and require large amounts of starting material. Simpler and more efficient tag-based affinity chromatography methods have since been developed for purifying CCT from human cells [20, 27] and yeast cells [28]. However, these rely on introducing an affinity tag into one of the CCT subunits, which may interfere with complex function.

An elegant approach enabling ligand-specific affinity chromatography for the purification of unmodified endogenous CCT has been applied to yeast [4] and human cell lines [29]. This strategy uses a high-affinity binding partner of CCT, a protein from the PhLP family of co-chaperones, into which an affinity tag is inserted to facilitate purification of the CCT:co-chaperone complex. However, these methods require either pre-purification of the co-chaperone [4] or transient co-expression in the cells from which CCT is purified [29]. An additional chromatographic step is also necessary to separate CCT from the co-chaperone. Here, we present a simplified version of this strategy for the purification of endogenous human CCT from Expi293F, a HEK293 derivative optimized for high-density suspension culture. Using the Sleeping Beauty transposon system [30, 31], we generated a stable cell line that inducibly expresses the PhLP1 co-chaperone engineered with a C-terminal Twin-Strep-tag

[32] (PhLP1-TST). Upon induction with doxycycline, CCT:PhLP1-TST complexes are formed. Purification is achieved in a single step using Strep-Tactin XT chromatography [33], with CCT specifically eluted from the column using a NaCl-based buffer. Compared to existing CCT purification methods, this approach offers significant advantages in terms of scalability, speed, and simplicity, while maintaining high purity and yield.

## 2 Materials

### *2.1 Common Equipment*

1. Cell culture $CO_2$ incubator.
2. Refrigerated centrifuge.
3. $CO_2$ incubator shaker.
4. Sterile cell culture laminar flow hood safety level II.
5. Inverted microscope.
6. Peristaltic pump.
7. Magnetic stirrer and stirring bars.
8. Gradient mixer.

### *2.2 Production of Plasmids*

1. Luria Broth Base (Miller's LB Broth Base): 170 mM NaCl, pH: 7.5, 1% (w/v) tryptone, and 0.5% (w/v) yeast extract.
2. Autoclave 125 mL Miller's LB Broth Base with 1.5% (w/v) agar, bacteriological grade.
3. Cool to 60°C in a water bath.
4. Aseptically add carbenicillin to 50 mL for a final concentration of 100 μg/mL and pour two ≈25 mL Petri dishes.
5. Pour two additional Petri dishes with 35 μg/mL chloramphenicol.
6. Allow plates to solidify for 12–16 h at room temperature in the dark.
7. Store wrapped at 4°C for up to 4 weeks.
8. Streak bacterial cultures (pSBtet-RP on carbenicillin plates; pCMV(CAT)T7-SB100 on chloramphenicol plates; available at Addgene) using a sterile loop or pipette tip.
9. Incubate inverted overnight (16–18 h) at 37°C.
10. Inoculate a single, well-isolated colony for each plasmid into 5 mL LB broth containing the appropriate antibiotic (100 μg/mL carbenicillin for pSBtet-RP or 35 μg/mL chloramphenicol for pCMV(CAT)T7-SB100).

11. Grow cultures overnight (16–18 h) at 37°C with orbital shaking (200–250 rpm).
12. Purify plasmid DNA using the QIAGEN Plasmid Mini Kit or similar, according to the manufacturer's instructions.
13. Quantify purified DNA using a NanoDrop or Qubit fluorometer to confirm yield and purity (*see* **Note 1**).

### 2.3 Construction of the Sleeping Beauty Vector

1. PhLP1-TST: Synthetic DNA containing the PhLP1-Twin-Strep-tag fusion protein coding sequence and the 5′ and 3′ hybridization sequences required for Gibson assembly. Resuspend the DNA in 10 mM Tris-HCl, pH 8.0 at 20 ng/μL and store at −20°C (*see* **Note 2**).
2. Oligonucleotides for PCR amplification: EBV-rev (GTGGTTTGTCCAAACTCATC); LNCX (AGCTCGTT TAGTGAACCGTCAGATC).
3. DreamTaq Hot Start PCR Master Mix or a similar hot start DNA polymerase and required solutions for PCR.
4. Restriction endonucleases: *Cla*I and *Nco*I.
5. Antarctic Phosphatase.
6. Gibson Assembly Cloning Kit (New England Biolabs).
7. Antibiotics: Carbenicillin Disodium Salt and Chloramphenicol.
8. QIAGEN Plasmid Mini Kit or other kit to purify plasmid DNA from small bacterial cultures.
9. QIAGEN MinElute Reaction Cleanup Kit or any other kit to purify small DNA from reactions (70 bp-4 kB).

### 2.4 Cell Lines and Maintenance

1. Expi293F cells.
2. HyClone peak expression medium (*see* **Note 3**).
3. Complete DMEM: DMEM, high glucose, GlutaMAX, pyruvate supplemented with 10% (v/v) Fetal Bovine Serum (*see* **Note 3**).
4. TrypLE Express or any trypsine-like enzyme, animal-free.
5. Dimethyl sulfoxide (DMSO), cell culture grade to freeze cells.
6. Thermo Scientific Nunc Cryogenic Tubes.
7. Corning CoolCell FTS30 Cell Freezing Vial Container.
8. Nunc EasYDish dishes or any cell plate made of noncytotoxic, nonpyrogenic gamma-irradiated polystyrene plastic, free of RNase, DNase, and human DNA, of 100 mm and 150 mm diameter.
9. Cell Drop BF automated cell counter.

### 2.5 Transfection of Expi293F Cells

1. PEI 25k solution: 1 mg/mL PEI 25k. Add 100 mg of PEI to 95 mL of ultra-pure water. While stirring, bring the pH of the solution to 2.0 by slowly adding hydrochloric acid. Once dissolved, carefully adjust the pH to 7.0 using 10 N sodium hydroxide. Bring the final volume to 100 mL with ultra-pure water, then sterilize the solution by filtration through a 0.22 μ m filter. Aliquot and store at −20°C. Once thawed, do not use aliquots older than 2 months.
2. 50 mg/mL normocin antibiotic. Sterile filtered.
3. 12-well Multi-well Cell Culture Plate.

### 2.6 Puromycin Selection

1. 10 mg/mL puromycin for selection. Seterile filtered.

### 2.7 Induction of PhLP1-TST Expression

1. Doxycycline monohydrate solution: 10 mg/mL doxycycline monohydrate in ultra-pure water. Sterilize by filtration through a 0.22 μm filter. Store at −20°C.
2. 10× PBS: 100 mM $Na_2HPO_4$ pH 7.4, 17.6 mM $KH_2PO_4$, 1.37 M NaCl, 27 mM KCl.

### 2.8 Isolation of CCT by Chromatography on Strep-Tactin XT 4Flow High-Capacity Columns

1. Potter-Elvehjem PTFE pestle and glass tube, 30 mL.
2. Biotin stock solution: 100 mM biotin in ultrapure water.
3. Protease inhibitors.
4. BioLock (IBA Lifesciences).
5. DNAse I solution: 20 mM Tris-HCl, pH 7.6, 10 mg/mL DNAse I, 50 mM NaCl, 2 mM $MgCl_2$, 2 mM $CaCl_2$, 1 mM DTT, and 50% (v/v) glycerol. Store in aliquots at −20°C.
6. RNAse A solution: 10 mM HEPES pH 7.2, 5 mg/mL RNAse A, 20 mM NaCl, 0.1% Triton X-100, and 50% (v/v) glycerol. Store in aliquots at −20°C.
7. Ethylenediaminetetraacetic acid (EDTA) solution: 0.5 M EDTA, pH 8.
8. Whatman GD/X sterile PES syringe filters.
9. Sterile gauze.
10. Strep-Tactin XT 4Flow high-capacity 0.2 mL column.
11. 5× Phosphate buffer: 250 mM $NaH_2PO_4{\cdot}H_2O$, pH 8.0. Adjust the pH to 8.0 with NaOH. Use ultra-pure water. Filter the solution through a 0.22 μm filter to sterilize and remove any particulates.
12. Lysis buffer: 1× phosphate buffer, 103 mM NaCl, 5 mM $MgCl_2$, 1 mM EDTA, 10% (v/v) glycerol, 0.5% Triton × -100, 10 mM 2-mercaptoethanol. Add protease inhibitors immediately before use.

13. EDTA lysis buffer: lysis buffer, 10 mM EDTA. Add protease inhibitors immediately before use.
14. Wash buffer: 1× phosphate Buffer, 103 mM NaCl, 5 mM $MgCl_2$, 1 mM EDTA, 10% (v/v) Glycerol, 10 mM 2-Mercaptoethanol. Add 2 mM ATP and protease inhibitors immediately before use.
15. NaCl Elution buffer: 1× phosphate buffer, 750 mM NaCl, 5 mM $MgCl_2$, 1 mM EDTA, 10% (v/v) glycerol, 10 mM 2-mercaptoethanol. Add 2 mM ATP and protease inhibitors immediately before use.
16. Biotin Elution buffer: 1× phosphate buffer, 103 mM NaCl, 5 mM $MgCl_2$, 1 mM EDTA, 10% (v/v) glycerol, 10 mM 2-mercaptoethanol, 50 mM Biotin. Add protease inhibitors immediately before use.
17. Buffer W: 10 mM Tris-HCl, pH 8.0, 150 mM NaCl, 1 mM EDTA.
18. Amicon Ultra-4 Centrifugal Filter 100 kDa.

### *2.9 Characterization of the Purification Process*

1. Bovine serum albumin standard ampules, 2 mg/mL.
2. 4× SDS-PAGE sample buffer: 50 mM Tris-HCl, pH 6.8, 8% SDS, 40% glycerol, 0.1% bromophenol blue, 400 mM DTT.
3. 4× Stacking buffer: 500 mM Tris-HCl, pH 6.8, 0.4% SDS.
4. 4× Resolving buffer: 1.5 M Tris-HCl, pH 8.8, 0.4% SDS.
5. Acrylamide bis-acrylamide solution: 40% acrylamide:bis-acrylamide (37.5:1).
6. Ammonium persulfate solution (APS): 10% ammonium persulfate (electrophoresis grade).
7. N,N,N′,N′-Tetramethyl-ethylenediamine (TEMED (electrophoresis grade)).

### *2.10 SDS-PAGE Mini-Gels*

To prepare 12% SDS-PAGE gels, 0.75 mm thick:

1. Combine 3.56 mL of 60% glycerol, 2 mL of 4x Resolving buffer, 2.4 mL of 40% Acrylamide/Bis-acrylamide, and 4 μL of TEMED to prepare the resolving gel solution.
2. Prepare the stacking gel solution by mixing 2.478 mL of ultra-pure water, 1 mL of 4x Stacking buffer, 0.5 mL of 40% Acrylamide/Bis-acrylamide, and 2 μL of TEMED.
3. Add 40 μL of freshly prepared APS to the resolving gel solution, mix thoroughly, and pour the solution into the assembled gel molds.
4. Immediately after, add 20 μL of freshly prepared 10% APS to the stacking gel solution, mix, and pour gently on top of

the unpolymerized resolving gels to fill the remaining space. Insert the desired combs.

5. Allow the gels to polymerize for at least 30 min at room temperature before use.

### 2.11 Gradient SDS-PAGE Mini-Gels

For the preparation of two 10%–12.5% gradient SDS-PAGE mini-gels (0.75 mm thick):

1. Prepare the 10% resolving gel solution by combining 2.467 mL of ultra-pure water, 1.25 mL of 4× Resolving buffer, 1.25 mL of 40% Acrylamide/Bis-acrylamide, and 3 μL of TEMED.
2. For the 12.5% resolving gel solution, combine 2.165 mL of 60% glycerol, 1.25 mL of 4x Resolving buffer, 1.563 mL of 40% Acrylamide/Bis-acrylamide, and 2 μL of TEMED.
3. The stacking gel solution should be prepared as described previously.
4. To pour a gel, set up a clean, dry gradient mixer on a magnetic stirrer with a stir bar in the outlet chamber.
5. Connect the mixer outlet to the inlet of a peristaltic pump and a 25G needle at the end of the outlet tubing.
6. Adjust the pump to a speed of approximately 1 mL/min.
7. With the communication between the chambers blocked, add 2 mL of the 12.5% resolving gel solution mixed with 10 μL of fresh 10% APS to the stir-bar chamber.
8. Add 2 mL of the 10% resolving gel solution thoroughly mixed with 12 μL of fresh 10% APS into the other chamber.
9. Unblock the communication and start the pump.
10. Monitor the pouring of the gel, keeping the needle near the gel surface to avoid distortion of the gradient.
11. Once the gel has reached the desired height, stop the pump.
12. Immediately flush the system with water to prevent polymerization.
13. Overlay the top of the gradient gel with 1 mL of water-saturated 2-butanol.
14. Clean and dry the gradient mixer before proceeding to cast the second gel.
15. After the resolving gel has polymerized for at least 30 min at room temperature, pour the stacking gel as described for the 12% gels, insert the combs, and allow it to polymerize for at least 30 min before use.

## 3 Methods

As shown in Fig. 1, this protocol is to be performed into two major phases. Phase 1 includes Sects. 3.1–3.3, and describes the construction of a stable cell line that inducibly expresses the target protein, PhLP1-TST. These three protocols must be performed sequentially, and the process can be paused between protocols.

Phase 2 includes Sects. 3.4–3.8, and covers the large-scale amplification of this cell line, the induction of PhLP1-TST, and the harvesting of the induced cells (Sect. 3.4), the isolation of CCT from the induced cells (Sects. 3.5 and 3.6), and the analysis of the purification process (Sects. 3.7 and 3.8).

### *3.1 Construction of the Sleeping Beauty Vector (pSBtet.PhLP1-TST) (see Note 2)*

1. Digest 2 μg of plasmid pSBtet-RP (Fig. 2b) with 20 units each of *Cla*I and *Nco*I in 50 μL of 1× rCutsmart buffer for ≥1 h at 37°C.

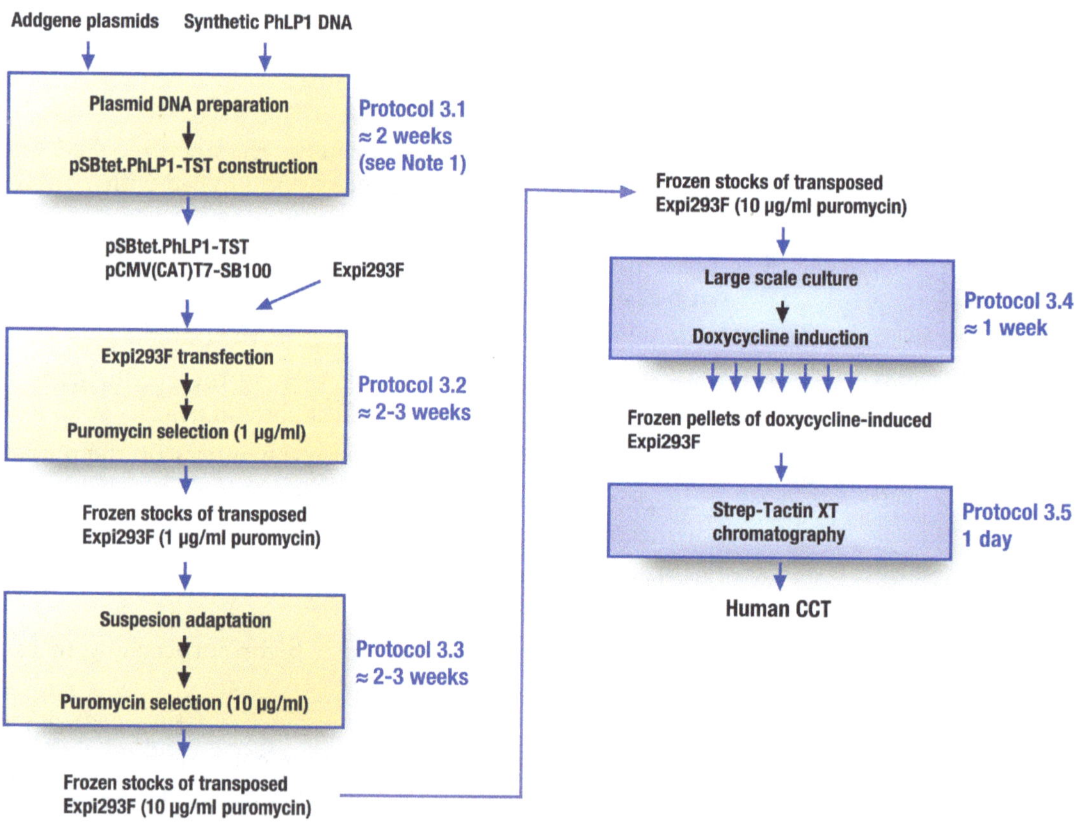


**Fig. 1** Outline of the CCT purification workflow. The figure illustrates the different steps in the process, the relevant protocol associated with each step, and its estimated duration. The inputs and outputs of each step are indicated. The steps on the left led to the construction of the Expi293 cell line transposed with the PhLP1-TST gene. Using this cell line, the processes indicated on the right led to the purification of CCT

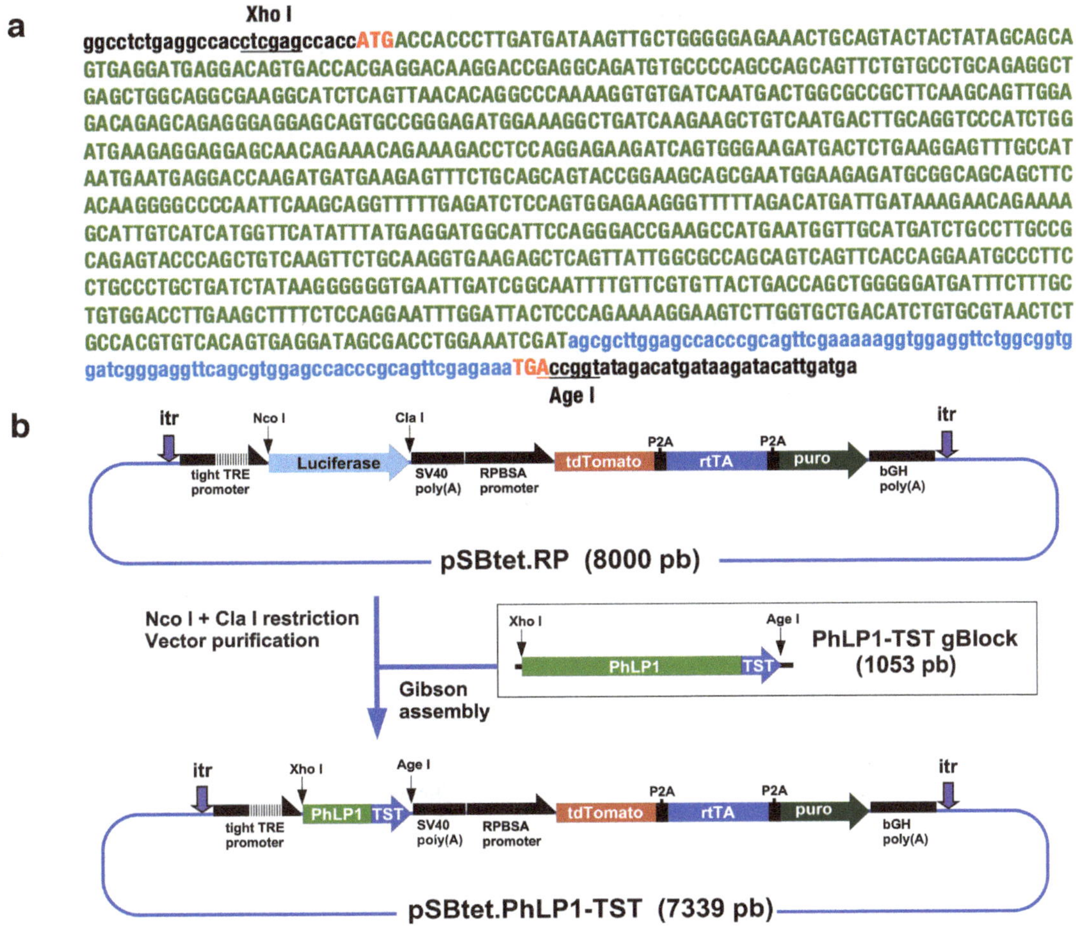


**Fig. 2** Construction of the pSBtet.PhLP1-TST plasmid. (**a**) The complete sequence of the synthetic DNA. The PhLP1 coding sequence is shown in green, fused to a Twin-Strep tag sequence shown in blue. Sequences required for hybridization during Gibson assembly are indicated in black, along with unique *Xho*I and *Age*I restriction sites. (**b**) Main features of the pSBtet.RB plasmid vector and outline of the cloning process. The synthetic DNA is cloned via Gibson assembly between the *Nco*I and *Age*I restriction sites, resulting in the formation of the pSBtet.PhLP1-TST plasmid

2. Verify digestion by loading 6 μL of the reaction, 200 ng of the undigested plasmid, and a DNA ladder onto a 1% agarose gel. Expected fragment sizes: 6329 bp (vector backbone) and 1677 bp (luciferase coding sequence).
3. Add 5 μL of 10X Antarctic Phosphatase buffer and 1 μL of Antarctic Phosphatase to the reaction. Incubate at 37°C for 30 min.
4. Inactivate enzymes by heating the reaction at 80°C for 20 min.
5. Run the entire digestion reaction on a 1% agarose gel.

6. Excise the 6329 bp vector backbone band, minimizing UV exposure to prevent DNA damage.
7. Purify the excised vector band using the QIAquick Gel Extraction Kit or similar, according to the manufacturer's protocol.
8. Quantify the purified vector DNA using a NanoDrop or Qubit fluorometer.
9. Store at −20°C.
10. Assemble 100 ng of digested pSBtet-PB vector with 20 ng of synthetic PhLP1-TST DNA using the Gibson Assembly Cloning Kit (Fig. 2b) or similar, following the manufacturer's protocol.
11. Identify positive colonies by colony PCR using EBV-rev and LNCX oligonucleotides with DreamTaq Hot Start PCR Master Mix, or a comparable reagent. Positive colonies yield a 1158 bp band; vector colonies (pSBtet-RP) yield a 1825 bp band.
12. Inoculate at least three putative positive colonies and purify plasmid DNA as previously indicated.
13. Quantify purified DNA using a NanoDrop or Qubit fluorometer to confirm yield and purity.
14. Confirm positive colonies by Sanger sequencing with EBV-rev and LNCX oligonucleotides, or preferably, by whole-plasmid sequencing.

### 3.2 Stable Cell Line Generation with the Sleeping Beauty Transposon System (see Note 4)

1. Expi293F cells are seeded at $0.5 \times 10^6$ cells/mL in vented shaker flasks and cultured in Hyclone Peak Expression Media at 125 rpm, 37°C, with 8% $CO_2$.
2. Cultures are routinely passaged every 2–3 days by diluting the cell suspension to $0.5 \times 10^6$ cells/mL when cell density is between $3 \times 10^6$ and $5 \times 10^6$ cells/mL.
3. Expi293F cells readily grow as adherent cultures when maintained in complete DMEM. Cells are cultured in tissue culture flasks in a humidified incubator at 37°C with 5% $CO_2$. Cells are passaged when they reach approximately 70%–80% confluence (usually every 3 days) by detaching with TrypLE Express solution and diluting at a 1:10 ratio into fresh complete medium.
4. For long-term storage, cells are harvested and centrifuged at 300 x*g* for 5 min.
5. Cell pellets are then resuspended at a density of $1 \times 10^7$ cells/mL in freezing medium consisting of 45% fresh complete growth medium, 45% conditioned medium (from Expi293F cultures), and 10% dimethyl sulfoxide (DMSO).

6. Aliquot $1 \times 10^7$ cells into cryovials, transfer to a controlled-rate freezing device, and store at −80°C for 24 h before transferring to liquid nitrogen for long-term storage.
7. Suspension-adapted cells are thawed in pre-warmed media at a concentration of $3–5 \times 10^5$ cells/mL and grown until cell density reaches $\geq 1 \times 10^6$ cells/mL with a viability ≥90%.
8. After cells have reached this threshold, they are cultured as indicated previously.
9. Allow at least 3 passages after thawing before using the cells for expression.
10. Adherent cells are thawed in 25 mL of pre-warmed complete DMEM and seeded in a 150 mm plate.
11. The following day, replace the medium with fresh pre-warmed complete DMEM, and maintain the cells as described previously.
12. Allow at least 3 passages after thawing before using the cells for transfection.
13. The day before transfection, seed a 12-well plate with $2.5 \times 10^5$ Expi293F cells per well in 1 mL of complete DMEM.
14. Seed a minimum of 3 wells.
15. On the day of transfection, replace the medium with fresh complete DMEM.
16. In a sterile 1.5 mL microcentrifuge tube, combine 728 ng of the pSBtet.PhLP1-TST plasmid with 72 ng of the pCMV(-CAT)T7-SB100 plasmid (*see* **Note 5**).
17. Adjust the volume to 25 μL with 150 mM NaCl.
18. For the puromycin selection control, prepare a separate tube containing only 72 ng of the pCMV(CAT)T7-SB100 plasmid in 25 μL of 150 mM NaCl.
19. In a sterile 1.5 mL microcentrifuge tube, add 7.2 μg of PEI and adjust the volume to 75 μL with 150 mM NaCl.
20. Incubate for 5 min at room temperature.
21. Add 25 μL of the PEI solution to the tube containing the plasmid DNA and to the control tube prepared in Step 17.
22. Briefly vortex for 3 s immediately after each addition.
23. Incubate for 20 min at room temperature to allow PEI/DNA complexes to form.
24. Add 50 μL of the prepared complexes to each designated well of the 12-well plate.
25. Gently mix to evenly distribute the complexes. Incubate the plate for 24 h at 37°C.

26. After 24 h, carefully remove the cell supernatant.
27. Add 1 mL of fresh complete DMEM with 100 μg/mL of Normocin.
28. Incubate until the cell monolayer reaches approximately 80%–90% confluence (typically within 24–48 h).
29. Detach the cells in each well and resuspend them in 10 mL of complete DMEM containing 1 μg/mL puromycin.
30. Transfer the cell suspension to a 100 mm culture dish.
31. Incubate the cells at 37°C.
32. Monitor the cells daily using an inverted microscope.
33. Replace the medium with fresh complete DMEM containing 1 μg/mL puromycin every 48 h.
34. Continue until all cells in the control dish (without pSBtet. PhLP1-TSTplasmid) have died, typically within 2–4 days.
35. During this period, discrete puromycin-resistant colonies should develop in the experimental dish (*see* **Note 6**).
36. Once the control cells are eliminated, detach puromycin-resistant cells.
37. Replate them in a fresh 100 mm culture dish containing 10 mL of complete DMEM with 1 μg/mL puromycin.
38. Culture until the cells reach 80%–90% confluence (usually 1–4 days) (*see* **Note 7**).
39. Detach the cells and resuspend them in 10 mL of complete DMEM containing 1 μg/mL puromycin.
40. Seed a 150 mm culture dish with 3 mL of this cell suspension, then add an additional 22 mL of DMEM containing 1 μg/mL puromycin.
41. Incubate until cells reach 80%–90% confluence (typically 2–3 days).
42. Detach the cells from the 150 mm plate and resuspend them in 5 mL of fresh complete DMEM.
43. Count the cells using a 1:5 dilution of the suspension. Set aside $1 \times 10^7$ cells for suspension adaptation (see Sect. 3.3).
44. Prepare frozen stocks with the remaining cells as described in Steps 3–5.

### 3.3 Adaptation to Suspension Culture and Selection of Cells Resistant to 10 μg/mL of Puromycin

1. Dilute the $1 \times 10^7$ cells reserved from Step 5 of the previous section to a total volume of 20 mL using Peak Expression Media containing 2% fetal bovine serum (FBS) and 1 μg/mL puromycin.
2. Transfer the suspension to a 125 mL vented Erlenmeyer flask and incubate in a shaker incubator at 37°C and 125 rpm

until the cell density reaches >$3 \times 10^6$ cells/mL (3–4 days). Initial viability may be below 80% but should rapidly increase to ≥ 90% (*see* **Note 8**).

3. Split the suspension culture into fresh Peak Expression Media containing 10 μg/mL puromycin, adjusting the density to $0.5 \times 10^6$ cells/mL.
4. Continue culturing by diluting the culture back to $0.5 \times 10^6$ cells/mL when cells reach a density >$3 \times 10^6$ cells/mL (approximately every 3 days), until viability consistently reaches ≥ 90% (typically 3 passages).
5. Split the cells into 50 mL of Peak Expression Media containing 10 μg/mL puromycin at a density of $0.5 \times 10^6$ cells/mL in a 250 mL vented Erlenmeyer flask.
6. Culture until cells reach >$3 \times 10^6$ cells/mL.
7. Prepare frozen stocks of these 10 μg/mL puromycin-resistant cells as described in Sect. 3.2 (Steps 3–5).
8. Maintain a 50 mL culture of the selected cells to proceed to the induction of the PhLP1 expression.

### 3.4 Induction of PhLP1-TST Expression

Cells are maintained in 50 mL cultures in 250 mL vented Erlenmeyer flasks in Peak Expression Media containing 10 μg/mL puromycin, seeded at $0.5 \times 10^6$ cells/mL, and passaged by diluting to this density once cell density reaches ≥$3 \times 10^6$ cells/mL. The expression of PhLP1-TST is induced when cells have reached a density of $2 \times 10^6$ cells/mL with a viability ≥ 90%. Achieve induction by adding doxycycline to a final concentration of 0.5 μg/mL every 24 h for a total of 48 h. The specific culture strategy depends on the availability of adequate vessels and shaker space. An example strategy for inducing 2 L of the 10 μg/mL puromycin-resistant cells is outlined below.

1. Day 0 (Friday): Seed 350 mL of Peak Expression Media containing 10 μg/mL puromycin with $0.5 \times 10^6$ cells/mL in a 1 L vented Erlenmeyer flask.
2. Day 3 (Monday): Count cells and dilute the culture to 2 L with Peak Expression Media at $0.5 \times 10^6$ cells/m in two 2 L vented Erlenmeyer flasks. Continue incubation.
3. Day 5 (Wednesday): Cells should be at approximately $2 \times 10^6$ cells/mL with ≥ 90% viability. Initiate induction by adding doxycycline to a final concentration of 0.5 μg/mL (*see* **Note 9**).
4. Day 6 (Thursday): Add a second dose of doxycycline to 0.5 μg/mL.
5. Day 7 (Friday): Cells should be at least $5 \times 10^6$ cells/mL with ≥ 90% viability. Count the cells and collect them by centrifugation at 500 ×*g* for 15 min at 4°C. Resuspend the cell

pellet in ice-cold 1× PBS. Distribute into aliquots containing ≈ $2.5 \times 10^9$ cells, pellet the cells by centrifugation at $1000 \times g$ for 5 min at 4°C. Remove the supernatant, freeze the pellets using liquid nitrogen, and store them at −80 for downstream processing.

### *3.5 Preparation of the Cytoplasmic Fraction for the Latter Isolation of CCT*

This protocol is designed for the isolation of CCT from $5 \times 10^8$–$2.5 \times 10^9$ cells using a 0.2 mL Strep-Tactin XT 4 Flow high-capacity column. It can be scaled up using larger columns and proportional buffer volumes for each step. To scale down below $5 \times 10^8$ cells, it is recommended to use MagStrep Strep-Tactin XT beads.

1. Retrieve an aliquot of $2.5 \times 10^9$ cells (approximately 8 g) from −80°C storage and place it on ice until thawed.
2. Place a 30 mL Dounce homogenizer on ice.
3. Add 15 mL of ice-cold Lysis buffer to the thawed cell pellet.
4. Gently disaggregate the cell pellet into a crude cell suspension using gentle vortexing or pipetting.
5. Transfer the suspension to the Dounce homogenizer and gently perform 5 push-and-pull strokes with the pestle to mechanically aid cell lysis.
6. Transfer the lysate to a 50 mL conical tube and adjust the volume to 25 mL with Lysis buffer.
7. Add 12.5 U of BioLock and 50 μL of DNase I (10 mg/mL) and RNase A (5 mg/mL) (*see* **Note 10**).
8. Incubate with gentle agitation for 20 min at 4°C. Adjust to 25 mM EDTA by adding 1.3 mL of 0.5M EDTA, pH 8.0. Mix gently and incubate 10 min at 4°C.
9. Take a 25 μL sample of the lysate into a 1.5 mL microcentrifuge tube and set aside at 4°C.
10. Centrifuge the lysate for 30 min at 25,000 x $g$ at 4°C.
11. Carefully transfer the supernatant (the soluble cytoplasmic fraction) to a new 50 mL conical tube, passing it through double-layered gauze to remove any remaining debris.
12. Resuspend the pellet in 20 mL of 1× PBS and disaggregate by vortexing or pipetting.
13. Sonicate the suspension with three 10 s pulses at 30% amplitude to homogenize.
14. Take a 25 μL sample into a 1.5 mL microcentrifuge tube and set aside at 4°C.
15. Filter the soluble fraction sequentially through a 4.5 μm filter and a 0.22 μm filter.
16. Take a 25 μL sample of the final filtered solution into a 1.5 mL microcentrifuge tube and set aside at 4°C.

### 3.6 Isolation of CCT by Strep-Tactin XT Column Chromatography

1. Prepare a 0.2 mL Strep-Tactin XT 4Flow high-capacity resin column on an appropriate chromatography stand, maintaining a temperature of 4°C.
2. Remove the top cap from the column, then twist off the lower cap, allowing the storage buffer to flow through.
3. Wash the column with 0.5 mL of 1xW buffer, followed by 0.5 mL of Lysis buffer containing 10 mM EDTA.
4. Add the filtered soluble fraction to the column and allow it to completely flow through by gravity (*see* **Note 11**).
5. Collect the entire flow-through.
6. Add 250 μL of Lysis buffer containing 10 mM EDTA and collect it with the flow-through.
7. Take a 10 μL sample into a 1.5 mL microcentrifuge tube and set aside at 4°C.
8. Wash the column with 1 mL of Wash buffer, collecting 0.25 mL fractions.
9. Transfer the column to room temperature and immediately elute CCT from the column using 6 mL of NaCl Elution buffer pre-equilibrated to room temperature.
10. Collect the first four 0.25 mL fractions, and a fifth 5 mL fraction.
11. Concentrate the 5 mL fraction to 250 μL using an Amicon Ultra-4 Centrifugal Filter 100 kDa following the manufacturer's instructions.
12. If desired, elute PhLP1-TST from the column using 1.5 mL of Biotin Elution buffer. Collect 0.5 mL fractions.
13. Regenerate the column by washing it successively with 1 mL of W buffer, 3 mL of freshly prepared 100 mM NaOH, and 3 mL of W buffer.
14. Overlay the column with 2 mL of 1×W buffer, and store at 4°C (*see* **Note 12**).

### 3.7 Characterization of the Purification Process

1. Prepare samples for SDS-PAGE: collect the 25 μL samples set aside at 4°C in Sect. 3.5.
2. Add 50 μL of ultrapure water and 25 μL of 4X SDS-PAGE sample buffer.
3. Mix and heat at 95°C for 5 min.
4. Centrifuge at 14000 x*g* for 5 min at room temperature (*see* **Note 13**).
5. Take 12 μL aliquots of the washing and elution samples, and add 4 μL of 4X SDS-PAGE sample buffer.
6. Mix, heat, and centrifuge as before.

7. Run a 10-well 12% SDS-PAGE gel loading 2.5 μL of the samples from Sect. 3.5 and 4 μL of the samples from the washing step (Fig. 3a).
8. Run two 10%–12.5% SDS-PAGE gradient gels loading 5 μL of the samples from the elution steps alongside with BSA as a mass standard. This will determinate the concentration of the eluted proteins (Fig. 3b, c and *see* **Note 14**).

### 3.8 Confirmation of CCT Isolation by Transmission Electron Microscopy (TEM)

1. Coat carbon-coated 300 mesh copper grids by glow-discharge at 25 mA for 15 s.
2. Place 5 μL of the sample on a clean piece of hydrophobic material.
3. Place 5 μL of 2% uranyl acetate nearby.
4. Place the carbon side of the grid onto the sample for 1 min at room temperature.
5. Blot gently to dry the grids with high-quality filter paper.

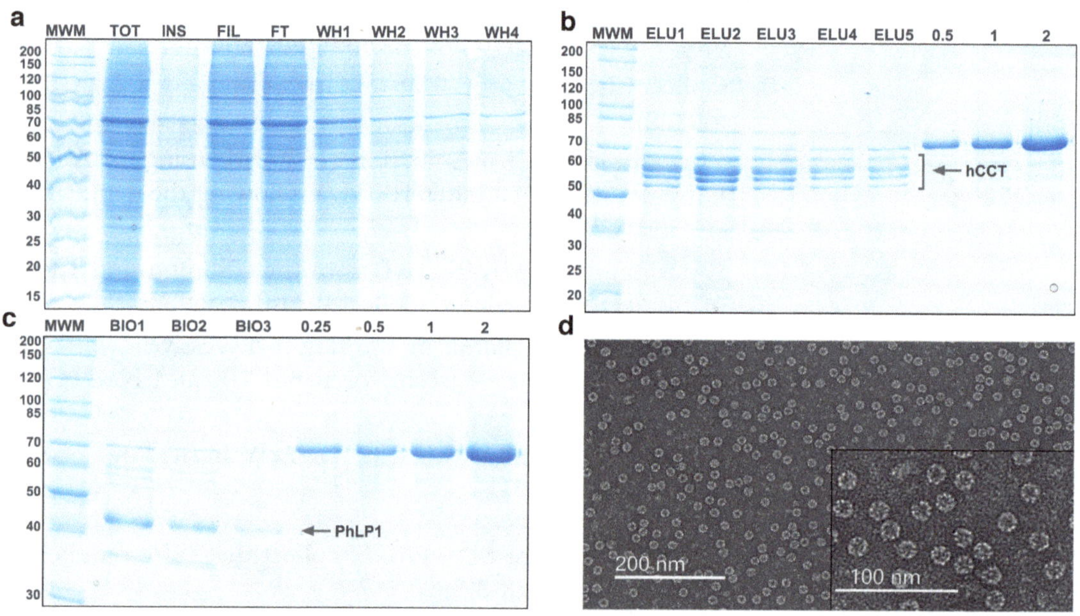


**Fig. 3** Characterization of the purification process. (**a**) 12% SDS-PAGE analysis of the subcellular fractionation and column washing steps. Samples (2.5 μL) from lysed extract (TOT), insoluble material after centrifugation (INS), soluble extract (FIL), and column flow-through (FT) are resolved along with 10 μL samples from the column washing steps (WH1-4). Samples are run alongside a protein molecular weight ladder, whose sizes are indicated. (**b**) 10–12.5% gradient SDS-PAGE analysis of the CCT-containing fractions eluted from the column using the NaCl elution buffer. Samples (5 μL) are run along with BSA as a mass standard, whose quantities are indicated in micrograms. (**c**) 10–12.5% gradient SDS-PAGE analysis of the PhLP1-containing fractions eluted from the column with the biotin elution buffer. Samples (5 μL) are run along with BSA as a mass standard. (**d**) Negatively stained sample (≈0.1 μM) of fraction ELU2 observed by transmission electron microscopy confirms the oligomeric nature of the purified CCT

6. Place the carbon side of the grid onto 2% uranyl acetate for 1 min at room temperature.
7. Blot gently the grids with high-quality filter paper.
8. Left the grids to air dry before imaging.
9. Proceed to TEM to visualize CCT complexes (Fig. 3d).

## 4 Notes

1. Use a miniprep kit that produces endotoxin-free plasmid DNA. We recommend verifying the sequence of the plasmids by whole-plasmid sequencing, or at least verifying the integrity of the plasmids by agarose gel electrophoresis.
2. Although we present a protocol for Gibson assembly of the synthetic PhLP1-TST DNA into the pSBtet-RP vector, for laboratories lacking specialized reagents or molecular biology expertise, outsourcing the cloning process may be more practical and economical. Numerous commercial providers offer cost-effective services to deliver synthetic DNA pre-cloned into the desired vector. Figure 2 includes the information necessary to design the relevant synthetic DNA.
3. We generally avoid using antibiotics in our cell cultures, except for the brief period after transfection and before puromycin selection. Continuous antibiotic use can hinder cell growth and mask potential contamination.
4. In our laboratory, we perform the construction of the Sleeping Beauty transposed cell lines in adherent cultures of Expi293 because our particular setup allows for a higher throughput and easier monitoring of the selection process. Also, we initially select for cells resistant to 1 μg/mL of puromycin to maintain a sample with the widest possible representation in the number of integrated transposons. We then select for cells resistant to 10 μg/mL of puromycin since we have experimentally verified that induction of cell lines selected under these conditions for 48 h with 0.5 μg/mL of doxycycline produces adequate levels of PhLP1.
5. Although plasmid DNA purified following standard procedures is usually free of bacterial contamination, it is advisable to heat the plasmid preparations for 20 min at 80°C to minimize bacterial contamination during transfection.
6. Transposed cells constitutively express tdTomato fluorescent protein; thus, the selection process and the enrichment of transposed cells can be easily followed by fluorescence

microscopy. Alternatively, transposed cells could be selected using fluorescence-activated cell sorting (FACS).

7. Discrete colonies of cells will develop after day 2 of selection and grow rapidly. If the number of transposed cells is low, few colonies will grow, and it will take more time to reach 80% confluence. To avoid the formation of colonies with excessive confluence at their center, after 4 days of selection, detach the cells and replate in a 100 mm dish at $2–4 \times 10^4$ cells/cm$^2$.
8. The objective is to obtain a unicellular suspension of cells. It is normal to observe the formation of clumps of several cells, which will disappear in the second or third passage. If excessive clumping is observed, transfer the cell suspension to a 50 ml conical tube and allow the cells to settle for 10 min. Recover the upper 75% of the culture, count the cells, and seed at $3 \times 10^5$ cells/mL.
9. It is convenient to monitor the induction of the PhLP1 protein during the first induction experiment by western blot, or as a troubleshooting step if purification fails later on. To do this, collect $0.5 \times 10^6$ cells of the culture to be induced before the addition of doxycycline, centrifuge at 500 xg for 5 min, and resuspend in 30 μL of PBS 1×. Add 10 μL of 4× SDS-PAGE sample buffer. Heat at 95° during 5 min. Centrifuge at 14,000 × *g* for 5 min at room temperature. Take similar aliquots at 24 h post-induction, and from the harvested cells at 48 h. Perform a western blot with 4 μL of the samples using Strep-Tactin conjugated with horseradish peroxidase at a 1:1000 dilution. PhLP1-TST appears as a ≈ 44 kDa band induced in the 24 h and 48 h post-induction lanes.
10. BioLock is a convenient semi-purified form of avidin that prevents biotin and biotinylated proteins from binding to Strep-Tactin XT. HEK293 cells contain approximately 1 μg of biotin per $2 \times 10^8$ cells. At least 1 U of BioLock should be added for each $2 \times 10^8$ cells.
11. The Wet Fred set facilitates the unsupervised application of large lysate volumes to gravity columns.
12. Properly regenerated columns can be reused up to five times. To verify the activity of the regenerated column, apply 3 mL of 1× buffer R (100 mM Tris/HCl, pH 8.0, 150 mM NaCl, 1 mM EDTA, 1 mM HABA (2–[4′-hydroxy-benzeneazo] benzoic acid)) (2–1002-100; IBA Lifesciences). An active column should display a uniform intense red color, whose intensity indicates the column activity status. After the verification of the column status, remove the bound HABA following the regeneration protocol (Step 13 in Sect. 3.6).
13. To reduce the viscosity of lysate caused by genomic DNA in total and insoluble extract samples, sonicate the samples

at 10%–20% amplitude using a sonicator equipped with a small tip, applying two to three short pulses of 5 s each. Alternatively, pass the lysate through a narrow-gauge needle (25–27 gauge) five to ten times to shear the DNA.

14. The expected yield of this protocol is around 0.5 mg of human endogenous CCT per $2.5 \times 10^9$ cells. This corresponds to approximately 120,000 CCT complexes per cell. Assuming that a cell contains $1.3 \times 10^6$ CCT complexes [1] this equals to a yield of approximately 10%.

## Acknowledgments

This research was supported by the Spanish Ministry of Science and the Spanish Research Agency through the grant PID2022-137175NB-I00 (AEI/FEDER, UE) and AEI/10.13039/501100011033, through the "Severo Ochoa" Program for Centers of Excellence in R&D (CEX2023-001386-S) to J.M.V. and J.C. The support of the grant CaixaResearch Health 2023 (HR23-00837) to J.M.V. is also acknowledged. In addition, J.C. acknowledges the support of a grant from Fundación Ramón Areces.

## References

1. Hein MY, Hubner NC, Poser I, Cox J, Nagaraj N, Toyoda Y, Gak IA, Weisswange I, Mansfeld J, Buchholz F, Hyman AA, Mann M (2015) A human interactome in three quantitative dimensions organized by stoichiometries and abundances. Cell 163(3):712–723. https://doi.org/10.1016/j.cell.2015.09.053
2. Willison KR (2018) The substrate specificity of eukaryotic cytosolic chaperonin CCT. Philos Trans R Soc Lond B Biol Sci 373(1749):20170192. https://doi.org/10.1098/rstb.2017.0192
3. Bueno-Carrasco MT, Cuéllar J (2018) Mechanism and function of the eukaryotic chaperonin CCT. In: eLS, John Wiley & Sons, Ltd., Chichester, UK, pp 1–9. https://doi.org/10.1002/9780470015902.a0028208
4. Leitner A, Joachimiak LA, Bracher A, Mönkemeyer L, Walzthoeni T, Chen B, Pechmann S, Holmes S, Cong Y, Ma B, Ludtke S, Chiu W, Hartl FU, Aebersold R, Frydman J (2012) The molecular architecture of the eukaryotic chaperonin TRiC/CCT. Structure 20(5):814–825. https://doi.org/10.1016/j.str.2012.03.007
5. Archibald JM, Logsdon JMJr, Doolittle WF (2000) Origin and evolution of eukaryotic chaperonins: phylogenetic evidence for ancient duplications in CCT genes. Mol Biol Evol 17(10):1456–1466. https://doi.org/10.1093/oxfordjournals.molbev.a026246
6. Kubota H, Hynes G, Carne A, Ashworth A, Willison K (1994) Identification of six Tcp-1-related genes encoding divergent subunits of the TCP-1-containing chaperonin. Curr Biol 4(2):89–99. https://doi.org/10.1016/s0960-9822(94)00024-2
7. Joachimiak LA, Walzthoeni T, Liu CW, Aebersold R, Frydman J (2014) The structural basis of substrate recognition by the eukaryotic chaperonin TRiC/CCT. Cell 159(5):1042–1055. https://doi.org/10.1016/j.cell.2014.10.042
8. Melville MW, McClellan AJ, Meyer AS, Darveau A, Frydman J (2003) The Hsp70 and TRiC/CCT chaperone systems cooperate in vivo to assemble the von Hippel-Lindau tumor suppressor complex. Mol Cell Biol 23(9):3141–3151. https://doi.org/10.1128/MCB.23.9.3141-3151.2003
9. Cuéllar J, Martín-Benito J, Scheres SH, Sousa R, Moro F, López-Viñas E, Gómez-Puertas P, Muga A, Carrascosa JL, Valpuesta JM (2008) The structure of CCT-Hsc70 NBD suggests a mechanism for Hsp70 delivery of substrates to the chaperonin. Nat Struct Mol Biol

15(8):858–864. https://doi.org/10.1038/nsmb.1464
10. Gestaut D, Roh SH, Ma B, Pintilie G, Joachimiak LA, Leitner A, Walzthoeni T, Aebersold R, Chiu W, Frydman J (2019) The chaperonin TRiC/CCT associates with prefoldin through a conserved electrostatic interface essential for cellular proteostasis. Cell 177(3):751–765. https://doi.org/10.1016/j.cell.2019.03.012.e15
11. Vainberg IE, Lewis SA, Rommelaere H, Ampe C, Vandekerckhove J, Klein HL, Cowan NJ (1998) Prefoldin, a chaperone that delivers unfolded proteins to cytosolic chaperonin. Cell 93(5):863–873. https://doi.org/10.1016/s0092-8674(00)81446-4
12. Park J, Kim H, Gestaut D, Lim S, Opoku-Nsiah KA, Leitner A, Frydman J, Roh SH (2024) A structural vista of phosducin-like PhLP2A-chaperonin TRiC cooperation during the ATP-driven folding cycle. Nat Commun 15(1):1007. https://doi.org/10.1038/s41467-024-45242-x
13. Han W, Jin M, Liu C, Zhao Q, Wang S, Wang Y, Yin Y, Peng C, Wang Y, Cong Y (2023) Structural basis of plp2-mediated cytoskeletal protein folding by TRiC/CCT. Sci Adv 9(11):eade1207. https://doi.org/10.1126/sciadv.ade1207
14. Stirling PC, Srayko M, Takhar KS, Pozniakovsky A, Hyman AA, Leroux MR (2007) Functional interaction between phosducin-like protein 2 and cytosolic chaperonin is essential for cytoskeletal protein function and cell cycle progression. Mol Biol Cell 18(6):2336–2345. https://doi.org/10.1091/mbc.e07-01-0069
15. Martín-Benito J, Bertrand S, Hu T, Ludtke PJ, McLaughlin JN, Willardson BM, Carrascosa JL, Valpuesta JM (2004) Structure of the complex between the cytosolic chaperonin CCT and phosducin-like protein. Proc Natl Acad Sci USA 101(50):17410–17415. https://doi.org/10.1073/pnas.0405070101
16. Shen PS, Willardson BM (2025) Protein folding by the CCT/TRiC chaperone complex. Curr Opin Struct Biol 91:102999. https://doi.org/10.1016/j.sbi.2025.102999
17. Brehme M, Voisine C, Rolland T, Wachi S, Soper JH, Zhu Y, Orton K, Villella A, Garza D, Vidal M, Ge H, Morimoto RI (2014) A chaperome subnetwork safeguards proteostasis in aging and neurodegenerative disease. Cell Rep 9(3):1135–1150. https://doi.org/10.1016/j.celrep.2014.09.042
18. Yam AY, Xia Y, Lin HT, Burlingame A, Gerstein M, Frydman J (2008) Defining the TRiC/CCT interactome links chaperonin function to stabilization of newly made proteins with complex topologies. Nat Struct Mol Biol 15(12):1255–1262. https://doi.org/10.1038/nsmb.1515
19. Nollen EA, Garcia SM, van Haaften G, Kim S, Chavez A, Morimoto RI, Plasterk RH (2004) Genome-wide RNA interference screen identifies previously undescribed regulators of polyglutamine aggregation. Proc Natl Acad Sci USA 101(17):6403–6408. https://doi.org/10.1073/pnas.0307697101
20. Kelly JJ, Tranter D, Pardon E, Chi G, Kramer H, Happonen L, Knee KM, Janz JM, Steyaert J, Bulawa C, Paavilainen VO, Huiskonen JT, Yue WW (2022) Snapshots of actin and tubulin folding inside the TRiC chaperonin. Nat Struct Mol Biol 29(5):420–429. https://doi.org/10.1038/s41594-022-00755-1
21. Zeng C, Han S, Pan Y, Huang Z, Zhang B, Zhang B (2024) Revisiting the chaperonin T-complex protein-1 ring complex in human health and disease: a proteostasis modulator and beyond. Clin Transl Med 14(2):e1592. https://doi.org/10.1002/ctm2.1592
22. Ghozlan H, Cox A, Nierenberg D, King S, Khaled AR (2022) The TRiCky business of protein folding in health and disease. Front Cell Dev Biol 10:906530. https://doi.org/10.3389/fcell.2022.906530
23. Frydman J, Nimmesgern E, Erdjument-Bromage H, Wall JS, Tempst P, Hartl FU (1992) Function in protein folding of TRiC, a cytosolic ring complex containing TCP-1 and structurally related subunits. EMBO J 11(13):4767–4778. https://doi.org/10.1002/j.1460-2075.1992.tb05582.x
24. Feldman DE, Spiess C, Howard DE, Frydman J (2003) Tumorigenic mutations in VHL disrupt folding in vivo by interfering with chaperonin binding. Mol Cell 12(5):1213–1224. https://doi.org/10.1016/s1097-2765(03)00423-4
25. Gao Y, Thomas JO, Chow RL, Lee GH, Cowan NJ (1992) A cytoplasmic chaperonin that catalyzes beta-actin folding. Cell 69(6):1043–1050. https://doi.org/10.1016/0092-8674(92)90622-j
26. Knee KM, Sergeeva OA, King JA (2013) Human TRiC complex purified from HeLa cells contains all eight CCT subunits and is active in vitro. Cell Stress Chaper 18(2):137–144. https://doi.org/10.1007/s12192-012-0357-z
27. Collier MP, Moreira KB, Li KH, Chen YC, Itzhak D, Samant R, Leitner A, Burlingame A, Frydman J (2021) Native mass spectrometry analyses of chaperonin complex TRiC/CCT reveal subunit N-terminal processing and re-association patterns. Sci Rep 11(1):13084. https://doi.org/10.1038/s41598-021-91086-6
28. Pappenberger G, McCormack EA, Willison KR (2006) Quantitative actin folding

reactions using yeast CCT purified via an internal tag in the CCT3/gamma subunit. J Mol Biol 360(2):484–496. https://doi.org/10.1016/j.jmb.2006.05.003

29. Cuéllar J, Ludlam WG, Tensmeyer NC, Aoba T, Dhavale M, Santiago C, Bueno-Carrasco MT, Mann MJ, Plimpton RL, Makaju A, Franklin S, Willardson BM, Valpuesta JM (2019) Structural and functional analysis of the role of the chaperonin CCT in mTOR complex assembly. Nat Commun 10(1):2865. https://doi.org/10.1038/s41467-019-10781-1
30. Mátés L, Chuah MK, Belay E, Jerchow B, Manoj N, Acosta-Sanchez A, Grzela DP, Schmitt A, Becker K, Matrai J, Ma L, Samara-Kuko E, Gysemans C, Pryputniewicz D, Miskey C, Fletcher B, VandenDriessche T, Ivics Z, Izsvák Z (2009) Molecular evolution of a novel hyperactive sleeping beauty transposase enables robust stable gene transfer in vertebrates. Nat Genet 41(6):753–761. https://doi.org/10.1038/ng.343
31. Kowarz E, Löscher D, Marschalek R (2015) Optimized sleeping beauty transposons rapidly generate stable transgenic cell lines. Biotechnol J 10(4):647–653. https://doi.org/10.1002/biot.201400821
32. Schmidt TG, Batz L, Bonet L, Carl U, Holzapfel G, Kiem K, Matulewicz K, Niermeier D, Schuchardt I, Stanar K (2013) Development of the Twin-Strep-tag® and its application for purification of recombinant proteins from cell culture supernatants. Protein Expr Purif 92(1):54–61. https://doi.org/10.1016/j.pep.2013.08.021
33. Schmidt TGM, Eichinger A, Schneider M, Bonet L, Carl U, Karthaus D, Theobald I, Skerra A (2021) The role of changing loop conformations in Streptavidin Versions engineered for high-affinity binding of the strep-tag II Peptide. J Mol Biol 433(9):166893. https://doi.org/10.1016/j.jmb.2021.166893

Chapter 3

# Integrated Protocol for γTuRC Expression and Purification for Microtubule Nucleation and Cryo-EM Analysis

Fabian Zimmermann and Marina Serna

## Abstract

Microtubule nucleation is a crucial process in cell biology that describes the formation of a new microtubule from free tubulin. As one of the key steps in establishing organized microtubule networks, microtubule nucleation is fundamental to the roles of the microtubule cytoskeleton in mediating intracellular transport, mitotic spindle formation, and cellular morphology. At the core of this process is the γ-tubulin ring complex (γTuRC), a multi-protein complex that serves as a template for microtubule nucleation, thus ensuring spatial and temporal control of microtubule growth. Understanding the molecular mechanisms that govern microtubule nucleation is essential for elucidating how cells dynamically organize their cytoskeleton and how dysregulation of this process contributes to pathologies such as cancer and neurodegenerative diseases.

In this method paper, we describe a straightforward and accessible approach to study microtubule nucleation without the need for advanced instrumentation. This includes a reliable and reproducible assay for assessing the activity of γTuRC and its regulatory factors that is suitable for a wide range of research settings. Detailed protocols for the recombinant production and purification of γTuRC, along with in vitro nucleation assays to assess its activity and regulation, are provided. Furthermore, we outline sample preparation for high-resolution structural analysis by cryo-electron microscopy (cryo-EM). Together, these methodologies provide a comprehensive framework for probing the mechanisms underlying γTuRC-templated microtubule nucleation and its regulation.

**Keywords** In vitro microtubule nucleation, Gamma-tubulin ring complex, Microtubule nucleation, Cryo-EM sample preparation, Fluorescence microscopy

## 1 Introduction

Microtubules are tube-like cytoskeletal filaments that undergo continuous assembly and disassembly cycles in response to cellular signals, a property known as dynamic instability [1]. To organize the microtubule cytoskeleton, cells restrict the formation of new microtubules, called microtubule nucleation, to microtubule organizing centers (MTOCs). Control over when and where new microtubules form is crucial to ensure the proper arrangement of microtubule networks

Noa B. Martín-Cófreces et al. (eds.), *The Cytoskeleton*, Methods in Molecular Biology, vol. 3070,
https://doi.org/10.1007/978-1-0716-5515-3_3,

that build the mitotic spindle, mediate the positioning of organelles, and coordinate intracellular transport and cellular motility.

Microtubule nucleation at MTOCs depends predominantly on the γ-tubulin ring complex (γTuRC). γTuRC is a multi-subunit protein complex that acts as a template for initiating αβ-tubulin polymerization into microtubules [2]. This 41-subunit complex provides a template that stabilizes early assembly intermediates of multiple tubulin dimers to facilitate formation of a 13-protofilament microtubule lattice, ensuring spatial and temporal control of microtubule growth [3–6]. However, despite significant insights into γ TuRC structure, important questions regarding the precise molecular mechanism underlying γTuRC-mediated nucleation remain open [7], highlighting the need of robust in vitro microtubule nucleation assays to understand the roles of individual γTuRC subunits and γ TuRC-binding regulators.

In this protocol, we describe a simple and accessible method to examine the microtubule nucleation activity of the γTuRC [8, 9]. First, we provide a detailed description of human γTuRC expression and purification strategies based on the baculovirus/insect-cell system and affinity chromatography, respectively [8, 9]. Second, we detail a straightforward in vitro microtubule nucleation endpoint assay based on fluorescence microscopy [8, 9]. This method allows monitoring the formation of microtubule filaments over time, providing a quantitative measure of nucleation activity. While methods for more precise determination of kinetic parameters have been developed [10], they usually require non-standard, specialized instrumentation and more extensive technical expertise. Instead, our approach, detailed here, can be performed using standard laboratory equipment by any researcher with basic biochemical training, making it widely accessible. Over the past years, visualization of the γTuRC structure using high-resolution cryo-electron microscopy (cryo-EM) has emerged as a powerful tool in elucidating γTuRC-mediated microtubule nucleation [5, 7]. Thus, we additionally provide a protocol optimized for cryo-EM sample preparation and imaging of γTuRC.

By offering relatively simple and reproducible methods, we aim to complement existing structural and biochemical protocols to facilitate broader research into microtubule regulation, contributing to a more comprehensive understanding of cytoskeletal organization.

## 2 Materials

### 2.1 Expression of the γTuRC in Insect Cells

1. Sf9 cells (Thermo Fisher Scientific) were maintained as suspension culture (*see* **Notes 1 and 2**).
2. Media suitable for cell culture of Sf9 cells (*see* **Note 3**).

3. Cell culture incubator. Cultures are maintained at 27°C with 120 rpm orbital shaking, without humidity or $CO_2$ control.
4. Plastic culture vessels for suspension culture. We use polycarbonate Erlenmeyer flasks with vented caps (commonly used capacity: 125 mL, 500 mL) and polypropylene bottles (2 L capacity) for larger culture volumes as described [11] (*see* **Note 4**).
5. Bacmid for co-expression of γTuRC subunits and RUVBLs, containing full-length coding sequences of γ-tubulin (TUBG1), MZT1, MZT2B, GCP2, GCP3, GCP4, GCP5, GCP6, RUVBL1, and RUVBL2 [8] (*see* **Note 5**). Recombinant GCP6 contains a C-terminal Twin-Strep-tag preceded by an Ala-Ser linker and an HRV (human rhinovirus) 3 C cleavage site, and GCP5 contains a C-terminal mTag-BFP followed by a $(Gly)_5$-linker, a TEV-protease cleavage site, and an ALFA-tag. These bacmids are generated by Tn7 transposition of pBIG2-based constructs into the EMBacY baculovirus genome [12] (*see* **Note 6**).
6. Polyethylenimine (PEI) linear 40 kDa transfection reagent (*see* **Note 7**).
7. Sterile phosphate-buffered saline (PBS) solution.
8. Sterile syringe filters with 0.2 μm pore size.
9. Biological safety cabinet for sterile cell culture.
10. Benchtop microcentrifuge, capable of reaching 14,000–16,000 × *g*, compatible with 1.5 and 2.0 mL microcentrifuge tubes.
11. Preparative centrifuge, capable of reaching 4,000–6,000 × *g*, compatible with 15 and 50 mL conical tubes.

### 2.2 Purification of the γTuRC from Insect Cells

1. γTuRC resuspension buffer (A-buffer): 50 mM Tris-HCl pH 8.0, 200 mM NaCl, 1 mM $MgCl_2$, 1 mM EGTA, 2 mM β-mercaptoethanol, 0.1% IGEPAL CA-630. Prepare and store at 4°C. Immediately before use, supplement with 2× EDTA-free protease inhibitors (Roche), and 0.1 mM GTP per gram of cell pellet (*see* **Note 8**).
2. γTuRC equilibration buffer (B-buffer): 50 mM Tris-HCl pH 8.0, 200 mM NaCl, 1 mM $MgCl_2$, 1 mM EGTA, 2 mM β-mercaptoethanol. Store at 4°C.
3. γTuRC first washing buffer (C-buffer): 50 mM Tris-HCl pH 8.0, 200 mM NaCl, 1 mM $MgCl_2$, 1 mM EGTA, 2 mM β-mercaptoethanol, 0.1 mM GTP. Store at 4°C.
4. γTuRC second washing buffer (D-buffer): 50 mM HEPES-KOH pH 7.5, 200 mM NaCl, 1 mM $MgCl_2$, 1 mM EGTA, 2 mM β-mercaptoethanol, 0.1 mM GTP. Store at 4°C.

5. γTuRC 3C-elution buffer (E-buffer): 50 mM HEPES-KOH pH 7.5, 200 mM NaCl, 1 mM $MgCl_2$, 1 mM EGTA, 2 mM β -mercaptoethanol, 0.1 mM GTP, supplemented with 3 C protease (10 μg per 0.1 mL of resin). Prepare fresh before use.
6. Sucrose (molecular biology grade).
7. 5× γTuRC sucrose gradient buffer (F-buffer): 250 mM HEPES-KOH pH 7.5, 1 M NaCl, 5 mM $MgCl_2$, 5 mM EGTA. Store at −20°C.
8. Gravity flow columns with 5–10 mL capacity, suitable for manual purification steps.
9. Dounce tissue grinder for mechanical lysis of insect cells.
10. Additional sample additives: Avidin to block residual biotin-binding sites, Universal Nuclease (Pierce) or Denarase (c-LEcta) to reduce viscosity from nucleic acids, and 3 C protease (IRB Barcelona Protein Expression core facility) for tag cleavage during elution.
11. Strep-Tactin XT chromatography resin (IBA Lifesciences) for high-affinity purification via the Twin-Strep tag (*see* **Note 9**).
12. Open-top thin-wall centrifuge tubes with 5 mL capacity for sucrose gradient ultracentrifugation.
13. Ultracentrifuge Optima MAX-XP with MLS-50 rotor (Beckman Coulter) or equivalent capable of reaching high g-forces necessary for sucrose gradient separation.
14. SDS-PAGE and Western-Blot systems for assessing the purity and identity of the purified γTuRC sample.
15. UV–vis spectrophotometer to determine the concentration of the γTuRC sample.

### 2.3 In Vitro Microtubule Nucleation Assay

1. Tubulin sample (*see* **Note 10**): Commercially available porcine brain tubulin, both rhodamine-labeled and unlabeled [13].
2. γTuRC sample: Sucrose-gradient purified γTuRC (see Sect. 3.2).
3. Nucleation assay buffer: 80 mM PIPES-KOH pH 6.9, 2 mM $MgCl_2$, 0.5 mM EGTA, and 1 mM GTP. Prepare fresh and keep on ice.
4. Nucleation fixation buffer: 80 mM PIPES-KOH pH 6.9, 2 mM $MgCl_2$, 0.5 mM EGTA, 50% glycerol, and 0.1% glutaraldehyde. Prepare fresh to ensure effective fixation.
5. PCR tubes for incubating the nucleation reactions at controlled temperatures.
6. Bradford reagent for protein concentration measurements.
7. Microscopy glass slides and 12 mm round glass coverslips for mounting samples.

8. PCR thermocycler to maintain controlled temperature conditions.
9. Fluorescence microscope equipped with appropriate excitation/emission filter for rhodamine and a 1.4–NA 63× oil immersion objective to visualize microtubules.

***2.4 Sample Preparation for Cryo-EM Studies of the γTuRC Complexes***

1. γTuRC sample: Strep-affinity purified γTuRC sample (see Sect. 3.5).
2. γTuRC biotin elution buffer (G-buffer): 57 mM HEPES-KOH pH 7.6, 200 mM NaCl, 1 mM $MgCl_2$, 1 mM EGTA, 2 mM 2-mercaptoethanol, 1 mM GTP, supplemented with 70 mM biotin. Prepare fresh before use.
3. Glow discharge plasma cleaner.
4. Cryo-EM grids: R1.2/1.3 300–square mesh copper holey-carbon grids (Quantifoil) are commonly used (*see* **Note 11**).
5. Grid boxes suitable for safe storage and handling of vitrified grids.
6. Vitrobot Mark IV (Thermo Fisher Scientific) for automated plunge-freezing under controlled humidity and temperature.
7. Cryogenic gases: Liquid nitrogen for cooling and storage, and liquid ethane as a cryogen for vitrification. Ensure high-purity gases to minimize ice contamination.

## 3 Methods

***3.1 γTuRC Baculovirus Generation***

1. Prepare a 15–25 mL suspension culture of Sf9 cells at $0.8 \times 10^6$ cells/mL for γTuRC baculovirus production.
2. Dilute purified bacmid DNA (*see* **Note 12**) in sterile-filtered PBS and mix with PEI (pH 7.0) at a 1:2 (w/w) DNA:PEI ratio. Typically, we use 1 μg of bacmid DNA per mL of Sf9 culture (*see* **Note 13**).
3. Incubate the DNA:PEI mixture at room temperature for 25 min to allow formation of DNA:PEI complexes.
4. Add the DNA:PEI mixture dropwise to the Sf9 cell culture and incubate at 27°C and 120 rpm for 5 days.
5. Monitor the fraction of yellow fluorescent protein (YFP)–positive Sf9 cells under a fluorescence microscope. Harvest the $P_0$ virus (supernatant fraction) when >95% of viable cells are YFP-positive by centrifugation at 300 x*g* for 5 min (*see* **Note 14**).
6. Filter the collected supernatant through a sterile 0.2-μm syringe filter to remove cell debris and store the $P_0$baculovirus stock at

4°C. No further amplification steps are typically required (*see* **Notes 15 and 16**).

### 3.2 γTuRC Expression in Insect Cells Using Baculovirus

1. Prepare suspension culture of Sf9 cells at $1 \times 10^6$ cells/mL for γTuRC expression. We routinely use expression culture volumes between 1–1.5 L.
2. Infect the Sf9 cell culture by adding the $P_0$ virus (see Sect. 3.1) at a 1:100 (v/v) ratio, i.e., 10 mL $P_0$ virus per liter of Sf9 suspension culture.
3. Culture the infected cells at 27°C and 120 rpm orbital shaking for 68–72 h, monitoring both YFP expression, as a marker for infection efficiency, and cell viability (*see* **Note 17**).
4. Harvest the cells by centrifugation at 300 x*g* for 5 min at room temperature.
5. Wash the collected pellet with PBS. The pellet may be used immediately for purification or snap-frozen in liquid nitrogen and stored at −80°C until use.

### 3.3 Purification of the γTuRC from Insect Cells

1. Resuspend the cell pellets containing overexpressed γTuRC with the tagged $GCP6^{3C\text{–}Twin\text{-}Strep}$ subunit in 5 mL of A-buffer per gram of cell pellet.
2. Lyse the cells using a dounce tissue grinder with 20 strokes on ice.
3. Clarify the cell extract by centrifugation at 20,000 x*g* for 25 min at 4°C.
4. Supplement the cleared cell extract with 2.5 mg avidin and 625 U Universal Nuclease per 10 mL of cell extract. Incubate for 5 min on a tube roller mixer at 4°C.
5. Prepare 2 mL of settled Strep-Tactin XT resin per 40 mL of cell extract. Add 4 mL of 50% slurry resin into a gravity column and allow the storage solution to drain by gravity.
6. Wash the resin with 5 resin volumes of B-buffer, followed by two resin volumes of A-buffer. Keep the resin in the A-buffer until sample loading to prevent resin drying.
7. Load the incubated cell extract into the pre-equilibrated Strep-Tactin XT resin and bind γTuRC by gravity flow. Collect the flow through the fraction containing the unbound material.
8. Wash the resin sequentially with two resin volumes of A-Buffer, three resin volumes of C-buffer and ten resin volumes of D-buffer to remove contaminants.
9. Elution of γTuRC can be done either by 3C protease cleavage or by biotin elution (see Sect. 3.5 and **Note 18**). For protease cleavage, resuspend resin in one resin volume D-Buffer. We

add in-house produced 3C protease at 10 μg per 100 μL resin and incubate with slight agitation for up to 4 h at 4°C.

10. Prepare ultracentrifuge tubes with 4.2 mL of a 10%–35% linear sucrose gradient in 1x F-buffer for density gradient purification (*see* **Note 19**).
11. Per gradient, layer 600 μL of the collected γTuRC elution fraction gently on top of the 10%–35% sucrose-gradient and centrifuge at 50,000 rpm for 3 h at 4°C in a MLS-50 rotor using an Optima MAX-XP ultracentrifuge (Beckman Coulter) with acceleration setting 9, and deceleration setting 7.
12. Remove top fraction (600 μL), then collect 350 μL sucrose gradient fractions by careful pipetting from top to bottom (*see* **Note 20**).
13. Identify gradient fractions containing γTuRC by SDS-PAGE, and Western blot if necessary (*see* **Note 21**).
14. Combine γTuRC-containing fractions, centrifuge at 16,100 × *g* for 10 min at 4°C in a benchtop centrifuge, and determine the sample concentration by measuring absorbance at 280 nm. Blank the spectrophotometer with the identical fractions from a control sucrose gradient run with D-Buffer alone.
15. Keep the purified γTuRC on ice for immediate use, or snap-freeze aliquots in liquid nitrogen and store at −80°C.

### 3.4 γTuRC in Vitro Microtubule Nucleation Assay

1. Mix unlabeled porcine brain αβ-tubulin with Rhodamine-labeled αβ-tubulin at a 10:1 molar ratio in the nucleation assay buffer to prepare the tubulin stock.
2. Spin the tubulin mix at 50,000 rpm in a TLA-55 rotor (Beckman Coulter) for 10 min at 4°C (*see* **Note 22**) to remove tubulin aggregates.
3. Determine tubulin concentration using a Bradford assay (*see* **Note 23**). While maintained on ice, dilute tubulin with nucleation assay buffer to a concentration of 13.4 μM.
4. Prepare 10 μL nucleation reactions in PCR tubes. We typically mix 1 μL of γTuRC (*see* **Note 24**) or 1 μL of control gradient fraction (see Step 15 in Sect. 3.3), and 9 μL of tubulin. This will result in a final assay concentration of 12 μM tubulin. The nucleation reaction is prepared and maintained on ice (*see* **Note 25**).
5. Set a PCR thermocycler to 37°C and prewarm the PCR tubes each with 9 μL fixation buffer for each assay time point.
6. For a "0-min" sample, mix 1 μL of the nucleation reaction with 9 μl of 37°C prewarmed fixation buffer. Mix well and

maintain the fixed sample at RT and protected from light until further processing.

7. Transfer the rest of the reaction to a PCR thermocycler set to 37°C to start the microtubule polymerization.
8. Take 1 μL aliquots at desired time points (*see* **Notes 26 and 27**).
9. After the last time point, pipette 1 μL of fixed sample on a microscope glass slide and cover it with 12-mm round coverslips.
10. Image rhodamine-labeled microtubules using a fluorescence microscope equipped with a rhodamine-compatible filter set and a 1.4–NA 63× oil immersion objective.
11. Quantify tubulin polymerization by counting microtubules per field of view using ImageJ software.

### 3.5 Sample Preparation for Cryo-EM Studies of γTuRC

1. Elute γTuRC sample from the Strep-Tactin XT resin (see Step 9 in **Sect.** 3.2) by incubating with 1.1 resin volume of the γ TuRC biotin elution buffer (G-buffer) for 3 h at 4°C. Chill an Eppendorf centrifuge to 4°C for later.
2. Assess sample purity by SDS-PAGE and determine its concentration by measuring absorbance at 280 nm (*see* **Note 28**).
3. Prepare 2.5-5.0 μL aliquots of purified γTuRC and immediately snap-freeze in liquid nitrogen. Store at −80°C.
4. Apply a 20 mA glow discharge in air (0.1 mbar) to the holey carbon grids for 40 s to render the grids hydrophilic.
5. Thaw a γTuRC aliquot immediately before use and spin it at 16,100 x*g* for 5 min at 4°C. Transfer the supernatant fraction to a new tube.
6. Dilute the γTuRC in cold D-buffer to a final concentration of 0.8-1 μM (*see* **Note 29**). Keep on ice.
7. Set the Vitrobot IV to 4°C and 90% humidity.
8. Liquefy ethane into a Vitrobot cryogen cup cooled by liquid nitrogen (*see* **Note 30**).
9. Apply 2.5 μL of sample to the grid pre-loaded into the Vitrobot chamber.
10. Immediately blot away excess buffer by setting the Vitrobot parameters to 3.0 s blotting time and −10 blot force (*see* **Note 31**), and automatically plunge-freeze grids in liquid ethane.
11. Transfer the vitrified grid to a grid box immersed in liquid nitrogen to prevent both sample warming and ice contamination.
12. Store vitrified cryo-EM grids in liquid nitrogen until use for data collection.

13. Cryo-EM image acquisition conditions entirely depend on the specific aims of the structural study. However, reference settings include the use of a 300–kV Titan Krios G4 electron microscope (Thermo Fisher Scientific) equipped with a BioQuantum energy filter (Gatan) and a K3 direct electron detector operated in counting mode. Nominal magnification of the microscope may be set to yield a pixel size between 0.5 and 1.0 Å/pixel (*see* **Note 32**), with a defocus range of −1.5 to−3.5 μm for optimal image contrast (*see* **Note 33**).

## 4 Notes

1. Sf9 cell cultures are maintained in serum-free media at 27°C and 120 rpm in capped Erlenmeyer plastic flasks. Cell density is maintained between 0.5–4 × $10^6$ cells/mL, and cell size and viability are monitored regularly at each passage to ensure optimal growth conditions. We recommend the use of prewarmed culture medium.
2. Protein expression efficiency may decrease with extended passaging; therefore, it is advisable to use fresh, low-passage Sf9 cultures for optimal yields. We avoid passing more than 20 times. After dilution of cells in fresh medium, we let cultures recover for 3–4 h before proceeding to the next step.
3. Although several commercial media are available, we use serum-free media Sf-900 III SFM (Gibco) supplemented with 12.5 U/mL penicillin and 12.5 mg/L streptomycin in all steps described.
4. Polycarbonate and polypropylene culture vessels can be cleaned, sterilized, and reused over several years to reduce plastic waste.
5. For TUBG1, MZT1, MZT2B, GCP2, GCP3, GCP4, GCP5, GCP6, we use cDNA with natural codons. For RUVBL1 and RUVBL2, we use cDNA codon-optimized for insect cell expression as previously described [14].
6. EMBacY bacmids have integrated an enhanced yellow fluorescence protein (YFP)-encoding gene that serves as a surrogate marker for successful expression of γTuRC subunits [15]. While monitoring YFP fluorescence enables rapid assessment of infection efficiency, phenotypic criteria such as cell size and refractile vacuolar content can also be considered.
7. PEI stock solution is prepared with $H_2O$ mQ at 1 mg/mL, pH is adjusted to 7.0, and the solution is sterilized by filtration before aliquoting, snap frozen, and stored at −80°C.

8. GTP is a critical component in the in vitro microtubule nucleation assay. It is highly recommended to use high-purity (≥99%) GTP to avoid contaminants that may interfere with tubulin polymerization. Aliquots of GTP stock (100 mM, pH 7.0) should be stored at −80°C and avoid freeze-thaw cycles, which might degrade the nucleotide.
9. γTuRC yields vary between affinity purification resins. While Strep-Tactin XT (IBA-lifesciences) can be used [9], we recommend Strep-Tactin XT 4Flow (IBA-lifesciences) [8].
10. The reliability and reproducibility of the in vitro microtubule nucleation assays highly rely on the quality of the tubulin preparation. Impurities or contaminants can significantly affect microtubule polymerization dynamics; therefore, using high-quality purified tubulin is critical for establishing controlled baseline conditions that allow for precise assessment of nucleation factors. Tubulin purification through repeated cycles of polymerization-depolymerization [13] ensures the removal of unwanted nucleation seeds and results in a preparation with a well-defined nucleation profile. We have used rhodamine-tubulin from Cytoskeleton (cat# TL950M).
11. Grid selection for γTuRC cryo-EM sample preparation is influenced by multiple factors, including vitrification protocols, the specific vitrification device used, and sample characteristics. The vitrification parameters described here have been optimized for Quantifoil R1.2/1.3 300–square mesh copper holey-carbon grids; however, alternative grid types (e.g., gold grids or grids with a continuous carbon support layer) may be employed depending on experimental needs.
12. High-quality bacmid preparation is essential for efficient Sf9 cell transfection. While several bacmid preparation protocols and kits are available, we use the Xtra BAC kit (Macherey Nagel). Bacmid concentration was determined by measuring absorbance at 260 nm, and purity was confirmed by 260/280 and 260/280 nm absorbance ratios, which should typically be ≥1.8 and ≥2.0, respectively. We recommend preparing single-use bacmid aliquots at a bacmid concentration between 1–2 μg/μL, which are snap-frozen and kept at −80°C for long-term storage. Bacmids prepared this way reproducibly yield highly-quality baculovirus for γTuRC expression.
13. The optimal ratio of DNA:PEI complexes for γTuRC production was empirically optimized, and we recommend a ratio of 1:2 (DNA:PEI).
14. $P_0$ baculovirus is typically harvested when infection signs become apparent, such as increased cell size and the presence of increased vacuolar cell content. These phenotypic changes are generally observed between 3–5 days post-transfection.

While this can be explored, for γTuRC expression harvesting at day 5 post-transfection consistently yields optimal results.

15. The method described here for γTuRC baculovirus generation is based on [16]. In our hands, the most reproducible expression results are obtained using $P_0$ without additional rounds of virus amplification.
16. Baculovirus stocks are ideally prepared freshly and used immediately for large-scale Sf9 cell infections. However, if storage is necessary, we recommend maintaining the baculovirus stock in sealed vials to minimize contamination risk, protected from light at 4°C for up to two months. Proper storage conditions help preserve the viral infectivity over time.
17. Infected insect cells are typically harvested 68–72 h post-infection. Extending the incubation beyond 72 h may lead to increased cell death and protein aggregation, negatively impacting γTuRC purification yield and quality.
18. Both ways of γTuRC elution yield comparable purity. While protease cleavage is slightly purer, it comes at the expense of longer elution time with an overall lower yield. 3 C protease activity may vary between providers and batches. We advise to test the available 3 C protease batch in small-scale purifications and optimizing the 3C protease final concentration and incubation time.
19. Variations in gradient formation may lead to inconsistent separation of samples during centrifugation. We produce reproducible linear gradients using a manual gradient mixer. Alternatively, automated gradient mixers may be employed. Gradient quality may be assessed by analyzing sedimentation of protein standards such as aldolase (158 kDa, 7 *S*) and thyroglobulin (669 kDa, 19 *S*) on a control gradient in the same centrifuge run.
20. Fraction collection may be a challenging step in the purification protocol due to the risk of disturbing the gradient and cross-contaminating the adjacent sucrose gradient layers. Here, we employed manual pipetting from the top using P1000 pipette tips with cut-off tips and low aspiration speeds. Alternatively, gradient fractionators or fraction collectors (e.g., BioComp Gradient Master) can be used for automated collection.
21. For Western blotting, the following antibodies were successfully used: rabbit polyclonal antibodies against GCP2, GCP3, GCP4, GCP5, GCP6, MZT1, and MZT2 [17, 18]; and anti-γ-tubulin mouse monoclonal antibody (Sigma-Aldrich, cat# T6557).

22. Purified tubulin preparations are prone to aggregation, a feature that can compromise their polymerizing activity during the nucleation assays. To minimize this, freeze-thaw cycles should be avoided. Protein aggregates that are commonly formed upon sample thawing may be efficiently removed by sample centrifugation as described. Once prepared, process tubulin as quickly as possible. We try to start nucleation assays within 30 min after thawing tubulin.
23. Tubulin concentration critically determines microtubule nucleation. In microtubule polymerization assays focused on studying the nucleation process, it is important to maintain the tubulin concentration at the level where no microtubule growth is observed under the described conditions in the absence of γTuRC, thereby ensuring physiologically relevant results.
24. We routinely use γTuRC concentration of 5–15 nM in nucleation assays. However, different concentrations can be tested to analyze concentration-dependent effects on nucleation efficiency.
25. This in vitro nucleation assay serves as a versatile platform for screening drugs and other regulatory factors that influence microtubule nucleation and dynamics. In addition to assessing the role of the γTuRC, various proteins and small molecules that might modulate microtubule assembly can be evaluated using this assay, enabling analysis of their mechanisms and potential therapeutic applications.
26. We have empirically determined that collecting time points up to 5–10 min reliably captures microtubule nucleation dynamics at the described tubulin concentrations. However, the time course can be adjusted based on experimental needs, especially when assessing microtubule nucleation regulatory factors.
27. If several assay conditions are tested in parallel, we recommend a successive starting time delay of each assay condition by a fixed amount of time. This ensures that the time points are exactly met throughout the assay conditions that are compared. In our hands, a 30 sec consecutive delay between each assay condition is sufficient for sample fixation and storage.
28. Using StrepTactin-XT 4Flow, a typical purification yields approximately 5 mg γTuRC per 1.5 L of infected Sf9 cells after the affinity purification step.
29. The concentration of the γTuRC sample is one of the key parameters to consider during vitrification. We have consistently obtained a high density of monodisperse particles in cryo-EM micrographs using the described range of γTuRC concentrations. While particle distribution also depends on

ice thickness, we have observed that when the γTuRC concentration falls below 0.5 μM, the protein complexes tend to adhere preferentially to the holey-carbon support film. In such cases, the use of a thin continuous carbon layer can improve particle distribution across the holes. It is important to adjust the vitrification parameters accordingly, particularly blot time and blot force, to prevent sample dryness and ensure optimal ice thickness.

30. We have successfully vitrified γTuRC and γTuRC-containing samples using both liquified ethane and ethane-propane mixtures as cryogen agents. Ensuring the cryogen is maintained at the proper temperature (just above its freezing point) and free from ice contamination is essential to achieving optimal vitrification conditions. Careful handling during both liquefaction and vitrification minimizes ice contaminants.
31. Vitrobot settings such as blotting time and blot force are highly dependent not only on the sample and grid type but also on the specific Vitrobot unit. Variations in the Vitrobot calibration and alignment of the blotting pads can significantly affect the reproducibility of the process. Therefore, these parameters should be empirically optimized for each experiment.
32. Nominal magnification typically employed for cryo-EM studies of the γTuRC balances between the number of particles captured per field of view, which is also determined by the sample concentration achieved, and the pixel size on the detector.
33. The recommended defocus range typically provides optimal image contrast for large complexes such as the γTuRC (approximately 30 nm in diameter). However, the ideal defocus range is also influenced by factors such as ice thickness. Therefore, it is essential to adjust the defocus range not only based on the sample's specific characteristics but also taking into account the experimental conditions.

## Acknowledgments

The methods described here were developed in the laboratories of Jens Lüders (IRB Barcelona, Spain) and Óscar Llorca (CNIO-Instituto de Salud Carlos III), and we gratefully acknowledge their support and contributions. During that period, research in J.L.'s lab was supported by grants PGC2018-099562-B-I00, PID2021-127603NB-I00, and RED2022-134792-T (Ministerio de Ciencia, Innovación y Universidades/FEDER); grant 2021 SGR 01280 (Agencia de Gestión de Ayudas Universitarias y de Investigación); and

IRB Barcelona intramural funds. Work in O.L.'s lab was funded by the Agencia Estatal de Investigación (AEI/10.13039/501100011033) and Ministerio de Ciencia e Innovación (PID2020-114429RB-I00). F.Z. was supported by a fellowship from the "La Caixa" Foundation (ID 100010434, fellowship code LCF/BQ/DI17/11620020).

## References

1. Desai A, Mitchison TJ (1997) Microtubule polymerization dynamics. Annu Rev Cell Dev Biol 13:83–117. https://doi.org/10.1146/annurev.cellbio.13.1.83
2. Moritz M, Braunfeld MB, Guenebaut V, Heuser J, Agard DA (2000) Structure of the gamma-tubulin ring complex: a template for microtubule nucleation. Nat Cell Biol 2(6):365–370. https://doi.org/10.1038/35014058
3. Brito C, Serna M, Guerra P, Llorca O, Surrey T (2024) Transition of human gamma-tubulin ring complex into a closed conformation during microtubule nucleation. Science 383(6685):870–876. https://doi.org/10.1126/science.adk6160
4. Vermeulen BJ, Bohler A, Gao Q, Neuner A, Zupa E, Chu Z, Wurtz M, Jakle U, Gruss OJ, Pfeffer S, Schiebel E (2024) Gamma-TuRC asymmetry induces local protofilament mismatch at the RanGTP-stimulated microtubule minus end. EMBO J 43(10):2062–2085. https://doi.org/10.1038/s44318-024-00087-4
5. Gao Q, Vermeulen BJA, Wurtz M, Shin H, Erdogdu D, Zheng A, Hofer FW, Neuner A, Pfeffer S, Schiebel E (2024) The structure of the gamma-TuRC at the microtubule minus end - not just one solution. Bioessays 46(9):e2400117. https://doi.org/10.1002/bies.202400117
6. Dendooven T, Yatskevich S, Burt A, Chen ZA, Bellini D, Rappsilber J, Kilmartin JV, Barford D (2024) Structure of the native gamma-tubulin ring complex capping spindle microtubules. Nat Struct Mol Biol 31(7):1134–1144. https://doi.org/10.1038/s41594-024-01281-y
7. Mammri L, Conduit PT (2024) Cryo-EM structures of gamma-TuRC reveal molecular insights into microtubule nucleation. Nat Struct Mol Biol 31(7):1004–1006. https://doi.org/10.1038/s41594-024-01345-z
8. Serna M, Zimmermann F, Vineethakumari C, Gonzalez-Rodriguez N, Llorca O, Luders J (2024) CDK5RAP2 activates microtubule nucleator gammaTuRC by facilitating template formation and actin release. Dev Cell 59(23):3175–3188. https://doi.org/10.1016/j.devcel.2024.09.001
9. Zimmermann F, Serna M, Ezquerra A, Fernandez-Leiro R, Llorca O, Luders J (2020) Assembly of the asymmetric human gamma-tubulin ring complex by RUVBL1-RUVBL2 AAA ATPase. Sci Adv 6(51):eabe0894. https://doi.org/10.1126/sciadv.abe0894
10. Consolati T, Henkin G, Roostalu J, Surrey T (2022) Real-time imaging of single gammaTuRC-mediated microtubule nucleation events in vitro by TIRF microscopy. Methods Mol Biol 2430:315–336. https://doi.org/10.1007/978-1-0716-1983-4_21
11. Rieffel S, Roest S, Klopp J, Carnal S, Marti S, Gerhartz B, Shrestha B (2014) Insect cell culture in reagent bottles. MethodsX 1:155–161. https://doi.org/10.1016/j.mex.2014.08.006
12. Weissmann F, Petzold G, VanderLinden R, Huis In 't Veld PJ, Brown NG, Lampert F, Westermann S, Stark H, Schulman BA, Peters JM (2016) biGBac enables rapid gene assembly for the expression of large multisubunit protein complexes. Proc Natl Acad Sci USA 113(19):E2564–E2569. https://doi.org/10.1073/pnas.1604935113
13. Vallee RB (1986) Reversible assembly purification of microtubules without assembly-promoting agents and further purification of tubulin, microtubule-associated proteins, and MAP fragments. Methods Enzymol 134:89–104. https://doi.org/10.1016/0076-6879(86)34078-3
14. Willhoft O, Bythell-Douglas R, McCormack EA, Wigley DB (2016) Synergy and antagonism in regulation of recombinant human INO80 chromatin remodeling complex. Nucleic Acids Res 44(17):8179–8188. https://doi.org/10.1093/nar/gkw509
15. Trowitzsch S, Bieniossek C, Nie Y, Garzoni F, Berger I (2010) New baculovirus expression tools for recombinant protein complex production. J Struct Biol 172(1):45–54. https://doi.org/10.1016/j.jsb.2010.02.010
16. Scholz J, Suppmann S (2017) A new single-step protocol for rapid baculovirus-driven protein production in insect cells. BMC Biotechnol 17(1):83. https://doi.org/10.1186/s12896-017-0400-3

17. Cota RR, Teixido-Travesa N, Ezquerra A, Eibes S, Lacasa C, Roig J, Luders J (2017) MZT1 regulates microtubule nucleation by linking gammaTuRC assembly to adapter-mediated targeting and activation. J Cell Sci 130(2):406–419. https://doi.org/10.1242/jcs.195321

18. Teixido-Travesa N, Roig J, Luders J (2012) The where, when and how of microtubule nucleation - one ring to rule them all. J Cell Sci 125(19):4445–4456. https://doi.org/10.1242/jcs.106971

# Chapter 4

# Isolation of Centrosomal Fractions for Proteomic Studies in Lymphocytes

**Marta Lozano-Prieto, Francisco Javier Chichón and Francisco Sánchez-Madrid**

## Abstract

Centrosomes are essential microtubule-organizing centers that coordinate spatial cues for cellular polarization and organelle trafficking. In T lymphocytes, centrosomes translocate toward the immunological synapse to orchestrate polarized secretion and sustain signaling with the antigen-presenting cell. Understanding the dynamic composition of the centrosome during lymphocyte activation is crucial for uncovering the molecular mechanisms behind immune responses. Here, we describe a detailed protocol to isolate centrosome-enriched fractions from human T lymphoblasts under near-physiological conditions, optimized for downstream applications such as proteomic or structural analyses.

**Keywords** Centrosomes, Lymphocytes, Proteomics

## 1 Introduction

The centrosome is a non-membranous organelle composed of a pair of centrioles surrounded by the pericentriolar matrix (PCM) [1]. It functions as the main microtubule-organizing center (MTOC) in animal cells, having a key role not only in global cellular organization but also in biological processes that require cellular polarization or/and intracellular transport (e.g., T cell activation) [2]. In CD4 T lymphocytes, centrosome repositioning toward the immunological synapse (IS) upon antigen recognition is a hallmark of effective activation [3–5]. Proteomic studies have uncovered that centrosomes undergo a significant change in their protein composition upon T cell receptor (TCR) stimulation. For instance, cytosolic CCT chaperonin, among others, increases in centrosome fractions from activated T lymphocytes, regulating the ability of T cells to respond to TCR/CD28 stimulation and co-stimulation, respectively [6].

Noa B. Martín-Cófreces et al. (eds.), *The Cytoskeleton*, Methods in Molecular Biology, vol. 3070,
https://doi.org/10.1007/978-1-0716-5515-3_4,

These findings highlight the importance of isolating intact centrosome fractions from T lymphocytes to enable confident proteomic and functional analysis results. Traditional centrosome isolation methods often rely on cytoskeletal disruptors. For example, nocodazole and cytochalasin D disrupt microtubules or actin filaments, respectively, inhibiting their polymerization [7–9]. These agents help increase centrosome recovery by detaching microtubules and actin-bound organelles. However, these drugs interfere with T cell activation and centrosome integrity by altering the preservation of post-translational modifications, the dynamic recruitment of chaperones like CCT, and activation-dependent protein-protein interactions [6, 10, 11]. This chapter presents a protocol that is optimized for human T lymphoblasts and other cell types to preserve centrosomes under near-physiological conditions, allowing accurate results from downstream functional and high-resolution mass spectrometry (MS) assays.

## 2 Materials

### 2.1 Cells, Culture Reagents, and Equipment

1. Freshly collected human peripheral blood or buffy coats.
2. Sodium chloride sterile: 0.9% NaCl.
3. Sterile Ficoll Histopaque or an equivalent lymphocyte separation medium.
4. Cell culture medium for lymphocytes: RPMI 1640 medium supplemented with 1 mM glutamine, nonessential amino acids, 25 mM Hepes, 10% fetal bovine serum (FBS).
5. Phytohemagglutinin (PHA), Staphylococcal Enterotoxin E (SEE).
6. Human recombinant Interleukin-2 (IL-2).
7. High-speed centrifuge for high volumes. A rotor with swinging buckets may be required.
8. Ultracentrifuge and a rotor with swinging buckets.
9. Ultracentrifuge tubes for 38.5 mL and 14 mL.
10. Incubator set at 37°C with 5% $CO_2$.
11. Non-treated culture-grade plastic dishes.
12. Sterile 50 mL polypropylene conical tubes.
13. Pipettes and sterile filter tips (P10, P200, P1000).
14. Hemocytometer or automated cell counter.
15. Syringes set for homogenization: 21 G and 25 G sterile needles.
16. 70 μm nylon mesh.

### 2.2 Latex Microbeads for Cell Activation (Optional)

Latex microbeads (6.4 μm) are conjugated with activating monoclonal antibodies against human CD3ε and against human CD28 (5 μg/mL and 1.66 μg/mL, respectively), or human γ -globulin (100 μg/mL). To coat the microbeads:

1. Add about $10^7$ latex microbeads to 450 μL of carbonate buffer to stick antibodies to the microbeads (*see* **Note 1**).
2. Add IgGs (either control or activating).
3. Incubate with gentle movement for 2 h at room temperature. Alternatively, it can be left at 4°C overnight.
4. Centrifuge at 5000 rpm for 5 min in a benchtop.
5. Wash twice with PBS.
6. Block with 1 mL of 1% BSA in PBS for 1 h at room temperature with gentle movement.
7. Wash twice with PBS.
8. Store the microbeads in incomplete cell culture medium, in about 200 μL at 4°C up to 2 weeks. Do not use beyond 2 weeks, in general (they can be contaminated and/or release the Ab).

### 2.3 Buffers and Solutions

1. Phosphate-buffered saline (PBS): 137 mM NaCl, 2.7 mM KCl, 10 mM $Na_2HPO_4$, and 1.8 mM $KH_2PO_4$, pH 7.4.
2. Tris-buffered saline (TBS) 1×, sterile and filtered: 50 mM Tris-HCl pH 7.4, 150 mM NaCl.
3. PIPES buffer: 10 mM PIPES, pH 7.2. Adjust pH with KOH
4. Lysis buffer: 1 mM Tris-HCl, pH 7.4, 0.5% v/v NP-40, 0.1 mM $MgCl_2$, 1 mM PMSF, protease and phosphatase inhibitor cocktails.
5. Gradient buffer: 10 mM PIPES, pH 7.2, 0.1% v/v Triton X-100, 0.1% v/v β-mercaptoethanol.
6. Sucrose solutions: prepare (w/w) 20%, 30%, 40%, 50%, and 70% sucrose solutions in gradient buffer and filter through 0.22 μm filters (*see* **Note 2**).

### 2.4 Handling and Validation of Fractions

1. Liquid nitrogen is ready to freeze fractions.
2. SDS-PAGE and Western blot equipment and reagents.
3. Anti-γ-tubulin primary antibody.
4. Secondary HRP-conjugated antibody.
5. Chemiluminescent imaging system.

## 3 Methods

### 3.1 Primary Human T Lymphocytes Isolation from Buffy Coat Preparations

1. Dilute 1:3 freshly collected human blood in sterile saline.
2. Carefully layer the diluted blood onto an equal volume of Ficoll in a centrifuge tube.
3. Centrifuge the sample without brake at room temperature at $880 \times g$ for 30 min. Three layers will become visible: a yellow-coloured plasma (top layer), a transparent layer containing Ficoll solution (middle layer), and a red-coloured layer of anucleate cells and polymorphonuclear leukocytes (bottom layer).
4. Carefully collect the mononuclear cells from the interphase between the top and middle layer using a sterile thin pipette in two 50 mL tubes (*see* **Note 3**).
5. Wash the collected cells in saline solution by spinning once at 390 g for 10 min with reduced or no brake.
6. Discard the supernatant carefully to avoid loss of cells.
7. Resuspend the cells gently and add 50 mL saline solution.
8. Perform five additional washes by centrifuging at 390 g for 5 min (*see* **Note 4**).
9. Plate the cell suspension onto untreated plastic dishes for at least 30 min at 37°C to allow adherence of monocytes.
10. Collect the suspended cells and discard the adhered monocytes (*see* **Note 5**).
11. Culture the purified lymphocytes in RPMI 1640 medium supplemented with 10% FBS at $2 \times 10^6$ cells per mL.

### 3.2 Primary Human T Lymphocytes Expansion

1. Add PHA (0.1 μg/mL) for 24 h or SEE (0.01 μg/mL) for 48 h (*see* **Note 6**).
2. Change the culture medium to the new complete RPMI 1640 medium supplemented with 10% FBS.
3. Add IL-2 (50–100 U/mL) every 2 days. Maintain cell density at $1–2 \times 10^6$ cells/mL and culture cells for 7–8 days.

### 3.3 Optional: T Lymphocyte Activation

1. Add latex microbeads (6.4 μm in diameter) conjugated to anti-CD3ε mAb (5 μg/mL) and anti-CD28 mAb (1.66 μg/mL) or Dynabeads™ for 15–20 min. Cell:bead ratio should be approximately 1:1 (*see* **Note 7**).
2. For control T lymphocytes (e.g., not TCR activated cells), add latex microbeads conjugated to γ-globulin (100 μg/mL) for 15–20 min in a 1:1 cell:bead ratio.
3. Immediately chill cells on ice to halt activation.

### 3.4 Optional: Application to Other Cell Lines

Although this protocol is optimized for primary human T lymphoblasts, it is also applicable to tissue as calf thymus [12]. Also, to other immortalized cell lines such as human embryonic kidney 293 cells (HEK293) and Jurkat E6-1 (JK) [13].

For these lines, no stimulation is required. For HEK293, cells can be scraped from cell culture plates and continue in Step 2 from Sect. 3.5 (*see* **Note 8**). For JK or other cells growing in suspension, cultured cells can be processed directly as T lymphoblasts, as described below.

### 3.5 Cell Lysis

The present protocol is based on the osmotic lysis of cells, supplemented with a low concentration of detergent to facilitate the process. In addition, protease inhibitors are included to prevent endogenous proteolytic degradation (*see* **Note 9**).

1. Centrifuge cells at $310 \times g$ for 10 min at 4°C. Discard supernatant.
2. Resuspend the pellet in cold filtered 1×TBS and centrifuge cells again at $310 \times g$ for 10 min at 4°C. Discard supernatant.
3. Lyse pellet in cold lysis buffer ($2 \times 10^6$-cells/mL) for 10 min on ice with gentle shaking (*see* **Note 10**).
4. Centrifuge at $24{,}000 \times g$ for 10 min at 4 °C.
5. Collect supernatant and filter through a 70 μm nylon mesh.
6. Add 20 mL of PIPES buffer per liter of lysate.
7. Reserve an aliquot as "Lysate 1" control (*see* **Note 11**).

### 3.6 Centrosome Enrichment

1. Centrifuge filtered lysate at $24{,}000 \times g$ for 1 h at 4°C. Apply a moderate deceleration rate (*see* **Note 12**).
2. A white centrosomal layer should form at the tube bottom. Gently resuspend this layer in 2 mL of gradient buffer.
3. Homogenize gently by passing the sample through a 21 G needle 3 times.
4. Homogenize gently by passing through a 25G needle 3 times.
5. Bring the volume up to 12 mL with gradient buffer.
6. Add 1 mL of 20% sucrose solution to increase density.
7. Reserve an aliquot as "Lysate 2" control (*see* **Note 13**).

### 3.7 Centrosome Isolation: First Sucrose Gradient

1. Prepare the following discontinuous sucrose gradient in ultracentrifuge tubes: 1 mL 70%, 2 mL 50%, 2 mL 40%, 2 mL 30%, and 2 mL 20% (w/w). See Fig. 1A and **Notes 14–18**.
2. Carefully layer the homogenized sample on top of the gradient (above the 20% sucrose layer) and fill the tube with gradient buffer if necessary.

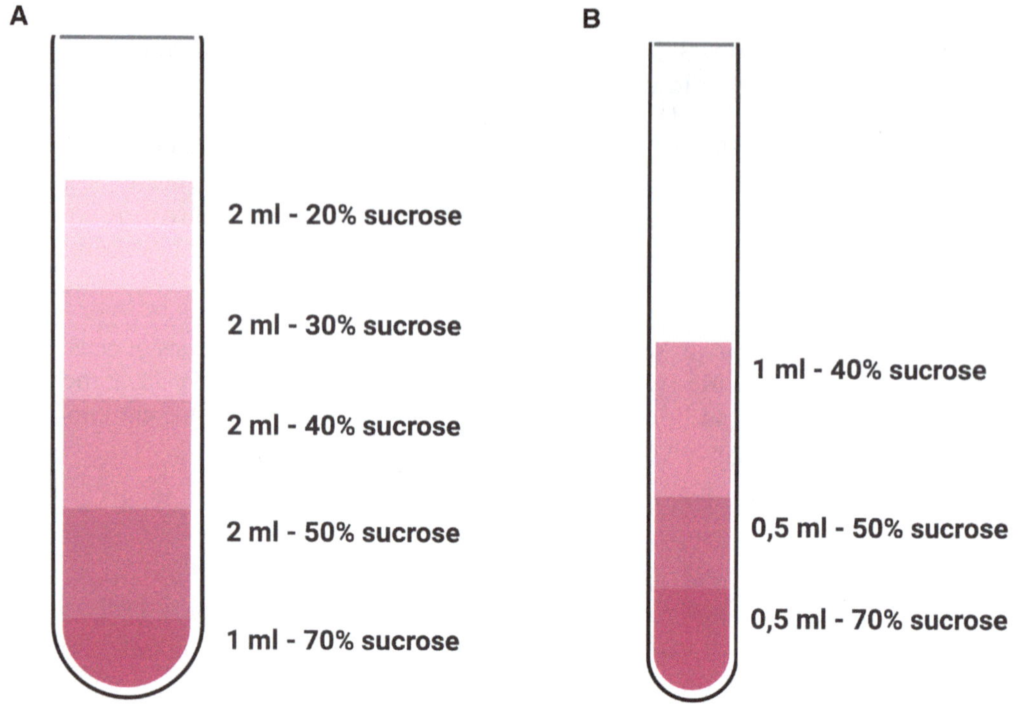


**Fig. 1** Scheme of the two sucrose gradients used for centrosome isolation. (**A**) First sucrose gradient layers (70%, 50%, 40%, 30%, and 20%, w/w) used in Sect. 3.7. (**B**) Second sucrose gradient layers (70%, 50%, and 40%, w/w) used in Sect. 3.8

3. Centrifuge at 32,500 × *g* for 1 h at 4°C with low acceleration rate and no brake.
4. Aspirate carefully to the 20% layer meniscus.
5. Collect 1 mL fractions (approx. 8; *see* **Note 19**).
6. Pool the 50% and 70% fractions and add gradient buffer up to 6 mL.
7. Reserve an aliquot of this latter diluted fraction as "Lysate 3" control (*see* **Note 20**).

### *3.8 Centrosome Isolation: Second Sucrose Gradient*

1. Prepare the following discontinuous sucrose gradient in ultra-centrifuge tubes: 0.5 mL 70%, 0.5 mL 50% and 1 mL 40% (w/w). See Fig. 1B and **Notes 14–18, and 21**.
2. Carefully layer the 6 mL of the pooled fraction on top of the gradient (e.g., the 40% sucrose layer) and fill the tube with gradient buffer if necessary.

3. Centrifuge at 120,000 × *g* for 1 h at 4°C with low acceleration rate and no brake.
4. Aspirate gently to the 40% meniscus.
5. Fractionate carefully: collect 2 fractions of 400 μL and the remaining into aliquots of 200 μL (*see* **Note 19**).

### 3.9 Sample Handling and Validation of Fractions

Upon validation through Western blot (Fig. 2), the γ-tubulin enriched fractions through this protocol can be subjected to different techniques, such as proteomic(LC-MS/MS)analysis to assess their content [6], or cryo-electron microscopy (cryo-EM) to assess centrosome ultrastructure [12], as in Fig. 3. Indeed, the centrosomes isolated through this protocol are functional and can be used to polymerize microtubules in vitro [6].

1. Store 30 μL aliquots of samples at 4°C for Western blot (WB) validation.
2. Freeze the remaining fractions in liquid nitrogen and store at −80°C until use (*see* **Note 22**).
3. Use samples "Lysates 1–3" as controls. See Fig. 2 and **Notes 11, 13, and 20**.
4. Identify the fractions containing the centrosomes by Western blot with antibodies anti-γ-tubulin (*see* **Note 23**).

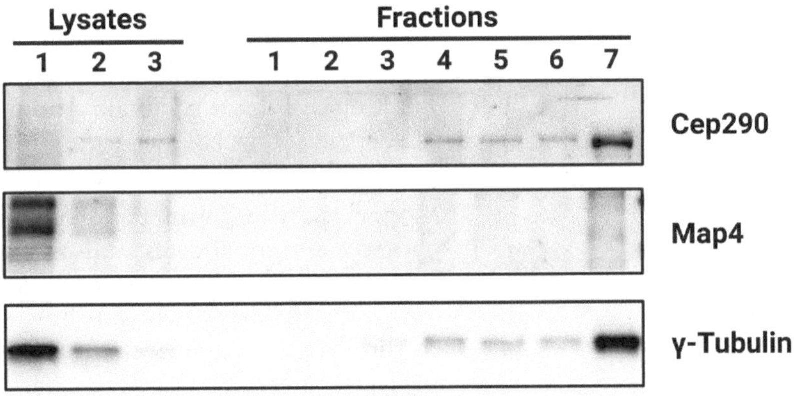


**Fig. 2** Western blot validation of centrosome isolation. Resting human T lymphoblasts were processed through the entire centrosome isolation protocol. Control lysates 1–3 and final gradient fractions [1–7] were collected (left to right wells). γ-Tubulin, Cep290, and Map4 were analyzed by Western blot as markers of centrosome proteins to validate their enrichment across the gradient. Fractions correspond to increasing sucrose percentages

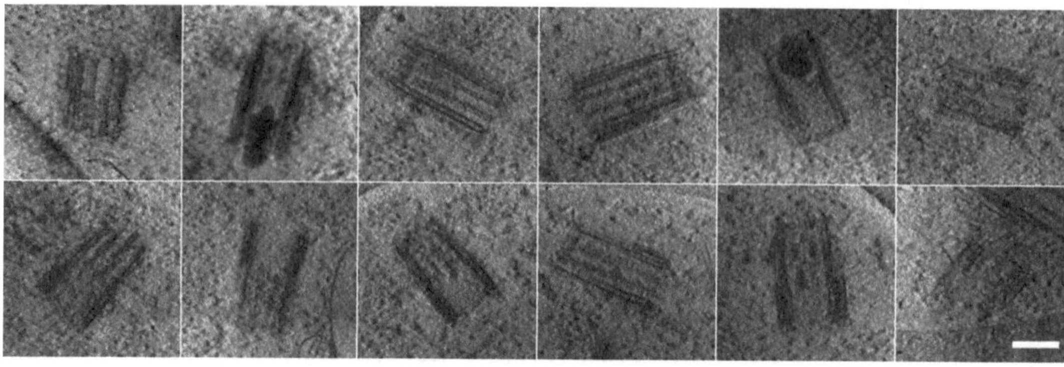

**Fig. 3** Representative central sections of centrioles isolated from calf thymus, obtained by cryo-electron tomography. Multiple protein complexes are visible surrounding the centrioles, and distinct internal densities can be observed within the centriolar lumen, highlighting the high quality of the centrosome purification. Scale bar, 200 nm

## 4 Notes

1. Dynabeads™ human CD3/CD28 T cell activators (Thermo Fisher Scientific, USA) or equivalent can also be used.
2. Weigh the required amount of sucrose in a tube, then carefully add the gradient buffer to the same tube to reach the required weight for the solution (e.g., a 20% sucrose solution contains 2 g of sucrose in 10 g of solution, regardless of the volume). Place the tube in a cold room at 4°C with gentle agitation overnight to obtain a homogeneous solution. Filter all sucrose solutions except the 70% sucrose solution, as it is too dense to filter.
3. Mononuclear cells from peripheral blood include mainly lymphocytes and monocytes, and usually appear as an additional layer at the interphase between the top and middle layer of the gradient. These cells can show red aggregates, including erythrocytes (bottom layer).
4. If a large number of erythrocytes are recovered with the mononuclear cells (indicated by an intensely red cell suspension), an additional lysis step can be performed on these cells. For this purpose, treat the cell pellet with 5–10 mL of sterile water for 5–10 s and then immediately add PBS or saline solution up to 50 mL to avoid general membrane damage to nucleated cells. After centrifugation, the isolation of mononuclear cells can continue.
5. This adherence step may be repeated twice to achieve more lymphocyte purity.

6. This step is performed to increase the number of T lymphocytes analyzed per donor, since thousands of millions of cells are required for optimal recovery of centrosomes.
7. This step can be useful to detect differences between activated and non-activated T lymphocytes. However, this will reduce the number of cells per donor and per sample for the isolation protocol.
8. The number of HEK293 cells required for optimal recovery of centrosomes is about $5 \times 10^8$ cells, corresponding to 5 x p150 cell culture plates.
9. The time to perform these steps is critical for centrosome integrity. Phosphatase inhibitors can be added if this kind of post-translational modification is to be studied in the isolated fractions. Cells will not be quiescent in the cell culture; therefore, interphase and mitotic centrosomes (spindle) can be recovered.
10. Approximately $1 \times 10^9$ cells (in the case of T lymphoblasts or JK cells, 1 L of $10^6$ cells/mL culture) are sufficient to proceed with centrosome isolation and to yield enough material for both proteomic and Western blot validation analyses.
11. Reserve an aliquot for "Lysate 1" control. This sample will serve as a baseline control for subsequent validation by Western blot, allowing comparison of total protein input before any gradient separation and helping to identify loss or enrichment during purification.
12. Usually, centrifuges include a point scale, where one is the slowest and nine is the fastest deceleration; level 5 is recommended in these cases.
13. Reserve an aliquot for "Lysate 2" control. This second control for Western blotting will allow a comparative assessment of centrosome enrichment versus initial Lysate 1 as well as an intermediate-stage validation for Lysate 3 and the final fractions.
14. Allow sucrose to dissolve slowly under gentle agitation at 4°C for at least 2 h, ensuring complete dissolution.
15. Normally, 70% is extremely viscous. It is possible to mark the 70% meniscus using a buffer, and then replace it with 70% sucrose until the mark.
16. Cut off the micropipette tip to facilitate the aspiration of the desired 70% volume.
17. Since the tubes usually used here are 38.5 mL due to large volumes of lysates, a rotor including swinging buckets such as the SW28 rotor (Beckman coulter) or equivalent will be needed. Swinging buckets are mandatory here.

18. Mark the menisci for future fraction collection. This facilitates accurate identification of gradient regions and consistent retrieval of centrosome fractions.
19. The first time that the protocol is performed, all the gradients can be aliquoted and examined by western blotting to determine the performance of the procedure and the potential distribution of different proteins of interest.
20. Reserve an aliquot for "Lysate 3" control. This control is used along with Lysates 1 and 2 to evaluate purification performance and centrosome proteins (e.g., γ-tubulin, CEP290) localization shifts to the final fractions.
21. The tubes used in this part of the protocol are usually of about 14 mL; a rotor including swinging buckets, such as the SW40 Ti rotor (Beckman Coulter) or equivalent, is required for this step of the protocol
22. Snap-freeze 80 μL aliquots in liquid nitrogen and store at −80°C until proteomics analysis. Usually, centrosomes are enriched in the last four fractions. For tubulin polymerization in vitro, 5 uL aliquots contain enough centrosomes to serve as seeds for a single reaction.
23. Using multiple centrosome markers (e.g., γ-tubulin, Map4, Cep290) for Western blot validation is recommended. While γ-tubulin is a well-established marker for centrosomes, including additional markers such as Map4 (a microtubule-associated protein) and Cep290 (a centriole and transition zone-associated protein), it provides a more comprehensive and robust validation of the centrosome enrichment.

## Acknowledgments

This work was supported by the Spanish Ministry of Science and Innovation, Agencia Estatal de Investigación, by competitive grant PID2023-149541OB-I00, by "La Caixa" Foundation under the project code LCF/PR/HR23/52430018, by Comunidad de Madrid through P2022/BMD7209 (INTEGRAMUNE), through the "Severo Ochoa" Programme for Centres of Excellence in R&D(CEX2023-001386-S), and a grant from the Asociación Española Contra el Cáncer (AECC; PRYCO223002PEIN). MLP is supported by an FPI fellowship (PRE2021-097478). We would like to acknowledge CryoEM Facility (CNB-CSIC), through the "Severo Ochoa" Program for Centers of Excellence in R&D(CEX2023-001386-S).

## References


1. Conduit PT, Wainman A, Raff JW (2015) Centrosome function and assembly in animal cells. Nat Rev Mol Cell Biol 16(10):611–624. https://doi.org/10.1038/nrm4062
2. Sancho D, Vicente-Manzanares M, Mittelbrunn M, Montoya MC, Gordón-Alonso M, Serrador JM, Sánchez-Madrid F (2002) Regulation of microtubule-organizing center orientation and actomyosin cytoskeleton rearrangement during immune interactions. Immunol Rev 189:84–97. https://doi.org/10.1034/j.1600-065x.2002.18908.x
3. Martín-Cófreces NB, Robles-Valero J, Cabrero JR, Mittelbrunn M, Gordón-Alonso M, Sung CH, Alarcón B, Vázquez J, Sánchez-Madrid F (2008) MTOC translocation modulates IS formation and controls sustained T cell signaling. J Cell Biol 182(5):951–962. https://doi.org/10.1083/jcb.200801014
4. Stinchcombe JC, Majorovits E, Bossi G, Fuller S, Griffiths GM (2006) Centrosome polarization delivers secretory granules to the immunological synapse. Nature 443(7110):462–465. https://doi.org/10.1038/nature05071
5. Martín-Cófreces NB, Baixauli F, Sánchez-Madrid F (2014) Immune synapse: conductor of orchestrated organelle movement. Trends Cell Biol 24(1):61–72. https://doi.org/10.1016/j.tcb.2013.09.005
6. Martin-Cofreces NB, Chichon FJ, Calvo E, Torralba D, Bustos-Moran E, Dosil SG, Rojas-Gomez A, Bonzon-Kulichenko E, Lopez JA, Otón J, Sorrentino A, Zabala JC, Vernos I, Vazquez J, Valpuesta JM, Sanchez-Madrid F (2020) The chaperonin CCT controls T cell receptor-driven 3D configuration of centrioles. Sci Adv 6(49):eabb7242. https://doi.org/10.1126/sciadv.abb7242
7. Vasquez RJ, Howell B, Yvon AM, Wadsworth P, Cassimeris L (1997) Nanomolar concentrations of nocodazole alter microtubule dynamic instability in vivo and in vitro. Mol Biol cell 8(6):973–985. https://doi.org/10.1091/mbc.8.6.973
8. Casella JF, Flanagan MD, Lin S (1981) Cytochalasin D inhibits actin polymerization and induces depolymerization of actin filaments formed during platelet shape change. Nature 293(5830):302–305. https://doi.org/10.1038/293302a0
9. Shoji K, Ohashi K, Sampei K, Oikawa M, Mizuno K (2012) Cytochalasin D acts as an inhibitor of the actin-cofilin interaction. Biochem Biophys Res Commun 424(1):52–57. https://doi.org/10.1016/j.bbrc.2012.06.063
10. Bornens M, Moudjou M (1999) Studying the composition and function of centrosomes in vertebrates. Methods in Cell Biol 61:13–34. https://doi.org/10.1016/s0091-679x(08)61973-1
11. Gomez TS, Billadeau DD (2008) T cell activation and the cytoskeleton: you can't have one without the other. Adv Immunol 97:1–64. https://doi.org/10.1016/S0065-2776(08)00001-1
12. Busselez J, Chichón FJ, Rodríguez MJ et al (2019) Cryo-electron tomography and proteomics studies of centrosomes from differentiated quiescent thymocytes. Sci Rep 9(1):7187. https://doi.org/10.1038/s41598-019-43338-9 Published 2019 May 10
13. Lozano-Prieto M, Camafeita E, Jorge I, Laguillo-Gómez A, Barrero-Rodríguez R, Devesa CA, Pertusa C, Calvo E, Sánchez-Madrid F, Vázquez J, Martin-Cofreces NB (2024) In-gel protein digestion using acidic methanol produces a highly selective methylation of glutamic acid residues. J proteom 304:105229. https://doi.org/10.1016/j.jprot.2024.105229

Check for updates

# Chapter 5

# Method for Analyzing Nucleoskeletal Components from the Isolated Nucleus

**Marina Armesto, Christopher P. Toseland and Javier Redondo-Muñoz**

## Abstract

Multiple skeletal components, including actin and associated proteins, are located in both the cytoplasm and the nucleus. Here, we describe a methodology to study nuclear actin and its associated proteins through a comprehensive approach. First, we focus on nuclear isolation as a singular step to separate and distinguish skeletal components within the nucleus. Nuclear actin organization can be regulated by using pharmacological agents (e.g., latrunculin, jasplakinolide) or genetic tools, such as mutants of nuclear actin (NLS-actinR62D). These approaches allow us to analyze the nuclear morphology, protein localization, and structural changes. Together, the development of advanced super-resolution microscopy, live-cell imaging, and biochemical approaches continues to refine our understanding of nuclear actin structure and dynamics.

**Keywords** Actin, Chromatin, Nuclear isolation, Lamin, Epigenetic changes

## 1 Introduction

Actin is a well-characterized and highly conserved cytoskeletal protein associated with cellular structure, motility, and intracellular trafficking. Recent decades have revealed a crucial role for actin within the nucleus, where it participates in essential processes [1], such as chromatin remodeling, chromosome reorganization, transcriptional regulation, DNA repair, and RNA processing [2–7]. In many instances, these roles are also linked to myosin proteins interacting with actin [8, 9]. Nuclear actin exists in monomers, short polymers, filaments, and rods, with its dynamics associated with a suite of actin-binding proteins (ABPs). While the presence of actin in the nucleus was initially controversial due to challenges in visualizing polymerized nuclear actin, advanced imaging and biochemical techniques have now firmly established its presence.

Actin-binding proteins (ABPs) play a fundamental role in modulating nuclear actin dynamics and function. Well-characterized

Noa B. Martín-Cófreces et al. (eds.), *The Cytoskeleton*, Methods in Molecular Biology, vol. 3070,
https://doi.org/10.1007/978-1-0716-5515-3_5,

cytoplasmic ABPs such as cofilin, profilin, Arp2/4, and N-WASP also exhibit nuclear functions, regulating actin filament dynamics [10–14]. Additionally, the formin family of actin nucleators, such as mDia1, has been shown to promote nuclear actin polymerization in response to cellular stress [15]. It has been reported that some other ABPs control and regulate epigenetic changes and the chromatin structure [16, 17]. Nuclear isolation provides a fundamental tool for analyzing the mechanical features of the nucleus and distinguishing cytosolic from nuclear components [18, 19], even when this step might promote nuclear shrinking [20]. Specific and non-specific methods exist for the manipulation of nuclear actin levels and organization. For example, this can be achieved with drug treatments such as latrunculin and jasplakinolide, which also impact cytoskeletal filaments and promote changes that are observed in isolated nuclei [21]. Specific nuclear perturbation occurs through overexpression of exportin-6 to deplete actin from the nucleus, or overexpression of a polymerization-deficient mutant of nuclear actin (NLS-actinR62D).

Despite significant progress, many questions remain regarding the structural organization of nuclear actin filaments, their precise regulatory mechanisms, and their functional implications in different cellular contexts. Here, we describe methods for analyzing actin and other cytoskeletal components and their functions within the nucleus of cells. We detail procedures for visualizing nuclear actin and associated proteins using conventional wide-field fluorescence microscopy and assessing nuclear mechanical properties. Additionally, we outline approaches for quantifying nuclear morphology, protein distribution, and structural changes using freely available image analysis software. These methods provide a comprehensive toolkit for investigating the roles of cytoskeletal elements in nuclear structure and function.

## 2 Materials

### 2.1 Solutions

1. RPMI 1640 medium with L-glutamine, 25 mM HEPES.
2. Opti-MEM I Medium.
3. Fetal bovine serum (FBS).
4. Poly-L-lysine solution: 10 μg/mL in sterile water.
5. Fixative solution: 4% formaldehyde in PBS.
6. Permeabilization buffer: 0.5% Triton X-100 in PBS.
7. Blocking solution: 10% fetal bovine serum, 0.1% Triton X-100 in PBS.
8. Buffer A: 10 mM HEPES, 10 mM KCl, 1.5 mM $MgCl_2$, 0.34 M sucrose, 10% (v/v) glycerol, 7. 1 mM DTT.
9. 0.5% Triton X-100 in PBS.

10. Protease inhibitors.
11. Phosphate-buffered saline (PBS), pH 7.5.
12. Tris Buffered Saline (TBS), pH 8.0.

### 2.2 Chemicals and Constructs

Several proteins can be used for immunofluorescence visualization of nucleoskeleton (see Sect. 1). Antibodies anti-lamin B1 and MLCK are used in this protocol.

1. Different cDNA constructs for nuclear actin are used [5]: NLS-YFP-WT (wild type actin), NLS-YFP-R62D (R actin mutant that inhibits actin polymerization), NLS-YFP-S14C (S actin mutant that stabilizes polymerized actin) (*see* **Note 1**).
2. Lipofectamine 2000.
3. Latrunculin-B (Lat-B).
4. Hoechst 33342 solution.

### 2.3 Equipment

1. Inverted DMi8 microscope (Leica) with ACS-APO 63x NA 1.30 glycerol immersion objective (*see* **Note 2**).
2. Centrifuge capable of 1400 × *g*.
3. $CO_2$ incubator set at 5% $CO_2$ and 37°C for cell culture.

### 2.4 Software and Image Analysis Tools

1. ImageJ software (National Institute of Health).
2. Leica software for 3D reconstructions.

## 3 Methods

### 3.1 Cell Culture

1. Culture the human Jurkat (CVCL_0367) cell line (*see* **Note 3**) in RPMI 1640 medium with L-glutamine, 25 mM HEPES, and 10% FBS and maintained in 5% $CO_2$ and 37°C.
2. Prepare mechanically transformed (TR) Jurkat cells for transient expression of actin mutants in the nucleus: Plate $8 \times 10^5$ TR Jurkat cells in 500 μL of antibiotic-free RPMI per well of a 24-well plate.
3. Prepare DNA-Lipofectamine Complex Formation in sterile tubes. Use separate tubes for each construct. Dilute 500 ng of NLS-YFP-WT (wild type actin), NLS-YFP-R62D (R actin) or NLS-YFP-S14C (S actin) DNA plasmids in 50 μL of Opti-MEM I Medium.
4. Mix gently.
5. In separate tubes, prepare 1 μL of Lipofectamine 2000 in 50 μL of Opti-MEM I Medium for each construct (*see* **Note 4**). Incubate for 5 min at room temperature.

6. Mix both solutions: the diluted DNA and the diluted Lipofectamine 2000 gently and incubate for 20 min at room temperature.
7. Add the 100 μL of DNA-Lipofectamine complexes to each well containing TR Jurkat cells.
8. Mix gently. Incubate the cells at 37°C in a $CO_2$ incubator for 18–48 h.
9. Optionally, change the medium after 4–6 h (*see* **Note 5**).

#### 3.2 Nuclear Isolation

1. Collect cells from culture conditions and transfer them to a sterile tube.
2. Resuspend the cell pellet in Buffer A (usually 150–200 μL of Buffer A for $1 \times 10^6$ cells) (*see* **Note 6**).
3. Mix harshly.
4. Incubate the cell suspension on ice for 5 min (*see* **Note 7**).
5. Centrifuge the lysed cell suspension at $1400 \times g$ for 5 min at 4°C (*see* **Note 8**).
6. Carefully aspirate and discard the supernatant.
7. Wash the nuclear pellet in Buffer A and centrifuge it again at $1400 \times g$ for 5 min at 4°C.
8. Gently resuspend the nuclear pellet in PBS.
9. Use the isolated nuclei for subsequent experiments as required.

#### 3.3 Immunofluorescence Staining of Isolated Nuclei

1. Coat glass slides with 10 μg/mL poly-L-lysine solution for 1 h at room temperature or overnight at 4°C (*see* **Note 9**).
2. Drain excess liquid from the coverslip.
3. Seed isolated nuclei onto coated slides and spin the samples at 1000 rpm for 2 min.
4. Allow isolated nuclei to sediment for 5 min at room temperature.
5. Fix cells with 4% formaldehyde for 10 min at room temperature.
6. Wash samples with TBS three times (*see* **Note 10**).
7. Permeabilize cells with 0.5% Triton X-100 in PBS for 5 min at room temperature.
8. Wash samples with PBS three times.
9. Block samples in blocking solution for 30 min.
10. Incubate with primary antibodies (1:100 dilution) for 1 h at room temperature (*see* **Note 11**).
11. Wash samples with PBS three times.

12. Incubate with secondary antibodies (1:200 dilution) for 1 h at room temperature (*see* **Note 12**).
13. Wash samples with PBS three times.
14. Stain samples with Hoechst 33342 (1 μg/mL) for 10 min at room temperature (*see* **Note 13**).
15. Wash samples with PBS and then with distilled water.
16. Mount slides with the appropriate mounting medium. Examples are included in Figs. 1 and 2.

### 3.4 Analysis of Lamin B1 Distribution Upon Actin Polymerization

Disruptions of the actin polymerization can lead to aberrant nuclear morphology, including the formation of invaginations. These invaginations are characterized by the presence of nuclear envelope markers, which are typically found at the nuclear periphery (see Fig. 1).

1. Actin polymerization can be inhibited by latrunculin B treatment (1 μg/mL for 1 h). Additionally, the actin construct NLS-YFP-R62D (R actin) is a nonpolymerizing β-actin mutant that recapitulates the effect of latrunculin B treatment (see Sect. 1 and Sect. 3.3).
2. To visualize nuclear lamin B1 localization, use the specific antibodies. Perform microscopy analysis, for example, using an inverted DMi8 microscope (Leica) with ACS-APO 63× NA 1.30 glycerol immersion objective (*see* **Note 14**).
3. Acquire z-stack images using appropriate excitation and emission wavelengths (*see* **Note 15**). We use a focus interval of 1 μm, although different intervals can be employed.
4. Export images in an appropriate format for the analysis (i.e., tiff format).

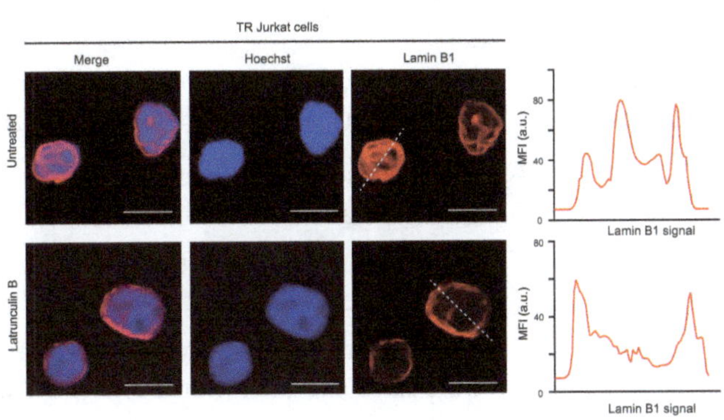


**Fig. 1** TR Jurkat cells were cultured in the presence or absence of latrunculin B (1 μg/mL) for 1 h at 37°C. Then, cells were seeded on poly-L-lysine-coated coverslips, fixed, and analyzed by confocal microscopy. Bar 10 μm. Line plots show the signal profile of lamin B1 from longitudinal sections of the nuclei observed. Bar 10 μm

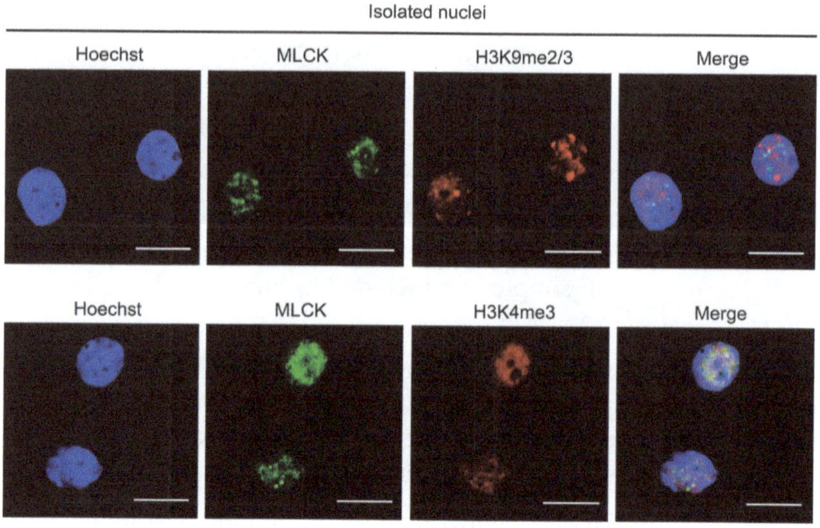


**Fig. 2** Isolated nuclei from Jurkat cells were seeded on polylysine-coated coverslips. Samples were fixed, permeabilized, and stained using specific antibodies to visualize the chromatin, MLCK, and H3K9me2/3 or H3K4me3. The middle section of the nucleus was analyzed by confocal microscopy. Bar 10 μm

5. In Fiji, make a maximum-intensity projection of the z stack using Image > Stacks > Z Project.
6. Select and draw a random line ROI perpendicular to the nuclear axis, and Analyze > Plot Profile. Both ends of the line should extend into the background outside the cell.
7. A Plot window should appear showing *Gray Value*(pixel value) versus *Distance*.
8. Save data (click Save on the Plot window) or copy and paste into a spreadsheet application such as Excel (More> Copy on the Plot window).

### 3.5 Detection of Skeletal Proteins into the Nucleus

As discussed above, several actins and ABPs have been described in the nucleus. To study them through microscopy:

1. Use Isolated nuclei from Jurkat cells seeded and fixed as indicated above (see Sect. 3.3).
2. To visualize skeletal components, use the specific antibodies and perform microscopy analysis.
3. Acquire z-stack images at a focus interval of 1 μm using appropriate excitation and emission wavelengths (*see* **Note 16**).
4. In Fiji, use the z-stack from the mid-section of the nucleus. This corresponds to the widest cross-section of the nucleus between the nuclear poles.
5. Define the ROI to measure. The ROI selected depends on the nuclear localization of the protein of interest. Using the

entire nucleus allows for determining the nuclear distribution or global nuclear intensity of the protein.

6. Select *Analyze > Set Measurements.* Select the parameters of interest by checking them (such as Area, Mean Gray Value, Integrated Density, etc.).
7. Choose *Analyze > Measure* and save the measurements that appear in the window for posterior analysis. Data can also be pasted into a spreadsheet application such as Excel.

## 4 Notes

1. The protocol for adherent cells may differ from cells growing in suspension. Please follow the instructions of the manufacturer.
2. Other microscopes can be used for this kind of image acquisition. User will need to adjust the acquisition parameters in these cases.
3. The Jurkat cell line is a widely used human T lymphocyte cell line derived from acute T cell leukemia (ATCC).
4. Optimization of DNA/Lipofectamine 2000 ratio may be necessary. Usually ratios of 1:2 to 1:3 (DNA, μg: Lipofectamine 2000 μl) work correctly for most cell lines.
5. The manufacturer explains that changing the medium after 4–6 h might enhance the transfection.
6. The volume of Buffer A buffer used should be adjusted based on the number of cells processed. We usually employ 100 μl for $2 \times 10^6$ of cells.
7. Keep all solutions and samples on ice throughout the procedure to maintain nuclear integrity.
8. Avoid over-pipetting when resuspending the nuclear pellet to prevent damage to the nuclei.
9. Apply some pressure to allow the entire coverslip to be covered by poly-L-lysine.
10. Wash the samples in TBS buffer to neutralize the formaldehyde more efficiently after the fixation step and increase signal-to-noise ratio.
11. Optimization of incubation times and concentration of some antibodies may be necessary.
12. Some antibodies may upgrade their performance by incubating them at 4°C overnight. A mix of primary antibodies can be used after optimization if secondary antibodies do not react with each other or if they are conjugated to fluorochromes. Highly cross-absorbed antibodies are recommended.

13. Include additional points to ensure there is no cross-reactivity between antibodies and to avoid unwanted labeling of background.
14. DNA stains such as DAPI or SYBR green can be used.
15. Nuclei can be observed without the optical sectioning ability of a confocal microscope, and some measurements can be performed with an epifluorescence microscope. However, for other methods, a confocal system is required to allow higher resolution.
16. If imaging two proteins, use sequential line scanning to avoid fluorescence bleed through.

## Acknowledgments

The authors thank the Microscopy Unit of Instituto de Investigación Biosanitaria Gregorio Marañón (IiSGM) for assistance with confocal analyses. This research was supported by UKRI-MRC (MR/M020606/1), UKRI-BBSRC (BB/X008460/1), and Science and Technology Facilities Council (19130001) to C.P.T.; and grants from the Ministerio de Ciencia e Innovación (MICINN) Agencia Estatal de Investigación (AEI) (PID2023-150232OB-I00, AEI/10.13039/501100011033) to J.R.M.

### References

1. Ulferts S, Lopes M, Miyamoto K, Grosse R (2024) Nuclear actin dynamics and functions at a glance. J Cell Sci 137:jcs261630
2. Olave IA, Reck-Peterson SL, Crabtree GR (2002) Nuclear actin and actin-related proteins in chromatin remodeling. Annu Rev Biochem 71:755–781
3. Chuang CH, Carpenter AE, Fuchsova B, Johnson T, de Lanerolle P, Belmont AS (2006) Long-range directional movement of an interphase chromosome site. Curr Biol 16:825–831
4. Hofmann WA, Stojiljkovic L, Fuchsova B, Vargas GM, Mavrommatis E, Philimonenko V, Kysela K, Goodrich JA, Lessard JL, Hope TJ, Hozak P, de Lanerolle P (2004) Actin is part of pre-initiation complexes and is necessary for transcription by RNA polymerase II. Nat Cell Biol 6:1094–1101
5. Hari-Gupta Y, Fili N, Dos Santos Á, Cook AW, Gough RE, Reed HCW, Wang L, Aaron J, Venit T, Wait E, Grosse-Berkenbusch A, Gebhardt JCM, Percipalle P, Chew TL, Martin-Fernandez M, Toseland CP (2022) Myosin VI regulates the spatial organisation of mammalian transcription initiation. Nat Commun 13:1346
6. Wei M, Fan X, Ding M, Li R, Shao S, Hou Y, Meng S, Tang F, Li C, Sun Y (2020) Nuclear actin regulates inducible transcription by enhancing RNA polymerase II clustering. Sci Adv 6:eaay6515
7. Hurst V, Shimada K, Gasser SM (2019) Nuclear actin and actin-binding proteins in DNA repair. Trends Cell Biol 29:462–476
8. Viita T, Kyheröinen S, Prajapati B, Virtanen J, Frilander MJ, Varjosalo M, Vartiainen MK (2019) Nuclear actin interactome analysis links actin to KAT14 histone acetyl transferase and mRNA splicing. J Cell Sci 132:jcs226852
9. Gawor M, Lehka L, Lambert D, Toseland CP (2025) Actin from within - how nuclear myosins and actin regulate nuclear architecture and mechanics. J Cell Sci 138:JCS263550
10. Shahid-Fuente IW, Toseland CP (2023) Myosin in chromosome organisation and gene expression. Biochem Soc Trans 51:1023–1034
11. Ulferts S, Lopes M, Miyamoto K, Grosse R (2024) Nuclear actin dynamics and functions at a glance. J Cell Sci 137:jcs261630

12. Oma Y, Harata M (2011) Actin-related proteins localized in the nucleus: from discovery to novel roles in nuclear organization. Nucleus 2:38–46
13. Yoo Y, Wu X, Guan JL (2007) A novel role of the actin-nucleating Arp2/3 complex in the regulation of RNA polymerase II-dependent transcription. J Biol Chem 282:7616–7623
14. Weston L, Coutts AS, La Thangue NB (2012) Actin nucleators in the nucleus: an emerging theme. J Cell Sci 125:3519–3527
15. Baarlink C, Wang H, Grosse R (2013) Nuclear actin network assembly by formins regulates the SRF coactivator MAL. Science 340:864–867
16. Wang P, Dreger M, Madrazo E, Williams CJ, Samaniego R, Hodson NW, Monroy F, Baena E, Sánchez-Mateos P, Hurlstone A, Redondo-Muñoz J (2018) WDR5 modulates cell motility and morphology and controls nuclear changes induced by a 3D environment. Proc Natl Acad Sci USA 115:8581–8586
17. Downing TL, Soto J, Morez C, Houssin T, Fritz A, Yuan F, Chu J, Patel S, Schaffer DV, Li S (2013) Biophysical regulation of epigenetic state and cell reprogramming. Nat Mater 12:1154–1162
18. Pagliara S, Franze K, McClain CR, Wylde G, Fisher CL, Franklin RJM, Kabla AJ, Keyser UF, Chalut KJ (2014) Auxetic nuclei in embryonic stem cells exiting pluripotency. Nat Mater 13:638–644
19. Guilluy C, Osborne LD, Van Landeghem L, Sharek L, Superfine R, Garcia-Mata R, Burridge K (2014) Isolated nuclei adapt to force and reveal a mechanotransduction pathway in the nucleus. Nat Cell Biol 16:376–381
20. Dahl KN, Engler AJ, Pajerowski JD, Discher DE (2005) Power-law rheology of isolated nuclei with deformation mapping of nuclear substructures. Biophys J 89:2855–2864
21. de Lope-Planelles A, González-Novo R, Madrazo E, Peralta-Carrero G, Cruz Rodríguez MP, Zamora-Carreras H, Torrano V, López-Menéndez H, Roda-Navarro P, Monroy F, Redondo-Muñoz J (2023) Mechanical stress confers nuclear and functional changes in derived leukemia cells from persistent confined migration. Cell Mol Life Sci 80:316

Check for updates

# Chapter 6

# In Vitro Analysis of Profilin 1 Binding to Recombinant Actin

**Leticia Villadangos, David Escot and Juan M. Serrador**

## Abstract

Profilin 1 (PFN1) is a key actin-binding protein (ABP) involved in cytoskeletal dynamics, regulating actin polymerization and filament remodeling. While native actin has been extensively studied, recombinant actin produced through biotechnological approaches remains less explored due to solubility and folding challenges. Previously, we assessed the effects of S-nitrosylation on actin binding, providing insights into post-translational modifications that regulate actin function. In the present study, we provide a comprehensive protocol to produce recombinant β-actin using an in vitro transcription-translation system supplemented with chaperonin CCT to analyze its interaction with PFN1. Our findings highlight the utility of recombinant actin for studying ABPs and their regulatory mechanisms, offering a cost-effective alternative for structural and functional analyses.

**Keywords** Recombinant β-actin, Profilin 1, CCT, S-nitrosylation, In vitro transcription-translation

## 1 Introduction

Actins are a highly conserved family of proteins found in all eukaryotic cells that play an essential role in cellular motility, division, and intracellular transport, among many other important physiological functions [1, 2]. In non-muscular cells, globular (G-) and filamentous (F-) actin exist in a dynamic equilibrium of polymerization/depolymerization. Human actin is a 375 amino acid protein of 42 kD that can be roughly divided into two major groups: muscle and cytoplasmic. The six known actin isoforms, classified as α-, β-, and γ-actin, are expressed in specific spatio-temporal cellular patterns and share the capacity to interact with a plethora of regulatory actin-binding proteins (ABPs), such as profilin, which fosters ATP binding to actin, polymerization, and treadmilling from the barbed ends of F-actin [3, 4]. Muscular actins (α-skeletal muscle actin, α-cardiac actin, and α- and γ-smooth muscle actin) are associated with fiber contraction, whereas β- and γ-cytoplasmic actin are ubiquitously expressed in cells and support internal mechanical forces to maintain cell shape, movement, transport, junctions, and polarity [2]. However, despite

Noa B. Martín-Cófreces et al. (eds.), *The Cytoskeleton*, Methods in Molecular Biology, vol. 3070,
https://doi.org/10.1007/978-1-0716-5515-3_6,

their similarities, β-actin is the most highly expressed isoform in non-muscular cells.

The actin cytoskeleton is a target of a wide variety of post-translational modifications (PTMs) [5, 6]. Acetylation, phosphorylation, glutathionylation, and ADP-ribosylation are among the most commonly detected PTMs on actin. However, actin is also highly sensitive to oxidation by reactive oxygen and nitrogen species (ROS and RNS, respectively) [5]. β-actin contains six redox-sensitive cysteines, of which S-nitrosylation at Cys374 is known to regulate the equilibrium between G- and F-actin and the reorganization of the actin cytoskeleton [7]. Purified actin from animal, yeast, or plant cells has been extensively used in vitro to investigate the function of ABPs. In contrast, biotechnologically produced recombinant actin has been much less studied, despite the advantages that an abundant and less expensive source of recombinant actin would provide for actin structural and dynamic studies, including those focused on the characterization and function of non-natural actin mutants. This is due to the fact that correct solubility and folding of recombinant actin expressed in *E. coli* or prokaryotic cell-free systems is a major challenge. In this regard, studies using an in vitro transcription-translation protein expression system have demonstrated that the eukaryotic chaperonin CCT facilitates the folding actin into its native conformation [8, 9].

This chapter describes methods to analyze the interaction of profilin-1 (PFN1) with recombinant actin and its regulation by S-nitrosylation. This approach allows the analysis of actin-ABP interactions using purified CCT to obtain recombinant actin mutants from a cell-free transcription-translation system reconstituted from *E. coli* translation components, providing procedures for studying actin regulation by PTM.

## 2 Materials

### 2.1 Equipment

1. Standard cell culture incubator at 37°C in 5% $CO_2$ humidified atmosphere.
2. Orbital shaker platform.
3. Thermoblock.
4. Benchtop microcentrifuge.
5. Protein electrophoresis and transfer system.

### 2.2 Reagents

1. Chaperonin CCT purified from bovine testis tissue as previously reported [9].
2. PUREexpress in vitro protein synthesis kit (*see* **Note 1**), containing Solution A: tRNAs and small molecules, including amino acids and rNTPs, and Solution B: ribosomes and T7 RNA polymerase, translation factors, aminoacyl-tRNA synthetases, and energy regeneration enzymes.

3. Plasmids for expression of 6XHis-tagged mCherry-β-actin: Wild type (WT), 3C (C257S+C272S+C285S) mutant, and 4C (C257S+C272S+C285S+C374S) mutant (*see* **Note 2**).
4. 25 mM S-nitroso-cysteine (L-Cys-NO), pH 6.8, freshly prepared as in [10]. Add 1 ml of 50 mM sodium nitrite (containing 100 μL of 1 N HCl) to 50 mM of L-cysteine; this will result in a stable 25 mM stock solution of approximately pH 3.0. Dilute the solution in alkalinized saline to increase the pH of L-Cys-NO to approximately pH 6.8.
5. DNase I covalently linked to CNBr-activated Sepharose beads: Prepare beads by coupling 20 mg of DNase I to 1 ml of BrCN-Sepharose. Wash DNase I-Sepharose beads three times with 1 mL of 10 mM Tris-HCl, pH 8.0, containing 0.05% NP-40, 30 mM NaCl, 0.2 mM CaCl2, 0.5 mM PMSF, and 0.1% ovalbumin.
6. $Ni^{2+}$-charged resin.
7. G-buffer: 5 mM Tris-HCl, pH 8.0, 0.2 mM ATP, 0.2 mM $CaCl_2$.
8. Imidazole.
9. Triton X-100.
10. Acrylamide/Bis-acrylamide solution: 40% Mix, 29:1.
11. 50 mM DTT reducing Laemmli loading buffer.
12. Human PFN1 monoclonal antibody.
13. mCherry polyclonal antibodies.
14. Coomassie blue staining solution (*see* **Note 3**).

### 2.3 Cells

1. Jurkat T cells as a biological source of human PFN1 (*see* **Note 4**).

## 3 Methods

### 3.1 Production and Analysis of Soluble Human β-Actin

1. Thaw aliquots of solutions A and B of the PUREexpress in vitro protein synthesis kit to perform the needed total reaction volume (*see* **Note 5**).
2. Prepare a 25 μL reaction mix for β-actin synthesis in a nuclease-free 0.5 mL microcentrifuge tube, adding components in the following order (*see* **Note 6**):

   a. 12.5 μL of Solution A (containing amino acids and rNTPs).
   b. 5 μL of Solution B (containing ribosomes and translation factors).
   c. 1 μL of 20 U/μL RNase inhibitor (to prevent mRNA degradation).

d. Add nuclease-free water to a final volume of 25 μL.
e. 3 μM purified CCT.
f. 250 ng of the tagged β-actin expression plasmids.

3. Mix and collect the reaction volume by briefly centrifuging to collect the contents at the bottom of the tube.
4. Incubate the reaction mix at 37°C for 1 h.
5. After 1 h, place the reaction on ice to stop the translation process.
6. To assess β-actin synthesis, take a 3 μL aliquot, mix with 2× Laemmli loading buffer containing 50 mM DTT, and incubate at 85°C for 5 min to denature and reduce proteins (*see* **Note 7**).
7. Load the sample onto an SDS-PAGE minigel and perform electrophoresis at 4°C in a cold chamber, applying a constant current of 20 mA (stacking gel) and 50 mA (running gel).
8. Once the proteins have been separated, stain the gel with 25 mL Coomassie blue for at least 30 min under shaking. Transfer the gel to distilled water and wash it two to three times for 20 min to remove any background staining.
9. Identify translated β-actins by SDS-PAGE followed by transference to nitrocellulose for Western blot analysis using mCherry-specific pAbs (*see* **Note 8**) (Fig. 1A).

### 3.2 Assessment of Functional in Vitro Expressed Recombinant β-Actin

#### 3.2.1 Check the Solubility of In Vitro Translated WT β-Actin and Its Corresponding C3 and C4 Mutants

1. *Centrifuge the remaining 22* μL of in vitro expressed β-actins in a microfuge at 13,000 rpm for 30 min at 4°C.
2. Separate the supernatant (soluble fraction) from the pellet (insoluble fraction) and transfer them into new microcentrifuge tubes.
3. Resuspend the pellet in 1× Laemmli buffer.
4. Mix 3 μL of the soluble fraction with 2× Laemmli loading buffer.
5. Perform SDS-PAGE electrophoresis to visualize soluble and insoluble β-actin by Coomassie blue staining and Western Blotanalysis (Fig. 1B).
6. To estimate the concentration of soluble β-actin, perform a quick semi-quantification by visual comparison of β-actin staining and the staining of increased amounts of BSA (0.5, 1, 2, and 4 μg) in an SDS-PAGE electrophoresis gel (*see* **Note 9**).

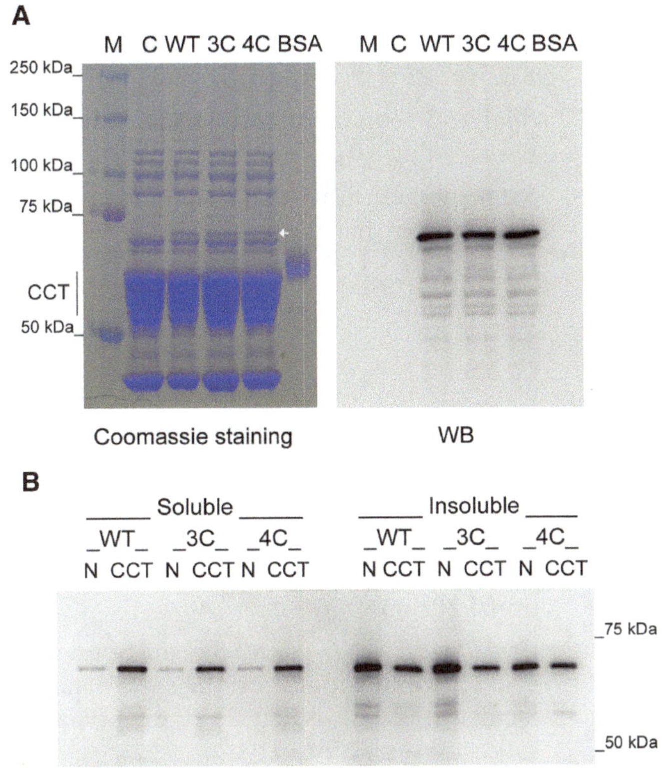


**Fig. 1** In vitro synthesis of soluble His-tagged recombinant WT-, 3C-, and 4C-mCherry-β-actin from purified *E. coli* translation components supplemented with CCT. (**A**) Coomassie staining of gels containing the total protein fraction (left) and immunoblot (WB) using mCherry-specific polyclonal antibodies (right) was performed in parallel. Control "C," CCT-containing *E. coli* translation components without the DNA template. The arrowhead points to synthesized actin. BSA (1.5 μg) was used as a loading control. (**B**) Western blot analysis of recombinant WT-, 3C-, and 4C-mCherry-β-actin in soluble and insoluble fractions from *E. coli* translation components in the absence (N) or presence (CCT) of chaperonin. Detection of recombinant WT actin and 3C and 4C mutants was performed with mCherry-specific rabbit polyclonal antibodies. M, molecular weights. (Reproduced from [6])

#### 3.2.2 Check Binding of In Vitro Translated β-Actin to DNase I

1. *Dilute 0.2* μg of soluble β-actin in 0.8 mL G-buffer containing 0.1% Triton X-100 in 1.5 mL microcentrifuge tubes.
2. *Add 10*μL of a 50% slurry of 5 mg/mL DNase I-Sepharose beads.
3. *Incubate soluble* β-actin and DNase I-Sepharose beads at room temperature for 1 h on the orbital shaker.
4. *Collect DNase I-Sepharose beads by centrifugation at 3000 rpm for 1 min.*
5. *Wash samples three times with 0.8 mL of 0.5% Triton X-100 buffer G.*

6. *Discard supernatants and resuspend DNase I-bound* β-actin in 30 μL of reducing Laemmli buffer and heat the samples for 5 min at 85°C.
7. *Analyze* β-actin binding to DNase I by SDS-PAGE electrophoresis loading 10 μL and 20 μL of samples.
8. Stain the gel with Coomassie blue and perform immunoblotting with mCherry-specific Abs as described in Sect. 3.1, Step 8 and Sect. 3.1, Step 9.

### 3.3 Profilin-β-Actin Interaction Assay

#### 3.3.1 Binding of In Vitro Translated β-Actin to $Ni^{2+}$-Charged Resin

1. *In 0.5 mL microcentrifuge tubes, mix 0.5* μg of recombinant 6XHis-tagged β-actins with 20 μL of 50% slurry $Ni^{2+}$-charged resin in 150 μL of buffer G and incubate them for 30 min at 4°C.
2. *Add 150* μM Cys-NO and incubate the reaction for an additional 20 min.
3. *Centrifuge Cys-NO-treated* β-actin-bound resin at 3000 rpm and 4°C for 1 min and discard the supernatant. Immediately, add β -actin-containing Jurkat T cell extracts as indicated in the next subsection (*see* **Note 10**).

#### 3.3.2 Binding of In Vitro Translated β-Actin to PFN1

1. *The day before the assay, culture 20 mL of Jurkat T cells (300,000/ mL) in a 75 cm²* flask at 37°C.
2. *Harvest* $6 \times 10^5$ cells per condition and centrifuge them at 6000 rpm and 4°C for 1 min. Remove supernatants, wash once in buffer G, and add 1 mL of buffer G containing 0.5% Triton X-100, 0.5% BSA, and 0.5 mM PMSF for 30 min. Cell lysis must be performed by continuous shaking in a cold chamber.
3. *Centrifuge cell lysates at 13,000 rpm and 4°C for 30 min and recover soluble supernatants (soluble protein fraction).*
4. *Add 0.8 mL of supernatants from Jurkat T cell extracts to a 1.5 mL microcentrifuge tube containing the 20* μL of 6XHis-tagged β -actin $Ni^{2+}$-charged resin prepared in Sect. 3.3.1.
5. *Incubate them by shaking at 4°C for 30 min (see* **Note 11**).
6. *Centrifuge tubes at 3000 rpm and 4°C for 1 min. Wash the resin twice with 0.8 mL G-buffer containing 0.1% Triton X-100 and the same buffer plus 5 mM imidazole.*
7. *Resuspend 6XHis-tagged* β-actin resin-bound protein complexes in 30 μL of reduced Laemmli loading buffer.
8. *Analyze PFN1 binding to* β-actin by SDS-PAGE and immunoblotting with PFN1 mAb (Fig. 2) (*see* **Note 12**).

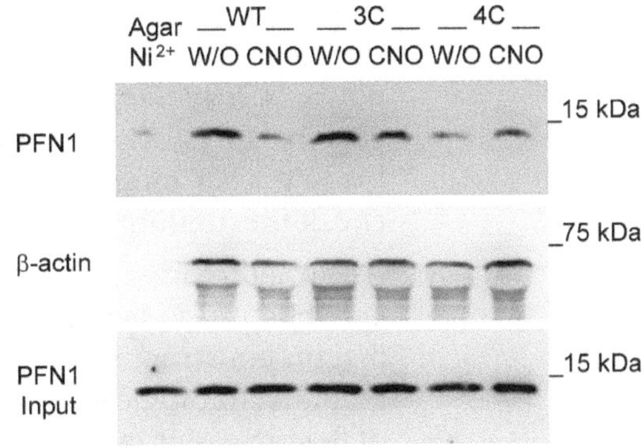


**Fig. 2** Binding of PFN1 from Jurkat T cell extracts to His-tagged WT-, 3C-, and 4C-mCherry-β-actin coupled to agarose-$Ni^{2+}$ and pretreated/S-nitrosylated or not with 100 μM Cys-NO (CNO). Agarose-$Ni^{2+}$ was used as a control (Agar$Ni^{2+}$). Detection of recombinant actins is performed with mCherry-specific rabbit polyclonal antibodies. Molecular weights are shown on the right

## 4 Notes

1. Although several in vitro transcription-translation kits are commercially available, we have chosen the Pure Express System (New England BioLabs) for its performance, cost-effectiveness, and usability.
2. All β-actin versions are cloned into the T7 promoter-containing pET-His6TEV-LIC vector [6].
3. Protein staining can be performed using Quick Coomassie Stain (Neo Biontech), a colloidal Coomassie-containing reagent that only requires distilled water to remove background staining.
4. This procedure employs Jurkat T cell extracts to obtain PFN1. Although this cell line serves as a convenient source of PFN1, other human cell lines can be used, as PFN1 is ubiquitously expressed in different tissues. A comprehensive list of PFN1-expressing cell types is available on The Human Protein Atlas (https://www.proteinatlas.org/ENSG00000108518-PFN1).
5. 25 μL of total reaction volume scaled up or down to maintain the corresponding proportion of solutions A and B as in Sect. 3.1, Step 2.
6. The order of reagent addition in this section is critical to maintain proper buffer conditions and improve translation efficiency. Add solution B to solution A and mix the β-actin plasmid template at the end.

7. Depending on the features of the plasmid used as a template, the sample reaction to be analyzed by SDS-PAGE electrophoresis may vary. Large plasmid templates are more difficult to express than smaller ones, reducing protein yield. The loading of a range of sample volumes might be helpful to check the amount of synthesized β-actin.
8. As an alternative procedure to detect recombinant β-actin, anti-His antibodies can be used in Western blotting. A tagged β-actin construct is essential to generate recombinant β-actin by biotechnological methods. This tag makes the purification of β-actin easier by affinity chromatography. The recombinant β-actin obtained in this procedure has two tags, a 6XHis tail and a mCherry tag that facilitates the identification of β-actin in the cell source of PFN1 by Western blot and immunofluorescence. If the synthesis of untagged recombinant β-actin is required, 6XHis can be released from mCherryβ-actin by incubation with Tobacco Etch virus (TEV) protease. A TEV proteolysis site is localized between both tags. Regarding the release of mCherry from β-actin, additional protease sites for synthetic biology, such as TEV, thrombin, or Tobacco Vein Mottling Virus (TVMVp), can be included in the plasmid expression vector by recombinant DNA technology.
9. More precise semi-quantitative estimations of in vitro-translated β-actin in reaction samples can be performed by densitometric analysis of protein bands on pictures from Coomassie blue-stained SDS-PAGE electrophoresis gels using an image analysis program (e.g., NIH ImageJ).
10. For binding assays between PFN1 and recombinant β-actin, this procedure uses Jurkat T cell extracts as sources of PFN1. However, some μg of recombinant PFN1 in buffer G can be used as an alternative. PFN1 can be obtained from these cells by poly-L-Lys affinity column purification as described in [11]. Alternatively, commercially PFN1 is available (Cytoskeleton, Inc.).
11. Optimization of β-actin binding to PFN1 requires checking of Jurkat T cell extract titers. To maximize results, add different volumes of Jurkat T cell extracts to 6XHis-mCherry-β -actin-bound resin. Similarly, the incubation time can be optimized to improve the PFN1-β-actin interaction and signal-to-noise ratio. Test different incubation times to enhance PFN1 binding assays.
12. A more detailed study of the interaction between β-actin and PFN1 may require a quantitative estimation of proteins in immunoblots by densitometric analysis. Perform densitometric quantifications with an image analysis program (e.g., NIH ImageJ) and normalize PFN1 binding to β-actin with respect

to PFN1 binding to control $Ni^{2+}$-resin, and represent the mean±SD and their statistical significance as the fold induction with respect to Cys-NO non-treated WT-mCherry-β -actin.

## Acknowledgments

This manuscript was supported by grant PID2022-140155OB-I00 to J.M.S. L.V. was supported by Formación de Personal Universitario Program (FPU23/00674), Ministry of Science, Innovation and Universities and Fondo Europeo de Desarrollo Regional (FEDER) and Fondo Social Europeo (FSE). CBMSO is a Severo Ochoa Center of Excellence (grant CEX2021-001154-S, funded by MICIN/AEI/10.13039/501100011033) and receives an institutional grant from Fundación Ramón Areces.

### References

1. Pollard TD, Cooper JA (2009) Actin, a central player in cell shape and movement. Science 326(5957):1208–1212. https://doi.org/10.1126/science.1175862
2. Pollard TD (2016) Actin and actin-binding proteins. Cold Spring Harb Perspect Biol 8(8):a018226. https://doi.org/10.1101/cshperspect.a018226
3. Terman JR, Kashina A (2013) Post-translational modification and regulation of actin. Curr Opin Cell Biol 25(1):30–38. https://doi.org/10.1016/j.ceb.2012.10.009
4. Ampe C, Van Troys M (2016) Mammalian actins: isoform-specific functions and diseases. Handb Exp Pharmacol 235:1–37. https://doi.org/10.1007/164_2016_43
5. Rouyere C, Serrano T, Fremont S, Echard A (2022) Oxidation and reduction of actin: origin, impact in vitro and functional consequences in vivo. Eur J Cell Biol 101(3):151249. https://doi.org/10.1016/j.ejcb.2022.151249
6. Garcia-Ortiz A, Martin-Cofreces NB, Ibiza S, Ortega A, Izquierdo-Alvarez A, Trullo A, Victor VM, Calvo E, Sot B, Martinez-Ruiz A, Vazquez J, Sanchez-Madrid F, Serrador JM (2017) eNOS S-nitrosylates beta-actin on Cys374 and regulates PKC-theta at the immune synapse by impairing actin binding to profilin-1. PLOS Biol 15(4):e2000653. https://doi.org/10.1371/journal.pbio.2000653
7. Povarova OI, Sulatskaya AI, Kuznetsova IM, Turoverov KK (2012) Actin folding, structure and function: is it a globular or an intrinsically disordered protein? In: Najman S (ed) Current frontiers and perspectives in cell biology. IntechOpen, Rijeka, Croatia. https://doi.org/10.5772/35293
8. Frankel S, Condeelis J, Leinwand L (1990) Expression of actin in *Escherichia coli*. Aggregation, solubilization, and functional analysis. J Biol Chem 265(29):17980–17987. https://doi.org/10.1016/S0021-9258(18)38260-7
9. Stemp MJ, Guha S, Hartl FU, Barral JM (2005) Efficient production of native actin upon translation in a bacterial lysate supplemented with the eukaryotic chaperonin TRiC. Biol Chem 386(8):753–757. https://doi.org/10.1515/BC.2005.088
10. Jourd'heuil D, Gray L, Grisham MB (2000) S-nitrosothiol formation in blood of lipopolysaccharide-treated rats. Biochem Biophys Res Commun 273(1):22–26. https://doi.org/10.1006/bbrc.2000.2892
11. Sohn RH, Chen J, Koblan KS, Bray PF, Goldschmidt-Clermont PJ (1995) Localization of a binding site for phosphatidylinositol 4,5-bisphosphate on human profilin. J Biol Chem 270(36):21114–21120. https://doi.org/10.1074/jbc.270.36.21114

Chapter 7

# In Vitro Production of Full-Length Myosin 2 and Visualization of Actin Filament Motility by Total Internal Reflection Microscopy

**Rafael Pérez-Díaz, Hugo Ramos-Solano, Marina Garrido-Casado, Vanessa C. Talayero, Johannes N. Greve and Miguel Vicente-Manzanares**

## Abstract

In vitro protein production constitutes a valid resource to generate data regarding the structure and function of myosin 2. This technique can be combined with proteins, molecular complexes and small molecules to study myosin II conformation, binding to actin and the effect of various myosin and actin modulators. Here, we describe a straightforward protocol to produce and purify NM2 (non-muscle myosin 2) in *Spodoptera frugiperda* Sf9 cells. We also provide an example of produced protein use in an in vitro actin motility assay using TIRF microscopy to visualize myosin 2-driven F-actin motion and quantify it using ImageJ. This type of analysis can be implemented in various ways and become useful for drug screenings and characterization of novel, disease-causing mutations.

**Keywords** In vitro myosin 2 production, Sf9 cells, Actin, TIRF microscopy, ImageJ

## Abbreviations

| | |
|---|---|
| cryo-EM | cryo-electron microscopy |
| ELC | Essential light chain |
| RLC | Regulatory light chain |
| TIRF | Total internal reflection fluorescence |

## 1 Introduction

Myosin 2 is one of the best characterized proteins at a structural and biochemical level, including high resolution cryo-EM structures in the enzymatically inactive state [1–4]. Myosin 2 interacts with F-actin and displays ATPase activity. These activities can be leveraged to characterize kinetic differences between paralogs [5–7]. Its ability to

Noa B. Martín-Cófreces et al. (eds.), *The Cytoskeleton*, Methods in Molecular Biology, vol. 3070,
https://doi.org/10.1007/978-1-0716-5515-3_7,

form filaments has been used in vitro to characterize specific features of their filamentation ability [8, 9].

Myosin 2 production in *E.coli* is not feasible due to the complexity of the coiled-coil, which prevents bacteria from producing it properly. Thus, the major system of choice is its expression in Sf9 cells. Sf9 cells can produce a good yield (mg) of full length myosin 2 [10], which can be used in downstream applications such as filament formation, cryo-EM, and in vitro actin motility. This protocol describes the entire pipeline of myosin 2 production, from cloning to purification, and its use in an in vitro actin motility assay measured using high resolution total internal reflection microscopy (TIRF). Here, we present two examples, NM2-A from human origin and Myo2 from *S.pombe*. While essentially equivalent function-wise, *S.pombe* Myo2 does not require regulatory light chain (RLC) phosphorylation to support strong actin motility [11].

## 2 Materials

For Sect. 2.1, tissue culture grade and sterile technique are required for all the materials used. For Sects. 2.2 and 2.3, all the solutions need to be prepared using ultrapure water. Specific instructions for storage and expiration are provided for each reagent.

### 2.1 Cell Culture and Propagation

1. Cell line of choice. Sf9 cells from *Spodoptera frugiperda* (DMSZ, ACC 125) are most commonly used [12] (*see* **Note 1**).
2. Sterile Erlenmeyer flasks of four sizes: 100 mL, 500 mL, 1 L, and 2 L (*see* **Note 2**).
3. Disposable or sterilizable Erlenmeyer caps. Hirschmann Silicosen stoppers can be sterilized in a conventional autoclave.
4. Specific media: Spodopan (*see* **Note 3**).
5. Media supplementation reagents. Fetal bovine serum (requires testing) and 100× poloxamer 188 (*see* **Note 4**).
6. Common tissue culture materials, including a tissue culture hood, temperature-controlled incubator, and temperature-controlled shaker ($CO_2$ control is not required).
7. Common disposable tissue culture materials: serological pipettes, tips, automatic pipettes.

### 2.2 Gene Cloning

1. pFastBac (*see* **Note 5**).
2. pFastDual (*see* **Note 6**).
3. cDNAs for myosins heavy and light chains (*see* **Note** 7).
4. Specific primers for cDNAs (*see* **Note 8**).

5. Cloning-grade DNA polymerase (*see* **Note 9**).
6. Common molecular biology reagents: cloning-grade agarose, restriction enzymes, DNA phosphatase (e.g., SAP), T4 DNA ligase, ampicillin agar plates.
7. DH5α or other conventional *E.coli* competent strain.

### 2.3 Bacmid Production

1. Gene of interest cloned in pFastBac (see Sect. 2.2).
2. Baculovirus-specific, *E.coli* bacmid production strains, specifically DH10B or DH10EMBacY.
3. Common microbiology media and reagents (LB, agar plates).
4. Antibiotics for selection: ampicillin (100 mg/mL in water), kanamycin (50 mg/mL in water), tetracyclin (10 mg/mL in water), gentamycin. Ampicillin is used for pFastBac propagation, the others for bacmid production (see below).
5. X-gal solution: 20 mg/mL Xgal, DMSO.
6. IPTG solution: 400 mM IPTG. Use distilled water, sterile filter and store at −20°C until use.

### 2.4 Essential and Regulatory Light Chains (ELC and RLC) Production in Bacteria

1. pGEX-6P1 or similar (for ELC/RLC expression in bacterial cells).
2. *E.coli* BL21 cells.
3. Common microbiology media and reagents (LB, agar plates, etc.).
4. Lysis buffer 1: 20 mM Tris-HCl pH 7.4, 0.45 M NaCl, 1 mM PMSF, 10 μg/mL leupeptin, 2 μg/mL aprotinin, 1 mg/mL lysozyme.
5. Wash buffer 1: 20 mM Tris-HCl pH 7.4, 0.45 M NaCl, 1 mM PMSF.
6. Gluthatione-agarose washed with Wash buffer 1.
7. Reduced L-Glutathione, powder.
8. Elution buffer 1: 20 mM Tris-HCl pH 8.5, 0.3 M NaCl, 50 mM Glutathione, 0.1% Triton X-100, 10 mM DTT (*see* **Note 10**).
9. Dialysis buffer: 25 mM Hepes pH7.4, 200 mM NaCl, 2 mM $MgCl_2$.
10. Prescission protease (*see* **Note 11**).

### 2.5 Sf9 Transfection and Baculovirus Propagation for Protein Production

1. Materials described in Sect. 2.1.
2. Suitable transfection reagent (*see* **Note 12**).

### 2.6 Protein Purification

1. Lysis buffer 2: 50 mM Hepes pH 7.4, 300 mM NaCl, 4 mM $MgCl_2$, 0.3 mM EDTA, 0.3 mM EGTA, 2 mM ATP, 7 mM β-mercaptoethanol, 10 mM imidazole (*see* **Note 13**), 1.5% Triton X-100, protease inhibitors.
2. Suitable purification resin. For these examples, $Ni^{2+}$-NTA-agarose (ABT 6BCL-NTANi-X).
3. Wash buffer 2: 25 mM Hepes pH 7.4, 300 mM NaCl, 4 mM $MgCl_2$, 50 mM imidazole (*see* **Note 13**), 1 mM DTT.
4. Elution buffer 2: 25 mM Hepes pH 7.4, 200 mM NaCl, 2 mM $MgCl_2$, 0.2–0.6 M imidazole (*see* **Note 13**), 1 mM DTT, 2 μM ELC (*see* **Note 14**), 2 μM RLC (*see* **Note 14**).
5. SEC buffer: 25 mM Hepes pH 7.4, 200 mM NaCl, 2 mM $MgCl_2$, 0.1 mM EGTA, 1 mM DTT. Prepare a minimum of 500 mL and filter it through 0.22 μM filters.
6. Suitable protein concentration determination assay such as BCA or Bradford.
7. Conventional protein electrophoresis system and reagents, including Coomassie staining/destaining.
8. Vivaspin 6 tubes, 10 kDa.
9. FPLC device.
10. HiLoad 16/600 Superdex 200 pg column.
11. Trehalose, powder.

### 2.7 Actin Purification (see Note 15)

1. Rabbit muscle acetone powder.
2. G-buffer: 10 mM Tris-HCl pH 7.8, 0.2 mM $CaCl_2$, 0.5 mM ATP, 0.5 mM DTT).
3. KCl.
4. $MgCl_2$.
5. Sucrose, powder.

### 2.8 Actin Polymerization and Labeling

1. G-actin from Sect. 3.7.
2. G-buffer from Sect. 2.7.
3. 10× polymerization buffer: 100 mM Tris-HCl pH 7.8, 1 M KCl, 50 mM $MgCl_2$, 5 mM EGTA, 1 mM DTT, 1 mM ATP.
4. Fluorescent phalloidin derivative (e.g., CST 8940).

### 2.9 Flow Cell Preparation for In Vitro Actin Motility Measurements Using Total Internal Reflection Fluorescence (TIRF) Microscopy

1. 1.5 H, 22 × 22 mm square coverslips (*see* **Note 16**).
2. Mounting slides.
3. Thin, double-sided tape (*see* **Note 17**).
4. Nitrocellulose paper (*see* **Note 18**).
5. Pentyl (amyl) acetate (Merck 227471), used as solvent for nitrocellulose paper (*see* **Note 18**).

#### 2.10 Myosin II Phosphorylation for In Vitro Actin Motility Measurements Using Total Internal Reflection Fluorescence (TIRF) Microscopy

1. Myosin 2 (from Sect. 3.6).
2. Calmodulin (*see* **Note 19**).
3. MLCK (*see* **Note 20**).
4. ELC (*see* **Note 14**).
5. RLC (*see* **Note 14**).

#### 2.11 In Vitro Actin Motility Assay for Measurements Using Total Internal Reflection Fluorescence (TIRF) Microscopy

1. Glucose oxidase solution: 100 μg/mL glucose oxidase in Aassay buffer 1 (see below).
2. Catalase solution: 100 μg/mL catalase in Assay buffer 1.
3. Methyl cellulose (15 cP) 0.5% methyl-cellulose 15 cP in Assay buffer 2 (*see* **Note 21**).
4. Polymerized actin obtained in (Sect. 3.7).
5. IVMA buffer: 25 mM HEPES pH 7.4, 50 mM KCl, 5 mM $MgCl_2$, 1 mM EGTA.
6. Phosphorylation buffer: 25 mM MOPS pH7.4, 50 mM KCl, 5 mM $MgCl_2$.
7. Assay buffer 1: IVMA buffer, 0.5 mg/mL BSA, 10 mM DTT.
8. Assay buffer 2: Assay buffer 1, 5 mg/mL glucose, 100 μg/mL glucose oxidase and 100 μg/mL catalase.
9. Assay buffer 3: Assay buffer 2, 0.5% methylcellulose and 4 mM ATP.
10. Assay buffer 4: Assay buffer 1, 4 mM ATP.
11. Microscope setup, with appropriate laser sources and asset to maintain physiological temperature (*see* **Note 22**).

#### 2.12 Image Analysis and Quantification

1. ImageJ (NIH) or Fiji.
2. A graphics representation program (e.g., Excel, Graph Pad, SPSS).

## 3 Methods

#### 3.1 Cell Culture and Propagation

The following steps need to be carried out in tissue culture laminar flow hoods following sterile techniques.

1. Thaw at 37°C (1–2 min) a frozen vial of Sf9 cells (1–2 mL).
2. Resuspend in a conical 15 mL tube (Falcon-type) in 10 mL of Spodopan medium (*see* **Note 23**).
3. Count total cells in the vial.

4. Seed in a round tissue culture dish at a density of $1–2 \times 10^6$ cells/mL. Depending on the number of cells, multiwell plates can be used.
5. Cells are cultured at 27°C–29°C in a conventional incubator (*see* **Note 24**).
6. Check viability daily in a brightfield microscope (*see* **Note 25** and Fig. 1).

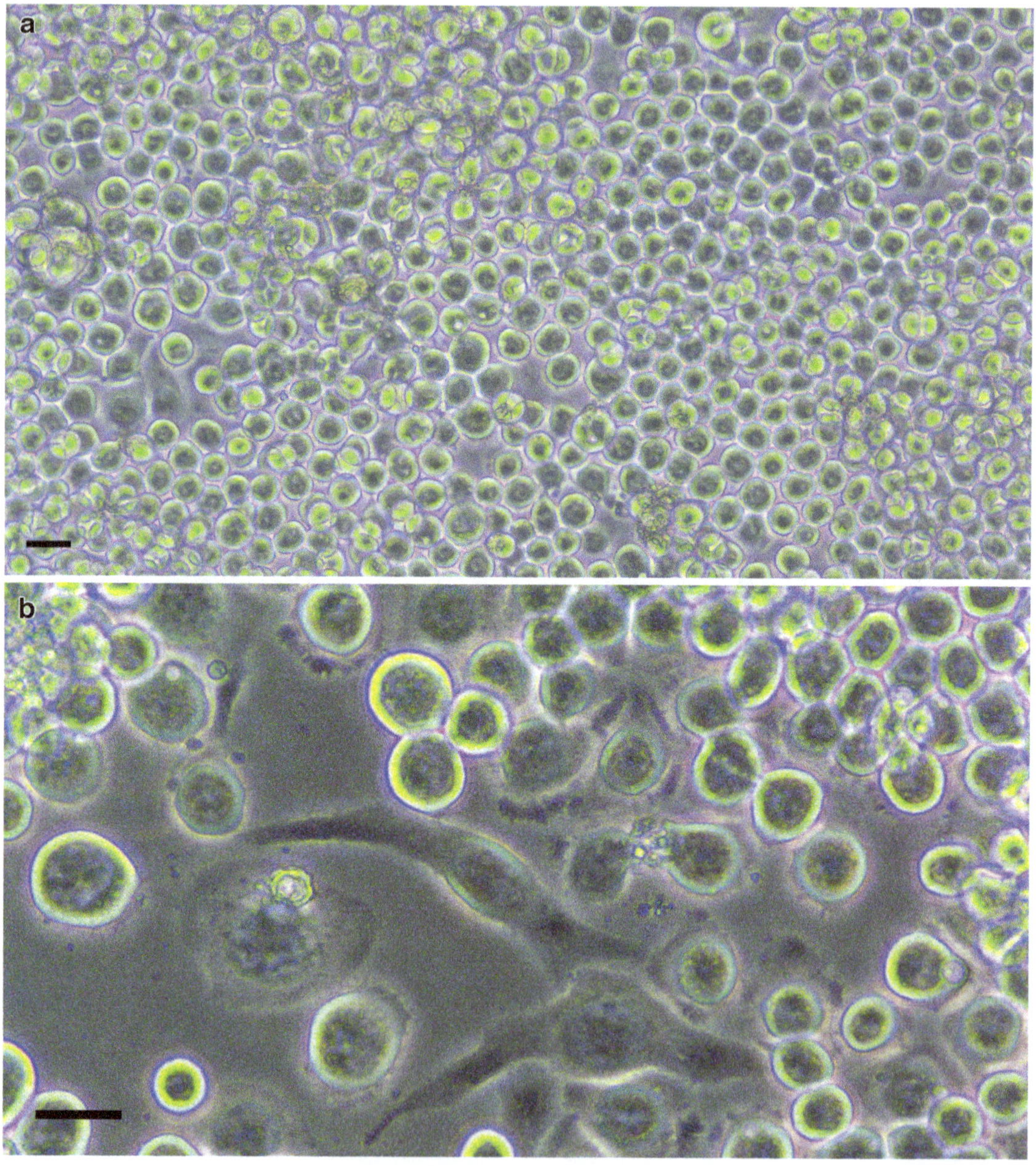


**Fig. 1** Sf9 cells under 10× (**a**) and 20× (**b**) magnification. Note adhered, spread cells in (**b**). Bar, 20 μm

7. Remove cells from the dish by mechanical means when the monolayer reaches 90% confluence (*see* **Note 26**).
8. Count cells and seed at 1–2 × $10^6$ cells/mL.
9. For suspension cultures, follow the indications on Table 1.
10. Incubate at 27°C–29°C (see Step 5) with shaking (*see* **Note 27**).
11. Divide the cells every 2–3 days after counting (*see* **Note 28**).
12. Cells can be frozen at 10–15 × $10^6$/mL in 90% FBS + 10% DMSO and preserved in liquid nitrogen for >10 years.

### 3.2 Gene Cloning

1. For de novo cloning, analyze the sequence of the desired gene using conventional online tools. The combination of NIH gene resources with Benchling (http://www.benchling.com) is a good, free option. Identify restriction targets that are not contained in the gene and obtain primers that flank the gene with those sites. $T_m$ should be between 55°C and 65°C without counting the overhangs.
2. Use free online tools, e.g., https://eu.idtdna.com/pages/tools/oligoanalyzer?returnurl=%2Fcalc%2Fanalyzer, or other suitable resource, to identify and prevent possible primer dimer or self-annealing problems.
3. Amplify the cDNA of the desired target using a high-fidelity DNA polymerase (see Sect. 2.1 and **Note 29**).
4. Incubate the amplicon and pFastBac (5 μg) in separate tubes with the selected restriction enzymes.
5. Gel-purify the amplicon and the vector.
6. Dephosphorylate the vector (SAP).
7. Ligate following common molecular biology procedures.
8. Transform into a suitable *E.coli* competent strain, e.g., DH5α.
9. Produce highly purified plasmid and sequence as outlined in Sect. 2.2.
10. Sequencing of final product is recommended (*see* **Note 30**).

**Table 1**
**Volumes, number of cells, and container sizes for Sf9 propagation**

| Volume (mL) | Number of cells | Erlenmeyer size |
|---|---|---|
| 10–30 | $10^7$–6 × $10^7$ | 100 mL |
| 25–100 | 2.5 × $10^7$–$10^8$ | 500 mL |
| 100–250 | $10^8$–2.5 × $10^8$ | 1 L |
| >250 | 2.5 × $10^8$– ... | 2 L |

### 3.3 Bacmid Production (see Note 31)

1. Transform DH10B or DH10EMBacY competent cells.
2. As a starting point, use 2 μg of plasmid per aliquot.
3. Incubate for 30 min on ice.
4. Heat shock the cells at 42°C for 45 s.
5. Add 1 mL SOC.
6. Incubate under moderate stirring (150 rpm) for 4 h.
7. Extend of Bac-agar plates containing LB/agar supplemented with 50 μg/mL kanamycin, 7 μg/mL gentamycin, 10 μg/mL tetracyclin, 100 μg/mL, X-gal and 40 μg/mL IPTG.
8. Incubate for 48–72 h.
9. Collect 4–6 white colonies and 1–2 blue colonies (*see* **Note 32**).
10. Grow colonies in 4 mL LB + 50 μg/mL kanamycin.
11. Produce bacmid DNA (*see* **Note 33**).
12. Verify that the purified bacmid contains the insert of interest by PCR (*see* **Note 34**).
13. Analyze amplicons in agarose gels: Expected results are 0.3 kbp for blue colonies (non-transposed), 2.3 kbp for empty pFastBac transposed, and supplementary 2.3 kbp insert size for inserted, transposed colonies. For pFastDual, sizes are 2.5 kbp for empty pFastDual and supplementary 2.5 kbp size of both inserts combined for inserted, transposed colonies.
14. Amplify the culture to 40–50 mL of LB+kanamycin by using the rest of the positive colonies.
15. Purify using a Large construct kit or by lysis with potassium acetate (*see* **Note 35**).
16. Once resuspended in water, sterilize by passing the DNA solution through a 0.22-μm filter syringe (*see* **Note 36**).

### 3.4 ELC/RLC Production in Bacteria and Purification

The production protocol is the same for ELC and RLC, and therefore, is not repeated here for each of the plasmids.

1. Transform *E.coli* BL21 cells with pGEX-6P1 or similar plasmid encoding the selected light chain.
2. Pick 1 colony and expand into 40 mL of liquid LB + ampicillin.
3. Incubate overnight at 37°C under continuous stirring.
4. Add the culture from Step 3 in to 400 mL LB + 100 μg/mL ampicillin.
5. Incubate at 37°C under continuous stirring.
6. Measure OD600 each hour. When Optical Density at 600 nm is in a [0.4–0.6] range, add 0.3 mM IPTG and incubate for 6 h at 25°C.

7. Collect by centrifugation (4000 × *g*, 15 min, 4°C).
8. Discard supernatant, resuspend in 30 mL cold Lysis buffer 1.
9. Disrupt by sonication, four cycles of 30 s, 30% amplitude, and allowing 2–3 min in between cycles.
10. Centrifuge the lysate (20,000 × *g*, 30 min, 4°C). Discard pellet.
11. Incubate the lysate with 1 mL prewashed glutathione-agarose with Wash buffer 1.
12. Incubate end-over-end at 4°C for 2 h.
13. Wash the agarose pellet five times with 10 mL Wash buffer 1.
14. Elute 3× (1 h, 4°C, end-over-end) with 1 mL of Elution buffer 1.
15. Pool eluates. Proceed to Step 16 or 17.
16. Dialyze against Dialysis buffer and proceed to Step 20.
17. Incubate for 4 h at 4°C with GSTSelectable marker-tagged Prescission protease as indicated by the manufacturer. Incubate again overnight.
18. Pool the eluates and incubate with 100 μL of prewashed gluthatione-agarose with Wash buffer 1 to remove Prescission protease. Incubate end-over-end at 4°C for 2 h.
19. Dialyze against Dialysis buffer.
20. Measure protein levels at 280 nm, considering that $\varepsilon_{RLC} = 5960\ M^{-1}\varepsilon cm^{-1}$, $M_W(RLC) = 20456.89$ Da; and $\varepsilon(ELC) = 6085\ M^{-1}\varepsilon cm^{-1}$, $M_W(ELC) = 16930.05$ Da.
21. Flash-freeze in PCR tubes at a 50–100 μM concentration.

### 3.5 Sf9 Transfection and Baculovirus Propagation for Protein Production

1. Add 10–20 μg bacmid (see Sect. 4.3 for details) to 100 μL Spodopan medium (no FBS) and 5 μL Fugene to 100 μL medium (no FBS) and incubate for 15 min.
2. Mix the two batches and incubate for 30 min.
3. Add 200 μL to Sf9 cells in a well of a 6-well plate ($0.5 \times 10^6$ cells in 2.5 mL).
4. Incubate for 96 h at 27°C–29°C.
5. Observe with a fluorescence microscope; >50% of the cells should be fluorescent if EMBacY or DH10B cells with a fluorescent insert, are used (*see* **Note 37** and Fig. 2).
6. Resuspend cells in well and add the entire well, including floating cells, to 20 mL shake culture ($10^6$ cells/mL).
7. Incubate for 96 h. Check the cells for fluorescence. Centrifuge the culture and store the supernatant at 4°C (P1) (*see* **Note 38**).

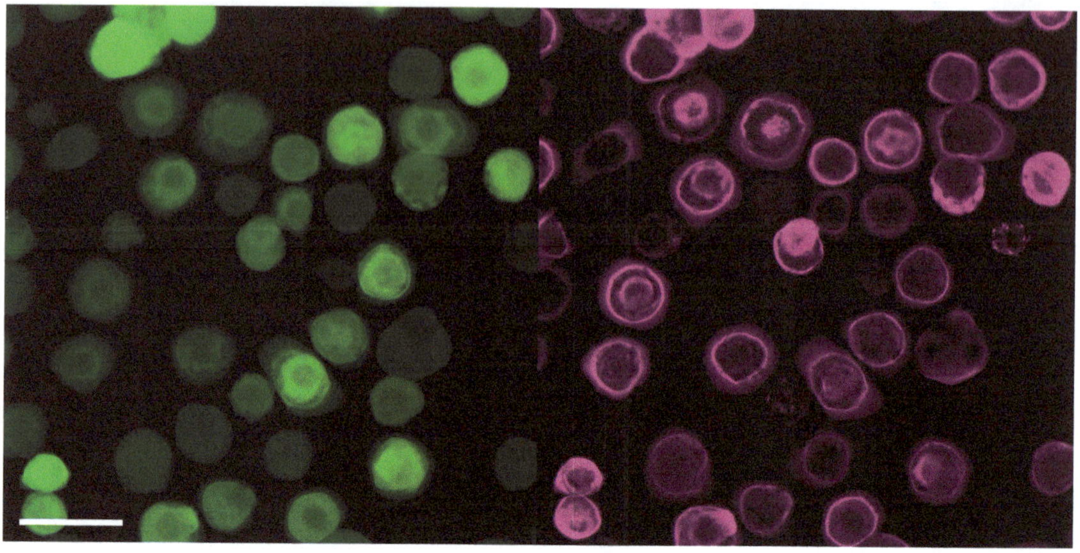

**Fig. 2** Expression of YFP (left) and Halotag (right) in Sf9 cells infected with FLAG-Halotag-nmMHC2-A wild-type full length. Cells were allowed to adhere to glass cover slides for 24 h, then incubated with 1:3000 dilution of JF635 (Janelia Farm), fixed and observed under a fluorescence microscope. Bar, 50 μm

8. For over-extended shelf-life, add 10% DMSO to the medium, mix well, aliquot, and conserve at −80°C.
9. Infect 200 mL (*see* **Note 39**) shake culture ($10^6$ cells/mL) with 4 mL of P1.
10. Incubate culture for 72 h.
11. Check the cells for fluorescence. Centrifuge the culture and store the supernatant at 4°C (P2) (*see* **Note 38**).
12. For over-extended shelf-life, add 10% DMSO to the medium, mix well, aliquot, and conserve at −80°C.
13. Infect 0.5–2 L of shaking culture with 10–40 mL P2 (for myosin 2, this means 10–40 mL of P2 for the heavy chain and 10–40 mL of pFastDual for both light chains) and incubate for 72 h.
14. Check the cells for fluorescence, centrifuge the culture and use the pellet(s) directly for protein preparation (next section) or store dry at −80°C. Each pellet should be from 400–500 mL of cell culture.

### 3.6 Protein Purification

For this example, we will use Ni-NTA beads to purify a 6×His-tagged myosin 2 heavy chain + ELC/RLC. All protocols are similar for the rest of required purified proteins for visualization.

1. Resuspend pellets in 10 mL cold Lysis buffer 2.

2. Raise the volume to 70–80 mL of Lysis buffer 2.
3. Sonicate in a glass beaker, at 40%, five cycles of 1 min each, on ice. Use a wide-bore sonication tip.
4. Centrifuge at 30,000 × *g* for 15 min at 4°C.
5. Reserve the supernatant; take 10 μL sample from supernatant for sampling.
6. In the meantime, wash Ni-NTA beads using dd$H_2O$. Ni-NTA beads are stored as 50% slurry. Centrifuge at 900 × *g* for 7 min at 4°C and add dd$H_2O$ to obtain a 50% mix. Centrifuge at 900 × *g* for 7 min at 4°C. Add 10 mL Lysis Buffer 2 to equilibrate the Ni-NTA beads. Centrifuge at 900 × *g* for 7 min at 4°C. Add Lysis Buffer 2 to obtain a 50% slurry.
7. The supernatant from Step 5 is mixed with the Ni-NTA beads (1.5 mL/L culture), incubated with end-over-end rotation for 2 h at 4°C.
8. Centrifuge the supernatant/beads mix after the incubation at 900 × *g* for 7 min at 4°C.
9. Wash beads twice with Wash buffer 2.
10. Load the beads on a gravity flow column (approx. 50 mL column).
11. Elute with 15 mL Elution buffer 2. Collect 5–6 fractions of 3 mL.
12. Use 5–10 μL of each fraction to perform a protein determination assay and identify the more enriched fraction.
13. Verify protein isolation through SDS-PAGE by using 10 μ L of each fraction. Stain proteins with Coomassie or similar to identify enriched fractions and observe if degradation is present.
14. Pool fractions containing the protein (normally 2–5 fractions).
15. Concentrate using Vivaspin tubes at 4°C to less than 1 mL according to the manufacturer's instructions.
16. Equilibrate the size exclusion column. Use approximately 120 mL filtered SEC buffer.
17. Load column onto an AKTA FPLC system. Run SEC column and elute in approximately 50–60 fractions of 1.2 mL for 4 h or overnight at 4°C.
18. Collect fractions, run SDS-PAGE to verify isolation and concentrate clean fractions using Vivaspin.
19. Determine concentration at 280 and 320 nm.
20. Subtract the 320 nm measurement from the 280 nm. Supplement protein solution with 10% (w/v) trehalose as a preservative (*see* **Note 40**).

21. Make 10 μL aliquots in PCR tubes and store inside 50 mL falcon tubes. Final Concentration should be 5–10 μM.
22. Flash-freeze the aliquots in liquid $N_2$. Store at −80°C until use.

### 3.7 Actin Purification (Optional)

1. Mix 5 g acetone powder with 100 mL G-buffer in a 250 mL beaker for 20 min at 4°C. Use a magnetic stirrer.
2. Filter suspension through gauze into a 500 mL beaker. Discard the first round of filtrated medium.
3. Collect the wet mass of powder, set it on a new 250 mL beaker, and add 50–70 mL of G-buffer.
4. Stir for 20 min at 4°C.
5. Filter again through the gauze.
6. Repeat Steps 3–5, 4–5 times until 200–250 mL filtrate have been collected.
7. Centrifuge at 40,000 x*g* for 30 min.
8. Discard pellet and increase salt concentration to 100 mM KCl and 5 mM $MgCl_2$ in the supernatant to trigger actin polymerization.
9. Incubate at RT for 3 h, and then move to 4°C overnight.
10. Increase concentration of KCl to 750 mM KCl to dissociate tropomyosin from actin filaments.
11. Stir slowly to complete dissolution.
12. Centrifuge for 3 h at 110,000 ×*g* at 4°C.
13. Dissolve pellets in fresh G-Buffer, e.g., by using a Dounce homogenizer.
14. Dialyze actin solution against 1 L G-Buffer for 3–5 days.
15. Add 3% sucrose to the actin solution, and determine concentration photometrically by measuring the absorbance at 280 and 320 nm. Use G-buffer to blank the spectrophotometer to subtract the effect of ATP or measure at 290 nm.
16. Aliquot at a concentration as close as possible to 100 μM.
17. Snap freeze actin in liquid nitrogen and store at −80°C.

### 3.8 Actin Polymerization and Labeling

1. Measure actin aliquots again at 290 and 320 nm. Calculate actual concentration, considering that ε (actin at 290 nm) = 26,600 $M^{-1} \times cm^{-1}$.
2. Dilute actin to 11 μM with G-buffer, and then to 10 μM with 10× polymerization buffer (*see* **Note 41**).
3. Allow it to polymerize for 2–3 h at room temperature.
4. Dilute polymerized actin to 2 μM in 1× polymerization buffer and add 3 μM labeled phalloidin. Store the phalloidin-stabilized actin and the remains of the 10 μM F-actin stock

(unlabeled stock) in the fridge overnight (*see* **Note 42**). The remaining unlabeled stock will be diluted to 2 μM on the day of data acquisition. We will need approximately 10 μL of this 2 μM dilution for blocking dead heads, but again, this depends on the width of the channel.

### 3.9 Flow Cell Preparation for In Vitro Actin Motility Measurements Using Total Internal Reflection Fluorescence (TIRF) Microscopy

1. Dilute 7 mg of nitrocellulose paper in 700 μL (1% w/v) pentyl (amyl) acetate. Incubate under rotation at 4°C overnight.
2. Clean coverslips (compressed airflow works best).
3. Fill a beaker to the edge with milliQ water.
4. Clean residual dust of the water surface by dragging a sheet of Parafilm over it.

**Fig. 3** Image corresponding to the description of coverslip coating with nitrocellulose

5. Drop one 10 μL drop of dissolved nitrocellulose (first step of this section) on the surface. It will form a thin film.
6. Overlay clean coverslips on the film. Keep them separate and away from the edges (Fig. 3).
7. Cut the nitrocellulose layer between the coverslips using a surgical scalpel.
8. Immerse the coverslip in water and let it dry with the nitrocellulose face up under flow.

### 3.10 Myosin 2 Phosphorylation for In Vitro Actin Motility Measurements Using Total Internal Reflection Fluorescence (TIRF) Microscopy

1. In a PCR tube, mix the components in Table 2.
2. Incubate for 30 min at 30°C.
3. Dilute the entire reaction with half a volume of IVAM buffer (in the example, add 50 μL).

### 3.11 In Vitro Actin Motility Assay Using Total Internal Reflection Fluorescence (TIRF) Microscopy

1. Dilute labeled actin (2 μM) to 30 nM in IVMA buffer.
2. Dilute unlabeled actin (10 μM) to 2 μM with IVMA buffer.
3. Prepare Assay buffers 1–4 freshly each time. The following amounts are needed: buffer 1250 μL per flow cell; buffers 2–4, 40 μL per flow cell of each.
4. Set up a glass slide with two symmetrical strips of double-sided adhesive tape, forming a 1–2 mm channel in the middle (Fig. 4).

**Table 2**
**Reagents and volumes for Myosin 2 phosphorylation in vitro**

| Reagent | Final concentration | Example stock solutions | Volume (for 100 μL) according to the example stock concentrations |
|---|---|---|---|
| Myosin 2 | 1.5 μM | 10 μM | 15 μL |
| $CaCl_2$ | 1 mM | 100 mM | 1 μL |
| DTT | 1 mM | 100 mM | 1 μL |
| ATP | 1 mM | 100 mM | 1 μL |
| Calmodulin | 0.2 μM | 10 μM | 1 μL |
| MLCK | 1 mg/mL | 20 mg/mL | 5 μL |
| ELC | 2 μM | 200 μM | 1 μL |
| RLC | 2 μM | 200 μM | 1 μL |
| Phosphorylation buffer | | | 74 μL |
| **TOTAL** | | | **100 μL** |

5. Peel off the outer side of the tape and adhere a nitrocellulose-coated square coverslip. The coated side faces down the slide (*see* **Note 43**).
6. Holding the flow cell vertically, place 10 μL of phospho-myosin II in the channel. It is essential that the solution enters the channel evenly and all of it is soaked.
7. Incubate for 3 min at room temperature. This will adhere the myosin II molecules to the nitrocellulose via the coiled-coil.
8. Flush 30 μL of Assay buffer 1 (wash).
9. Flush 10 μL of unlabeled actin and incubate for 3 min at room temperature. The unlabeled actin binds to all myosin II molecules, active or inactive ("dead").

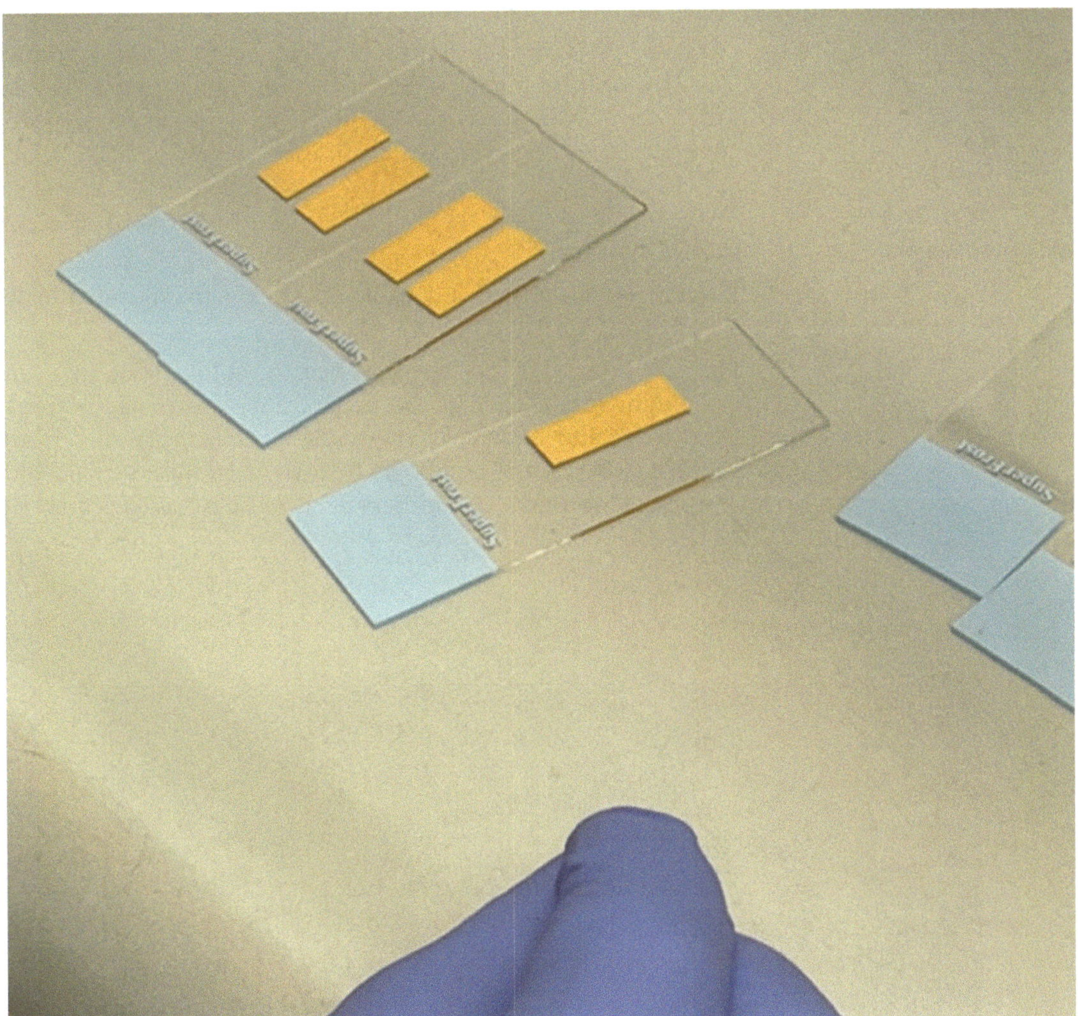

**Fig. 4** Slides showing the channel flanked by double-sided adhesive tape

10. Flush 30 μL of Assay buffer 4 (*see* **Note 44**).
11. Flush 30 μL of Assay buffer 1 (wash).
12. Flush 10 μL of labeled actin (see sect. 4.8). Incubate for 3 min at room temperature. Labeled actin will bind to active heads only.
13. Flush 30 μL of Assay buffer 1 (wash).
14. Flush 30 μL of Assay buffer 2 (wash).
15. Flush 30 μL of Assay buffer 3 (wash).
16. Seal the edges of the flow cell with petroleum jelly.
17. Move the slide into the microscope stage.
18. Set up lasers, 63× or 100× objective (resolution will be important in the analysis step) and TIRF angle to 100 nm. Focus on the center of the channel. The field should display clear filaments (Fig. 5a).
19. Acquire 90–125 total frames, allowing 2–5 s between frames (3–10 min total imaging time). If a motorized stage is available, up to four fields can be acquired simultaneously. This needs to be noted in the quantification stage (see Sect. 3.12).

### 3.12 Image Analysis and Quantification

1. Analysis is based on the wrMTrck (worm track) plugin for ImageJ/Fiji [13].
2. The software automatically will track each filament individually over time.
3. The raw movie (Fig. 5a) needs to be thresholded (Fig. 5b), and the plugin interface (Fig. 6) programmed with data regarding the size and velocity of the tracks. Essential parameters are minSize (>30 pixel$^2$), minTrackLength (>10 frames), and FPS (frames per second). Parameters need to be tweaked according

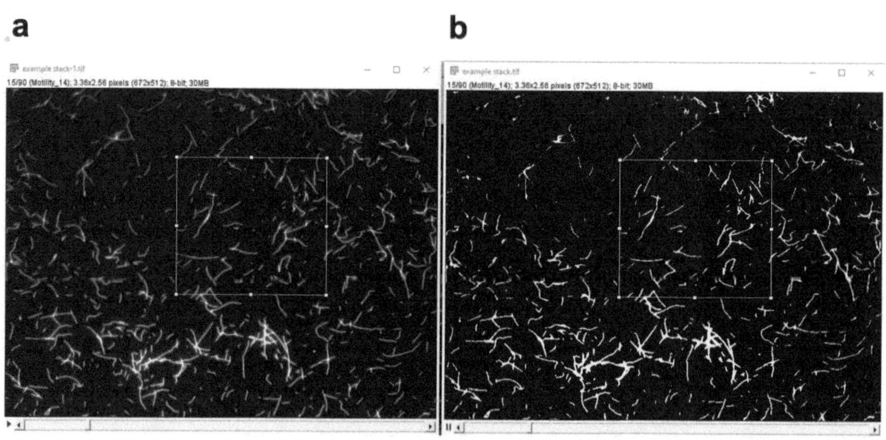


**Fig. 5** Raw (**a**) and thresholded (**b**) movie

to the features of the movie. The plugin returns a movie with labeled filaments (Fig. 7a), a track map (Fig. 7b), and a Results spreadsheet (Fig. 8) that contains essential information that can be translated into instantaneous speed with basic operations. For example, average velocity can be extracted from the distance travelled by a filament over the number of frames, considering the resolution of the image and the interval between frames.

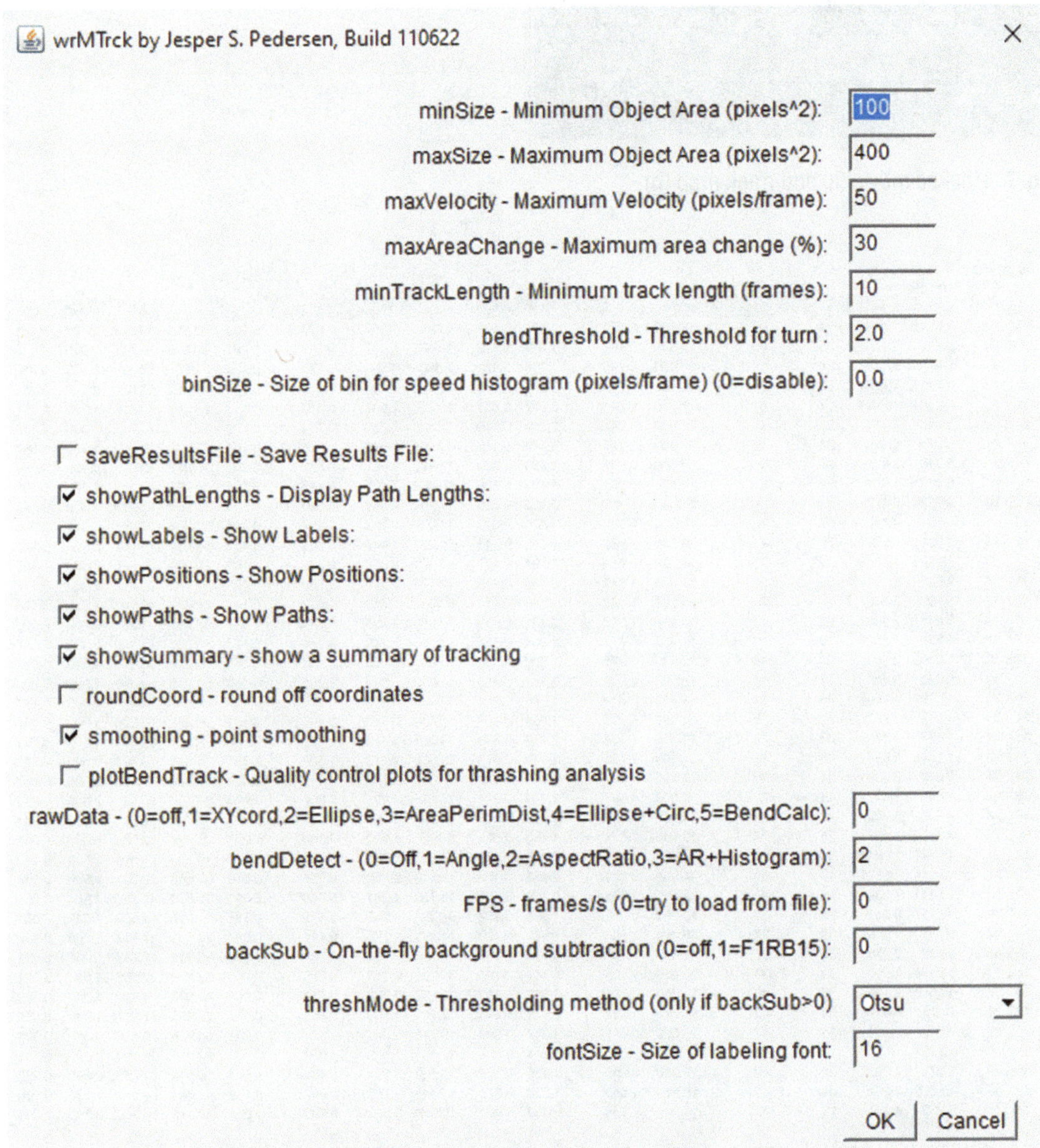


**Fig. 6** Interface of the wrMTrck plugin, developed by Jesper S. Pedersen

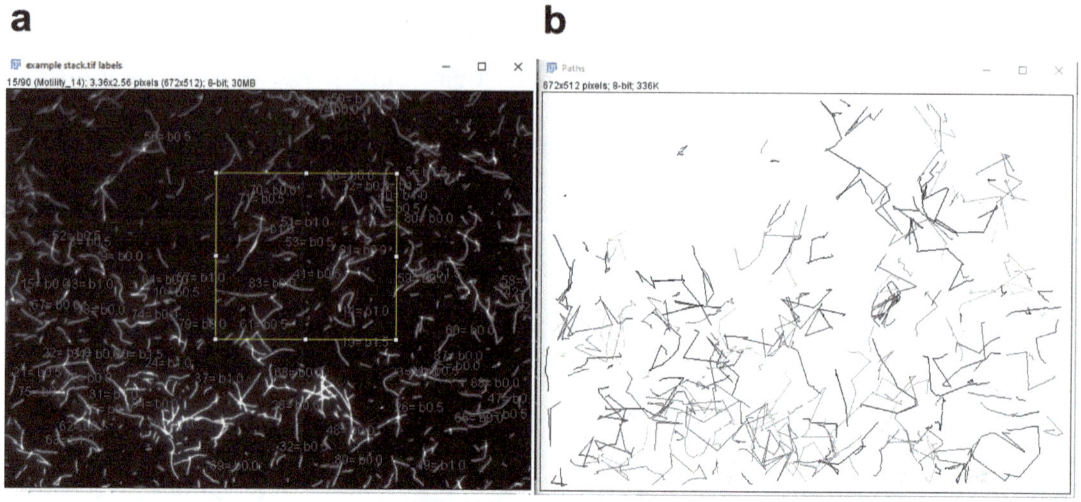


**Fig. 7** Tracked movie (**a**) and track map (**b**)

Results
File Edit Font Results

| | Track | Length | Distance | #Frames | 1stFrame | Time(s) | MaxSpeed | Area | sdArea | Perim | sdPerim | avgSpeed | BLPS | avgX | avgY | Bends | BBPS |
|---|---|---|---|---|---|---|---|---|---|---|---|---|---|---|---|---|---|
| 1 | 1 | 11.614 | 11.206 | 37 | 0 | 18.500 | 0.931 | 1.693 | 0.265 | 9.694 | 1.157 | 0.628 | 0.130 | 30.159 | 33.448 | 2.000 | 0.108 |
| 2 | 2 | 5.554 | 5.073 | 35 | 0 | 17.500 | 0.597 | 1.861 | 0.422 | 10.553 | 1.127 | 0.317 | 0.060 | 52.519 | 31.657 | 2.000 | 0.114 |
| 3 | 3 | 5.970 | 5.721 | 20 | 0 | 10.000 | 0.775 | 1.220 | 0.132 | 7.050 | 0.538 | 0.597 | 0.169 | 5.277 | 37.120 | 0.500 | 0.050 |
| 4 | 4 | 10.325 | 10.274 | 33 | 0 | 16.500 | 0.785 | 2.235 | 0.180 | 12.415 | 0.196 | 0.626 | 0.101 | 64.656 | 51.068 | 2.000 | 0.121 |
| 5 | 5 | 11.801 | 10.844 | 37 | 0 | 18.500 | 1.244 | 2.207 | 0.559 | 10.928 | 2.007 | 0.638 | 0.117 | 72.393 | 49.678 | 1.000 | 0.054 |
| 6 | 6 | 3.845 | 3.724 | 23 | 1 | 11.500 | 0.548 | 2.298 | 0.545 | 12.875 | 1.220 | 0.334 | 0.052 | 52.875 | 27.221 | 2.000 | 0.174 |
| 7 | 7 | 6.675 | 6.499 | 23 | 3 | 11.500 | 0.979 | 1.773 | 0.346 | 10.865 | 1.069 | 0.580 | 0.107 | 72.729 | 16.735 | 1.000 | 0.087 |
| 8 | 8 | 5.716 | 3.855 | 20 | 3 | 10.000 | 0.961 | 2.660 | 1.072 | 11.190 | 4.051 | 0.572 | 0.102 | 80.256 | 43.397 | 0.000 | 0.000 |
| 9 | 9 | 8.913 | 4.067 | 29 | 3 | 14.500 | 1.285 | 1.791 | 0.526 | 10.184 | 2.501 | 0.615 | 0.121 | 84.777 | 48.169 | 2.000 | 0.138 |
| 10 | 10 | 9.723 | 9.617 | 32 | 3 | 16.000 | 1.356 | 2.133 | 0.485 | 11.162 | 1.951 | 0.608 | 0.109 | 87.054 | 57.954 | 1.000 | 0.062 |
| 11 | 11 | 4.695 | 4.112 | 37 | 4 | 18.500 | 0.525 | 1.970 | 0.303 | 11.039 | 1.153 | 0.254 | 0.046 | 53.904 | 3.840 | 0.000 | 0.000 |
| 12 | 12 | 24.081 | 16.783 | 76 | 4 | 38.000 | 0.980 | 1.815 | 0.543 | 11.237 | 1.947 | 0.634 | 0.113 | 58.087 | 14.781 | 5.000 | 0.132 |
| 13 | 13 | 10.895 | 10.356 | 39 | 4 | 19.500 | 0.910 | 2.104 | 0.407 | 10.382 | 0.962 | 0.559 | 0.108 | 22.861 | 51.298 | 2.500 | 0.128 |
| 14 | 14 | 9.616 | 8.951 | 34 | 5 | 17.000 | 1.179 | 2.436 | 0.779 | 12.171 | 3.006 | 0.566 | 0.093 | 84.183 | 66.037 | 2.000 | 0.118 |
| 15 | 15 | 8.798 | 7.440 | 29 | 7 | 14.500 | 1.028 | 1.774 | 0.443 | 8.524 | 1.759 | 0.607 | 0.142 | 8.647 | 40.141 | 1.500 | 0.103 |
| 16 | 16 | 9.547 | 7.601 | 26 | 10 | 13.000 | 1.213 | 2.431 | 0.639 | 12.559 | 2.755 | 0.734 | 0.117 | 64.491 | 3.825 | 0.500 | 0.038 |
| 17 | 17 | 6.227 | 5.754 | 22 | 13 | 11.000 | 0.872 | 1.901 | 0.624 | 8.873 | 3.183 | 0.566 | 0.128 | 80.297 | 71.912 | 0.500 | 0.045 |
| 18 | 18 | 13.195 | 2.891 | 59 | 14 | 29.500 | 1.297 | 1.689 | 0.449 | 9.380 | 1.856 | 0.447 | 0.095 | 66.334 | 38.288 | 1.500 | 0.051 |
| 19 | 19 | 6.270 | 5.451 | 31 | 14 | 15.500 | 0.632 | 2.312 | 0.466 | 11.938 | 1.849 | 0.405 | 0.068 | 64.248 | 69.895 | 1.000 | 0.065 |
| 20 | 20 | 10.799 | 10.586 | 36 | 15 | 18.000 | 0.818 | 1.503 | 0.205 | 9.535 | 0.736 | 0.600 | 0.126 | 54.677 | 8.092 | 2.000 | 0.111 |
| 21 | 21 | 4.618 | 3.997 | 21 | 16 | 10.500 | 1.005 | 1.864 | 0.688 | 8.744 | 2.843 | 0.440 | 0.101 | 30.942 | 40.836 | 1.000 | 0.095 |
| 22 | 22 | 7.077 | 1.179 | 33 | 18 | 16.500 | 1.164 | 2.472 | 0.825 | 13.617 | 3.831 | 0.429 | 0.063 | 42.274 | 9.607 | 1.500 | 0.091 |
| 23 | 23 | 10.615 | 8.888 | 63 | 18 | 31.500 | 1.105 | 2.149 | 0.613 | 12.538 | 2.212 | 0.337 | 0.054 | 38.669 | 18.492 | 3.500 | 0.111 |
| 24 | 24 | 6.670 | 6.339 | 24 | 20 | 12.000 | 1.257 | 2.880 | 0.820 | 14.017 | 2.773 | 0.556 | 0.079 | 73.070 | 68.127 | 2.000 | 0.167 |
| 25 | 25 | 12.494 | 12.137 | 44 | 21 | 22.000 | 0.887 | 2.420 | 0.671 | 14.994 | 2.421 | 0.568 | 0.076 | 83.456 | 9.214 | 2.000 | 0.091 |
| 26 | 26 | 8.190 | 7.792 | 28 | 22 | 14.000 | 0.719 | 1.463 | 0.069 | 8.536 | 0.197 | 0.585 | 0.137 | 63.702 | 38.944 | 1.000 | 0.071 |
| 27 | 27 | 4.049 | 3.457 | 21 | 24 | 10.500 | 1.671 | 1.737 | 0.561 | 8.492 | 1.900 | 0.386 | 0.091 | 37.197 | 49.383 | 0.000 | 0.000 |
| 28 | 28 | 11.321 | 11.040 | 48 | 25 | 24.000 | 0.969 | 2.303 | 0.622 | 14.227 | 2.759 | 0.472 | 0.066 | 18.733 | 25.695 | 3.500 | 0.146 |
| 29 | 29 | 13.394 | 12.578 | 48 | 32 | 24.000 | 0.981 | 2.068 | 0.669 | 11.573 | 2.758 | 0.558 | 0.096 | 37.651 | 29.231 | 1.500 | 0.062 |
| 30 | 30 | 7.581 | 7.559 | 25 | 35 | 12.500 | 1.064 | 2.366 | 0.367 | 13.350 | 1.261 | 0.606 | 0.091 | 75.076 | 41.219 | 2.500 | 0.200 |
| 31 | 31 | 6.871 | 6.341 | 26 | 35 | 13.000 | 0.815 | 1.568 | 0.252 | 8.623 | 0.942 | 0.529 | 0.123 | 71.098 | 44.580 | 1.000 | 0.077 |
| 32 | 32 | 9.451 | 9.379 | 38 | 37 | 19.000 | 0.834 | 2.020 | 0.590 | 11.471 | 1.728 | 0.497 | 0.087 | 19.633 | 17.317 | 2.500 | 0.132 |
| 33 | 33 | 2.632 | 2.526 | 22 | 37 | 11.000 | 1.173 | 3.196 | 0.311 | 15.694 | 1.061 | 0.239 | 0.030 | 46.021 | 47.747 | 1.000 | 0.091 |
| 34 | 34 | 14.479 | 13.001 | 42 | 38 | 21.000 | 0.805 | 1.467 | 0.125 | 8.722 | 0.318 | 0.689 | 0.158 | 36.620 | 32.806 | 1.500 | 0.071 |
| 35 | 35 | 11.030 | 2.144 | 36 | 40 | 18.000 | 0.977 | 2.082 | 0.480 | 12.722 | 3.092 | 0.613 | 0.096 | 46.823 | 21.214 | 1.500 | 0.083 |
| 36 | 36 | 7.672 | 5.047 | 34 | 40 | 17.000 | 1.662 | 2.615 | 0.448 | 12.703 | 1.790 | 0.451 | 0.071 | 28.053 | 67.967 | 1.000 | 0.059 |
| 37 | 37 | 2.982 | 2.857 | 35 | 41 | 17.500 | 0.671 | 1.542 | 0.380 | 10.613 | 1.217 | 0.170 | 0.032 | 50.926 | 20.136 | 0.500 | 0.029 |
| 38 | 38 | 12.452 | 10.883 | 39 | 42 | 19.500 | 2.844 | 2.254 | 0.352 | 12.481 | 1.389 | 0.639 | 0.102 | 22.361 | 46.194 | 3.000 | 0.154 |
| 39 | 39 | 7.272 | 7.061 | 30 | 42 | 15.000 | 0.995 | 2.435 | 0.397 | 13.130 | 1.565 | 0.485 | 0.074 | 10.614 | 62.773 | 0.500 | 0.033 |
| 40 | 40 | 11.875 | 2.901 | 37 | 42 | 18.500 | 1.043 | 1.748 | 0.374 | 9.241 | 1.783 | 0.642 | 0.139 | 40.634 | 66.351 | 2.000 | 0.108 |
| 41 | 41 | 5.282 | 4.652 | 22 | 43 | 11.000 | 1.139 | 3.204 | 0.861 | 15.020 | 3.562 | 0.480 | 0.064 | 92.440 | 50.906 | 0.500 | 0.045 |

**Fig. 8** Results sheet from wrMTrck plugin in Fiji

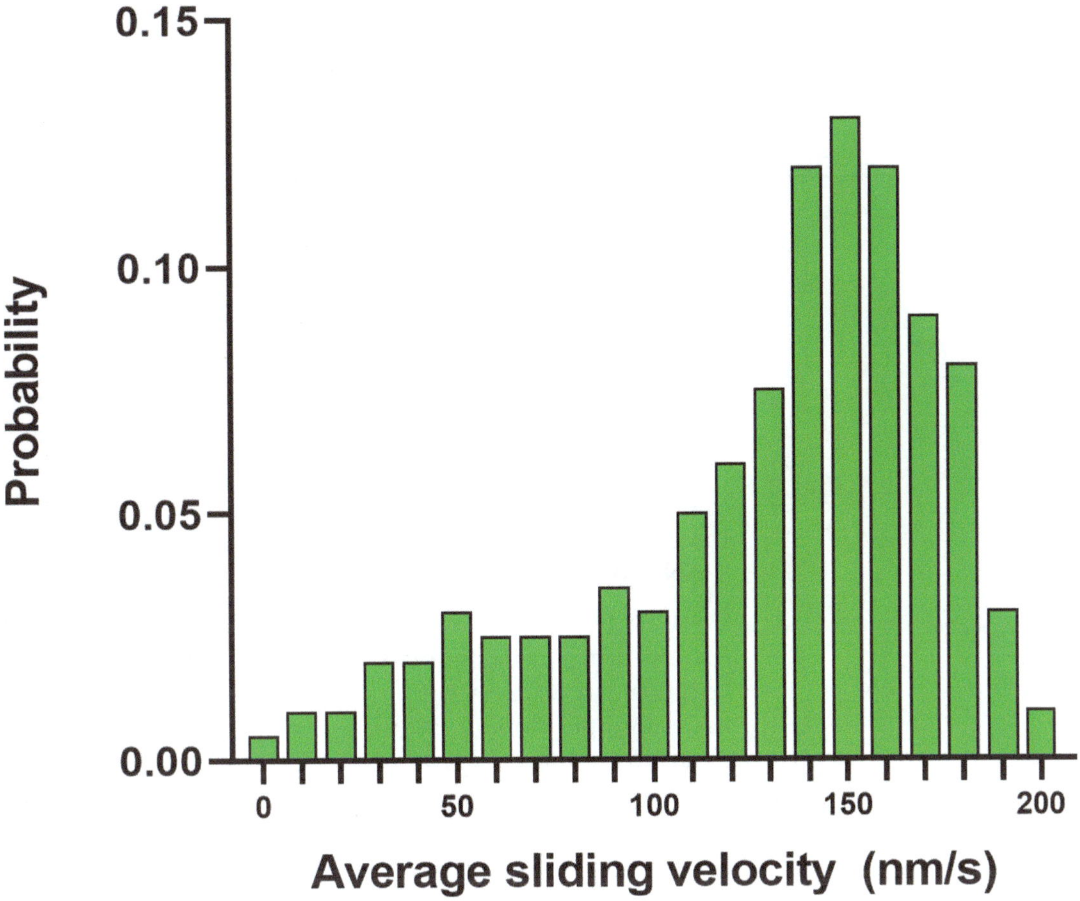


**Fig. 9** Probability distribution of NM2-A IVAM

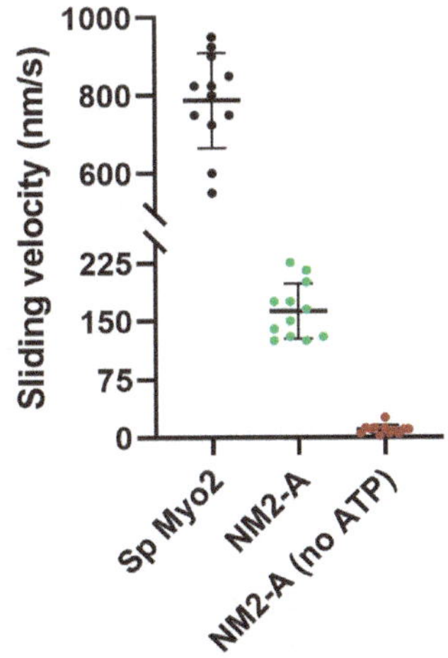


**Fig. 10** Average sliding velocity of *S. pombe* Myo2 (does not require rlc1 phosphorylation, as described in [11]), NM2-A wild-type and unphosphorylated NM2-A

4. That information needs to be translated into a spreadsheet for representation. Typical representation plots are shown in Figs. 9 and 10.

## 4 Notes

1. Several alternatives exist, for example, Sf21 cells. Sf21 cells are better suited to evaluate virus cytopathic effects, whereas Sf9 are easier and faster to amplify for protein production.
2. Two alternatives exist: borosilicate glass or single-use plastic. For convenience and to reduce the carbon footprint of our experiments, we use borosilicate glass Erlenmeyer.
3. Many alternatives exist, including Gibco's SFM Sf-900™ II or Insect-XPRESS™ from Lonza. We routinely use Spodopan (PAN Biotech).
4. Previously known as Pluronic F-68, Gibco 24040032.
5. Myosin heavy chains are cloned into this vector using conventional, restriction enzyme-based cloning. This vector has a multiple cloning site with plenty of options. It can be obtained from the Bac-to-Bac kit (Thermo 10359016).
6. ELC and RLC are cloned into this dual promoter vector using conventional, restriction enzyme-based cloning. This vector has a multiple cloning site with plenty of options (Thermo 10712024).
7. cDNAs can be obtained from different sources, as original gene sequence, commercial sources or other sources, for example, an expression plasmid.
8. These primers need to be designed specifically for each gene to be cloned, including suitable restriction overhangs for cloning into the expression vectors.
9. These enzymes can be obtained commercially; we prefer CloneAmp (Takara 639298).
10. Crucial step: Adjust pH to 8.5.
11. There are several commercial options. All of them work similarly.
12. We regularly use Fugene HD from Promega, but other alternatives, including homemade PEI solutions can be used.
13. Imidazole is not necessary for non-$Ni^{2+}$/$Co^{2+}$-based purifications. Imidazole in wash buffer needs to be titrated for optimal removal of contaminants without releasing the protein. 50 mM is a good starting point.

14. ELC and RLC are produced in BL21 *E. coli* cells, see Sects. 2.4 and 3.4.
15. G-actin can be purchased directly from Cytoskeleton Inc (AKL95).
16. There are several commercially available elements, but we prefer covers from Thorlabs (CG15CH).
17. We use Tesa 05338 (15 mm) tape for our assets.
18. The nitrocellulose from BioRad (1620113) is recommended.
19. Calmodulin can be acquired from Merck (cat. No. C4874) or produced in *E. coli* (protocol not provided).
20. Catalytically active MLCK can be acquired from Merck (cat. No. M9197) or produced in Sf9 cells (protocol not provided).
21. We recommend Merck (cat. No. M7027).
22. 37°C temperature is critical to observe actin polymerization.
23. If growth is slow, Spodopan or other suitable medium can be supplemented with 10% fetal bovine serum.
24. Although 27°C–29°C range is usually adequate, optimal temperature for cell grow needs to be determined for each laboratory condition.
25. When seeded on plastic, Sf9 cells are round and refringent under brightfield microscopy, with occasionally adhered and spread cells.
26. Pipetting the medium up and down until the monolayer comes off the plastic is enough to recover cells. Scraping is not used, as it damages the cells too much. Do not use trypsin or other chemical means.
27. Shaking should be energic but not violent. An orbital speed between 90–130 rpm is good in most incubators (this needs to be determined experimentally for each system). To reduce damage, non-ionic surfactant 1% poloxamer 188 (formerly Pluronic F-68) can be added to the cell culture medium.
28. Do not resuspend them at a number of cells below $10^6$ cells/mL. Cultures can be transiently cultured as concentrated as $10^7$ cells/mL.
29. A good source should provide good amplification provided the primers are designed appropriately. Data corresponding to Figs. 8 and 9 were made with non-muscle myosin 2-A, which included cloning of the heavy chain (*MYH9*, gene id. 4627) into pFastBac and RLC (*MYL12B*, gene id. 103910) and ELC (*MYL6*, gene id. 4637) in pFastDual. Example 2 also included data made with *S. pombe Myo2* (SPCC645.05c) in pFastBac and *rlc1* (SPAC926.03) and *cdc4* (SPAP8A3.08) in pFastDual.

30. DNA sequencing. In the past, this was carried out using Sanger sequencing with 800 bp-spaced vectors to visualize the entire gene. Whole plasmid nanopore-based sequencing is replacing this method.
31. Bacmids are produced independently for pFastBac containing the myosin two heavy chain and pFastDual containing the ELC and RLC.
32. In general, transposed transformants are scarce (<1%). Transposed colonies are white, non-transposed colonies remain blue.
33. It is a good idea to save a bit of the initial culture for the post-verification steps. While Large-construct DNA purification columns can be used, simple lysis + potassium acetate purification is more cost-effective.
34. Using pFastBac and EMBacY cells, the purified bacmid is the template for PCR using the primers: pUC/M13 Forward 5′-CCCAGTCACGACGTTGTAAAACG-3′ and pUC/M13 Reverse, 5′-AGCGGATAACAATTTCACACAGG-3′. Any polymerase can be used for this purpose. Regular reaction conditions include: first denaturation step at 98°C for 5 min, 30 cycles including a denaturation step at 94°C for 30 s followed by an annealing Step at 55°C for 30 s and an extension step at 72°C for 3 min; a final extension Step at 72°C for 12 min is performed. The extension step in the cycles requires 3 min for inserts below 0.5 kbp. Over that, add 1 min per kbp.
35. DNA purity needs to be optimal (260/280 nm ratio = 1.8–2.0; 260/230 nm ratio = 2–2.2) for subsequent steps.
36. This is a critical step. There will be a significant loss of yield, but a lack of contaminants in this DNA is essential for the next steps.
37. In this example, the *MYH9* gene was fused with Halotag. Protein expression was visualized with Janelia Halo-derivative JF635 [14].
38. It is highly recommended to filter this supernatant through a 0.22-μm filter syringe for extended shelf-life.
39. P2 and P3 can be scaled up or down based on need or previous experience.
40. Use a 50% solution of trehalose prepared in the moment and dilute four times (to 10%). Addition of trehalose makes the protein less sensitive to freezing/thawing. After addition of the trehalose, determine the concentration again.
41. As an example, to 10 μL of actin from the freezer, add 15 μL of G-buffer and measure concentration. Consider a concentration of 22 μM. Add 25 μL of G-buffer for a final

concentration of 11 μM (total volume = 50 μL). Add 5.5 μL of 10× polymerization buffer. Final = 55 μL of 10 μM actin in 1× polymerization buffer.

42. Store polymerized actin in aliquots of 30–40 μL and concentrations from 100–200 μM. One of these aliquots is more than enough for 1 day of measurements. For the preparation of the labeled actin, use 12 μL of the 10 μM F-actin stock + 1.18 μL of labeled-Phalloidin stock (153 μM) + 46.8 μL of 1× polymerization buffer. A much smaller volume can also be used, as the 2 μM labeled stock is later used to prepare a stock with a working concentration of around 30 nM (10 μL of this 30 nM stock is required for each flow-cell, but this depends on the width of the flow-cell channel).

43. These are good for a week at 4°C, but they work better if freshly made.

44. The ATP in buffer 4 will detach the unlabeled actin from the active myosin heads, while it will have no effect on the dead myosin II heads. Therefore, the inactive myosin heads are now blocked and will not interfere with the assay.

## Acknowledgments

The authors thank Jesper S. Pedersen for the free wrMTrck plugin for Fiji and Sarah Heissler and Jim Sellers for helpful discussions. This work was funded by PID2023-153018NB-I00 (MINECO) and STOP Ras cancers (AECC) to M.V-M. JNG is supported by the PREPARE program for medical scientists from Hannover Medical School. Authors' contributions: RPD, HRS, MGC, experimental work; VCT, protocol hands-on setup, data acquisition and analysis; JNG, experimental design and manuscript correction; MVM, funding acquisition, global design, data acquisition and analysis, manuscript writing and correction. Availability of data and material: Original data and unique plasmids and other materials will be made available by the authors upon reasonable request.

### References

1. Heissler SM, Arora AS, Billington N, Sellers JR, Chinthalapudi K (2021) Cryo-EM structure of the autoinhibited state of myosin-2. Science Adv 7(52):eabk3273. https://doi.org/10.1126/sciadv.abk3273
2. Scarff CA, Carrington G, Casas-Mao D, Chalovich JM, Knight PJ, Ranson NA et al (2020) Structure of the shutdown state of myosin-2. Nature 588(7838):515–520. https://doi.org/10.1038/s41586-020-2990-5
3. Yang S, Tiwari P, Lee KH, Sato O, Ikebe M, Padron R et al (2020) Cryo-EM structure of the inhibited (10S) form of myosin II. Nature 588(7838):521–525. https://doi.org/10.1038/s41586-020-3007-0
4. Bharda AV, Jung HS (2022) Review on the structural understanding of the 10S myosin II in the era of Cryo-electron microscopy. Appl Microsc 52(1):9. https://doi.org/10.1186/s42649-022-00078-x

5. Hu A, Wang F, Sellers JR (2002) Mutations in human nonmuscle myosin IIA found in patients with May-Hegglin anomaly and Fechtner syndrome result in impaired enzymatic function. J Biol Chem 277(48):46512–46517. https://doi.org/10.1074/jbc.M208506200
6. Kim KY, Kovacs M, Kawamoto S, Sellers JR, Adelstein RS (2005) Disease-associated mutations and alternative splicing alter the enzymatic and motile activity of nonmuscle myosins II-B and II-C. J Biol Chem 280(24): 22769–22775. https://doi.org/10.1074/jbc.M503488200
7. Johnson CA, Walklate J, Svicevic M, Mijailovich SM, Vera C, Karabina A et al (2019) The ATPase cycle of human muscle myosin II isoforms: adaptation of a single mechanochemical cycle for different physiological roles. J Biol Chem 294(39):14267–14278. https://doi.org/10.1074/jbc.RA119.009825
8. Billington N, Wang A, Mao J, Adelstein RS, Sellers JR (2013) Characterization of three full-length human nonmuscle myosin II paralogs. J Biol Chem 288(46):33398–33410. https://doi.org/10.1074/jbc.M113.499848
9. Pal K, Nowak R, Billington N, Liu R, Ghosh A, Sellers JR, et al (2020) Megakaryocyte migration defects due to nonmuscle myosin IIA mutations underly thrombocytopenia in MYH9-related disease. Blood 135(21):1887–1898. https://doi.org/10.1182/blood.2019003064
10. Sellers JR (1998) In vitro motility assay to study translocation of actin by myosin. Current protocols in cell biology. 00: 13.2.1 13.2.10. https://doi.org/10.1002/0471143030.cb1302s00
11. Pollard LW, Bookwalter CS, Tang Q, Krementsova EB, Trybus KM, Lowey S (2017) Fission yeast myosin Myo2 is down-regulated in actin affinity by light chain phosphorylation. Proc Natl Acad Sci USA. 114(35):E7236–E7244. https://doi.org/10.1073/pnas.1703161114
12. Kwiatkowska J, Stein E, Romanenko A, Plens-Gałąska M, Podsiadła-Białoskórska M, Szołajska E et al (2025) A beginners guide to Sf9 and Sf21 insect cell line culture and troubleshooting. Sci Rep 15(1):19907. https://doi.org/10.1038/s41598-025-99812-0
13. Nussbaum-Krammer CI, Neto MF, Brielmann RM, Pedersen JS, Morimoto RI (2015) Investigating the spreading and toxicity of prion-like proteins using the metazoan model organism *C. elegans*. J Vis Exp 95:52321. https://doi.org/10.3791/52321
14. Grimm JB, Muthusamy AK, Liang Y, Brown TA, Lemon WC, Patel R et al (2017) A general method to fine-tune fluorophores for live-cell and in vivo imaging. Nat Methods 14(10):987–94. https://doi.org/10.1038/nmeth.4403

# Chapter 8

## Reconstituting the Motility of Intraflagellar Transport Trains Ex Vivo

**Aditya Chhatre, Ludek Stepanek, Stefan Diez and Gaia Pigino**

### Abstract

Intraflagellar transport (IFT) is essential for the assembly, maintenance and function of most eukaryotic cilia. It acts as a bidirectional cargo transport system, powered by kinesin and dynein motors to navigate through the crowded ciliary environment. Here, we describe a method to isolate intact IFT trains from live cells and reconstitute their motility on synthetic microtubules in vitro. This system enables the controlled investigation of the biophysical properties of IFT trains, including their sensitivity to tubulin post-translational modifications. With minor adaptations, this approach can be extended to study other intracellular transport processes.

**Keywords** Intraflagellar transport, Kinesin, Dynein, Microtubules, In vitro motility assay, Cilia, Ex vivo single cell assay

## 1 Introduction

Cilia, also known as flagella, are microtubule-based organelles present on the surface of most eukaryotic cells, where they perform sensory and motility functions. Their internal cytoskeletal scaffold, the axoneme, consists of nine microtubule doublets, each comprising a complete A-tubule and an incomplete B-tubule. Intraflagellar transport (IFT) is a molecular motor-driven system critical for cilia assembly and function [1]. Anterograde IFT trains, powered by kinesin-2, travel along the B-tubules to the ciliary tip, where they unload their cargo and rearrange into dynein-driven retrograde trains, which return to the cell body along the A-tubules [2]. Such complex transport machinery requires precise organization, which remains challenging to study within the intricate environment of a living cell.

Many biochemical and biophysical properties of molecular motor systems can be dissected in vitro under controlled conditions [3],

**Supplementary Information** The online version contains supplementary material available at https://doi.org/10.1007/978-1-0716-5515-3_8

Noa B. Martín-Cófreces et al. (eds.), *The Cytoskeleton*, Methods in Molecular Biology, vol. 3070,
https://doi.org/10.1007/978-1-0716-5515-3_8,

for example, using techniques such as single molecule stepping assays [4], optical trapping [5], three-dimensional motility assays [6], controlled molecular crowding [7], and exposure to modified microtubules [8]. We sought to reconstitute IFT movement on synthetically polymerized microtubules with defined post-translational modifications to determine whether IFT train motility is influenced by such biochemical cues [9].

In vitro motility assays usually require purification of functional motor complexes in sufficient quantity. However, IFT is an elaborate system, composed of at least 22 different proteins [10], organized into A- and B- complexes [11], which polymerize to form long linear trains [12]. Despite significant progress [13, 14], complete reconstitution of functional IFT trains from purified components has not yet been achieved. We therefore adopted the ex vivo reconstitution approach. While IFT material can be isolated from cilia in bulk [15], its native architecture is not retained during the purification process [16] due to weak interactions between some of its subcomplexes [17]. To overcome this, we developed an assay to transfer intact, motile IFT trains directly from cilia onto synthetic microtubules. Our method leverages several convenient properties of the biflagellate alga *Chlamydomonasreinhardtii*, which we use as a source of IFT trains. A comprehensive range of fluorescently tagged or mutated strains is available for this model organism [18], and additional modifications can be introduced via established protocols [19]. Upon contact with a solid surface, *Chlamydomonas* transitions from swimming to gliding motility, tightly attaching its outstretched cilia to the substrate in a configuration suitable for IFT extraction. Moreover, the *Chlamydomonas* cell body is protected by a detergent-resistant cell wall, while the ciliary membrane is not. This allows selective extraction of the ciliary compartment without lysing the entire cell.

Overview of the reconstitution setup is illustrated in Fig. 1. To deliver intact IFT trains onto synthetic microtubules, living *Chlamydomonas* cells are allowed to glide on a coverslip coated with microtubules. Ciliary membranes are then selectively solubilized using detergent, enabling IFT trains to transfer directly onto the synthetic microtubules after detaching from the axonemal microtubule doublets. Because the gliding cells adhere to the surface via transmembrane proteins [20], strong detergent treatment (when introduced in bulk—using pipetting or microfluidics) can displace both the cell bodies and the released IFT trains before the trains have a chance to rebind. To overcome this issue, we developed a method for locally dispensing femtoliter volumes of detergent using a capillary micropipette. In the presence of ATP-containing motility buffer, some IFT trains detach from the demembranated axonemes and immediately rebind to the synthetic microtubules, where their motility can be observed (Fig. 2a). Some IFT trains do not detach immediately, allowing their motility to be observed on

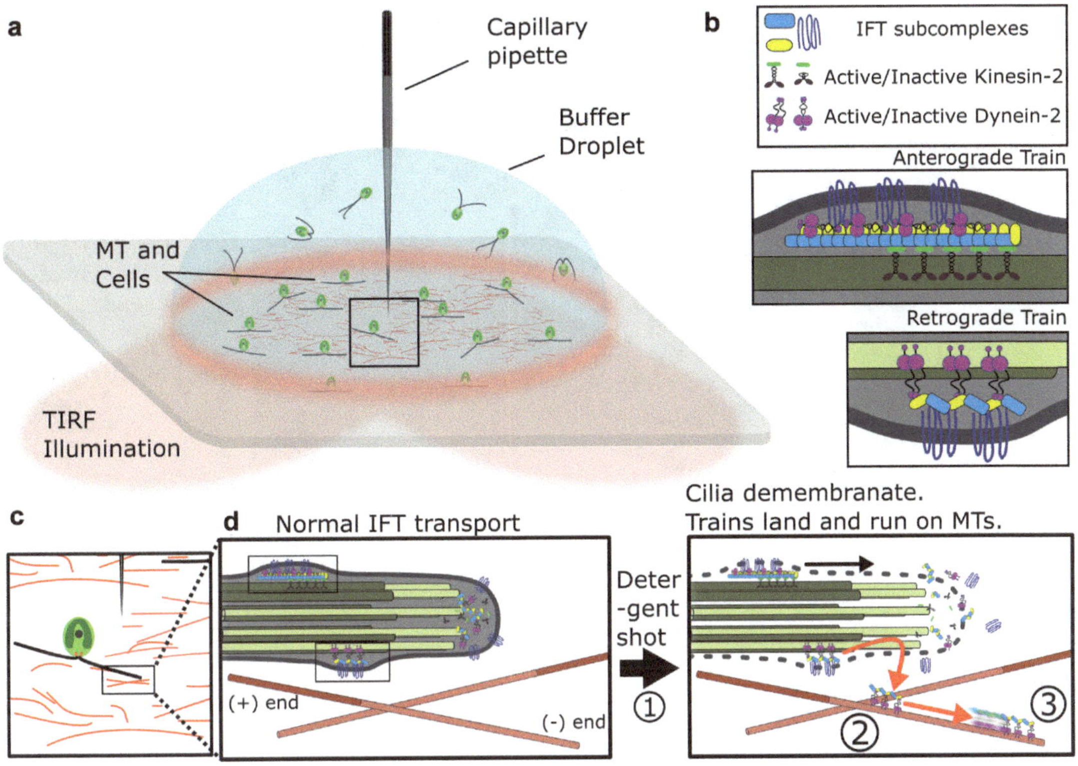


**Fig. 1** Overview of IFT train motility reconstitution setup. (**a**) *Chlamydomonas* cells gliding on synthetic microtubules adhered to a coverslip in a droplet of motility buffer on a TIRF microscope. A detergent-filled micropipette is vertically immersed in the buffer. (**b**) Legend showing components of IFT trains (top), and illustrative representations of an anterograde (middle) and retrograde train (bottom). (**c**) Illustration of relative positions of microtubules, cells, and capillary pipette.(**d**) Illustration of in vivo IFT train motility (left) and ex vivo IFT train motility on synthetic polarity-marked microtubules (right). ① Detergent shot from capillary pipette solubilizes ciliary membrane. ② IFT trains are released into the surrounding environment. Some motile trains continue to move (black arrow) on the demembranated cilia (parent axoneme), while ③ others land and move (red arrows) on the polarity-marked microtubules. (Reproduced from ref. [9] under CC BY 4.0)

membrane-free axonemes and enabling quantification of detachment kinetics (Fig. 3). As the detergent diffuses rapidly after ejection, its area of effect is confined to approximately 30 μm (see Fig. 4a, b). This localized treatment enables us to repeat the extraction procedure on neighboring cells within the same sample, significantly improving experimental throughput. The entire procedure is carried out on an inverted microscope, and images are recorded using total internal reflection fluorescence (TIRF) microscopy (see Movie 1). Analysis of the TIRF video sequences reveals that the mean velocity of IFT trains does not change during motility reconstitution (Fig. 2b), suggesting that the IFT architecture and function remain conserved.

In conclusion, we introduce an ex vivo method for reconstituting the motility of intact IFT trains extracted directly from detergent-lyzed cells. Our assay is applicable to any cell type or organism

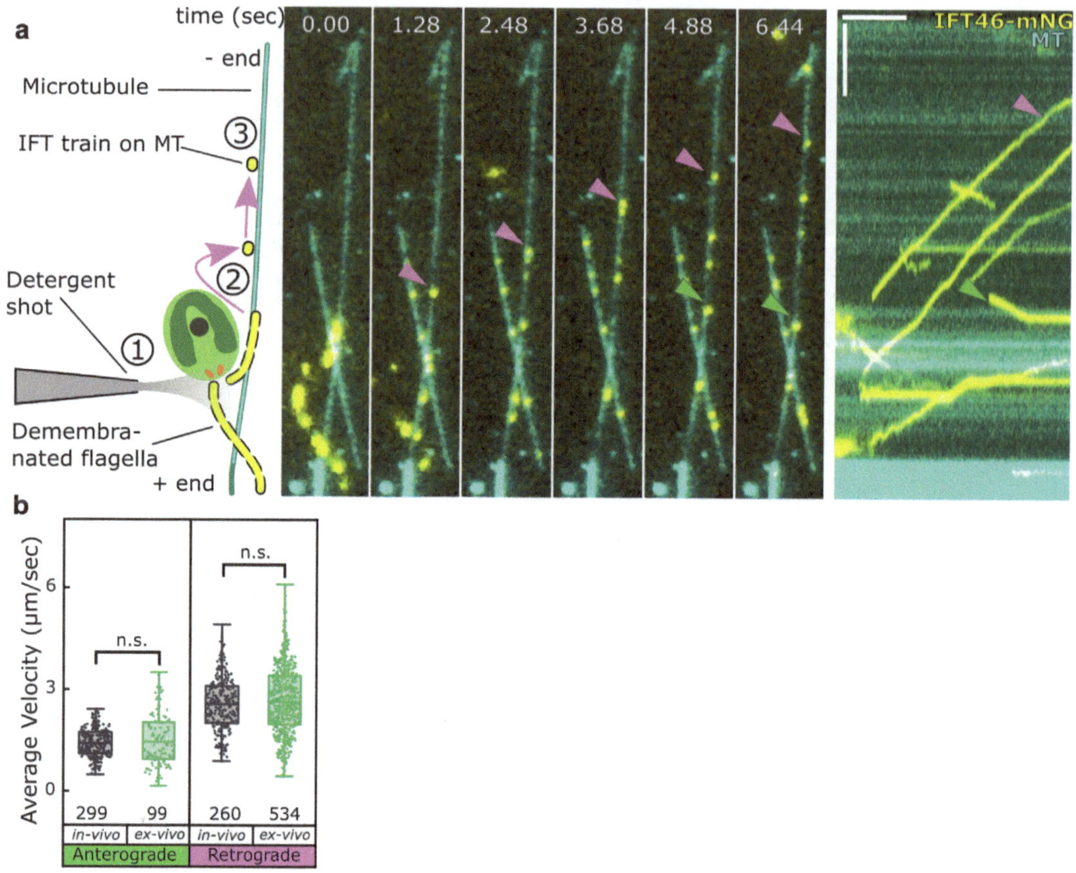


**Fig. 2** Observation and analysis if IFT trains during ex vivo motility reconstitution.(**a**) Left: Schematic of ① detergent shot, ② landing of IFT train, and ③ motion on microtubule. Right: Representative time-lapse montage and kymograph of motility reconstitution of IFT trains comprising mNeonGreen labeled IFT46. Timepoint 0.00 shows intact cilia (yellow) attached to synthetic microtubules (cyan). Upon detergent addition, reconstituted anterograde IFT trains (green arrowheads) can be seen moving toward the brightly labeled plus end of a microtubule, while retrograde IFT trains (magenta arrowheads) move toward the minus end. (**b**) Box plots of average velocities as measured from kymographs in (**a**), showing that the velocities of IFT trains remain conserved during motility reconstitution. Statistics by Student's two-tailed *t*-test. n.s., $p =$ not significant. Each point represents an individual train, from 50–70 cells across 6–7 biological replicates. Box bounds represent 25% and 75% quartiles. Line represents the median. Box whiskers represent maxima and minima. Horizontal bar, 5 μm, vertical bar, 2 s. (Adapted from [9] under CC BY 4.0)

that adheres to a coverslip and can be readily implemented in any laboratory with access to TIRF microscope and microinjection instrumentation. In complementary studies, microfluidic platforms have been introduced to analyze single-cell extracts, obtained through mechanical [21] or laser-induced [22, 23] cell disruption. All these targeted ex vivo approaches share several key advantages: they allow rapid reactivation of unstable molecular complexes or transient interactions, capture cell-to-cell variability, minimize input sample

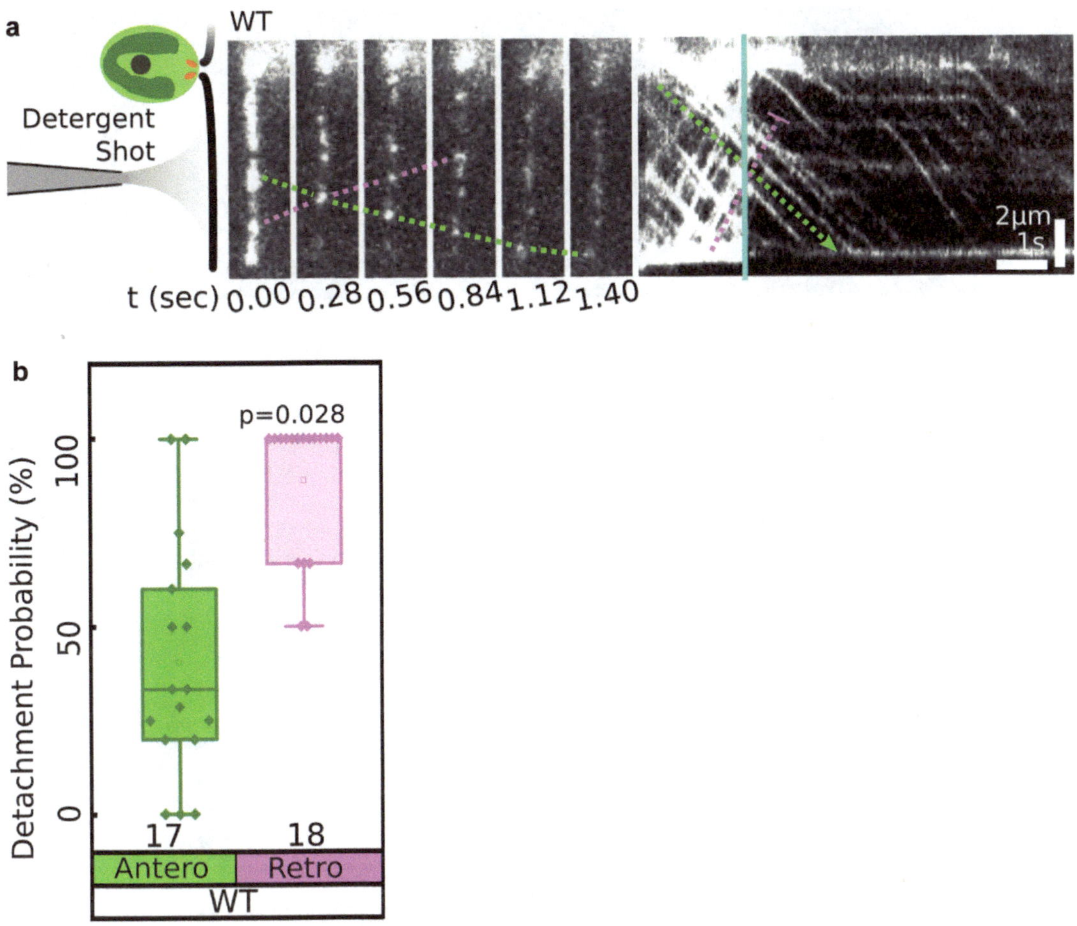


**Fig. 3** Observation and analysis of IFT trains detachment from parent axonemes. (**a**) Time lapse montage and kymograph of motility reconstitution of IFT trains comprising mNeonGreen labeled IFT46. Timepoint 0.00 shows intact cilia. Upon detergent addition (cyan line), retrograde train (dashed magenta segment) detaches quickly, while anterograde train (dashed green segment) continues unperturbed to cilia tip. (**b**) Calculated detachment frequency of anterograde or retrograde trains (see Sect. 3.4). Total individual cilia estimated as shown. Statistics by Two-sided Wilcoxon Signed Rank Test. *p*-values as shown. Each point represents an individual axoneme analyzed as shown, from three biological replicates each. Box bounds represent 25% and 75% quartiles. Line represents the median. Box whiskers represent maxima and minima. (Adapted from [9] under CC BY 4.0)

quantity, enable single-molecule sensitivity, and provide immediate experimental readouts. Together, these features facilitate a controlled simplification of the complex and dynamic intracellular environment, offering deeper insight into the organizational principles of living systems.

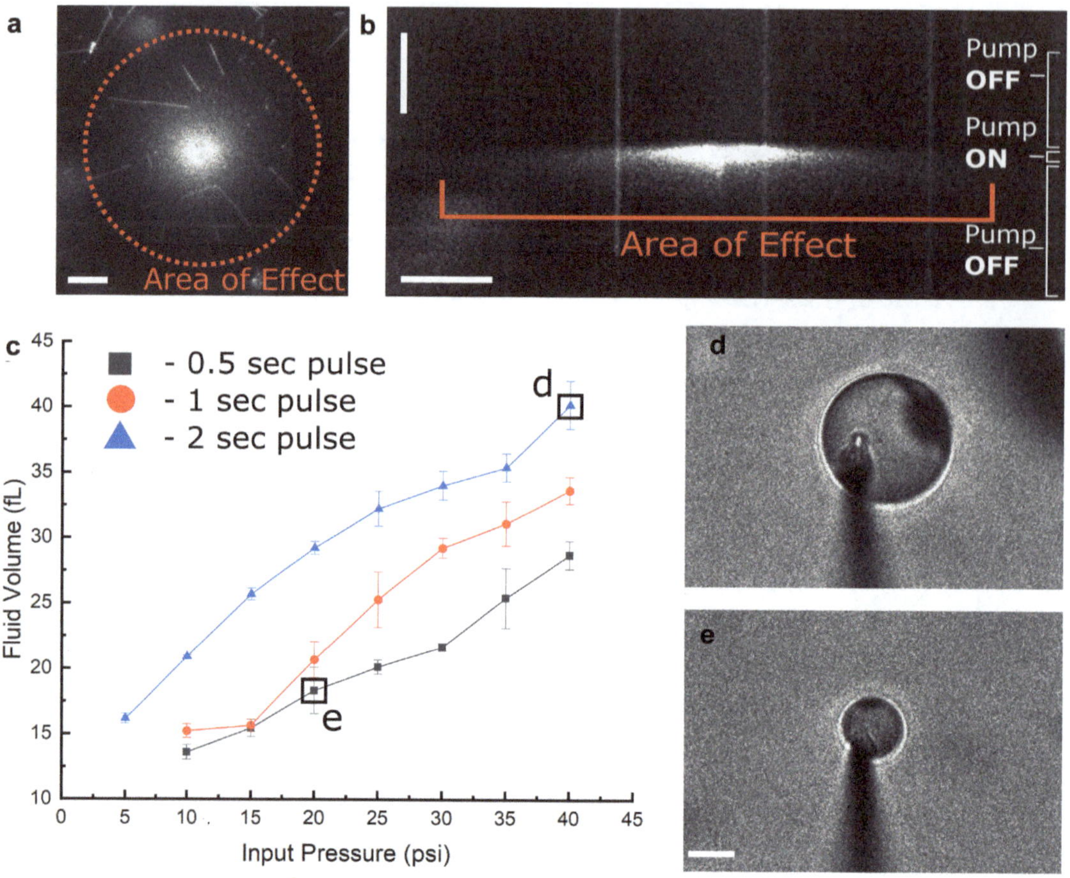


**Fig. 4** Calibration of the microinjector pump and capillary pipette. (**a**) Ejection of rhodamine solution to visualize its concentration profile and approximate area of effect across the field of view. Rhodamine-labeled microtubules are present in the field of view. (**b**) Kymograph of ejection dynamics (**c**) Volume of ejected fluid (femtoliter) as a function of pulse duration and input air pressure in the injector pump (psi). Volumes are estimated from radii of fluid spheres. Calibrations were performed across N = 3 repeats, average of 3–4 ejections per data point. (**d, e**) Representative images across the dynamic range of input pressures and pulse times. Horizontal bar, 5 μm, vertical bar, 2 s. (Adapted from [9] under CC BY 4.0)

## 2 Materials

### 2.1 Cell Culture

1. *Chlamydomonas* cell line expressing a fluorescently tagged IFT component of choice: several suitable cell lines are available from the Chlamydomonas Resource Center [18].
2. TAP salts: 1.5% $NH_4Cl$, 0.4% $MgSO_4$, 0.2% $KH_2PO_4$.
3. Phosphate solution: 28.8% $K_2HPO_4$, 14.4% $KH_2PO_4$.
4. Hunter's trace elements stock solution [24], available from the Chlamydomonas Resource Center: 0.134 mM $Na_2EDTA \bullet {}_2H_2O$, 0.077 mM $ZnSO_4 \bullet 7H_2O$, 0.184 mM $H_3BO_3$,

$0.026\,mM\ MnCl_2 \bullet 4H_2O$, $0.018\,mM\ FeSO_4 \bullet 7H_2O$, $0.007\,mM\ CoCl_2 \bullet 6H_2O$, $0.005\,mM\ CuSO_4 \bullet 5H_2O$, $0.0008\,mM\ (NH_4)_6 Mo_7O_{24} \bullet 4H_2O$.

5. TAP (Tris-Acetate-Phosphate) medium [25]: 0.02 M Trisma Base pH 7, 0.017 M $CH_3COOH$, 2.5% TAP salts (v/v), 0.1% Phosphate solution (v/v), 0.1% Hutner's trace elements stock solution (v/v) (*see* **Note 1**).
6. Culture cells in bottles aerated with sterile air under 12 h light/12 h dark cycles at 23°C.

### 2.2 Silanized Coverslips

1. Clean the coverslips (22 × 22 mm, #1.5 thickness coverslips) by 15 min sonication in 5% Mucasol diluted in water.
2. Rinse for 2 min with deionized water.
3. Incubate in 0.05% dichlorodimethylsilane diluted in 90% ethanol/10% methanol (v/v) for 60 min.
4. Rinse and sonicate 2 × 10 min with 100% methanol.
5. Sonicate 2× with deionized water.
6. Blow dry using nitrogen gas.
7. Store in an airtight container.

### 2.3 Capillary Micropipettes

1. Glass capillaries: borosilicate glass, with filament, thin wall, 1 mm outer diameter (e.g., BF100-75-15, Sutter Instruments).
2. Programmable pipette puller (e.g., P97 or P1000, Sutter Instruments).
3. Optimize the puller settings per manufacturer's instructions (*see* **Note 2**) to obtain suitable pipette parameters of 6–8 mm taper length and 2–3 μm tip size.
4. Inspect micropipettes using a low-magnification stereomicroscope. Pipette shoulders should not have significant asymmetries as this might cause inconsistencies in ejection volumes.
5. Confirm that pipettes dispense consistent liquid volumes in the range of 20–40 fL by using a pipette for injection pump calibration (see Sect. 2.7, Fig. 4).
6. Prepare a batch of 15–20 capillary micropipettes (length 20–40 mm).

### 2.4 Polarity-Marked Microtubules

To visualize the directionality of reconstituted IFT trains, synthetic microtubules are assembled with brightly labeled plus-ends. This is achieved by extending dim rhodamine labeled microtubule seeds with an elongation mix containing bright rhodamine labeled tubulin. Targeted labeling at the plus-end is ensured through N-ethylmaleimide modification of bright rhodamine tubulin.

1. Unlabeled tubulin: commercially available or purified from porcine brain [26].
2. Rhodamine-labeled tubulin: commercially available or labeled according to standard procedures [27].
3. BRB80 buffer: 80 mM PIPES, 1 mM $MgCl_2$, 1 mM EGTA, pH 6.8.
4. Dimly-labeled microtubule seeds: mix one part of rhodamine-labeled tubulin with nine parts of unlabeled tubulin for a final concentration 2 μM in BRB80, add GMPCPP to 1 mM, and incubate at 37°C for 30 min.
5. N-ethylmaleimide (NEM) tubulin: incubate 40 μM unlabeled tubulin with 0.4 mM NEM on ice for 10 min, and quench excess NEM with 20 mM DTT for 10 min. Use immediately.
6. Plus-end labeling mix: mix one part of rhodamine-labeled tubulin with three parts of unlabeled tubulin for a final concentration of 4 μM, 1 μM NEM-tubulin, 1 mM GTP, 4 mM $MgCl_2$ in BRB80. Prepare on ice.
7. Plus-end labeled microtubules: Warm the plus-end labeling mix at 37°C for 1 min, incubate with 1/10 volume of dimly-labeled microtubule seeds at 37°C for 60 min and stabilize by adding taxol to 10 μM. Harvest by spinning at 17,000 g for 15 min. Polarity-marked microtubules should be used on the same day.

### 2.5 Instrumentation for Reconstitution and Microscopy: TIRF Microscope

We have successfully validated the procedure using different models of microscopes, micromanipulator and microinjection pumps. For each instrument, we list the key features, followed by the specifications of our typical setup. The TIRF microscope must be configured for high-speed imaging of IFT moving along polarity-marked microtubules. Some features, such as motorized control and automatic focus compensation are beneficial, but not strictly necessary (*see* **Note 3**). Essential features are:

1. Single-molecule detection sensitivity.
2. ≥25 fps frame rate.
3. Simultaneous dual-color imaging.
4. Inverted stand with sufficient space on the sample side to allow micromanipulation. In particular, the condenser arm should be retractable to a sufficient degree. An enclosure box is generally not suitable.

### 2.6 Instrumentation for Reconstitution and Microscopy: Micromanipulator

The micromanipulator is required to hold the micropipette directly above the objective, in the center of the field of view. Distinct from typical microinjection setups, the pipette is held vertically, not obliquely (*see* **Note 4**). The following considerations are essential when choosing and installing the micromanipulator (*see* **Note 5**):

1. Coarse positioning in the range of centimeters and fine positioning with precision of 1 μm or smaller in X, Y, Z.
2. Fine positioning should be remotely controlled, electronic or hydraulic.
3. Custom mounting accessories might be needed to install the micromanipulator onto the microscope. Minimize mechanical loop to reduce micropipette vibrations.
4. Ensure that the manipulator cables or connectors do not hinder the objective or optical path in any way.
5. As the micromanipulator usually interferes with the condenser and the transmitted light path, it should be easily detachable from the microscope.

### 2.7 Instrumentation for Reconstitution and Microscopy: Microinjection Pump

The microinjection pump is used for femtoliter-scale delivery of detergent solution onto the sample. The adjustability of the following settings is essential (*see* **Note 6**):

1. Compensation pressure.
2. Ejection pressure.
3. Ejection pulse duration.
4. To calibrate ejection pressure and ejection pulse duration, inject the demembranation solution (see Sect. 2.8, Step 3) into a droplet of paraffin oil (Fig. 4c–e). Ejected volume can be calculated from the radius of the ejected droplet. Desired volume is 20–40 fL.
5. To validate the setup, eject rhodamine solution into HMDEK solution and observe the pulse kinetics using TIRF microscope. Visible rhodamine fluorescence should encompass an area of approximately 30 μm (Fig. 4a) and dissipate within approximately 1 s (Fig. 4b).

### 2.8 Reconstitution

1. HMDEK buffer: 30 mM HEPES, 5 mM $MgSO_4$, 1 mM DTT, 0.5 mM EGTA, 25 mM KCl, pH 7.2.
2. $Mg^{2+}$-ATP solution (100 mM stock).
3. Demembranation solution: 1% Igepal CA-630 (NP-40) detergent in HMDEK.
4. Teflon-tipped tweezers for coverslip handling.
5. Coverslip holder with screw clamps.
6. 1 mL Luer-Lock syringe with 27-gauge beveled needle.

## 3 Methods

### 3.1 Preparation of Cells and Microtubules

1. 3–4 days prior to the experiment, inoculate 100–250 mL of TAP medium with the *Chlamydomonas* cell strain of choice.
2. On the day of the experiment, check the cell culture quality: cell suspension should appear bright green and most cells should be swimming when examined with a microscope. Enough cells should adhere to solid surfaces in a gliding position.
3. Prepare polarity-marked microtubules and evaluate their density and labeling quality using a TIRF microscope. Repeat the preparation if microtubule density is low or polarity labeling is inadequate.
4. Aspirate 1–2 mL of the cell suspension from culture bottle (*see* **Note 7**).
5. Wash the cells thrice in HMDEK buffer.
6. Resuspend cells in 1 mL of HMDEK to a final concentration of approximately $10^6$–$10^7$ cells/mL (*see* **Note 8**).
7. Add ATP to the cell suspension to have a concentration of 1 mM in the final cell-microtubule mix. Take into account the volume of microtubule suspension added later (see Sect. 3.2, Step 12; *see* **Note 9**).
8. Keep the suspension on a slow rocker until use.

### 3.2 Setting Up the Microscope

1. Mount the micromanipulator on the microscope stage so that the pipette holder is perpendicular to the stage and directly above the objective (see Fig. 5).
2. Ensure that manipulator adjustment screws are near the center of their travel range before starting the experiment.
3. Mount a silanized coverslip in the coverslip holder. Avoid over tightening of the holder screws to prevent bending the coverslip.
4. Place the mount on the microscope stage.
5. Fill a syringe with approximately 0.5 mL of demembranation solution.
6. Backfill carefully a capillary micropipette with a few microliters of the demembranation solution by using a 27-gauge needle (*see* **Note 10**).
7. Ensure the solution reaches the tip of the pipette. This can be verified by observing a small meniscus around the pipette shoulder (*see* **Note 11**).
8. Immediately mount it in the pipette holder supplied with the microinjection pump (*see* **Note 12**).

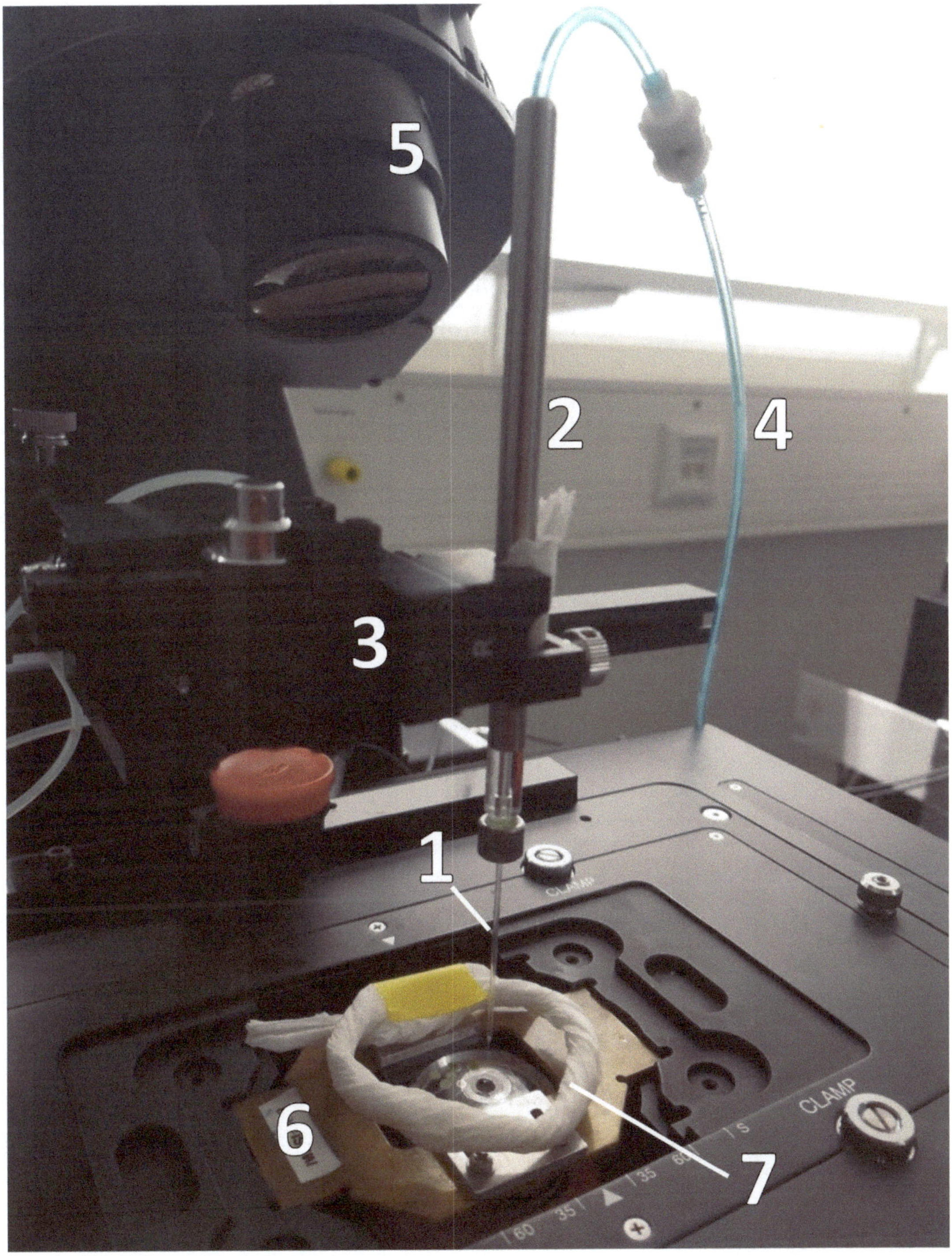


**Fig. 5** IFT train motility reconstitution setup mounted on a microscope stage. A capillary micropipette **(1)** inserted in a microinjection pipette holder **(2)**, held vertically by a hydraulic micromanipulator **(3)**, connected to a microinjection pump by tubing **(4)**. Transmitted illumination arm of the microscope **(5)** is retracted back to accommodate the assembly. Coverslip with the sample is mounted in a custom holder **(6)** and surrounded with a wet filter paper **(7)** to reduce evaporation (Reproduced from [28])

9. Mount the pipette holder in the micromanipulator so that the pipette tip is suspended 3–4 cm above the coverslip. Achieve this height by slowly sliding the pipette holder in the manipulator, not by moving the manipulator Z-axis screw (*see* **Note 13**).
10. Start a live video stream at 10× magnification. Turn on the transmitted lamp to illuminate the stage and the pipette tip.
11. Using the coarse manipulator, position the pipette tip directly over the objective. Coarse alignment can be performed looking at the pipette tip from sides (*see* **Note 14**).
12. Once it is approximately centered, slowly lower the pipette using the coarse or fine Z-axis screw (*see* **Note 15**).
13. The pipette tip or its shadow should now appear in the microscope field of view (FOV). Center it using the manipulator coarse X and Y adjustment screws.
14. Raise or lower the tip so that it just touches the surface (*see* **Note 16**). The final tip position should be 10–20 μm above the coverslip, or roughly 1–2 Chlamydomonas cell diameters.
15. Switch to an intermediate magnification objective (40× or 60×).
16. Recenter the pipette.
17. Switch to the TIRF objective (100× or 150×) and repeat the centering (*see* **Note 17**).
18. Place a small droplet (0.5–1 μL) of polarity-marked microtubules on the coverslip a few mm away from the pipette and overlay it slowly with 5–10 μL of the cell suspension in HMDEK-ATP buffer. The final ATP concentration should be 1 mM.
19. Bring the droplet into the FOV so that the pipette tip is immersed and approximately centered (*see* **Note 18**).
20. Recenter the pipette as needed.
21. Check for leakage of liquid from the pipette tip. If necessary, eliminate it by adjusting the compensation pressure on the pump.

### 3.3 IFT Train Motility Reconstitution on Synthetic Microtubules

1. Turn off the transmitted illumination. Focus the microtubules, cilia and IFT in TIRF mode.
2. Start a continuous video stream recording of 4–5 min duration.
3. Navigate to a field of view (FOV) with a suitable density of microtubules and ciliated cells. FOV of 50 μm × 50 μm containing 6–7 cilia is usually ideal for observing processive IFT runs (*see* **Note 19**).

4. Center the region of interest in the FOV.
5. Trigger the pump to eject a predetermined volume of detergent. The gliding cells will typically detach from the coverslip and begin ciliary motion, as the axonemal motility gets reconstituted along with IFT upon cilia demembranation.
6. Observe IFT trains as they land and move along polarity-marked microtubules.
7. Continue imaging on the same FOV for 3–5 s after demembranation to capture complete runs (see Movie 1).
8. While still imaging, move to a new FOV nearby.
9. Repeat Steps 3–8 until the stream acquisition finishes. The result is a long movie that shows multiple reconstitution events interleaved with segments showing the process of locating and focusing on the next area of interest (*see* **Note 20**).
10. Start over from Step 2 to acquire additional data.
11. Intermittently turn on transmitted light to check the pipette tip remains centered. If off-center or lost from FOV, repeat Steps 9–11, Sect. 3.2, before resuming.
12. Continue recording until the pipette becomes clogged (*see* **Note 21**) or the cells start dying (*see* **Note 22**).
13. Stop recording and begin with a new pipette and fresh cells at that point.
14. At the end of the session, raise the pipette, remove it from the holder, and discard it carefully with sharp waste (*see* **Note 23**).
15. Extract successful reconstitution events from the continuous recordings for further analysis in image analysis software such as FIJI/ImageJ [29]. For a typical analysis, see Fig. 2.

### 3.4 IFT Train Motility Reconstitution on Parent Axonemes

In addition to studying IFTmotility on reconstituted microtubules, it can also be reconstituted directly on demembraneted axonemes. This approach lets us assess IFT train behavior without interference from the flagellar membrane or cytoplasm. Further, measurement of detachment kinetics after demembranation reveals relative train affinity for the axoneme in-situ, which are illustrated in Fig. 3. These measurements are especially interesting for investigating how microtubule post-translational modifications, associated MAPs, or lattice defects in the microtubule doublets affect IFT motor binding and dynamics. IFT can be reconstituted on parent axoneme with few modifications of the protocol described above.

1. Completely exclude polymerized microtubules for this experiment, as the acquisition will focus on parent axonemes, i.e.,

demembranated axonemes from the cells that are exposed to detergent.

2. Start a continuous video stream recording of 4–5 min duration. Acquire a few seconds of normal in-vivo IFTmotility.
3. Proceed as for Steps 3–14 described in Sect. 3.3.
4. For analysis, include predemembranation frames to be able to efficiently trace back trains later during the analysis.
5. Calculation of Detachment Probability: For each train type on each parent axoneme, calculate "Retention fraction" (Rf) as,
6. Rf = Npa/Nc
7. where,
8. Npa = Number of trains per demembranated parent axoneme
9. Nc = Number of trains per cilium before demembranation
10. Disregarding newly injecting trains after demembranation (particularly anterograde trains) in the above evaluation, calculate Detachment Probability as,
11. Pd = (1 − Rf) × 100
12. Occasionally, the target cilium will change shape upon demembranation, due to non-native activation of axonemal dyneins. Trace and plot kymographs individually for each of these shapes during post-processing before taking a maximum Z-projection to obtain full temporal coverage (*see* **Note 24**).

## 4 Notes

1. Dissolve 2.42 g Tris base in 850 mL of deionized water. Add 25 mL of TAP salts. Add 1 mL of Phosphate solution. Add 1 mL of Hutner's trace elements [24]. Adjust pH to 7.0 using about 1 mL glacial acetic acid. Bring volume to 1 L with deionized water. Autoclave and store at room temperature.
2. Micropipette puller settings must be determined individually for each machine. For consistent results, adjustments may be necessary over time, particularly after replacing the heating filament. Always aim for pipettes with reasonable neck flexibility: rigid neck will shatter upon even light contact with a glass coverslip, while overly flexible neck will be deflected by the surface tension of the sample droplet and thus difficult to center.
3. The setup used in the method described here is: Ti2 Eclipse (Nikon, Japan) equipped with a 100×/1.49NA TIRF

objective with a galvo-driven orbital TIRF (iLas2, Gataca Systems, France) coupled to a multilaser combiner (VSLMS-MOT100, Visitron Systems, Germany). Multicolor, simultaneous acquisition is done with EMCCD cameras (iXon life, Oxford Instruments, UK) controlled by VisiView software 5.0 (Visitron Systems, Germany). Data are acquired as a continuous stream at 40 ms/frame, 1024 × 1024 pixels at a resolution of 0.086 μm/px.

4. Pipettes positioned at non-perpendicular angles allow the manipulator assembly to occupy less headspace over the stage. However, ejecting detergent at an obtuse angle can induce lateral flows, potentially displacing microtubules and reconstituted particles from the field of view (FOV).
5. We used MN-4 for coarse manipulation and MMO-4 for fine manipulation, both Narishige, Japan.
6. Our pump model: PV-850, World Precision Instruments, UK.
7. When using non-hatching *Chlamydomonas* mutants, aspirate the desired quantity and treat with autolysin for 30 min, and allow recovery in fresh TAP for 1–2 h prior to proceeding further.
8. Precise cell concentration matters less for this experiment. However, the number of cells adhered to the glass surface within each TIRF FOV during reconstitution is essential. This may depend on different factors, including the cell strain used, precise stage of the cell cycle, or surface modification of the coverslip.
9. For example, if 0.5 μL of the microtubule suspension is to be overlaid with 5 μL of the cell suspension, add 11 μL of 100 mM ATP to 989 μL of the cell suspension to have an intermediate concentration of 1.1 mM.
10. One microliter of detergent solution is sufficient for a full experiment. Typical ejection volumes are in the femtoliter range.
11. Sometimes the solution may not reach the pipette tip. If this occurs, gently tap the broad end or rub it with a small piece of paper to create frictional vibrations. This should help draw the solution down to the narrow tip.
12. Avoid carrying the pipette around after filling.
13. It is essential to keep the micromanipulator adjustment screws around the middle of their travel range during the preparatory phase. Their travel range is limited and reaching end stops during an experiment leads to tedious realignment.

14. Insufficient headspace over the microscope may prevent the manipulator from fitting when the condenser illumination arm is in proper position. In such cases, illuminate the stage while the condenser arm is retracted. Although this will result in oblique transmitted illumination, it should suffice for locating cells on the surface and centering the pipette tip.
15. Be careful! Do not lower it too quickly or too far. Doing either will break the tip against the coverslip.
16. A good way to confirm contact with the coverslip surface is to observe a slight buckling of the pipette shaft as it touches the glass surface.
17. Pre-emptively raise the pipette during centering to prevent it from breaking or scratching the coverslip. Lower it in a controlled manner using coarse Z-axis adjustment screws. This precaution might be especially important when using high-magnification oil-immersion objectives that physically contact the coverslip. Ensure objectives are not raised so high that they push the coverslip into the pipette tip.
18. Ensure that the pipette tip is fully immersed and not deflected sideways. Sideward deflection can result from interactions between the pipette neck and the surface of the hemispherical liquid drop. If this occurs frequently, the pipette neck may be too flexible and the puller program may require optimization to produce stiffer pipettes.
19. Gliding *Chlamydomonas* cells and microtubule density must be optimized according to experimental needs. Low microtubule density reduces the frequency of reconstituted IFT train runs. Very high density increases run frequency, but complicates tracking due to overlapping and crossing of trajectories.
20. This approach is more efficient than attempting to record only the reconstitution events, as switching between live mode and acquisition can take several seconds. Avoiding these interruptions not only streamlines the experimental workflow but also helps minimize phototoxic damage to the sample. Events of interest are later identified and extracted from the full recording.
21. After prolonged use, debris may accumulate and clog the pipette tip. Keep a small (5–6 μL) droplet of HMDEK buffer or distilled water for washing pipette tip next to the cell suspension, gently move the pipette back and forth within the cleaning droplet using fine adjustment screws to dislodge any contaminants. Discard the pipette if the issue is not resolved and restart the protocol with a new pipette.
22. Over time, repeated detergent ejections and buffer evaporation may reduce cell viability. Placing a small open humidity

chamber over the coverslip (see Fig. 5, Sect. 7) can reduce evaporation, although it may also limit the space available for pipette manipulation.

23. Always use a new pipette for each session.
24. In a few cases, especially at saturating ATP concentrations, demembranated cilium will beat and cause whole cell movement, making further analysis impossible. To overcome this, adapt the protocol to immobilize cells further by either lowering ATP concentration in final buffer or choosing a mutant with immobile cilium and an IFT reporter, such as *pf18::*d1bLIC-GFP (CC-4489 in Chlamydomonas Resource Center).

## Acknowledgments

We acknowledge all members of the Pigino and Diez labs for fruitful discussions. The project was supported by the Deutsche Forschungsgemeinschaft (DFG, German Research Foundation) under Germany's Excellence Strategy—EXC 2068—390729961 Cluster of Excellence, Physics of Life of TU Dresden (starting grant to A.C.). A.C. and S.D. were further supported by the TU Dresden. L.S. was supported by the INTER-ACTION grant LUAUS24182 by the Czech Ministry of Education, Youth and Sports. G.P. was supported by the Human Technopole, Milan, Italy.

### References

1. Kozminski KG, Johnson KA, Forscher P, Rosenbaum JL (1993) A motility in the eukaryotic flagellum unrelated to flagellar beating. Proc Natl Acad Sci USA 90:5519–5523. https://doi.org/10.1073/pnas.90.12.5519
2. Stepanek L, Pigino G (2016) Microtubule doublets are double-track railways for intraflagellar transport trains. Science 352:721–724. https://doi.org/10.1126/science.aaf4594
3. Howard J, Hudspeth AJ, Vale RD (1989) Movement of microtubules by single kinesin molecules. Nature 342:154–158. https://doi.org/10.1038/342154a0
4. Vale RD, Funatsu T, Pierce DW et al (1996) Direct observation of single kinesin molecules moving along microtubules. Nature 380:451–453. https://doi.org/10.1038/380451a0
5. Neuman KC, Nagy A (2008) Single-molecule force spectroscopy: optical tweezers, magnetic tweezers and atomic force microscopy. Nat Methods 5:491–505. https://doi.org/10.1038/nmeth.1218
6. Mitra A, Meißner L, Gandhimathi R et al (2020) Kinesin-14 motors drive a right-handed helical motion of antiparallel microtubules around each other. Nat Commun 11: 2565. https://doi.org/10.1038/s41467-020-16328-z
7. Leduc C, Padberg-Gehle K, Varga V et al (2012) Molecular crowding creates traffic jams of kinesin motors on microtubules. Proc Natl Acad Sci 109:6100–6105. https://doi.org/10.1073/pnas.1107281109
8. Vemu A, Garnham CP, Lee D-Y, Roll-Mecak A (2014) Generation of differentially modified microtubules using in vitro enzymatic approaches. In: Vale RD (Ed.)Eds. Methods in enzymology. Academic Press, 149–166
9. Chhatre A, Stepanek L, Nievergelt AP et al (2025) Tubulin tyrosination/detyrosination regulate the affinity and sorting of intraflagellar transport trains on axonemal microtubule doublets. Nat Commun 16: 1055. https://doi.org/10.1038/s41467-025-56098-0

10. Taschner M, Lorentzen E (2016) The intraflagellar transport machinery. Cold Spring Harb Perspect Biol 8:a028092. https://doi.org/10.1101/cshperspect.a028092
11. Lacey SE, Foster HE, Pigino G (2023) The molecular structure of IFT-A and IFT-B in anterograde intraflagellar transport trains. Nat Struct Mol Biol 30:584–593. https://doi.org/10.1038/s41594-022-00905-5
12. Jordan MA, Diener DR, Stepanek L, Pigino G (2018) The cryo-EM structure of intraflagellar transport trains reveals how dynein is inactivated to ensure unidirectional anterograde movement in cilia. Nat Cell Biol 20:1250–1255. https://doi.org/10.1038/s41556-018-0213-1
13. Taschner M, Lorentzen E (2016) Recombinant reconstitution and purification of the IFT-B Core Complex from Chlamydomonas reinhardtii. In: Satir P, Christensen ST (Eds.)Eds. Cilia: methods and protocols. Springer, New York, NY, 69–82
14. Mohamed MAA, Stepp WL, Ökten Z (2018) Reconstitution reveals motor activation for intraflagellar transport. Nature 557:387–391. https://doi.org/10.1038/s41586-018-0105-3
15. Sloboda RD (2016) Purification and localization of intraflagellar transport particles and polypeptides. In: Gavin RH (Ed.)Eds. Cytoskeleton methods and protocols: methods and protocols. Springer, New York, NY, 119–137
16. Mencarelli C, Mitchell A, Leoncini R et al (2013) Isolation of intraflagellar transport trains. Cytoskeleton 70:439–452. https://doi.org/10.1002/cm.21121
17. Petriman NA, Loureiro-López M, Taschner M et al. (2022) Biochemically validated structural model of the 15-subunit intraflagellar transport complex IFT-B. EMBO J 41:e112440. https://doi.org/10.15252/embj.2022112440
18. Chlamydomonas Resource Center. https://www.chlamycollection.org/. Accessed 22 May 2025
19. Nievergelt AP, Diener DR, Bogdanova A et al (2024) Protocol for precision editing of endogenous *Chlamydomonas reinhardtii* genes with CRISPR-Cas. STAR Protoc 5:102774. https://doi.org/10.1016/j.xpro.2023.102774
20. Bloodgood RA, Tetreault J, Sloboda RD (2019) The Chlamydomonas flagellar membrane glycoprotein FMG-1B is necessary for expression of force at the flagellar surface. J Cell Sci 132:jcs233429. https://doi.org/10.1242/jcs.233429
21. Dickinson DJ, Schwager F, Pintard L et al (2017) A single-cell biochemistry approach reveals PAR complex dynamics during cell polarization. Dev Cell 42(416–434):e11. https://doi.org/10.1016/j.devcel.2017.07.024
22. Sarıkaya S, Dickinson DJ (2021) Rapid extraction and kinetic analysis of protein complexes from single cells. Biophysical J 120:5018–5031. https://doi.org/10.1016/j.bpj.2021.10.011
23. Zimyanin V, Magaj M, Manzi NI, Yu CH, Gibney T, Chen YZ, Basaran M, Horton X, Siller K, Pani A, Needleman D, Dickinson DJ, Redemann S (2026) Chromokinesin Klp-19 regulates microtubule overlap and dynamics during anaphase in C elegans J Cell Sci 13:jcs.264066. https://doi.org/10.1242/jcs.264066. Epub ahead of print. PMID: 42439517
24. Hutner SH, Provasoli L, Schatz A, Haskins CP (1950) Some approaches to the study of the role of metals in the metabolism of microorganisms. Proc Am Phil Soc 94:152–170
25. Gorman DS, Levine RP (1965) Cytochrome f and plastocyanin: their sequence in the photosynthetic electron transport chain of *Chlamydomonas reinhardi*. Proc Natl Acad Sci 54:1665–1669. https://doi.org/10.1073/pnas.54.6.1665
26. Castoldi M, Popov AV (2003) Purification of brain tubulin through two cycles of polymerization–depolymerization in a high-molarity buffer. Protein Expr Purif 32:83–88. https://doi.org/10.1016/S1046-5928(03)00218-3
27. Hyman A, Drechsel D, Kellogg D et al (1991) Preparation of modified tubulins. Methods Enzymol 196:478–485. https://doi.org/10.1016/0076-6879(91)96041-o
28. Chhatre A (2023) Ex-vivo reconstitution of intraflagellar transport (IFT) train motility. Dissertation, Technische Universität Dresden. https://nbn-resolving.org/urn:nbn:de:bsz:14-qucosa2-900655
29. Schindelin J, Arganda-Carreras I, Frise E et al (2012) Fiji: an open-source platform for biological-image analysis. Nat Methods 9:676–682. https://doi.org/10.1038/nmeth.2019

# Chapter 9

# Planar Immunological Synapses Formed by Human CD4 T Cells as In Vitro System to Study Actin and Tubulin Dynamics Through TIRF Microscopy

Álvaro Gómez-Morón, Marta Lozano-Prieto and Noa B. Martín-Cófreces

## Abstract

The rapid and transient rearrangement of the actin cytoskeleton and the microtubule network toward the antigen contact site is essential for immunological synapse (IS) formation in T cells. This process can be reproduced in vitro using activating antibodies against the T-cell receptor and costimulatory molecules. During the IS formation, the actin cytoskeleton is remodeled into a dense peripheral F-actin ring, accompanied by polarization of the centrosome, which acts as the major microtubule-organizing center at the synapse. Both processes rely on tightly regulated cycles of polymerization and depolymerization of cytoskeletal filaments. In this chapter, we present a live-cell, super-resolution microscopy approach based on Total Internal Reflection Fluorescence (TIRF), using planar IS as an in vitro model to investigate actin and tubulin dynamics in primary human CD4 T cells.

**Keywords** Actin, Tubulin, Cytoskeleton, TIRF microscopy, T lymphocytes, Immunological synapse

## 1 Introduction

The immunological synapse (IS) is a specialized cell–cell contact site established between a T lymphocyte and an antigen-presenting cell (APC) displaying cognate MHC–peptide complexes. It provides a signaling platform, where multiple signals are integrated to drive T cell activation and the initiation of immune responses [1, 2]. The establishment of the IS is a highly dynamic process that requires the integration of spatial, mechanical and biochemical cues that are essential for specific adaptive immune responses [3, 4]. Actin and tubulin dynamics at the IS sustain signaling by generating forces that regulate downstream signaling molecules, while also directing organelle and vesicle polarization toward the IS [5]. These dynamic rearrangements are regulated in space and time by a complex network of signaling pathway and cytoskeleton-associated proteins [6]. Live-cell

Noa B. Martín-Cófreces et al. (eds.), *The Cytoskeleton*, Methods in Molecular Biology, vol. 3070,
https://doi.org/10.1007/978-1-0716-5515-3_9,

super-resolution microscopy is a suitable approach to elucidate the molecular dynamics underlying this process [7, 8].

This chapter takes advantage of these biological events to use human primary CD4 T cells transfected with plasmid encoding fluorescent actin and tubulin cytoskeleton proteins as an in vitro model to study actin and tubulin dynamics in a physiological context. T cells activated on coverslip coated with anti-CD3 plus anti-CD28 antibodies form a flat lamella while they spread onto the activating surface. This two-dimensional pseudo-IS allows accurate imaging of cytoskeletal processes through specific microscopy techniques, such as Total Internal Reflection Fluorescence microscopy (TIRFm). TIRFm uses a specific mode of sample illumination to excite fluorophores near the adherent cell surface (within ~100 nm), corresponding to the plasma membrane and the cortical cytoskeleton at the pseudo-IS. This illumination system is based on an evanescent wave produced when the light beam falls at a shallow angle greater than the critical angle into a glass coverslip at its interface with the cell aqueous environment. The beam light is then totally internally reflected and forms an evanescent wave due to the electromagnetic fields generated at the interface between the cover glass and the adhered cell plasma membrane [9]. Because the evanescent wave decreases exponentially from the interface, the illumination does not penetrate deeply into the cell. Therefore, the illumination provided by the TIRFm imaging is restricted to the neighboring region to the plasma membrane, providing a high axial resolution for imaging. In this region of the activated T cell, the cortical cytoskeleton undergoes rapid and tightly regulated cycles of polymerization and depolymerization that drive IS formation [10]. Using this protocol, we demonstrate how planar IS formed by primary human $CD4^+$ T cells can be used as a model system to investigate actin and tubulin dynamics under near-physiological stimulation conditions. This approach further enables the quantitative analysis of pharmacological compounds that modulate signaling pathway involved in cytoskeletal regulation.

## 2 Materials

### 2.1 Common Disposables

1. Micropipettes (0.2–2 μL, 1–20 μL, 20–200 μL, 200–1000 μL) and sterile filter tips.
2. Nonpyrogenic sterile serological pipettes (5, 10, and 25 mL).
3. 50 mL sterile polypropylene conical tubes.
4. 1.5 mL low-binding microcentrifuge tubes.
5. 24- and 6-well plates for cell culture.
6. Neubauer hemocytometer or automated cell counter.

### 2.2 Common Equipment

1. Cell culture incubator for 37°C and 5% $CO_2$ ambient.
2. Sterile cell culture laminar flow hood safety level II (BSL2).
3. Cooled bench top centrifuge.
4. Cell Nucleofector System for transfection (*see* **Note 1**).

### 2.3 Buffers and Reagents

1. Lymphocyte separation medium: any commercial density gradient media (1.077 g/mL) compatible with live lymphocytes (*see* **Note 2**).
2. Saline solution: 154 mM NaCl. Sterilize before use.
3. Hank's Balanced Salt Medium (HBSS): 4.166 mM $NaHCO_3$ pH:6.9, 0.337 mM $Na_2HPO_4$, 0.44 mM $KH_2PO_4$, 0.136 M NaCl, 5.365 mM KCl, 1.261 mM $CaCl_2$, 0.405 mM $MgSO_4$, 0.491 mM $MgCl_2$, 5.55 mM Glucose. Sterilize before use.
4. Cell culture medium: RPMI 1640, 100 mM Glutamine, 25 mM HEPES, 10% fetal bovine serum (FBS; v/v).
5. Transfection medium: Opti-MEM I (*see* **Note 3**).
6. Live-cell imaging medium: RPMI 1640 without phenol red, 25 mM HEPES, 2% FBS (v/v).
7. Poly-L-Lys solution: 50 μg/mL in deionized $H_2O$. Sterilize before use through 0.1–0.22 μm filters.
8. Fibronectin solution: 10 μg/mL in RPMI 1640. Sterilize before use through 0.1–0.22 μm filters.
9. Bicarbonate-carbonate buffer: 0.1 M $NaHCO_3$ pH: 9.6, 0.032 M $Na_2CO_3$.
10. Antibody coating mix: stimulating monoclonal antibody anti-human CD3ε (clones Hit3a, UCHT-1, OKT3) and monoclonal antibody anti-human CD28 (clone CD28.2) diluted in bicarbonate-carbonate buffer at a 3:1 ratio.
11. Human recombinant Interleukin-7 (rhIL-7).

### 2.4 Cells and Plasmids

1. Freshly collected human peripheral blood or buffy coat preparations.
2. Plasmids encoding fluorescent cytoskeleton components: EB3-GFP (kind gift from Dr. A. Akhmanova) to study the growth of microtubules through the observation of the incorporation of fluorescent EB3 to the end of microtubules and β-actin-mCherry to study actin retrograde flow (equilibrium between polymerization and depolymerization).

### 2.5 Equipment and Analytical Software for Microscopy

1. TIRF microscope equipped with an objective with maximal magnification and numerical aperture (NA) (*see* **Note 4**) coupled to a camera with high resolution.

2. Glass-bottom-18-well chambers for live-cell microscopy (*see* **Note 5**).
3. Image J freeware (https://imagej.net/ij).
4. Leica accompanying software (LAS-AF 2.6.0. build 7266).

## 3 Methods

### 3.1 Primary Human CD4 T Lymphocytes Isolation From Buffy Coat Preparations

1. Dilute 1:3 (v/v) the buffy coat cell preparation in sterile saline solution (*see* **Note 6**).
2. Carefully layer 30 mL of the diluted blood onto 15 mL of lymphocyte separation medium in a 50 mL centrifuge tube.
3. Centrifuge the two-layered gradient without brake at room temperature at 1800 rpm for 30 min.
4. Upon centrifugation, three layers become visible: a top layer containing yellow-colored plasma, a transparent middle layer containing lymphocyte separation medium, and a red-colored bottom layer containing anucleate cells and polymorphonuclear leukocytes.
5. Carefully recover the mononuclear cells from the interphase between the top and middle layer of each 50 mL centrifuge tube using a sterile Pasteur pipette in a new 50 mL centrifuge tube.
6. Wash the collected cells in saline solution by centrifuging once at 1200 rpm for 10–15 min.
7. Discard the supernatant carefully to avoid loss of cells (*see* **Note 7**).
8. Resuspend the cells gently and add 50 mL saline solution.
9. Perform five additional washes by centrifuging at 1100 rpm, 1200 rpm, 1300 rpm, 1400 rpm, and 1500 rpm for 5 min each.
10. Purify $CD3^+CD4^+$ cells by negative selection with magnetic beads-based methods (*see* **Note 8**).
11. Resuspend the freshly isolated CD4 T cells in HBSS or saline solution and keep at 4°C until nucleofection.

### 3.2 Nucleofection of Primary Human CD4 T Lymphocytes

1. Prewarm at 37°C the transfection mix (Opti-MEM I with the DNA or RNA included; *see* **Note 9**), the cuvettes, the post-transfection complete cell culture medium and the cell culture plates before nucleofection.
2. Count the cells. Use a maximum of $5 \times 10^6$ cells per nucleofection.
3. Spin $5 \times 10^6$ of CD4 T cells onto 1.5 mL tubes in a minicentrifuge (3500 rpm, 3 min, 4°C).

4. Aspirate all the supernatant in the tube.
5. Immediately, add 100 μL of the prewarmed transfection mix and resuspend carefully.
6. Instantly, add the cell mix to the prewarmed cuvette.
7. Directly nucleofect the cells: program U-04 from Nucleofector I.
8. Nucleofect each sample separately, recover the cells with Pasteur pipettes, and plate them immediately (*see* **Note 10**) onto the pre-warmed 6-well cell culture plates in complete medium supplemented with recombinant human IL-7 (10 ng/mL) (*see* **Note 11**). Use a 2.5–4 × $10^6$ cells/mL cell density for increased survival.
9. Incubate at 37°C and 5% $CO_2$ for 24 h until imaging.

### 3.3 Stimulatory Surface Preparation for TIRF Microscopy

1. The day before the experiment, pipette 40 μL of poly-L-Lys solution (10 μg/mL) per well of the glass-bottom-18-well chamber and incubate overnight (o/n) at 4°C. Alternatively, incubate for 1 h at 37°C (*see* **Note 12**).
2. Wash twice with 100 μL of sterile $dH_2O$ or HBSS.
3. Prepare the antibody coating mix (*see* **Note 13**).
4. Pipette 40–50 μL per well of diluted antibodies and incubate o/n at 4°C.
5. Wash twice with 100 μL of sterile $dH_2O$ or HBSS and add 150 μL of cell imaging medium per well and store at 37°C until use (*see* **Note 14**).

### 3.4 TIRFm Imaging of Actin Retrograde Flow and Microtubule TIPs

1. Prewarm the microscope chamber at 37°C and 5% $CO_2$ for at least 4 h before image acquisition (*see* **Note 15**).
2. Put a drop of prewarmed immersion oil onto the 100× objective and in the glass-bottom-18-well plate and place the plate with the adapter in the stage.
3. Proceed to the laser alignment (*see* **Note 16**).
4. Wash cells once with imaging medium and resuspend in imaging medium (1 × $10^7$ cells/mL). Prewarm cells at 37°C and 5% $CO_2$ for at least 15 min until used.
5. Add gently 20 μL (2 × $10^5$ cells) as little drops in a well of the precoated and prewarmed p18-well plate.
6. Let cells settle for 3 min for EB3-GFP tracking or record immediately after cell addition for actin retrograde flow analysis through β-actin-mCherry tracking.
7. Localize the transfected cells with the oculars and the epifluorescence mode of the microscope and place them at the center of the imaging area. Select cells with similar expression of the fluorescent tag (*see* **Notes 17 and 18**).

8. When the laser is aligned, select manually the direction of incidence of the laser to generate the evanescent wave (*see* **Note 19**).
9. Adjust the acquisition parameters for video recording. The intensity of the laser power should be enough to visualize the fluorescent proteins or probes but avoiding phototoxicity and photobleaching (*see* **Notes 20 and 21**).
10. Activate mode for simultaneous imaging if simultaneous recording is performed; TIRF microscopes allowing it have different options (*see* **Note 22**).
11. Acquire an initial image including EB3-GFP and/or β-actin-mCherry fluorescence and brightfield (BF) for localization of the cell shape. In the case of β-actin-mCherry tracking, this initial image is avoided if recording includes initial cell spreading.
12. Select the laser penetration (objective relative angle). Recommended laser penetration is 150 nm to detect the centrosome and microtubule dynamics and 90 nm to detect cortical actin cytoskeleton near to the plasma membrane (*see* **Note 23**).
13. Acquire frames every 300 ms for 3–4 min for EB3-GFP tracking and every 60–90 ms for 5–6 min to track the actin retrograde flow (*see* **Note 24**).
14. Acquire a final image including EB3-GFP or β-actin-mCherry fluorescence and brightfield (BF) for localization of proteins and characterization of the cell shape.

### 3.5 Image Analysis of β-Actin Clearing in β-Actin-mCherry Transfected Human Primary CD4 T Lymphocytes

All the data analysis described in this section can be performed with the Image J freeware.

1. Use 16-bit images in gray scale.
2. Open individual image sequences from TIRFm imaging with Image J software.
3. Adjust the Brightness and Contrast (Image > Adjust > Brightness/Contrast). These conditions must be used consistently and uniformly for all images processed. An image composition as in Fig. 1 will be obtained.
4. Crop time and XY dimensions from the gray image corresponding to β-actin-mCherry fluorescence to analyze only the desired time range and the cell of interest by making a ROI ("selection tool") and duplicating (Image > Duplicate) the image sequence.
5. Save as a.tiff file for analysis.

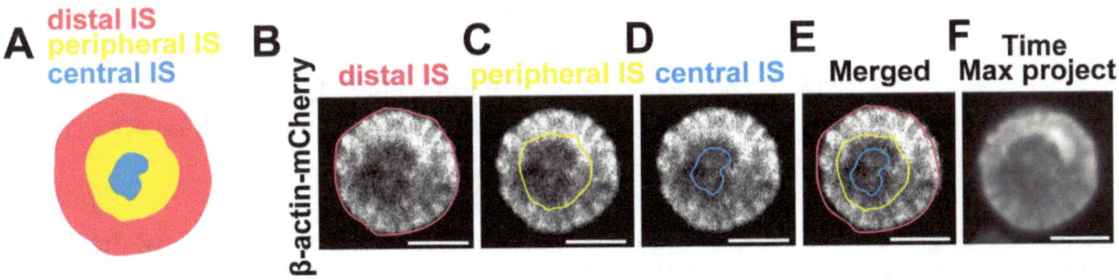


**Fig. 1** Image analysis of β-actin clearing in β-actin-mCherry transfected human primary CD4 T lymphocytes. Representative fluorescent image of a human CD4 T cell transfected with β-actin-mCherry (gray pseudocolor), spreading over a stimulatory surface. (**A–D**) ROIs used to calculate β-actin clearing are indicated: (**A**) scheme of the regions of the IS, (**B**) distal IS (magenta), (**C**) peripheral IS (yellow) and (**D**) central IS (blue). (**E**) shows the merged of all the areas used to calculate β-actin clearance. (**F**) shows the maximal projection of the time. Bar, 5 μm

6. Usually, the analysis of an image sequence expanding from 1–2 min for tracking is sufficient to have reproducible results.
7. To analyze β-actin clearing from the central IS (Fig. 1A), draw the external perimeter of the cell by using the "*polygon selection*" tool (Fig. 1B). Save the ROI in the ROI manager ("Ctrl + T).
8. Measure ("Ctrl + M") the area and the fluorescence in the ROI (*see* **Note 25**).
9. Draw a second polygon at the distal lamella end by using the "*polygon selection*" tool (Fig. 1C). Save the ROI in the ROI manager ("Ctrl + T").
10. Measure the area and the fluorescence in the ROI.
11. In mature synapses, a more central area devoid of β-actin appears.
12. Draw a third polygon surrounding the area devoid of β-actin by using the "*polygon selection*" tool (Fig. 1D). Save the ROI in the ROI manager ("Ctrl + T").
13. Measure the area and the fluorescence in the ROI.
14. Compare the area of the whole cell (Fig. 1B, magenta line), the limit between the central region of the IS and the peripheral IS (Fig. 1C, yellow line) and the area of the central IS (Fig. 1D, blue line). To see all the ROIs in the image (Fig. 1E), select the "*Show All*" and "*Labels*" options in the ROI manager.
15. A summatory of all the frames acquired along the time can be obtained as in Fig. 1F (Image > Stacks > Z project > Sum slices).
16. ROIs can be saved as.zip files for posterior analysis.

17. Measurements can be performed individually or collectively by using the "*Multimeasure option*". Export data as.txt files.
18. The percentage of fluorescence and the area occupied in ROIs containing diverse β-actin densities can be calculated.
19. Compile all datasets for statistical analysis.

### 3.6 Image Analysis of the Retrograde Flow Speed in β-Actin-mCherry Transfected Human Primary CD4 T Lymphocytes

1. To calculate the speed of the actin retrograde flow, draw a line across the radius of the cell toward the center by using the "*straight-line tool*" (Fig. 2A, red arrows).
2. Create a kymograph using the "*reslice tool*" inside the stacks function (Image > Stacks > Reslice).
3. Check the "*Avoid interpolation*" option. This kymograph is used to calculate the actin speed (Fig. 2B).
4. Draw a line following the signal of β-actin in the distal area of the synapse (Yellow line in Fig. 2B). The slope of this line corresponds to the speed of the particles tracked.
5. Measure the line selection ("Ctrl+M") and use the values in the resulting table: "Height" indicates the time used by a particle of β-actin-mCherry (magenta line in Fig. 2B) to travel the distance "Width" (red line in Fig. 2B). These values are given in pixels. The ratio Width:Height corresponds to the speed.
6. The angle yielded by the program (white line in Fig. 2B) corresponds to a negative angle between the line and the bottom of the image. The cotangent of this positive angle corresponds

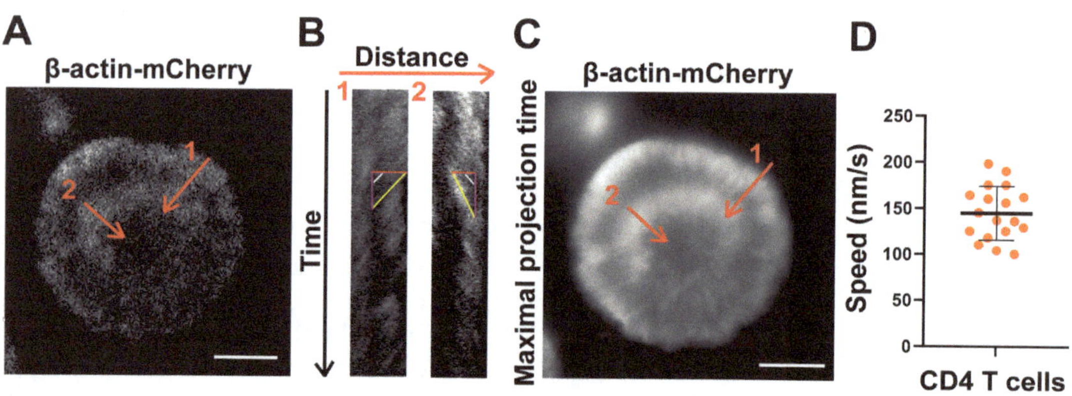


**Fig. 2** Image analysis of the speed of actin retrograde flow in β-actin-mCherry transfected human primary CD4 T lymphocytes.Representative fluorescent image of a human CD4 T cell transfected with β-actin-mCherry (gray pseudocolor), spreading over a stimulatory surface. (**A**) Red arrows represent the ROIs used to create the kymographs. (**B**) Kymographs of cell shown in (**A**) containing the lines and angle used for the speed calculation. The displacement of the actin signal (X axis) over the time (Y axis) is labeled. (**C**) shows the maximal projection of the time. (**D**) Graph representing the speed of the actin retrograde flow in nm/s. Dots represent individual cells analyzed ($n = 18$) and black lines the mean and the standard deviation (SD). Bar, 5 μm

to the slope of the line L (yellow line in Fig. 2B), which is the speed of a given particle of β-actin-mCherry.

7. Perform similar measurements 5–6 times for statistical calculation of the parameter in this kymograph.
8. Perform Steps 1–7 at least 3–4 times in different locations of the cell (*see* **Note 26**).
9. A summatory of all the frames acquired along the time can be obtained as in Fig. 2C (Image > Stacks > Z project > Sum slices).
10. Multiply the values of speed obtained by the size of the pixel in microns (if pixels are not square, use the corresponding value) and divide by the time corresponding to the frame interval of the image sequence (in general, 1 px will correspond to 30 ms for this frame interval between consecutive images). This information is compiled in the metadata accompanying the image sequence. In ImageJ, this information is available when opening the images for the first time, or by using the properties option inside the image menu (Image > Properties).
11. Now your data are expressed in normalized units, such as nm/s (Fig. 2D).
12. Pool data from different cells in different fields for each condition to be compared through statistical analysis.

### 3.7 Image Analysis of Microtubule Dynamics in EB3-GFP Transfected Human Primary CD4 T Lymphocytes

All the data analysis described in this section can be performed with the Image J freeware (*see* **Note 27**).

1. Open the file from TIRFm individually with Image J software.
2. Adjust the Brightness and Contrast (Image > Adjust > Brightness/Contrast).
3. Merge EB3-GFP and Brightfield channels (Image > Color > Make composite). An image composition as in Fig. 3A, B will be obtained.
4. A summatory of all the frames acquired along the time can be obtained as in Fig. 3C (Image > Stacks > Z project > Sum slices).
5. Open the Image J plugin "*TrackMate*" (Plugins > Tracking > TrackMate).
6. Usually, 30 s–2 min for tracking is sufficient to have reproducible results.
7. Select a Detector for accurately analyzing microtubule dynamics. For EB3-GFP labeled cells, use *LoG (Laplacian of Gaussian) Detector* (*see* **Note 28**).

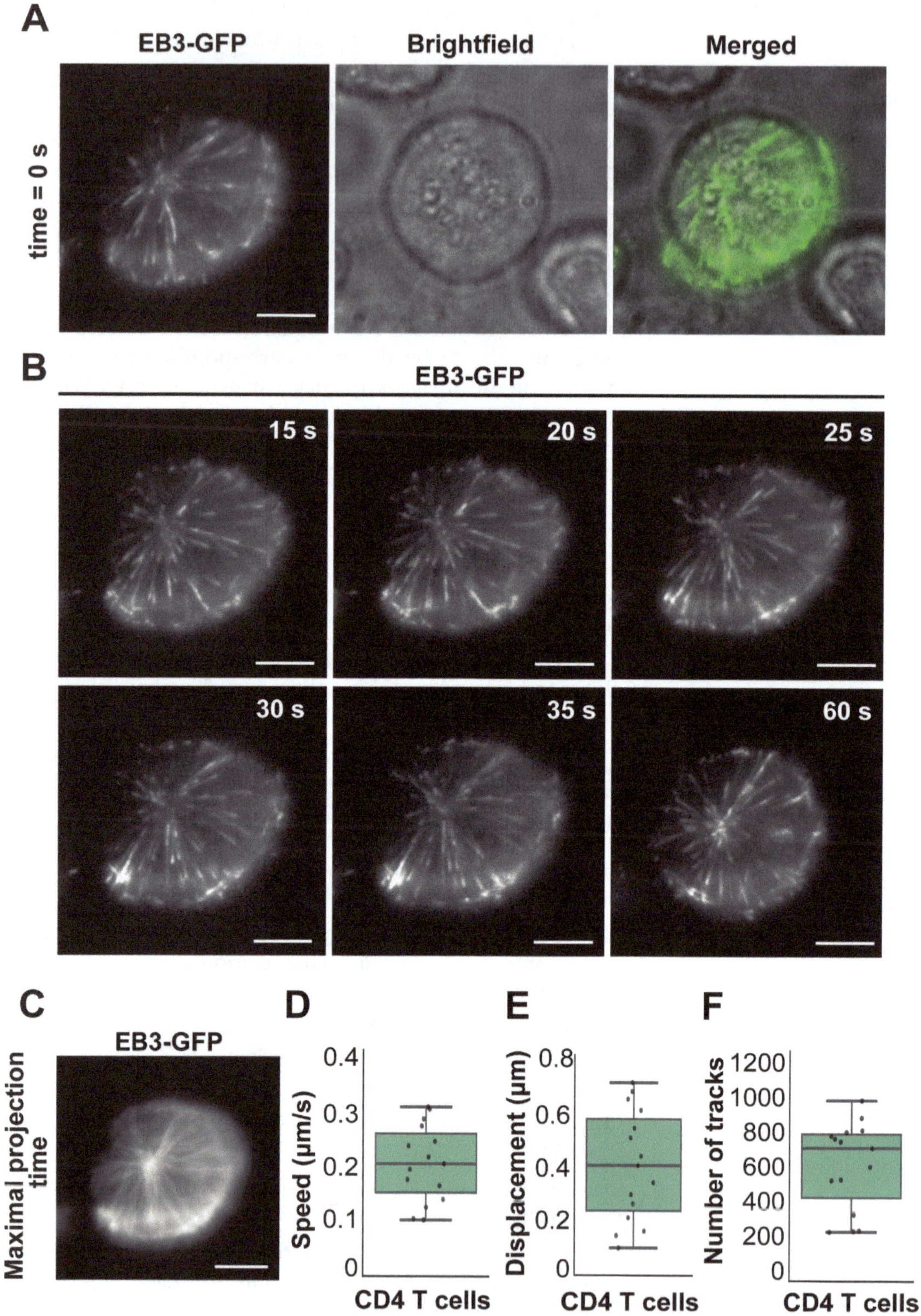


**Fig. 3** Microtubule growth and dynamics from TIRFm imaging in EB3-GFP transfected human primary CD4 T lymphocytes. (**A**) Fluorescence and brightfield images from time = 0 s of EB3-GFP (gray pseudocolor and green

8. Adjust the object diameter "*Estimated object diameter*" to detect the EB3-GFP comets (*see* **Note 29**), and "*quality threshold*" to exclude background noise (*see* **Note 30**).
9. Turn on "*Sub-pixellocalization*" option, to improve tip position accuracy and get more precise velocities.
10. Select a Tracker to detect microtubule plus-ends along the time. For EB3-GFP labeled cells, use *LAP tracker* (*see* **Note 31**).
11. Adjust LAP tracker parameters such as the "*linking max distance*," "*gap-closing max distance*," and "*gap-closing max frame gap*," considering the pixel size and time between frames used for image acquisition (*see* **Note 32**).
12. Calculate Spots, Edges, and Tracks to analyze the number of tracks, the displacement and the speed (Fig. 3D–F).
13. Export the results to.csv files.
14. Pool the data from different cells in different fields corresponding to the diverse conditions for statistical analysis.

### 3.8 (Optional) Post-Processing of TrackMate Outputs for EB3-GFP Microtubule Dynamics

To make the downstream analysis robust and fully reproducible, we provide an open-source Python pipeline that standardizes the. csv files exported by *TrackMate* plugin of Image J, aggregates per-cell metrics, and optionally performs non-parametric statistics, and produces publication-ready plots (Fig. 3D–F). Code and a detailed documentation specifying what each step computes is available in the repository instructions (https://github.com/MLozanoPrieto/trackmate-eb3-tirfm-analysis). Briefly, below we focus on how to run the pipeline to reproduce our analysis.

1. From each cell/video analyzed in Sect. 3.7, export *TrackMate* tables as comma-separated values and place the <CellBase>-tracks.csv and <CellBase>-edges.csv under a single input root organized as <ROOT>/<Experiment>/<Condition>/<Plate>/ (*see* **Notes 33** and **34**).
2. Clone the repository and install the listed Python dependencies (see the repository's README.md and requirements.txt). The pipeline is a single command-line program, trackmate_eb3_tirfm_analysis.py, which can be run from any working directory.

**Fig. 3 (Continued)**
in merged) expressing human primary CD4 T cell. (**B**) Fluorescence images from a time-lapse of an EB3-GFP (gray pseudocolor) expressing human primary CD4 T cell. Time interval is indicated in the figure. (**C**) Maximal projection of the time showing all the tracks analyzed in 3 min. (**D–F**) Graphs representing the (**D**) speed (**E**) displacement and (**F**) number of tracks of MT growth. Dots represent individual cells analyzed ($n = 15$) and black lines the mean and the standard deviation (SD). Bar, 5 μm

3. Execute the script with the input root (e.g., -i <ROOT> ).The program scans the directory tree, harmonizes variable CSV headers, and computes per-cell summaries from *-tracks.csv (e.g., speed, duration-based measures, linearity, and confinement descriptors) and per-track descriptors derived from *-edges.csv (medians/percentiles/IQR for speed and directional-change rate). Results are written to 1_METRICS/cell_metrics.csv and 1_METRICS/track_metrics_from_edges.csv (*see* **Note 35**).
4. Optionally, if exactly two conditions are present, rerun or invoke the script with the statistics flag (e.g., -s). The pipeline applies pooled Mann–Whitney *U* tests and a Van Elteren-type stratified analysis across experiments, followed by Benjamini–Hochberg FDR correction. Results are saved to 2_STATS/mannwhitney_all_metrics.csv, including effect sizes to aid interpretation.
5. Optionally, create publication-ready box-and-dot plots by condition and per-experiment figures with the plotting flag (e.g., –p). If statistics have been computed, the figures include significance brackets (uncorrected p and FDR-adjusted q) (Fig. 3D–F).
6. Upon completion, consult the time-stamped log file in logs, which records command-line arguments, parsing choices for CSV headers and encodings, and any warnings (e.g., empty or malformed files) (*see* **Note 36**).

## 4 Notes

1. Nucleofector is a specific electroporator from AMAXA with dedicated programs for different cell types, either primary or established cell lines. Other methods can be also used to transfect the cells to overexpress exogenous fluorescent proteins to track cytoskeleton dynamics through microscopy.
2. There are several commercially available media, such as Ficoll Histopaque or Biocoll.
3. The use of Opti-MEM I (GIBCO, Invitrogen) here allows excluding other commercial kits for transfection through nucleofection, which is a kind of electroporation. Dedicated 0.2 mL cuvettes are required, which are commercially available from different sources.
4. In our case, we use a Leica AM TIRF MC M system mounted on a Leica DMI 6000B fitted with a HCX PL APO 100×1.46 NA oil objective microscope, coupled to an Andor-DU8285 VP-4094 camera. This camera is specifically

good for detection of monomeric cherry fluorescent protein and allows 1004×1002 images (px) with frame lapse of 32/s.

5. 10 mm glass-bottom dishes can also be used for TIRFm imaging. The disadvantage of using this type of dish is that the laser of the TIRF microscope must be realigned each time the dish is changed, which increases the time required for acquisition and causes a loss of temperature when the doors are opened to change the plate. For this reason, glass-bottom-18-well chambers (e.g., iBidi) can be a better option for TIRFm imaging.
6. Each buffy coat preparation contains approximately 60 mL from 450 mL of complete blood, so 100 mL of saline solution would need to be added to dilute ~1:3 (v/v). Higher dilutions can be used at this stage, but this will increase the number of gradients required to process without great improvement in lymphocyte separation.
7. The first pellet obtained after the gradient separation of the mononuclear cells is not tightly adhered to the bottom of the tube and a high number of cells can be lost in this step.
8. A cocktail of antibodies combined with magnetic beads such as EasySep Human CD4$^+$ T Cell Isolation Kit from StemCell Technologies or the CD4 negative selection kit from Miltenyi Biotech is recommended to purify total CD4 T cells without promoting cell activation.
9. A maximum of 5 µL of DNA or RNA should be added to 100 µL of transfection medium. RNA concentration ranges 1–3 µM. Optimized range: 1.5–2.5 µL in 100 µL of Opti-MEM I or commercial Amaxa mix. To study actin and tubulin cytoskeletonsdynamics in the protocol described in this chapter, a plasmid encoding EB3-GFP and a plasmid that express low levels of β-actin-mCherry under the control of a speckle promoter, based on mutated CMV promoter, were used.
10. The cell viability is about 50% after the transfection procedure. To improve cell survival, plate all cells immediately after nucleofection in a single mix with warmed medium in a cell culture plate or flask. Do not keep cells in the nucleofection solution for more than 3–5 min, so this is critical to improve cell survival.
11. After nucleofection, T cells experience significant stress and viability loss. IL-7 is a critical cytokine for T cell survival and homeostasis, so it enhances its survival upregulating anti-apoptotic proteins (e.g., Bcl-2) but also supporting homeostatic proliferation without pushing them into exhaustion or terminal effector states, unlike IL-2, which is a signal for strong proliferation upon expression of its high-affinity receptor

CD25 and promotes cell differentiation and activation-induced cell death (AICD).

12. In some assays, human plasma fibronectin (10 μg/mL) can be used as an adhesive substrate. Fibronectin enhances T cell adhesion, through the interaction with $\alpha4\beta1$ (VLA-4, very late antigen-4) and $\alpha5\beta1$ (VLA-5, very late antigen-5) integrins. In these cases, a prior coating of poly-L-Lys can be also used. There is no need of coating with stimulatory antibodies if only these integrins are to be assayed (*see* **Note 13**).
13. Recombinant human ICAM-1 (1 μg/mL) can be included in the stimulatory surface cocktail to improve CD4 T cell-mediated adhesion through LFA-1 integrin. This protein is usually an Fc-fusion protein (fraction constant of immunoglobulins). Including ICAM-1 in the stimulatory surface cocktail enables studying different areas of actin cytoskeleton at the pseudo-IS than with TCR stimulation alone due to integrin adhesion. By implementing this, results similar to those obtained with the lipid bilayer model could be achieved [11].
14. Glass-bottom-18-well chambers or 10 mm glass-bottom dishes can be frozen at this step.
15. Since TIRFm is a super-resolution microscopy technique at the Z-axis, it is important that the temperature remains stable at 37°C to prevent thermal drift. Thermal drift in TIRF microscopy refers to small, unwanted movements of the sample or optical components caused by temperature changes over time. Even tiny fluctuations in temperature can lead to mechanical expansion or contraction of microscope internal components, resulting in image drift, a gradual shift in the observed field of view and loss of focus. In this chapter, images were acquired under a Leica AM TIRF MC M system mounted on a Leica DMI 6000B fitted with a HCX PL APO 100 × 1.46 NA oil objective microscope, coupled to an Andor-DU8285 VP-4094 camera.
16. Laser alignment in TIRF microscopes is required to ensure the formation of a precise evanescent field by adjusting the incident angle above the critical angle of the excitation beam at the glass–sample interface. Depending on the optical system, there are different procedures. In the case of the examples showed here, the alignment is performed by using the "automatic alignment" mode in the Leica accompanying software. The laser is directed through the back focal plane of a high magnification and high NA objective (usually a 100× objective, 1.4 NA or higher) and shifted laterally to achieve total internal reflection. Proper alignment confines excitation to a thin region near the coverslip, maximizing contrast and minimizing background fluorescence. Ideally, each laser

wavelength should be individually aligned to maintain uniform illumination and avoid chromatic mismatches, but this is not required in most experiments. Stable mechanical components and thermal equilibration are essential to preserve alignment accuracy during imaging.

17. Detection of fluorescence through TIRFm imaging is only possible very near the objective, where the emission derived from the evanescent field is detectable. The working distance range is between 70 and 300 nm. If the centrosome is not close to the glass surface, its fluorescence will not be detected. Indeed, the fluorescence from the EB3-enriched microtubules tips emerging are detected upon reaching the area occupied by the cortical cytoskeleton at the pseudo-IS. Therefore, the dynamics of polymerization of microtubules originating out of the lamella cannot be detected, such as those around the nucleus.
18. Cells can be sorted before imaging if the transfection efficiency is very low, in order to decrease the time used to find adequate cells for imaging and the microscope use. In addition, it is relevant to analyze cells with similar fluorescent protein expression, since the levels of over-expression of EB3 or actin can change the cytoskeleton dynamics.
19. To select the better laser incidence direction, it is necessary to consider the one with the best signal-to-noise ratio, while offering a homogeneous evanescent field.
20. In our hands, 2%–5% laser power is enough to visualize the probes and tags while avoiding phototoxicity and photobleaching. However, the conditions must be adjusted for each assay. Usually, the laser power required for visualization of fluorescent proteins indicates expression. In this set of experiments, although low levels of expression are required, most of the over-expressed proteins are localized at the immune synapse organized at the interface with the coverslip. This makes local concentration of fluorescent proteins high enough for good detection.
21. Frequent exposure times correspond to around 100–200 ms for EB3-GFP and 30–60 ms for β-actin-mCherry fluorescence.
22. Select the option "*hardware synchronized*" of Leica software to increase the acquisition speed, as well as dual filter for GFP and RFP, if you are using this microscope.
23. When recording simultaneously tubulin and actin dynamics, penetration depth of 150 nm is recommended together with 60 ms for time-lapse. If low signal for β-actin-mCherry is detected with a penetration depth of 90 nm, it can be changed to 150 nm, as this would increase the number of excited molecules, although it will decrease resolution in Z.

24. Acquire image sequences of similar size (equal number of pixels in regions of interest (ROIs)) and similar exposure times for different conditions.

25. Measuring the area of the distal IS, after fixing the cells at different times of interaction with the stimulatory surface, allows you to determine the spreading capacity of T lymphocytes after stimulation [12].

26. Optionally, the speed of retrograde actin flow can be calculated in a similar manner in the peripheral IS as in the distal IS, although other structures, such as actomyosin arches, are present in that region and must be considered.

27. Other software, such as IMARIS V.8.4 software (https://imaris.oxinst.com), can be used, following protocols described elsewhere [13].

28. The detector determines how your whole microtubule, or your microtubule ends or comets are identified in each frame. The best detector depends mainly on how your microtubules are labeled and imaged [e.g., EB1/EB3 microtubule plus-ends, fluorescently labeled tubulin, or fluorescently labeled Microtubule-Associated Proteins (MAPs)]. LoG (Laplacian of Gaussian) Detector is optimal for EB1/EB3 microtubule plus-end tracking, because EB1/EB3 comets are spot-like particles, so you can adjust the estimated blob diameter to match the size of your comet. If you are tracking whole labelled microtubules, such as Tubulin-GFP or Microtubule-Associated Proteins, such as Microtubule-associated protein-4 (MAP4)-GFP, you should use DoG (Difference of Gaussian) Detector, as it can better detect extended bright regions, unlike the LoG (Laplacian of Gaussian) Detector, which may not capture long or curved filaments, which is what happens when the entire filament is fluorescently labeled.

29. Usually 0.4–0.7 μm, with an average of 0.5 μm, is recommended for *estimated object diameter* value to properly detect the EB3 comets.

30. If your image is noisy or have a high background, you can apply a Gaussian blur filter before detection to help the LoG detector pick out real comets.

31. To analyze microtubule dynamics through microtubule plus-ends growing in EB3-GFP labeled cells, LAP tracker is optimal, since it is designed to link detections between consecutive frames optimally, and it can also detect comets that appear and disappear, since EB1/EB3 comets are moving spots that appear and vanish frequently. In this regard, LAP tracker can create new tracks when a new comet appears and end tracks when a comet disappears. This can be used to measure speed, growth lifetimes, and catastrophe frequencies.

32. For an acquisition of 300 ms per frame on a 100× objective and a pixel size of 60 nm/pixel, set: *Linking Max distance*: 0.5 μm; *Gap-closing Max frame gap*: 1–2 frames; *Gap-closing Max distance*: 1.0 μm. These settings will ensure the covering of very fast comets while keeping false links limited. If you detect many false links, lower linking distance or gap distance. If real tracks break often, increase gap frames or gap distance.
33. <CellBase>-spots.csv and <CellBase>-all-spots.csv files may be present in the folder but are ignored by the pipeline.
34. Ensure that *TrackMate* calibration (pixel size and frame interval) and output units are consistent across all cells and conditions. The pipeline trusts the numeric values written to CSV by *TrackMate*; it does not convert units.
35. To control the output location, provide an output base directory with -o <OUTPUT_BASE> to redirect the 1_METRICS/, 2_STATS/, and 3_PLOTS/ folders away from the input tree. This is useful for keeping raw exports separate from derived results.
36. Retain the log alongside the 1_METRICS/, 2_STATS/, and 3_PLOTS/ directories to ensure full reproducibility of the post-processing.

## Acknowledgments

This work was supported by grants from the Madrid Regional Government (S2022/BMD-7209-INTEGRAMUNE-CM) to NBMC, Obra Social Fundación la Caixa (LCF/PR/HR23/52430018) to NBMC, and grant funded by MCIN/AEI/10.13039/501100011033 (PID2022–141895OB-I00) to NBMC. AGM is supported by a PhD Fellowship from Madrid Regional Government (PIPF-2023/SAL-GL-30092). MLP is supported by a FPI fellowship (PRE2021-097478). We are thankful for the microscopy experimentation and analysis conducted at the Microscopy & Dynamic Imaging Unit, CNIC, ICTS-ReDib, cofunded by MCIN/AEI/10.13039/501100011033 and at the Videomicroscopy Facility of the IIS Princesa (Madrid, Spain), cofunded by IFEQ21/00085 and IFCS22/00014 from ISCIII and FEDER.

**Code Availability** The analysis pipeline is available on GitHub (https://github.com/MLozanoPrieto/trackmate-eb3-tirfm-analysis). Release v1.0.0 was used for the plots shown in this article.

## References


1. Monks CR, Freiberg BA, Kupfer H, Sciaky N, Kupfer A (1998) Three-dimensional segregation of supramolecular activation clusters in T cells. Nature 395(6697):82–86. https://doi.org/10.1038/25764
2. Grakoui A, Bromley SK, Sumen C, Davis MM, Shaw AS, Allen PM, Dustin ML (1999) The immunological synapse: a molecular machine controlling T cell activation. Science 285(5425):221–227. https://doi.org/10.1126/science.285.5425.221
3. Huse M (2025) Mechanoregulation of lymphocyte cytotoxicity. Nat Rev Immunol 25(9):680–695. https://doi.org/10.1038/s41577-025-01173-2
4. Dustin ML (2014) The immunological synapse. Cancer Immunol Res 2(11):1023–1033. https://doi.org/10.1158/2326-6066.CIR-14-0161
5. Martín-Cófreces NB, Baixauli F, Sánchez-Madrid F (2014) Immune synapse: conductor of orchestrated organelle movement. Trends in Cell Biol 24(1):61–72. https://doi.org/10.1016/j.tcb.2013.09.005
6. Gómez-Morón Á, González-Pinar A, Carrasco-Padilla C, Roda-Navarro P, Martín-Cófreces NB (2025) Exploring how cytoskeleton dynamics tunes T cell activation at the immunological synapse. European J Immunol 55(8):e70028. https://doi.org/10.1002/eji.70028
7. Balagopalan L, Sherman E, Barr VA, Samelson LE (2011) Imaging techniques for assaying lymphocyte activation in action. Nat Rev Immunol 11(1): 21–33. https://doi.org/10.1038/nri2903
8. Dustin ML, Depoil D (2011) New insights into the T cell synapse from single molecule techniques. Nat Rev Immunol 11(10):672–684. https://doi.org/10.1038/nri3066
9. Poulter NS, Pitkeathly WT, Smith PJ, Rappoport JZ (2015) The physical basis of total internal reflection fluorescence (TIRF) microscopy and its cellular applications. Methods Mol Biol 1251:1–23. https://doi.org/10.1007/978-1-4939-2080-8_1
10. Grigoriev I, Akhmanova A (2010) Microtubule dynamics at the cell cortex probed by TIRF microscopy. Methods Cell Biol 97:91–109. https://doi.org/10.1016/S0091-679X(10)97006-4
11. Dustin ML (2010) Insights into function of the immunological synapse from studies with supported planar bilayers. Curr Top Microbiol Immunol 340:1–24. https://doi.org/10.1007/978-3-642-03858-7_1
12. Blas-Rus N, Bustos-Morán E, Pérez de Castro I, de Cárcer G, Borroto A, Camafeita E, Jorge I, Vázquez J, Alarcón B, Malumbres M, Martín-Cófreces NB, Sánchez-Madrid F (2016) Aurora A drives early signalling and vesicle dynamics during T-cell activation. Nat Commun 7:11389. https://doi.org/10.1038/ncomms11389
13. Blas-Rus N, Bustos-Morán E, Sánchez-Madrid F, Martín-Cófreces NB (2017) Analysis of microtubules and microtubule-organizing center at the immune synapse. Methods Mol Biol 1584:31–49. https://doi.org/10.1007/978-1-4939-6881-7_3

Check for updates

# Chapter 10

# Experimental Tools and Techniques for Studying Microtubule Nucleation at the Centrosome and the Golgi Apparatus

Chiara Marcozzi, Maria P. Gavilán and Rosa M. Ríos

## Abstract

Microtubules are dynamic cytoskeletal filaments that organize intracellular space, facilitate vesicle and organelle transport, and are essential for cell division and polarity. They also contribute to signal transduction and the generation of mechanical forces. This wide range of functions depends on the ability of microtubules to form well-organized arrays. The spatial organization of these arrays is largely determined by specific subcellular sites—known as microtubule-organizing centers— where new microtubules are nucleated. The centrosome and the Golgi apparatus are the primary microtubule-organizing centers, and their microtubule nucleation activity is tightly regulated to fulfill cellular requirements. In this chapter, we present immunofluorescence techniques to study microtubule cytoskeleton, centrosome, and Golgi apparatus in both two- and three-dimensional cultures. We focus on fixation and pre-fixation extraction methods to enhance protein visualization and analysis. We also describe assays to investigate microtubule nucleation at the centrosome and the Golgi apparatus after microtubule depolymerization, along with methods to accurately quantify microtubules from each site. Finally, as a complementary live-cell imaging method, we explain how to visualize growing microtubules using fluorescent microtubule plus-end-binding proteins together with fluorescent markers for the centrosome or the Golgi apparatus. These approaches offer valuable tools to better understand microtubule organization and dynamics across different cellular contexts, providing insights into their roles in health and disease.

**Keywords** Microtubules, Centrosome, Golgi apparatus, Microtubule nucleation, Immunofluorescence, Live-cell imaging, Quantitative analysis

## 1 Introduction

Microtubules (MTs) are long, rigid polymers that form dynamic networks within the cell. These networks serve as tracks for intracellular transport and organelle positioning, and play essential roles in maintaining cell shape, establishing cell polarity, and enabling cell motility. During mitosis, MTs are indispensable for chromosome segregation, while in post-mitotic cells, they give rise to the axonemal structures

**Supplementary Information** The online version contains supplementary material available at https://doi.org/10.1007/978-1-0716-5515-3_10

Noa B. Martín-Cófreces et al. (eds.), *The Cytoskeleton*, Methods in Molecular Biology, vol. 3070,
https://doi.org/10.1007/978-1-0716-5515-3_10,

of cilia and flagella [1]. The arrangement of MT networks is mainly controlled by the locations where new MTs are formed, a process called MT nucleation. This process typically takes place at specialized structures globally known as MT-organizing centers (MTOCs [2]).

Traditionally, the centrosome has been considered the primary, if not the only, MTOC in animal cells [3]. However, over the past twenty years, the Golgi apparatus (GA) has emerged as a prominent alternative MTOC [4–6]. This new Golgi function has been reported across a wide range of organisms—from *Drosophila* to humans—and in diverse cell types, from undifferentiated progenitors to fully differentiated cells [2, 7]. MT nucleation at the centrosome occurs within the pericentriolar material (PCM), a dense protein matrix that surrounds the centrioles [8–10], whereas at the GA it is associated with the cis-Golgi cisternae [11].

To ensure that the MT cytoskeleton organization meets the structural and functional needs of each cell type throughout its lifespan, the nucleation activity of each MTOC is tightly regulated. In cycling cells, MT nucleation activities of the centrosome and of the GA exhibit inverse correlation throughout the cell cycle. During G1 and S phases, they actively nucleate MTs. At the G2/M transition, centrosomal nucleation markedly increases [12, 13], while GA-mediated nucleation sharply declines [14]. As cells progress into telophase, MT nucleation at the GA gradually resumes, whereas centrosomal nucleation activity substantially decreases.

During differentiation, as cells acquire tissue- or organ-specific functions, their MT networks reorganize according to defined differentiation programs. This remodeling—particularly prominent in cells that permanently exit the cell cycle—involves a reduction in centrosomal MTOC activity and a shift toward MT nucleation from the GA. For instance, neurons, striated muscle cells, and pancreatic β-cells retain minimal, if any, MTOC activity at the centrosome, and the GA assumes a predominant role as MTOC [15–17].

For decades, immunofluorescence (IF) alongside with biochemistry assays has been instrumental in advancing our understanding of MT organization and dynamics in cells, tissues, and organisms. More recently, the combination of IF and live-cell imaging with CRISPR/Cas9-based genome editing has enabled more in-depth analyses of the molecular machinery involved in the MT nucleation process. Key components of this mechanism such as the γ-TuRC, their recruiters (including Pericentrin, AKAP450, and CEP192/NEDD1), and cofactors or activators (such as ch-TOG, CDK5Rap2, and Myomegalin)—have been identified at the centrosome, at the GA, or at both structures, and have been shown to contribute to their MT nucleation activities [2, 18–23]. Despite these advances, many aspects remain poorly understood, particularly regarding the mechanisms underlying MT organization in specific cell types or during key biological processes like apoptosis and autophagy, and deserve further investigation.

This chapter provides a practical guide to widely used methods for labeling MTs, centrosomes, and the GA by IF in both two- (2D) and three-dimensional (3D) cultured cells. As an example of 3D models, we present MDCKII (Madin–Darby canine kidney type II) cyst grown on Matrigel, although our protocol can also be successfully applied to other 3D systems, such as Caco-2 (human colorectal adenocarcinoma) cysts and organoids derived from human colon crypts. This chapter also includes protocols for studying MT nucleation in both fixed and live cells. Centrosomal MT nucleation has been extensively studied across a wide range of cell types, while GA-associated MT nucleation is commonly investigated in hTERT–RPE1 cells (retinal pigment epithelial cells immortalized through the introduction of the human telomerase reverse transcriptase subunit; RPE1 hereafter). These non-transformed, nearly-diploid cells are particularly suitable for studying normal cellular processes and for gene editing. Other non-transformed cell lines, including MCF10A (breast epithelial cells), HDF (human dermal fibroblasts) or Nthy-ori 3-1 (normal thyroid epithelial cells) also exhibit strong GA-associated MT nucleation activity (our unpublished observations). In contrast, this activity is either absent or minimal in transformed cell lines such as HeLa or MCF7 (our unpublished observations). Finally, the chapter outlines strategies for the quantitative analysis of MT nucleation. Our goal is to offer protocols that serve both experts and, in particular, those at the beginning of their training in the field, such as students and postdocs.

## 2 Materials

### *2.1 Buffers and Reagents*

1. PBS: 137 mM NaCl, 2.7 mM KCl, 10 mM $Na_2HPO_4$, 1.76 mM $KH_2PO_4$, pH 7.4.
2. PHEM buffer: 25 mM HEPES, 60 mM PIPES, 10 mM EGTA, 2 mM $MgCl_2$, pH 6.8.
3. PBS-T: PBS containing 0.1% Tween-20.
4. PBS-T–BSA: PBS containing 0.1% Tween-20 and 3% Bovine Serum Albumin (BSA).
5. Dulbecco's Phosphate-Buffered Saline (DPBS) with calcium and magnesium.
6. Adhesion Metasilicate Solution (AMS): 0.64 μM disodium metasilicate, 0.013 μM sodium hexametaphosphate in ultrapure water. Leave on stirrer for several hours until the solution is completely clear, and filter through 0.22 μm filters.

### 2.2 Cell Culture

1. RPE1 cells are cultured in DMEM (Dulbecco's Modified Eagle Medium)/F12 supplemented with 10% Fetal Bovine Serum (FBS), 2 mM L-glutamine, penicillin (100 U/mL), streptomycin (100 μg/mL) (*see* **Note 1**).
2. MDCKII cells (canine epithelial-like cell line isolated from the distal renal tubule). MDCKII are cultured in MEM (Minimal Essential Medium) supplemented with 5% FBS, 2 mM L-glutamine, penicillin (100 U/mL), streptomycin (100 μg/mL) (*see* **Note 1**).
3. HEK293 cells (immortalized human embryonic kidney cells). HEK293 cells are cultured in DMEM (Dulbecco's Modified Eagle Medium) high-glucose medium supplemented with 10% FBS, 2 mM L-glutamine, penicillin (100 U/mL), and streptomycin (100 μg/mL (*see* **Note 1**).
4. Cells are maintained in a humidified incubator at 37 °C with 5% $CO_2$ and subcultivated at a 1:5 or 1:10 ratio (3 times a week).
5. 0.25% Trypsin-EDTA solution.
6. Forceps suitable for coverslip handling.
7. Coverslip mini-racks.

### 2.3 AMS-Treated Coverslips

Adherent cells for IF experiments are usually grown on glass coverslips, which provide a less supportive surface than plastic tissue culture dishes. To enhance cell adhesion and preserve cell morphology, we treat glass coverslips with AMS prior to seeding:

1. Pour 100 mL of AMS into a 500-mL glass beaker and bring to boil. To speed up this step, AMS can be pre-warmed in the microwave and subsequently put on a heated stirrer. When the solution reaches its boiling point, add 100 coverslips (we use 12-mm round coverslisps) and a magnetic stirrer to gently move the solution for 30 min. Cover the glass beaker with a loose-fitting lid.
2. While boiling, move coverslips inside the beaker every 10 min by circularly swirling the beaker using heat-resistant gloves. This will ensure that the buffer solution spreads equally on all coverslips and does not precipitate.
3. Following 30 min boiling, remove AMS and wash coverslips extensively (at least ten times) by swirling them in ultrapure water (*see* **Note 2**).
4. Boil coverslips in 100 mL ultrapure water for 30 min, swirling every 10 min as before.
5. Wash coverslips thoroughly again (at least ten times) with ultrapure water.

6. Remove water and leave coverslips in 100% ethanol. The protocol can be stopped at this point.
7. Carefully separate coverslips with forceps and let them air-dry on filter paper (*see* **Note 3**).
8. Place dried coverslips into a glass Petri dish.
9. Heat sterilize for 2 h at 200°C.
10. Store at room temperature (RT).

### 2.4 Immunofluorescence

1. Fixatives: prechilled (−20°C) ultrapure methanol or freshly prepared ultrapure 4% paraformaldehyde (methanol-free) in PBS.
2. Nocodazole.
3. Primary antibodies: Antibodies that we routinely use to stain the centrosome are anti-γ-tubulin (T6557, Sigma-Aldrich), anti-pericentrin (ab4448 or ab28144, Abcam), anti-CEP192 (A302-324A, Bethyl Laboratories), anti-ninein (A-R-H28, CurieCoreTech) anti-CAP350 ([24] or NB100-911, Novus Biologicals). For the cis-GA, we mainly use anti-GM130 (703794, Invitrogen or A-R-H53, CurieCoreTech, or 610822, BD Transduction Laboratories), anti-GMAP210 ([25] or 26456-1-AP, Proteintech), anti-golgin160 (21193-1-AP, Proteintech), or anti-giantin (A-R-H03, CurieCoreTech). For MTs, we use anti-α-tubulin (T6199, Sigma-Aldrich) and anti-EB1 (mouse, 610534, BD Transduction Laboratories).
4. Secondary antibodies: The secondary antibodies that we employ are purchased from Jackson ImmunoResearch Laboratories, Inc. and all are produced in donkey: Alexa Fluor 488 AffiniPure® Donkey IgG, Cy™3 AffiniPure® Donkey and Alexa Fluor 647 AffiniPure® Donkey IgG against the species of choice.
5. DAPI to stain nucleus.
6. Mounting media: N-propyl gallate or ProLong Gold-like.
7. Nail polish to seal mounted coverslips.

### 2.5 Plasmids for Lentivirus Production

When co-expression of two proteins of interest at similar levels is desired, we clone their cDNAs into a bi-cistronic lentiviral plasmid separated by a P2A cleavage site. The P2A sequence induces ribosomal "skipping" during translation, allowing approximate equimolar expression of the two distinct proteins. In addition, we clone cDNAs under the control of a doxycycline-inducible promoter that permits to finely tune expression levels and to prevent undesired effects caused by overexpression. Plasmids for lentivirus generation:

1. Bi-cistronic expression plasmid pCW57 (#71782, Addgene) as a backbone to clone the cDNAs encoding for proteins of interest.
2. Second-generation lentiviral packaging plasmid psPAX2 (#12260, Addgene).
3. VSV-G envelope expressing plasmid pMD2.G (# 12259, Addgene).

### 2.6 Lentivirus production

For lentivirus production, we use HEK293 cells, a suitable packaging host known for high transfection efficiency and for the ability to produce high-quality recombinant lentiviral particles. Preparation of lentiviral particle is as follows:

1. Coat a 10-cm dish with poly-D-lysine (2 μg/mL in PBS) and incubate for 1 h at 37°C in a humidified incubator.
2. Wash it twice with PBS and air-dry it in a laminar flow hood.
3. Seed $4 \times 10^6$ cells per 10 cm dish 18–24 h before transfection to obtain 70–80% cell confluency at the time of transfection.
4. 30 min before adding transfection reagents, replace the culture medium with complete fresh medium.
5. Co-transfect the expression plasmid (pCW57) containing the cDNAs coding for the proteins of interest, the packaging plasmid (psPAX2), and the envelope plasmid (pMD2.G) into HEK293 at a 3:2:1 ratio by plasmid mass using LipoD293 transfection reagent following manufacturer's instructions.
6. Move cells into a Biosafety Level 2 containment.
7. 72 h post-transfection, collect the cell culture medium where lentiviral particles will have been released.
8. To remove cell debris, centrifuge the collected medium at 2000 rpm for 20 min at 4°C.
9. Filter the supernatant through a 0.45-μm filter, then concentrate the viral particles by ultracentrifugation at 22,000 rpm for 1 h and 30 min at 4°C (e.g., using a Beckman Coulter ultracentrifuge).
10. Carefully remove supernatant without disturbing the viral pellet.
11. Resuspend the pellet in 200 μL PBS.
12. Prepare 20 μL aliquots and store them at −80°C for future use. Avoid multiple freeze–thaw cycles, as they reduce viral stability and infectivity.

### 2.7 Generation of RPE1-Derived Cell Lines Stably Expressing Proteins of Interest

Two common methods for ectopic protein expression are transfection and viral delivery. In RPE1 cells, however, low transfection efficiency [26] and high variability in expression levels among transfected cells limit the reliable assessment of physiological processes in vivo. For this reason, our recommended strategy is to generate stable cell lines expressing the protein(s) of interest through lentiviral transduction followed by antibiotic selection. In addition, the use of a lentiviral vector carrying a doxycycline-inducible promoter allows tighter control over expression levels. Although more laborious than direct transfection, this approach produces a higher proportion of RPE1 cells that inducibly express ectopic proteins. An outline of the workflow for generating these RPE1-derived cell lines is presented in Fig. 1.

1. On the day before infection, seed $2.5 \times 10^5$RPE1 cells into a single well of a six-well plate.
2. On the day of infection, move cells into a Biosafety level 2 laboratory, replace the culture medium with 1 mL of fresh complete medium, and add 20 μL of lentiviral preparation. To increase infection efficiency, polybrene can be added at a final concentration of 8 μg/mL. Polybrene neutralizes the negative charges of the cell surface and viral particles, facilitating closer contact between cell and virus, and enhancing viral absorption.
3. Incubate the cells in a humidified incubator at 37°C with 5% $CO_2$ to allow recovery and viral uptake.

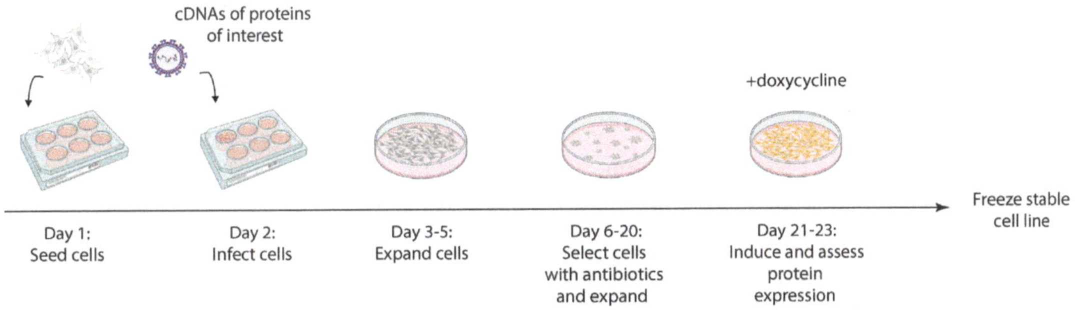


**Fig. 1** Workflow of stable cell line generation. Cells are seeded onto culture dishes and infected with lentivirus carrying the cDNA(s) of interest. After infection, cells are expanded to 70–80% confluency and subjected to antibiotic selection to enrich for stably transduced populations for approximately 10-14 days. Resistant colonies are pooled to generate a stable cell line, protein expression is induced with doxycycline, and infection efficiency and protein expression are assessed. After successful validation of protein expression, cells are cryopreserved in liquid nitrogen for future use. Schematic elements were designed using BioRender

4. 6h post-infection, replace the medium with 2mL of fresh complete medium.
5. Expand cells until they reach 70%–80% confluency.
6. Apply selection using an appropriate antibiotic to enrich for stably transduced cells.
7. When resistant colonies become visible, combine them to generate a stable cell line to use for subsequent assays.
8. Assess the absence of lentivirus in the cell culture medium using Lenti-X GoStix Plus. If no viral signal is detected, the cells may be safely returned to a Biosafety Level 1 laboratory.
9. Induce expression with doxycycline and assess infection efficiency and protein expression levels.
10. After successful validation of protein expression, freeze cells for long-term storage in liquid nitrogen.

### 2.8 Live Cell Imaging

1. RPE1 medium for time-lapse microscopy: DMEM/F12 without phenol red supplemented with 10% FBS, 2 mM L-glutamine, and penicillin (100 U/mL), streptomycin (100 μg/mL)or Leibovitz L-15 medium supplemented with 10% FBS, and penicillin (100 U/mL), streptomycin (100 μg/mL).
2. Time-lapse microscope slides.
3. Spinning Disk Confocal microscope.
4. Analysis software: ImageJ/Fiji (NIH). https://imagej.net/downloads; https://github.com/imagej/ImageJ

## 3 Methods

### 3.1 Visualization of Microtubule Cytoskeleton and Centrosomal and Golgi Apparatus Proteins by Immunofluorescence: 2D Cultured Cells

In this section, we will discuss our step-by-step procedure to perform IF staining, fixation options, and their applications, challenges, and solutions to enhance antibody labeling and signal detection. As mentioned before, we routinely use RPE1 cells although these protocols can be applied to most cultured cell lines. An example is shown in Fig. 2A.

1. Place treated coverslips onto a dish.
2. Add a small amount of medium to the coverslips, and firmly press coverslips down with a pipette tip to ensure good adherence of the coverslips to the dish. This will prevent cells from attaching to both coverslip surfaces.
3. Seed cells on the dish containing coverslips. Set up the IF experiment so that cells are at 70%–80% confluency on the day of the assay.

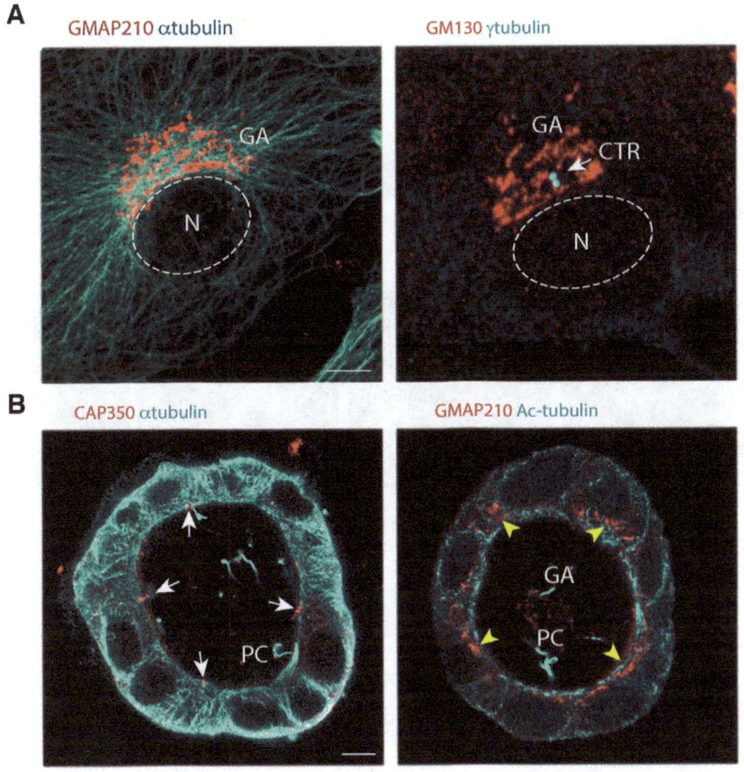


**Fig. 2** Immunofluorescence of 2D and 3D cell cultures. (**A**) Confocal images of RPE1 cells processed by IF following the protocol detailed in Sect. 3.1. On the left, cells were stained with antibodies against α-tubulin (cyan) and GMAP210 as a Golgi marker (red). On the right, cells were stained with γ-tubulin to label the centrosome (cyan) and GM130 (red) to mark the GA. Merged images are shown. Arrows indicate the centrosome (CTR). The white dashed line outlines the nucleus (N). Bar, 12 μm. (**B**) Confocal sections through the middle region of a representative MDCKII cyst. On the left, cysts were double-labeled for α-tubulin (cyan) and CAP350 (red) to mark MTs and the centrosome, respectively. On the right, cysts were stained with antibodies against acetylated tubulin (Ac-tubulin, cyan) and GMAP210 (red). Merged images are shown. Arrows point to the centrosome. Yellow arrowheads indicate Golgi elements. PC: primary cilium. Bar, 5 μm. These images illustrate the organization and distribution of MTs, GA, and centrosomes in both non-differentiated cells (**A**) and cells that have differentiated to form 3D structures (**B**)

4. To facilitate handling of glass coverslips prior to antibody incubation, it is recommended to use coverslip racks (mini-racks) that allow washing and fixation of multiple coverslips at the same time (as shown in Fig. 3).
5. Using forceps (preferably with pointed tips), take coverslips out of the dish and place them into the mini-rack.
6. Submerge the coverslip mini-rack into a container filled with pre-warmed PBS, drain any residual PBS by placing it on absorbent paper before introducing it into a container with the fixative solution of choice (*see* **Notes 4–6**).

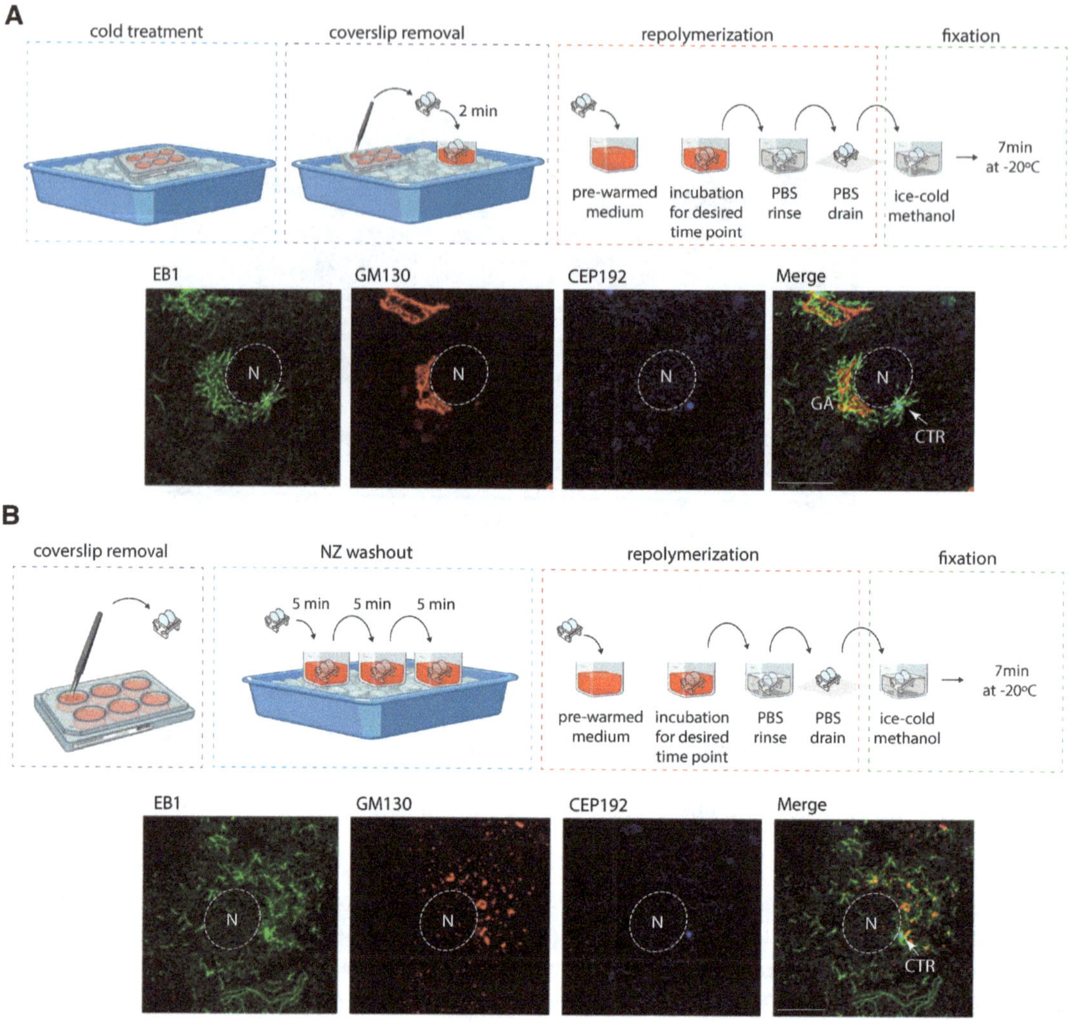


**Fig. 3** Analysis of MT nucleation after MT depolymerization. (**A**) Schematic representation of the MT nucleation assay following cold treatment. Cells are first incubated on ice for 1 h to depolymerize MTs. Coverslips are then placed on a mini-rack kept in a container with cold medium on ice. Next, the mini-rack is transferred to a container with pre-warmed medium at 30°C and incubated for the selected recovery time to allow MT regrowth. After incubation, the mini-rack is briefly rinsed in PBS, drained, and fixed in ice-cold methanol for 7 min at −20°C. Confocal images show MT nucleation regrowth after cold treatment. Cells were stained with EB1 (green), GM130 (red), and CEP192 (blue). Bar, 10 μm. Arrows indicate the centrosome (CTR). The white dashed line outlines the nucleus (N). (**B**) Stepwise illustration of the MT nucleation assay after NZ washout. Coverslips with NZ-treated cells are lifted and transferred into a mini-rack. To wash out NZ, mini-racks are sequentially incubated for 5 min in cold medium by transferring it through three separate containers of cold medium. After washing, the mini-rack is submerged in pre-warmed medium at 30°C for the selected time point, briefly rinsed in PBS, drained, and fixed in ice-cold methanol for 7 min at −20°C. Confocal images show MT regrowth from the centrosome and GA fragments in cells stained with EB1 (green), GM130 (red), and CEP192 (blue). Arrows indicate the centrosome (CTR). The white dashed line outlines the nucleus (N). Bar, 10 μ m. Schematic elements were designed using BioRender and OpenAI

7. If a previous step for permeabilization is required (*see* **Note 4**):
   - Prepare fresh PHEM buffer and PHEM buffer containing either 0.1% Triton X-100 or 0.1% saponin and 0.02% BSA.
   - Pre-warm buffers to 37°C.
   - Move coverslips into the mini-rack.
   - Rinse the mini-rack in PHEM buffer.
   - Incubate coverslips in PHEM buffer solution containing detergents for 40 s to 1 min.
   - Rinse the mini-rack in PHEM buffer.
   - Fix immediately using the desired fixation method.
8. If fixation with paraformaldehyde is performed (*see* **Note 5**):
   - Submerge the mini-rack containing coverslips into 4% PFA solution.
   - Incubate for 15 min at RT.
   - Rinse twice in PBS for 5 min each.
   - To allow antibody access to intracellular targets, permeabilize cells by incubating coverslips with 0.5% Triton X-100 in PBS for 10 min at RT.
   - Proceed with IF.
9. If fixation with methanol is performed (*see* **Note 6**):
   - Submerge coverslips in ice-cold methanol for 7 min at −20°C.
   - Rehydrate coverslips in PBS or PBS-T for 5–10 min.
   - Given that methanol also permeabilizes membranes, proceed directly with the IF blocking step.
10. After fixation, wash coverslips twice in PBS-T for 5 min each.
11. Add primary antibodies at working dilution prepared in PBS-T–BSA for 1 h at RT (Sect. 2.4).
12. Wash coverslips twice in PBS-T for 5 min each.
13. Add fluorescent dye-labeled secondary antibodies at predetermined working dilution in PBS-T–BSA, along with DAPI for nuclear staining at a final concentration of 0.1 μg/mL when needed (Sect. 2.4).
14. Incubate for 45 min at RT.
15. Wash coverslips twice in PBS-T for 5 min each.
16. Dehydrate coverslips in 100% ethanol for 2 min at RT. This step helps to preserve antibody labeling, reduce background, and stabilize samples for storage and imaging.

17. Air-dry coverslips on a filter paper with the cell surface facing up.
18. Prepare a glass slide with mounting medium compatible with the imaging technique of choice (one drop of mounting medium/coverslip). Gently place the coverslip on the drop with the cell surface facing down, avoiding air bubbles.
19. Remove extra mounting medium with a piece of paper and, if desired, seal the coverslips with nail polish.

### *3.2 Visualization of Microtubule Cytoskeleton and Centrosomal and Golgi Apparatus Proteins by Immunofluorescence: 3D Cell Cultures on Matrigel*

Epithelial cells can self-organize and polarize in vitro through cell–cell and cell–matrix interactions, forming 3D structures that recapitulate their in vivo organization within tissues. 3D cell culture models have become essential tools for studying complex biological processes such as epithelial morphogenesis, cell polarity, tissue self-organization, and organogenesis. A widely used approach for 3D cell cultures involves the use of basement membrane extracts derived from Engelbreth–Holm–Swarm (EHS) mouse sarcoma, such as Matrigel. Matrigel provides optimal conditions for establishing proper cell polarity, thereby promoting the formation of organized 3D structures.

The following protocol describes how to perform IF on MDCKII cells that, when grown on Matrigel, form hollow spheres known as cysts (see Fig. 2B). This system allows the analysis of basic cellular processes in a more physiologically relevant context than traditional 2D monolayer cultures. In addition to MDCKII cysts [24], the protocol is also suitable for other 3D cultured systems, such as Caco-2 cysts and organoids derived from human colon crypts grown in Matrigel.

1. For cyst formation, MDCKII cells growing in conventional tissue culture dishes are trypsinized, counted, and a single cell suspension at 1000 cells/mL is prepared in complete medium supplemented with 2% Matrigel (*see* **Note 7**).
2. 1 mL or 500 μL of cell suspension is plated on a four- or eight-well glass chamber slide pre-coated with 100% Matrigel. Select the most appropriate slide format depending on the number of conditions to assess.
3. The slides are kept in a humidified incubator at 37°C with 5% $CO_2$ for 5–7 days. Medium is replaced every 24–48 h with complete medium containing 2% Matrigel.
4. Unless otherwise specified, all subsequent IF staining steps are performed at RT with gentle agitation on an orbital shaker.
5. Carefully wash twice the chamber slides with DPBS containing calcium and magnesium for 5 min each to remove residual medium. Make sure that DPBS completely fills the chamber.
6. Fix cysts with 4% PFA in DPBS by gently swirling for 30 min. Adjust solution volume according to the size of the chamber

slides used: 500 μL is usually adequate for a four-well chamber slide and 250 μL for an eight-well chamber slide (*see* **Note 8**).

7. (Optional) To quench residual PFA, an additional incubation with 1% BSA and 0.1 M glycine in PBS for 30 min can be included after fixation.
8. Wash the cysts three times with DPBS for 5 min each.
9. Permeabilize PFA-fixed cysts with 0.5% Triton X-100 in DPBS for 15 min.
10. (Optional) Staining with some antibodies (e.g., γ-tubulin), may benefit from an additional 10 min incubation with 0.5% SDS in DPBS.
11. Wash cysts three times with DPBS for 5 min each.
12. Block non-specific binding sites by incubating with DPBS containing 0.025% saponin and 3% BSA (blocking solution) for 1 h.
13. Incubate samples with primary antibodies diluted in blocking solution overnight at 4°C (*see* **Note 9**). A volume of 125 and 250 μL of primary antibody solution should be enough for eight-well and four-well chambers, respectively. Ensure that Matrigel remains fully covered with antibody solutions while moving on the orbital shaker.
14. Wash the cysts three times with blocking buffer for 10 min each to remove unbound primary antibodies.
15. Incubate cysts with appropriate fluorophore-conjugated secondary antibodies prepared in blocking solution for 90 min at RT.
16. Wash the cysts three times with blocking buffer for 5 min each to remove unbound secondary antibodies.
17. Incubate with 1 μg/mL DAPI for 10 min when desired.
18. Extensively wash the samples with blocking buffer for 1 h, replacing the buffer with fresh solution every 20 min. A final wash with DPBS for 30 min is recommended.
19. Carefully lift the chamber wells upward to remove them from the slide.
20. Mount the resulting slide on a glass coverslip using ProLong Gold. Do not apply excessive pressure as this may damage the Matrigel.
21. Detailed imaging of MDCKII cysts requires the use of a Confocal Laser Scanning microscope. Select an appropriate Z-stack range and step size to capture the entire volume of the cyst.
22. Cysts with normal lumen formation can be identified by staining for apical markers such as actin and PAR3, which localize

to the lumen-facing side, and basolateral markers such as α -catenin and β-catenin, which are found at cell–cell junctions. Abnormal cysts typically show disorganized or multiple small lumens (*see* **Note 10**).

### 3.3 Microtubule Regrowth Assay After Ice-Induced Depolymerization

Several strategies can be used to assess the contribution of different organelles and cellular structures to MT nucleation. In fixed cells, these approaches typically rely on depolymerizing the MT network, then restoring conditions that permit MT regrowth. Careful analysis of the sites where new MTs assemble enables the identification of the structures involved in the nucleation process. MT depolymerization can be induced through various methods, each providing complementary insights into the mechanisms regulating MT nucleation.

Incubating cells on ice induces MT depolymerization [11, 27], likely due to conformational changes in tubulin dimers that reduce their ability to interact and assemble into stable structures at low temperatures [28]. Although cold treatment does not affect GA integrity or positioning, the pericentrosomal location of the GA makes it difficult to distinguish between centrosome- and Golgi-derived MTs. To overcome this limitation, it is advisable to analyze cells in which these organelles are spatially separated—a condition that naturally occurs during the S phase of the cell cycle [29]. Figure 3A depicts the experimental procedure and provides an example of the outcome.

1. Replace cell culture medium with pre-chilled complete medium.
2. Fill an ice bucket with ice.
3. Place cell culture dishes on ice and completely cover them with it.
4. Incubate for 1 h in order to ensure complete MT depolymerization. Shorter incubation times may not be sufficient to depolymerize a subpopulation of stable MTs (e.g., acetylated MTs) present in most cell types, and that typically colocalizes with the GA.
5. Prepare one container with cold medium and place it on ice. Prepare a second container with pre-warmed medium and keep it at RT.
6. Remove coverslips from the culture dishes while keeping them on ice, then place them into a mini-rack.
7. Submerge the mini-rack in a container filled with cold medium for 2 min.
8. Transfer the mini-rack to the container with medium at RT and incubate for various time points to monitor MT nucleation kinetics. Using mini-racks ensures that MT repolymerization begins simultaneously on all coverslips. For RPE1 cells, time points between 45 s and 3 min are optimal for visualizing

MT regrowth from both GA and the centrosome (Fig. 3A, top panel). MT repolymerization usually initiates earlier and more predominantly at the centrosome (*see* **Note 11**).

9. After each selected recovery time point, quickly rinse the coverslip mini-rack once in PBS and remove PBS excess by draining it on absorbent paper.
10. Transfer the mini-rack into a container with ice-cold methanol and incubate for 7 min at −20°C to fix cells.
11. Follow IF as outlined in Sect. 3.1. When analyzing MT repolymerization from both the centrosome and the GA, we recommend simultaneously labeling cells with antibodies against a cis-Golgi protein (e.g., GM130 or GMAP210), a PCM protein (e.g., γ-tub, pericentrin, CEP192) and either tubulin (α- or β-) or EB1/EB3 proteins, which bind to growing MT ends (Fig. 3A, bottom panel) (*see* **Notes 12–14**).

### 3.4 Microtubule Regrowth Assays After Nocodazole Treatment

Nocodazole, the most commonly used drug to depolymerize MTs, is a benzimidazole-based compound that binds to tubulin heterodimers, preventing their polymerization. Without new tubulin incorporation, MTs spontaneously depolymerize. NZ treatment also causes GA fragmentation into stacks of cisternae that disperse throughout the cytoplasm. Most of these stacks localize far from the centrosome, allowing a clear distinction between MT nucleation events occurring at the centrosome and those originating at the GA fragments [5, 11, 30]. Figure 3B illustrates the procedure and shows an example of the experiment outcome.

1. Replace culture medium with fresh medium containing 10 μM NZ.
2. Incubate cells at 37°C in a humidified $CO_2$ incubator for 2 h to induce complete MT depolymerization. Optionally, place culture dishes on ice for an additional 30 min to enhance depolymerization of highly stable MTs.
3. Prepare containers with cold medium and place them on ice. Three containers are required for each desired MT regrowth time point.
4. Wash cells in cold medium three times for 5 min each, by moving the mini-rack through the three containers with cold medium.
5. Transfer the mini-rack into a container with pre-warmed medium and keep it at RT for the selected time point. Optimal time points must be determined empirically for each cell type. In RPE1 cells, a regrowth time window between 2 min 30 s and 5 min generally allows clear visualization of distinct asters emerging from both the centrosome and the dispersed GA

fragments, before neighboring MT asters merge and hinder the assessment and quantification of MT nucleation.

6. Quickly rinse the coverslip mini-rack once in PBS and remove PBS excess by draining it on absorbent paper.
7. Fix in ice-cold methanol and proceed with IF protocol described in Sect. 3.1. As for ice-recovery experiments, we recommend to simultaneously label cells with antibodies against a cis-Golgi protein (e.g., GM130 or GMAP210), a PCM protein (e.g., γ-tub, pericentrin, CEP192) and either tubulin (α- or β-) or EB1/EB3 proteins, which bind to growing MT ends (Fig. 3B, bottom panel) (*see* **Notes 12–14**).

### 3.5 Quantitative Analysis of Microtubule Nucleation at the Centrosome

Quantification of MT nucleation at the centrosome is based on measuring EB1 intensity at the centrosome. A schematic representation of the procedure is shown in Fig. 4A.

1. Acquire fluorescent images for EB1 and a centrosomal marker such as α-tubulin or CEP192.
2. Export images in TIFF format and import them into ImageJ/Fiji (*see* **Notes 15** and **16**).
3. Split EB1 and centrosome marker images: Image > Color > Split Channels.
4. Open ROI Manager: Analyze > Tools > ROI Manager.
5. Using the centrosome marker image, draw a circular ROI around each centrosome using the oval ROI tool from ImageJ toolbar. Choose a regrowth time point at which MTs at the centrosome have just begun to form. Normally, we apply a 3-μm radius oval ROI (as shown in Fig. 4A).
6. Click 'Add' in the ROI Manager to save each selection.
7. Repeat for all centrosomes in the image.
8. In parallel, select the EB1channel image and apply Process > Subtract Background to reduce background noise (*see* **Note 17**).
9. Apply the previously saved ROIs to the background-corrected EB1 image.
10. Use the ROI Manager > Measure function to quantify the intensity of MT asters.
11. To ensure statistical robustness, analyze a sufficient number of images to obtain data for approximately 100 centrosomes per independent experiment.
12. Plot the results as box plots or scatter dot plots to visualize the distribution of EB1 intensities at centrosomes.

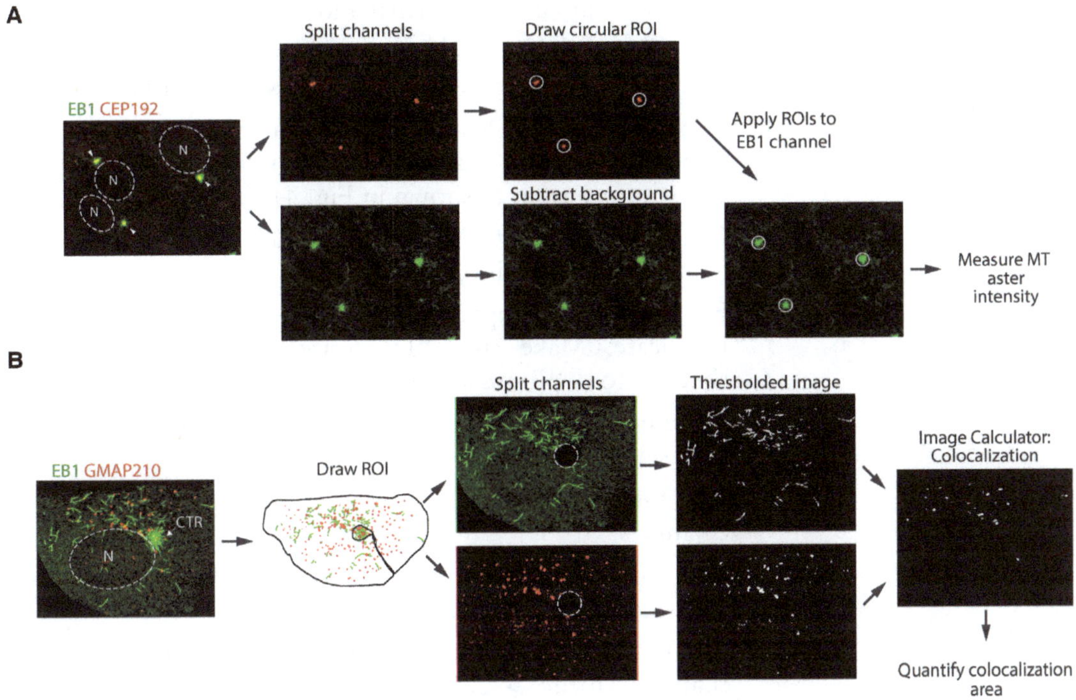


**Fig. 4** Quantitative analysis of MT nucleation at the centrosome and the GA. (**A**) Outline of the procedure used to quantify centrosomal MT nucleation after NZ washout in cells stained with CEP192 (red) to mark the centrosomes and EB1 (green) to label MT plus-ends. After splitting channels, circular ROIs are drawn around each centrosome in the CEP192 image and added to the ROI Manager. In parallel, background is subtracted from the EB1 image, and the saved ROIs are transferred from the ROI Manager to this image. Signal intensity corresponding to MT asters is then quantified using the Measure function. (**B**) Schematic representation of MT nucleation analysis from the GA. Cells were stained with GMAP210 (red) to label GA fragments and EB1 (green) to mark growing MT ends. A region of interest (ROI) is drawn to include the entire cell while excluding the centrosomes as indicated (dashed circle). Channels are then split, and background is subtracted before thresholding the images. Colocalization is calculated using the *Image Calculator* function, and the overlapping area is quantified. Arrows indicate the centrosome (CTR). The white dashed line outlines the nucleus

### 3.6 Quantitative Analysis of Microtubule Nucleation at the GA

MT nucleation at the GA is usually quantified by measuring the overlapping signals between growing MTs, detected by EB1 (or EB3), and NZ-induced Golgi elements stained with a cis-Golgi marker. Their colocalization serves as a proxy for newly formed MTs originating from the GA. Figure 4B provides a schematic overview of the procedure.

1. Acquire fluorescent images for EB1 and a cis-Golgi marker such as GM130 or GMAP210. Since the analysis relies on measuring overlap between Golgi fragments and MTs, using a medial- or trans-Golgi marker may underestimate colocalization values.

2. Export images in TIFF format and import them into ImageJ/ Fiji (*see* **Notes 15** and **16**).
3. Open ROI Manager: Analyze > Tools > ROI Manager.
4. Use the Polygon Selection Tool or Freehand Tool from ImageJ toolbar to outline individual cells and exclude the centrosomal aster, as shown in Fig. 4B (*see* **Note 18**).
5. Click Add in the ROI Manager for each selection, and repeat Steps 4 and 5 for all cells to analyze.
6. For each ROI, duplicate the image by selecting Image > Duplicate. Make sure "Duplicate Hyperstack" option is checked to duplicate all channels.
7. To remove centrosome signal, use Edit > Clear Outside for each selected ROI.
8. Split the channels to isolate each one: Image > Color > Split Channels.
9. Apply background subtraction to each channel: Process > Subtract Background (*see* **Note 17**).
10. Adjust the threshold to select the signal of interest for each channel: Image > Adjust > Threshold. Threshold values should be determined for each experiment dataset and consistently applied to all cells analyzed.
11. Convert each thresholded image to Binary Mask: Process > Binary > Convert to Mask.
12. Use "Image Calculator" to identify overlapping signals between EB1 and GA: Process > Image Calculator. Select EB1 mask as Image 1, choose the operation AND, and select GA mask as Image 2. This will generate a new image showing colocalization between the 2 masks.
13. To select colocalization signal go to Edit > Selection > Create.
14. Click "Add" in the ROI Manager to save the selection, then Measure to quantify the colocalization area.
15. Plot results as box plots or scatter dot plots to evaluate the variability of MT nucleation at the GA.
16. *(Optional)* To verify the accuracy of the colocalization measurement, merge the colocalization area image (from Step 12) with the original EB1 and GA channels. If images are not in the same bit depth format, convert them by selecting Image > Type > 8-bit, then merge with the colocalization image using Image > Color > Merge Channels (**Note 19**).

### 3.7 Live Cell Imaging to Assess Microtubule Nucleation In Vivo

MT dynamics can be monitored in vivo by time-lapse microscopy of fluorescently tagged EB1 or EB3 proteins. Beyond their association with MT tips, these proteins are also enriched at the centrosome, aiding its identification during live-cell imaging. It is worth reminding that overexpressed EB1 and EB3 tend to uniformly distribute along MTs and to stabilize them; thus, controlling their expression is crucial for this kind of experiment [31, 32]. Co-expression of GA or centrosome markers with EB1 or EB3 allows the simultaneous visualization of MT nucleation sites and the dynamic behavior of growing MTs.

As an example, we describe how to generate an inducible cell line stably expressing mRuby-EB3 and eGFP-GalT (β-1,4-galactosyltransferase) as a GA marker, how to validate and optimize their expression (see Fig. 5A for a scheme of the procedure), and how to perform live-cell imaging to study MT nucleation at the centrosome and the GA (Fig. 5B and Electronic Supplementary Movie 1). We have also generated an RPE1-derived cell line inducibly expressing EB1-eGFP and Centrin2-iRFP670 to monitor centrosomal MT nucleation activity (Fig. 5A). The same method and setup can be applied to co-express other Golgi or centrosome markers, provided that tagging does not affect the localization or function of the selected proteins.

Before setting up a time-lapse experiment to assess MT nucleation, it is important to establish optimal protein expression levels and to define imaging conditions, with the aim to achieve a high signal-to-noise ratio and to minimize photobleaching. These steps are essential for obtaining reliable, reproducible, and quantifiable data.

1. One day prior to imaging, seed $0.7 \times 10^5$ cells into each well of a four-well Ibidi slide.
2. Add doxycycline at a final concentration of 1 μg/mL to different wells at staggered time points (e.g., every 2 h), so that by the time of imaging, cells will have been induced for different durations.
3. Prior to filming, change the cell culture medium to pre-warmed complete DMEM/F12 without phenol red. If $CO_2$chamber is not available on the time-lapse microscope, complete Leibovitz L-15 medium, which is a $CO_2$-independent medium, can be used.
4. Examine the cells using a spinning disk confocal microscope to identify which doxycycline induction time provides optimal signal intensity while maintaining normal MT and GA dynamics. Low EB3 expression may hinder imaging due to signal weakness and rapid photobleaching, whereas excessively high levels can disrupt MT dynamics. Select cells with sufficient GalT fluorescence and, if possible, an extended Golgi morphology (*see* **Note 20**). Although expression levels will not be uniform even across cells at the same induction time, choose the condition where the majority of cells display adequate levels of both EB3 and GalT.

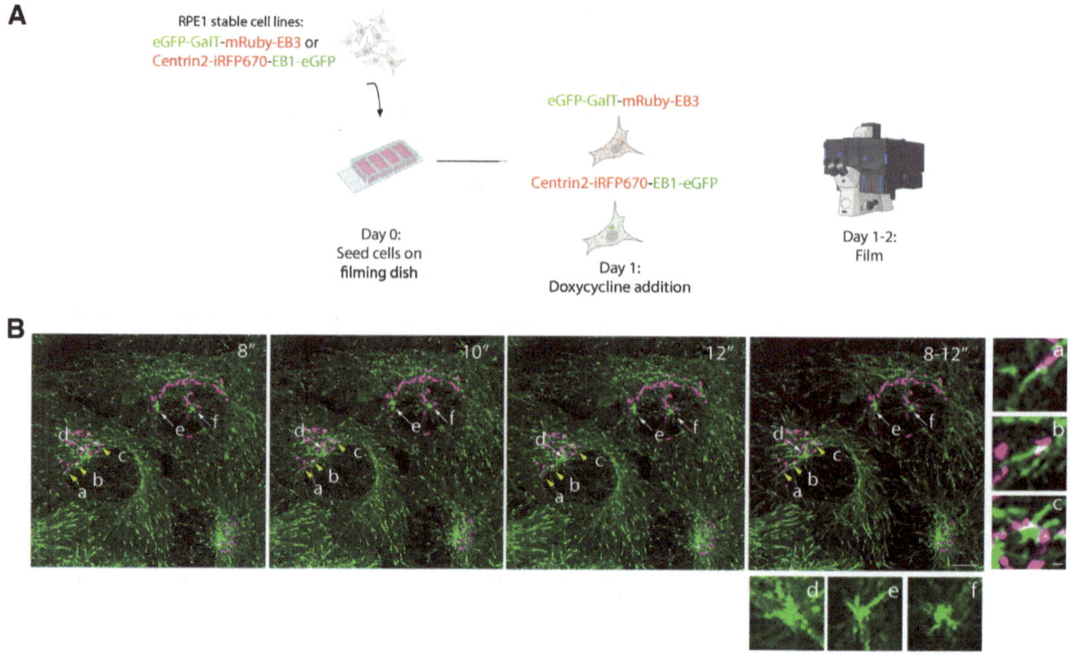


**Fig. 5** Workflow of MT nucleation analysis by live-cell imaging. (**A**) Cell lines stably expressing either eGFP-GalT-mRuby-EB3 or Centrin2-iRFP670-eGFP-EB1 are seeded onto dishes suitable for time-lapse microscopy. Protein expression is induced by doxycycline, and MT nucleation is subsequently monitored by time-lapse imaging using a spinning disk confocal microscope. Schematic elements were designed using BioRender. (**B**) RPE1 cells co-expressing eGFP-GalT and mRuby-EB3 were imaged every 2 s over a 1-min period (see Electronic Supplementary Movie 1). Z-stack projections at the indicated time points (8, 10, and 12 s panels) are shown. To better visualize the trajectories of growing MTs, a projection of EB3 signals over a three-frame interval was generated (8–12 s panel). Two representative cells are shown: on the left, a G1-phase cell in which both the centrosome and GA nucleate MTs; on the right, a G2/M-phase cell where duplicated centrosomes have separated to initiate spindle formation and actively generate new MTs, while the perinuclear fragmented GA retains residual MT-nucleating activity. Yellow arrowheads mark EB3 comets originating from the GA (see magnified panels a, b, and c on the right), and white arrows indicate EB3 comets emerging from the centrosome (see magnified panels d, e, and f at the bottom). For visualization purposes, the original fluorophore colors have been altered. Bar, 10 μm (main images); 2 μm (magnified insets)

5. Optimize imaging parameters on these cells—including laser power, exposure time, and binning settings—to capture images every 2 s over 1 min period. Set the number of Z-stacks and step size according to the signal distribution and imaging depth required to capture the structure of interest (*see* **Note 21**).
6. Repeat the image at least 15–20 cells per condition.
7. Export all acquired image sequences in .TIFF format and import them into ImageJ/Fiji (Fig. 5B and Electronic Supplementary Movie 1) to examine MT regrowth events.

## 4 Notes

1. RPE1 cells (ATCC CRL-4000) and HEK293 cells (ATCC CRL-1573) were obtained from the ATCC (American Type Culture Collection) collection. MDCKII cells (ECACC 00062107) were obtained from the ECACC (European Collection of Authenticated Cell Cultures).
2. When preparing large batches of coverslips, Steps a–c of the protocol should be performed with no more than 100 coverslips at a time, repeating the process as needed. After Step c, the treated coverslips can be transferred into a bottle containing ultrapure water. Subsequent batches can be added to the same bottle. Keep the bottle gently rotating on a roller while preparing the remaining coverslips. Once all coverslips have been washed, remove them from the bottle and proceed to Step d. It is not necessary to scale the water volume proportionally to the number of coverslips, but ensure that they are fully submerged.
3. Although stirring AMS while boiling helps prevent it from settling on the coverslips, a brown residue may still appear on some coverslips after air-drying. Each coverslip should be screened individually, and any with visible stains or spots should be discarded, as these may interfere with antibody staining and increase background noise.
4. In certain cases, IF staining can benefit of an extraction step prior to fixation using detergents to remove soluble cytoplasmic pool of the protein to examine, thereby facilitating its detection when associated with subcellular structures. When targeting cytoskeletal or centrosomal proteins, permeabilization with Triton X-100 is the most widely employed method. When preserving intracellular membranes is necessary, saponin should be the detergent of choice. Saponin targets cholesterol and specifically permeabilizes the plasma membrane in a concentration-dependent manner.
5. The overall aim of the fixation process is to preserve samples in a defined state at a specific time point, minimizing environmental alterations and sample degradation, while maintaining cellular architecture and the spatial relationships between cellular components. The choice of fixative depends on the structure to be examined, the type of assay being performed, and antibody requirements or compatibility. In some cases, fixation can be preceded by an extraction step (*see* **Note 4**). PFA is a cross-linker fixative that binds cellular contents and forms intra- and inter-molecular covalent links, thus

stabilizing cell morphology and architecture. It is widely used for visualizing membrane proteins and the actin cytoskeleton.

6. Organic solvents, such as methanol, are denaturating fixatives that work by displacing water and precipitating cellular components. Methanol is commonly used when staining MT cytoskeleton, the centrosome, and some Golgi-associated proteins, such as the Golgin family proteins.
7. For cyst formation, MDCKII cells are cultured on Matrigel (354230, Corning, Life Sciences) using four- or eight-well glass chamber slides (PEZGS0416, PEZGS0816, Millipore).
8. Please note that cysts should not be fixed in methanol, as they will detach from the Matrigel. Structures such as MTs, centrosome and even primary cilium can be clearly visualized by fixing in PFA and following the instructions outlined in this protocol.
9. Most of the antibodies indicated in Sect. 2.4 can also be used to stain the centrosome, the GA and MTs in PFA-fixed 3D cysts: anti-γ-tubulin (T6557, Sigma-Aldrich), anti-pericentrin (ab4448, Abcam), or anti-CAP350 [24] to stain the centrosome; anti-GMAP210 [25] for the cis-GA; anti-α-tubulin (T6199, Sigma-Aldrich); or anti-acetylated-tubulin (T7451, Sigma-Aldrich) for the MTs. To monitor proper cyst formation, we use antibodies for either the apical membrane, such as ZO-1 (40-2200, Invitrogen) or PAR3 (07-330, Millipore) or the basolateral membrane including α-catenin (C2081, Sigma-Aldrich) or β-catenin (C2206, Sigma-Aldrich). Labeling of F-actin with TRITC-Phalloidin (P1951, Sigma-Aldrich) allows simultaneous visualization of both apical membrane and cell adhesions.
10. Automated image analysis methods offer significant advantages for the analysis of 3D epithelial cysts. Deep learning algorithms such as CartoCell [33] are capable to accurately segment hundreds of cysts with minimal human intervention, effectively removing the bottleneck associated with labor-intensive manual data curation.
11. In our hands, pre-warming the medium to 30°C and then performing the MT polymerization step at RT represents the most practical experimental setup. Since recovery times are short, the medium temperature remains approximately stable throughout the experiment. Ice recovery experiments may also be performed incubating with medium pre-warmed to 37°C. However, it should be noted that under this condition recovery times range between 10 and 30 s. When establishing the experimental conditions, it is important to verify that MTs are completely depolymerized after cold treatment. To this end, it is recommended to fix a coverslip prior to the

regrowth step (time 0) and stain with the same antibody mixture used to assess MT nucleation.

12. At short repolymerization times, we do not observe substantial differences between α-tubulin or EB1/EB3 stainings. However, since the cytosolic pool of EB1/EB3 is less abundant than that of α-tubulin, growing MTs labeled with EB1/EB3 are easier to distinguish due to lower background noise. Alternatively, performing a pre-fixation extraction step—as described in **Note 4**— greatly reduces background from soluble tubulin and improves the signal-to-noise ratio.
13. Several combinations of commercially available primary antibodies are suitable for these experiments. Examples include:

a. anti-EB1 (mouse, 610534, BD Transduction Laboratories), anti-GM130 (rabbit, 703794, Invitrogen), and anti-ninein (human, A-R-H28, CurieCoreTech);
b. anti-EB1 (mouse, 610534, BD Transduction Laboratories), anti-CEP192 (rabbit, A302-324A, Bethyl Laboratories), and anti-GM130 (human, A-R-H53, CurieCoreTech);
c. anti-α-tubulin (mouse, T6199, Sigma-Aldrich), anti-pericentrin (rabbit, ab4448, Abcam), and anti-GM130 (human, A-R-H53, CurieCoreTech).

14. The ideal setup to minimize variability in assessing MT nucleation would be to plate cells from different experimental conditions on the same coverslip, enabling direct comparison. However, this is rarely feasible in practice. In our experience, such a setup is only possible when the conditions being compared can be reliably distinguished using an appropriate marker—for example, when comparing siRNA-treated versus untreated cells, knockout versus wild-type cells, or cells expressing a tagged protein versus untransfected ones. In these cases, a reliable proxy can be a specific antibody staining that allows clear identification of each condition—for instance, by detecting protein loss in siRNA-treated or knockout cells compared to wild type, or the presence of a tagged protein in transfected versus untransfected cells. However, a major pitfall of this approach is that triple labeling (as described in **Note 13**) is no longer possible. As a result, MT nucleation at the GA and the centrosome must be assessed separately.
15. Images for analysis can be exported as either 16-bit or 8-bit files, each with specific considerations. Sixteen-bit images are more suitable for quantitative measurements, such as mean fluorescence intensity, as they preserve subtle signal variations and enable more accurate thresholding and segmentation. In contrast, 8-bit images may still be appropriate for quantitative

analysis when the detected signal is clear and well-defined and are often adequate for qualitative assessments of parameters such as area, shape, and distance.

16. Our methodology is both simple and accessible to research laboratories, as it relies entirely on a free and open-source software (ImageJ/Fiji). Other commercial platforms such as Imaris (Bitplane) and MetaMorph (Molecular Devices, LLC) provide built-in colocalization plugins suitable for this analysis [21] and can be used.
17. Choose a rolling ball radius value to subtract background ensuring that no signal from structures of interest is removed.
18. NZ treatment may induce centriole separation, therefore when drawing an ROI to delineate cells with distant centrioles, both centriole asters should be excluded.
19. This step allows to verify that the majority of the colocalized signal has been accurately assigned to EB1 and the GA by the applied colocalization threshold.
20. It is important to select cells with appropriate and consistent ectopic protein levels to ensure reliable analysis across different experimental conditions.
21. To ensure accurate temporal resolution of EB1/EB3 dynamics, we acquire a single Z-stack of the GA using the GalT channel at the beginning of filming, assuming minimal or no GA variation over the 1-min imaging period.

## Acknowledgments

Work in Ríos lab is supported by FEDER/Ministerio de Ciencia e Innovación–Agencia Estatal de Investigación (PGC2018-095057-B-I00; PID2022-141680NB-I00) and Consejo Superior de Investigaciones Científicas (CSIC PIE Ref 202080E095).

## References

1. Alberts B et al (2022) Molecular biology of the cell, 7th edn. W. W. Norton & Company
2. Akhmanova A, Kapitein LC (2022) Mechanisms of microtubule organization in differentiated animal cells. Nat Rev Mol Cell Biol 23(8):541–558. https://doi.org/10.1038/s41580-022-00473-y
3. Bornens M (2021) Centrosome organization and functions. Curr Opin Struct Biol 66:199–206. https://doi.org/10.1016/j.sbi.2020.11.002
4. Chabin-Brion K, Marceiller J, Perez F, Settegrana C, Drechou A, Durand G, Pous C (2001) The Golgi complex is a microtubule-organizing organelle. Mol Biol Cell 12(7):2047–2060. https://doi.org/10.1091/mbc.12.7.2047
5. Efimov A, Kharitonov A, Efimova N, Loncarek J, Miller PM, Andreyeva N, Gleeson P, Galjart N, Maia AR, McLeod IX, Yates JR 3rd, Maiato H, Khodjakov A, Akhmanova A, Kaverina I (2007) Asymmetric CLASP-dependent

nucleation of noncentrosomal microtubules at the trans-Golgi network. Dev Cell 12(6):917–930. https://doi.org/10.1016/j.devcel.2007.04.002

6. Rios RM (2014) The centrosome-Golgi apparatus nexus. Philos Trans R Soc Lond B Biol Sci 369(1650):20130452. https://doi.org/10.1098/rstb.2013.0462
7. Sallee MD, Feldman JL (2021) Microtubule organization across cell types and states. Curr Biol 31(10):R506–R511. https://doi.org/10.1016/j.cub.2021.01.042
8. Lawo S, Hasegan M, Gupta GD, Pelletier L (2012) Subdiffraction imaging of centrosomes reveals higher-order organizational features of pericentriolar material. Nat Cell Biol 14(11):1148–1158. https://doi.org/10.1038/ncb2591
9. Woodruff JB, Wueseke O, Hyman AA (2014) Pericentriolar material structure and dynamics. Philos Trans R Soc Lond B Biol Sci 369(1650). https://doi.org/10.1098/rstb.2013.0459
10. Bornens M (2002) Centrosome composition and microtubule anchoring mechanisms. Curr Opin Cell Biol 14(1):25–34. https://doi.org/10.1016/s0955-0674(01)00290-3
11. Rivero S, Cardenas J, Bornens M, Rios RM (2009) Microtubule nucleation at the cis-side of the Golgi apparatus requires AKAP450 and GM130. EMBO J 28(8):1016–28. https://doi.org/10.1038/emboj.2009.47
12. Lee KS, Park JE, Ahn JI, Zeng Y (2021) Constructing PCM with architecturally distinct higher-order assemblies. Curr Opin Struct Biol 66:66–73. https://doi.org/10.1016/j.sbi.2020.09.013
13. Vasquez-Limeta A, Loncarek J (2021) Human centrosome organization and function in interphase and mitosis. Semin Cell Dev Biol 117:30–41. https://doi.org/10.1016/j.semcdb.2021.03.020
14. Maia AR, Zhu X, Miller P, Gu G, Maiato H, Kaverina I (2013) Modulation of Golgi-associated microtubule nucleation throughout the cell cycle. Cytoskeleton (Hoboken) 70(1):32–43. https://doi.org/10.1002/cm.21079
15. Kapitein LC, Hoogenraad CC (2015) Building the neuronal microtubule cytoskeleton. Neuron 87(3):492–506. https://doi.org/10.1016/j.neuron.2015.05.046
16. Lucas L, Cooper TA (2023) Insights into cell-specific functions of microtubules in skeletal muscle development and homeostasis. Int J Mol Sci 24(3):2903. https://doi.org/10.3390/ijms24032903
17. Bracey KM, Gu G, Kaverina I (2022) Microtubules in pancreatic beta Cells: convoluted roadways toward precision. Front Cell Dev Biol 10:915206. https://doi.org/10.3389/fcell.2022.915206
18. Roostalu J, Surrey T (2017) Microtubule nucleation: beyond the template. Nat Rev Mol Cell Biol 18(11):702–710. https://doi.org/10.1038/nrm.2017.75
19. Liu P, Wurtz M, Zupa E, Pfeffer S, Schiebel E (2021) Microtubule nucleation: the waltz between gamma-tubulin ring complex and associated proteins. Curr Opin Cell Biol 68:124–131. https://doi.org/10.1016/j.ceb.2020.10.004
20. Wu J, de Heus C, Liu Q, Bouchet BP, Noordstra I, Jiang K, Hua S, Martin M, Yang C, Grigoriev I, Katrukha EA, Altelaar AFM, Hoogenraad CC, Qi RZ, Klumperman J, Akhmanova A (2016) Molecular pathway of microtubule organization at the Golgi apparatus. Dev Cell 39(1):44–60. https://doi.org/10.1016/j.devcel.2016.08.009
21. Gavilan MP, Gandolfo P, Balestra FR, Arias F, Bornens M, Rios RM (2018) The dual role of the centrosome in organizing the microtubule network in interphase. EMBO Rep 19(11):e45942. https://doi.org/10.15252/embr.201845942
22. Choi YK, Liu P, Sze SK, Dai C, Qi RZ (2010) CDK5RAP2 stimulates microtubule nucleation by the gamma-tubulin ring complex. J Cell Biol 191(6):1089–1095. https://doi.org/10.1083/jcb.201007030
23. Ali A, Vineethakumari C, Lacasa C, Luders J (2023) Microtubule nucleation and gammaTuRC centrosome localization in interphase cells require ch-TOG. Nat Commun 14(1):289. https://doi.org/10.1038/s41467-023-35955-w
24. Gavilan MP, Arjona M, Zurbano A, Formstecher E, Martinez-Morales JR, Bornens M, Rios RM (2015) Alpha-catenin-dependent recruitment of the centrosomal protein CAP350 to adherens junctions allows epithelial cells to acquire a columnar shape. PLoS Biol 13(3):e1002087. https://doi.org/10.1371/journal.pbio.1002087
25. Infante C, Ramos-Morales F, Fedriani C, Bornens M, Rios RM (1999) GMAP-210, A cis-Golgi network-associated protein, is a minus end microtubule-binding protein. J Cell Biol 145(1):83–98. https://doi.org/10.1083/jcb.145.1.83
26. Ghetti S, Burigotto M, Mattivi A, Magnani G, Casini A, Bianchi A, Cereseto A, Fava LL (2021) CRISPR/Cas9 ribonucleoprotein-mediated knockin generation in hTERT-RPE1 cells. STAR Protoc 2(2):100407. https://doi.org/10.1016/j.xpro.2021.100407
27. Grimaldi AD, Fomicheva M, Kaverina I (2013) Ice recovery assay for detection of

Golgi-derived microtubules. Methods Cell Biol 118:401–415. https://doi.org/10.1016/B978-0-12-417164-0.00024-0
28. Li G, Moore JK (2020) Microtubule dynamics at low temperature: evidence that tubulin recycling limits assembly. Mol Biol Cell 31(11):1154–1166. https://doi.org/10.1091/mbc.E19-11-0634
29. Frye K, Renda F, Fomicheva M, Zhu X, Gong L, Khodjakov A, Kaverina I (2020) Cell cycle-dependent dynamics of the Golgi-Centrosome Association in motile cells. Cells 9(5):1069. https://doi.org/10.3390/cells9051069
30. Hurtado L, Caballero C, Gavilan MP, Cardenas J, Bornens M, Rios RM (2011) Disconnecting the Golgi ribbon from the centrosome prevents directional cell migration and ciliogenesis. J Cell Biol 193(5):917–933. https://doi.org/10.1083/jcb.201011014
31. Thoppil RJ, Sanders A, Kaverina I (2020) Detection of microtubule nucleation hotspots at the Golgi. Methods Mol Biol 2101:179–189. https://doi.org/10.1007/978-1-0716-0219-5_12
32. Ezquerra A, Viais R, Luders J (2020) Assaying microtubule nucleation. Methods Mol Biol 2101:163–178. https://doi.org/10.1007/978-1-0716-0219-5_11
33. Andres-San Roman JA, Gordillo-Vazquez C, Franco-Barranco D, Morato L, Fernandez-Espartero CH, Baonza G, Tagua A, Vicente-Munuera P, Palacios AM, Gavilan MP, Martin-Belmonte F, Annese V, Gomez-Galvez P, Arganda-Carreras I, Escudero LM (2023) CartoCell, a high-content pipeline for 3D image analysis, unveils cell morphology patterns in epithelia. Cell Rep Methods 3(10):100597. https://doi.org/10.1016/j.crmeth.2023.100597

Chapter 11

# Study of Unconventional Cytoskeleton Components Regulating Actin Contraction

**Jing Zhang, Weiwei Qin, Che Hu, Wei Miao and Jie Xiong**

## Abstract

The cytoskeleton, as the internal structural framework of the cell, is not only critical for maintaining cell shape but also plays a central role in regulating cell motility and other various cellular functions. Recent studies have shown that certain organisms rely on unconventional cytoskeletal components to carry out their unique biological processes. For example, studies on the single-celled eukaryotic predator *Lacrymaria* have revealed that this organism employs unconventional cytoskeletal structures to drive the rapid contraction of its "neck". In this study, we developed an integrated approach combining mass spectrometry (MS), immunofluorescence (IF), super-resolution imaging, subcellular structural analysis, and three-dimensional (3D) reconstruction. This methodology is not only well-suited for investigating *Lacrymaria* but also broadly applicable for exploring the presence and biological functions of unconventional cytoskeletal components in other eukaryotic systems.

**Keywords** Unconventional cytoskeleton, *Lacrymaria*, Mass spectrometry, Immunofluorescence, Super-resolution imaging, 3D reconstruction

## 1 Introduction

The cytoskeleton is a critical intracellular structure that maintains cell shape, supports cell motility, and facilitates organelle transport. Classically, it has been thought to consist primarily of microtubules, actin filaments, and intermediate filaments. However, recent studies have increasingly uncovered the complexity and diversity of the cytoskeletal system. In particular, the discovery of unconventional cytoskeletal components has provided new insights into the dynamic regulation and functional versatility of the cytoskeleton [1–4]. These unconventional cytoskeletal components include, but are not limited to, centrin, novel microtubule-associated giant proteins, and *Plasmodium*-like unconventional actin. They play essential roles in maintaining cellular function and regulating fundamental biological processes, thereby ensuring organismal survival and reproduction.

Noa B. Martín-Cófreces et al. (eds.), *The Cytoskeleton*, Methods in Molecular Biology, vol. 3070,
https://doi.org/10.1007/978-1-0716-5515-3_11,

To investigate unconventional cytoskeletal components, we selected the single-celled eukaryote *Lacrymaria*, known for its extraordinary dynamic shape-shifting, as the study subject. *Lacrymaria* is a predatory ciliate commonly found in freshwater environments such as lakes and ponds. Morphologically, the *Lacrymaria* cell can be artificially divided into three parts: the teardrop-shaped cell body, the elongated "neck" and the "head" (oral apparatus) from posterior to anterior [5, 6]. When active, *Lacrymaria* repeatedly extends and contracts its neck in search of prey [7–9]. In the extension-contraction cycle, the neck can extend up to 30 times the length of the body before retracting, enabling the cell to reorient its search direction with each cycle [5, 8, 10].

To capture the dynamic behavior of *Lacrymaria*, DeepLabCut [11], an AI-based tool for markerless pose estimation, was used to track the positions of the cell's head, the head-body boundary, and the tail. The trajectory of the head serves as a definitive path, effectively mapping the spatial region in which the cell is capable of capturing prey across a full 360° range. To elucidate the molecular composition of the neck, 5000 head-neck parts were isolated via microsurgery, enabling low-input protein mass spectrometry (MS) to identify neck-specific protein components. The analysis revealed that an actin-myosin system is involved in mediating the extreme neck deformation [2]. Neck deformation was also found to be influenced by different cations, as well as by actin and myosin inhibitors. Through immunofluorescence (IF), GST pull-down and AlphaFold2, we found myosin is a component of the *Lacrymaria* myoneme and colocalizes and interacts with centrin [2]. Through IF, transmission electron microscopy (TEM), and focused ion beam scanning electron microscopy (FIB-SEM), a giant protein facilitates the coordination between the microtubule cytoskeleton and centrin-myosin fiber, specifically in the neck [2]. *Plasmodium*-like actin in *Lacrymaria* by IF indicates highly dynamic short actin filaments involved in neck deformation [2].

Through these methods, MS has emerged as a powerful and versatile tool for protein detection, identification, and quantification, owing to its high accuracy and sensitivity [12]. Protein identification is typically performed by digesting proteins into peptides, ionizing them, and analyzing their mass-to-charge ratios [13, 14]. Then, the acquired spectra are matched to theoretical primary and secondary mass spectra derived from protein databases for confident identification [13]. This process involves protein sample preparation, protease hydrolysis, and MS analysis.

The GST pull-down assay is a robust in vitro method for detecting direct protein-protein interactions [15, 16]. When combined with MS, it is used to search for unknown proteins that may have an interaction relationship with the target protein [17, 18]. When combined with Western blotting (WB), it is possible to verify direct interactions between two known proteins. The principle of the method is that the probe protein is fused with GST via recombinant technology. The fusion protein binds to GSH immobilized on beads, thereby isolating proteins that interact with the fusion protein [15, 16].

IF technology, when integrated with high-resolution imaging techniques, plays a critical role in visualizing and analyzing cellular structures. The principle of this technology is based on the fact that fluorescent dyes absorb specific wavelengths of excitation light and transition to an excited state, then emit longer wavelength fluorescence to achieve target molecule localization, which can be combined with antibody-antigen reactions or direct dye use [19–21]. The experimental procedure includes sample preparation (fixation and permeabilization), blocking, primary antibody incubation, followed by incubation with secondary antibody or fluorescent dye, and image capture using confocal microscopy [22].

TEM and FIB-SEM are crucial technologies for revealing the ultrastructure of cells. TEM provides high-resolution images that clearly display the arrangement, diameter, and interconnections of structural components such as microtubules, actin filaments, and intermediate filaments [23, 24]. FIB-SEM is an advanced imaging technology that combines FIB and SEM techniques [25]. The system uses FIB for high-precision sectioning, while SEM enables high-resolution imaging. With the aid of image processing software, nanoscale three-dimensional (3D) reconstruction of the internal structure of the sample can be achieved [25]. Therefore, the combined application of TEM and FIB-SEM provides an effective strategy for investigating cellular structure, enabling a comprehensive understanding of the characteristics and spatial distribution of relevant cellular structures.

Conducting pharmacological inhibitor experiments is also an effective strategy for investigating cytoskeletal function by observing changes in cell morphology, motility, and internal structures, such as microtubules, actin filaments, and intermediate filaments [26–28]. Combining with IF and TEM, researchers can elucidate the dynamic roles of the cytoskeleton in maintaining cell structure, facilitating migration, regulating mechanotransduction, and coordinating intracellular transport [28]. Furthermore, the methods should be integrated with genetic manipulations and biochemical analyses to provide a comprehensive understanding of cytoskeletal mechanisms in physiological processes.

Through the application of these integrated techniques, we identified several unconventional cytoskeletal components in unicellular protists that play critical roles in regulating cell motility. We anticipate that these methodological advances will offer powerful tools for the broader cell biology community, enabling more systematic and efficient investigations. By streamlining experimental workflows and offering reproducible protocols, our approach aims to promote broader participation and collaboration, contributing valuable resources and insights into the structural and functional diversity of the cytoskeleton.

## 2 Materials

All solutions are prepared using ultrapure water. The materials already listed above will not be repeated in subsequent sections.

### 2.1 Mass Spectrometry

1. 1% Agar: Weigh 1.0 g agar into a 250 mL conical flask containing approximately 80 mL ultrapure water. Mix thoroughly, then adjust the final volume to 100 mL with ultrapure water. Autoclave at 121°C for 20 min (*see* **Note 1**).
2. Mouth pipette: Aspirator tube assemblies for calibrated microcapillary pipettes, Sigma, USA.
3. 10× PBS (final concentrations: 1.37 M NaCl, 0.1 M $Na_2HPO_4$, 0.018 M $KH_2PO_4$, 0.027 M KCl; pH 7.4): Weigh 80 g NaCl, 35.8 g $Na_2HPO_4{\cdot}12H_2O$, 2.4 g $KH_2PO_4$ into a 1 L beaker containing approximately 900 mL ultrapure water. Stir until fully dissolved, then adjust the volume to 1 L with additional ultrapure water. Store the solution at 4°C.
4. Lysis buffer: 30 mM Tris-HCl, 20 mM KCl, 2 mM $MgCl_2$, 1 mM phenylmethylsulfonyl fluoride (PMSF), and 0.1% Triton X-100 (*see* **Note 2**).
5. Sonicator: JY92-IIN, Ningbo Scientz Biotechnology Co., Ltd., Zhejiang, China.
6. Bicinchoninic acid (BCA) kit: Beyotime Institute of Biotechnology, Jiangsu, China.
7. High-pH reverse-phase high-pressure liquid chromatography: HPLC, C18, 40 μm, 60-Å pore size; Agilent Technologies.
8. Solvent A: Prepare a 0.1% formic acid (FA) solution by adding 10 μL FA to 10 mL ultrapure water in a 15 mL RNase-free centrifuge tube (*see* **Note 3**).
9. Solvent B [0.1% FA in 80% Acetonitrile (ACN)]: Add 80 mL ACN and 100 μL FA to a 100 mL graduated cylinder containing 10 mL ultrapure water. Mix well, and adjust the volume to 100 mL with ultrapure water (*see* **Note 3**).
10. 75 μm × 15 cm analytical column: C18, 3 μm, Thermo Fisher Scientific.
11. Q Exactive HF-X mass spectrometer: Thermo Fisher Scientific.

### 2.2 Western Blotting

1. LB medium: Weigh 5 g tryptone, 10 g yeast extract, and 5 g NaCl into a 1 L beaker containing approximately 800 mL ultrapure water. Mix well, then adjust the volume to 1 L with ultrapure water. Autoclave at 121°C for 20 min.
2. 1.5% Agar in LB medium: Add 1.5 g agar to 100 mL LB medium. Autoclave for 20 min (*see* **Note 4**).
3. 100 mg/mL Ampicillin: Weigh 1 g ampicillin sodium salt into a 15 mL RNase-free centrifuge tube. Add 8 mL ultrapure water and mix thoroughly until fully dissolved. Adjust the volume to 10 mL with ultrapure water. Sterilize the solution by passing through a 0.22 μm syringe filter. Aliquot into 1 mL portions and store at −20°C.
4. 0.5 M Isopropyl β-D-thiogalactoside (IPTG): Weigh 1.1915 g IPTG into a 15 mL RNase-free centrifuge tube. Add 8 mL ultrapure water and mix until fully dissolved. Adjust the volume to 10 mL with ultrapure water. Sterilize the solution by passing through a 0.22 μm syringe filter. Aliquot into 1 mL portions and store at −20°C (*see* **Note 5**).
5. Coomassie blue super fast staining solution (Beyotime Institute of Biotechnology, Jiangsu, China).
6. His-tag protein purification kit (Beyotime Institute of Biotechnology, Jiangsu, China).
7. NuPAGE bis–tris precast polyacrylamide gel: 4%–12%, Thermo Fisher Scientific (Invitrogen).
8. NuPAGE MES SDS running buffer: Thermo Fisher Scientific (Invitrogen).
9. Western blotting transfer buffer (0.025 M Tris, 0.192 M glycine, 20% methanol): Weigh 3.03 g Tris, 14.4 g glycine into a 1 L beaker containing approximately 700 mL ultrapure water. Stir until fully dissolved, then adjust the volume to 800 mL with ultrapure water. Then add 200 mL methanol, and mix thoroughly before use.
10. Polyvinylidene difluoride (PVDF) membrane (Merck Millipore, MA, USA).
11. WesternBright sirius HRP substrate kits (Advansta, CA, USA).
12. 1× PBS containing 0.1% Tween-20: Add 1 mL Tween-20 to 999 mL of 1× PBS and mix thoroughly (*see* **Note 6**).

### 2.3 GST Pull Down

1. 1× PBS: Add 100 mL 10× PBS to 900 mL ultrapure water. Mix thoroughly and store at room temperature (RT).
2. Glutathione Sepharose™ 4B (GE Healthcare Dharmacon, USA; *see* **Note 7**).

3. 1 M Tris-HCl, pH 8.0: Weigh 181.7 g Tris into a 1 L glass beaker containing approximately 800 mL ultrapure water. Stir thoroughly until dissolved and adjust pH to 8.0 with concentrated HCl (*see* **Note 8**). Adjust the final volume to 1 L with ultrapure water. After sterilization, store at RT.
4. Elution buffer: 50 mM Tris-HCl (pH 8.0), 10 mM GSH (*see* **Note 9**).
5. 2× SDS sample buffer: 20 mM Tris-HCl, pH 6.8, 100 mM dithiothreitol (DTT), 2% SDS, 20% glycerol, 0.016% bromophenol blue. Add 200 μL of 1 M Tris-HCl (pH 6.8), 1 mL of 20% SDS, 2 mL of glycerol, and 160 μL of 1% bromophenol blue in a 15 mL RNase-free centrifuge tube, adjust the volume to 9 mL with ultrapure water and mix thoroughly by vortexing until homogeneous. Aliquot the buffer into 0.9 mL portions without DTT and store at −20°C. Immediately before use, add 0.1 mL of freshly prepared 1 M DTT to 0.9 mL of the stock buffer (final volume 1 mL) and mix well. (*see* **Note 10**).

### 2.4 Inhibitor Treatment

1. 1% Agar: see Sect. 2.1.1.
2. 10 mM Latrunculin A (LAT-A): Add 23.7 μL DMSO to the bottle containing 100 μg of dry powder and vortex thoroughly. The solution is aliquoted and stored at −20°C (*see* **Note 11**).
3. 10 mM Cytochalasin B (CB): Add 208.5 μL DMSO to the bottle containing 1 mg of dry powder and vortex thoroughly. The solution is aliquoted and stored at −20°C (*see* **Note 11**).
4. 50 mM CK666: Add 1.68 mL DMSO to the bottle containing 25 mg of dry powder and vortex thoroughly. The solution is aliquoted and stored at −20°C (*see* **Note 11**).
5. 10 mM para-Nitroblebbistatin: Add 1.48 mL DMSO to the bottle containing 5 mg of dry powder and vortex thoroughly. The solution is aliquoted and stored at −20°C (*see* **Note 11**).

### 2.5 Immunofluorescence

1. 10% Triton X-100: Add 10 mL Triton X-100 to 90 mL ultrapure water. Mix thoroughly and store at RT (*see* **Note 12**).
2. 1× PBS with 0.5% Triton X-100: Add 5 mL 10% Triton X-100 to 95 mL 1× PBS. Mix thoroughly and store at RT.
3. 2× PHEM (final concentrations: 120 mM Pipes, 50 mM Hepes, 20 mM EGTA, 8 mM $MgSO_4$, pH 7.0): Weigh 18.14 g Pipes, 6.5 g Hepes, 3.8 g EGTA, 0.99 g $MgSO_4 \cdot 7H_2O$ into a 1 L beaker containing 450 mL ultrapure water. Stir to dissolve, then adjust the pH to 7.0 with 1.0 M KOH. Adjust the final volume to 500 mL with ultrapure water. Store at 4°C.

4. 1× PHEM: Add 500 mL 2× PHEM to 500 mL ultrapure water. Mix thoroughly and store at 4°C.
5. Paraformaldehyde (PFA) solution: 20% PFA, pH 7.4. Weigh 20 g PFA into a 250 mL beaker containing 80 mL ultrapure water. Heat the mixture to 60-65°C with continuous stirring using a magnetic stirrer until a milky white suspension forms. Add 1.0 M KOH dropwise until the solution becomes clear, indicating complete dissolution of PFA. Allow the solution to cool at RT, and use HCl to precisely adjust the pH back to 7.4. Then adjust the final volume to 100 mL with ultrapure water. Sterilize the solution by passing through a 0.22 μm syringe filter and store at −20°C (*see* **Note 13**).
6. Fixation solution A: 1× PHEM, 2% PFA, 0.5% Triton X-100. Add 5 mL 2× PHEM, 1 mL 20% PFA, 0.5 mL 10% Triton X-100 to a 15 mL RNase-free centrifuge tube containing 3.5 mL ultrapure water. Mix thoroughly and store at 4°C.
7. Fixation solution B: 1× PHEM, 2% PFA. Add 5 mL 2× PHEM, 1 mL 20% PFA into a 15 mL RNase-free centrifuge tube containing 4 mL ultrapure water. Mix thoroughly and store at 4°C.
8. Fixation solution C: 1× PHEM, 4% PFA. Add 5 mL 2× PHEM, 2 mL 20% PFA to a 15 mL RNase-free centrifuge tube containing 3 mL ultrapure water. Mix thoroughly and store at 4°C.
9. Fixation solution D: 4% PFA, 17% NaCl. Weigh 1.7 g NaCl into a 15 mL RNase-free centrifuge tube containing 5 mL ultrapure water, mix thoroughly until fully dissolved. Add 2 mL 20% PFA, mix well, and adjust the final volume to 10 mL with ultrapure water. Store at 4°C.
10. Blocking buffer: Add 0.3 g bovine albumin (BSA) to a 15 mL RNase-free centrifuge tube containing 5 mL 1× PBS, mix thoroughly until fully dissolved. Add 0.3 mL 10% Triton X-100, mix well, and adjust the final volume to 10 mL with 1× PBS.
11. 1 mg/mL DAPI: Add 5 mL ultrapure water to the bottle containing 5 mg dry powder. Vortex thoroughly until fully dissolved. Aliquot the solution and store at −20°C (*see* **Note 14**).
12. 1% Saponin: Weigh 0.1 g saponin into a 15 mL RNase-free centrifuge tube containing 8 mL ultrapure water. Mix thoroughly, then adjust the volume to 10 mL with ultrapure water.
13. Phalloidin (cat. GC15692): Dissolve the vial contents (1 mg) in 150 μL of anhydrous DMSO to prepare a 400× stock solution at a concentration of approximately 66 μM.

14. SiR–Actin (cat. CY-SC001): Dissolve the vial contents of SiR–Actin in 50 μL of anhydrous DMSO to make a 1 mM stock solution. The work solution is 1 μM.

### 2.6 TEM and FIB-SEM

1. 0.2 M PB: 0.2 M $NaH_2PO_4$, 0.2 M $Na_2HPO_4$, pH 7.4. Weight 31.2 g $NaH_2PO_4{\cdot}2H_2O$ into a 1 L beaker containing 900 mL ultrapure water and stir to dissolve. Adjust the volume to 1 L with ultrapure water to prepare a 0.2 M $NaH_2PO_4$ solution. Weigh 71.6 g $Na_2HPO_4{\cdot}12H_2O$ into another 1 L beaker containing 900 mL ultrapure water and stir to dissolve. Adjust the volume to 1 L with ultrapure water to prepare a 0.2 M $Na_2HPO_4$ solution. Mix 19 mL of 0.2 M $NaH_2PO_4$ with 81 mL of 0.2 M $Na_2HPO_4$ thoroughly to prepare 0.2 M PB. The solution pH is 7.4. Store at RT.
2. 0.1 M PB: 0.1 M $NaH_2PO_4$, 0.1 M $Na_2HPO_4$ , pH 7.4. Add 500 mL 0.2 M PB (pH 7.4) to a 1 L beaker containing 500 mL ultrapure water and mix thoroughly. Store at RT.
3. 1% osmium tetroxide ($OsO_4$): Weigh 0.1 g $OsO_4$ into a 15 mL RNase-free centrifuge tube. Add 10 mL sterilized water and mix thoroughly to dissolve. Divide into 1 mL portions and wrap the container with aluminum foil. Store the solution at 4°C in the dark (*see* **Note 15**).
4. Prepare ethanol solutions by mixing ethanol and ultrapure water in a 15 mL RNase-free centrifuge tube as follows. Mix thoroughly and store at RT (*see* **Note 16**). 30% Ethanol: Add 3 mL ethanol to 7 mL water. 50% Ethanol: Add 5 mL ethanol to 5 mL water. 70% Ethanol: Add 7 mL ethanol to 3 mL water. 80% Ethanol: Add 8 mL ethanol to 2 mL water. 90% Ethanol: Add 9 mL ethanol to 1 mL water.
5. Prepare ethanol-acetone mixtures in a 15 mL RNase-free centrifuge tube as described below. Mix thoroughly and store at RT (*see* **Note 17**). 3:1 Ethanol:Acetone: Add 9 mL ethanol and 3 mL acetone. 1:1 Ethanol:Acetone: Add 5 mL ethanol and 5 mL acetone. 1:3 Ethanol:Acetone: Add 3 mL ethanol and 9 mL acetone.
6. Prepare acetone and son812 epoxy resin mixtures in a 15 mL RNase-free centrifuge tube as follows. Mix thoroughly and store at RT (*see* **Note 17**). 1:1 Acetone:Son812 epoxy resin: Add 5 mL acetone and 5 mL son812 epoxy resin. 1:2 Acetone:Son812 epoxy resin: Add 4 mL acetone and 8 mL son812 epoxy resin.
7. 2% uranyl acetate: Weigh 2 g uranyl acetate into a 250 mL glass beaker containing 80 mL ultrapure water. Stir until fully dissolved, and then adjust the volume to 100 mL with ultrapure water. Sterilize the solution by passing through a 0.22 μ

m syringe filter, and store at 4°C in a brown bottle (*see* **Note 18**).

8. 3% lead citrate. Weigh 3 g lead citrate into a 250 mL glass beaker containing 80 mL ultrapure water. Stir thoroughly until dissolved, adjust pH to 12 with 1 M NaOH, then adjust the volume to 100 mL with ultrapure water. Sterilize the solution by passing through a 0.22 μm syringe filter, and store at RT in a brown bottle (*see* **Note 18**).

## 3 Methods

### *3.1 Search for Components of* Lacrymaria *Cytoskeleton Through Mass Spectrometry (MS)*

MS is a powerful method for protein identification, but its success depends on the quality of sample preparation. This protocol covers the entire workflow from cell culture to protein identification of the *Lacrymaria* head and neck region, which is essential for the dynamic extension-contraction behavior of the organism. The same approach-can be applied to related organisms.

1. For cell culture, add sterilized lake water to a plastic Petri dish containing a bottom layer of 1% solid agar. Then, introduce *Cyclidium* sp. and *Escherichia coli* as food sources [7, 8] (*see* **Note 19**).
2. When the cell density reaches approximately 30 cells/mL, individual cells are isolated using a mouth pipette. Wash the cells 2–3 times with sterilized lake water to remove debris and food, and then starve for 3–5 h.
3. Using a mouth pipette under a stereomicroscope (*see* **Note 20**), collect the head-neck parts from active cells (Fig. 1).
4. Immediately flash-freeze the collected head-neck parts in dry ice. After each batch of collection, transfer the samples to a −80°C freezer for long-term storage.
5. After harvesting a sufficient quantity of head-neck parts, take out all frozen samples and resuspend them in lysis buffer supplemented with 1× cOmplete proteinase inhibitor (*see* **Note 21**).
6. Sonicate the mixture on ice for 10 min using a cycle of 3 s on and 3 s off, at an output power of 135 W.
7. Centrifuge at 10,000 × *g* for 30 min at 4°C. Discard the pellet and collect the supernatant.
8. Measure protein concentration using a bicinchoninic acid (BCA) kit according to the manufacturer's instructions. Before protein identification, in-solution and in-gel protease digestion could be performed optionally (*see* **Note 22**).

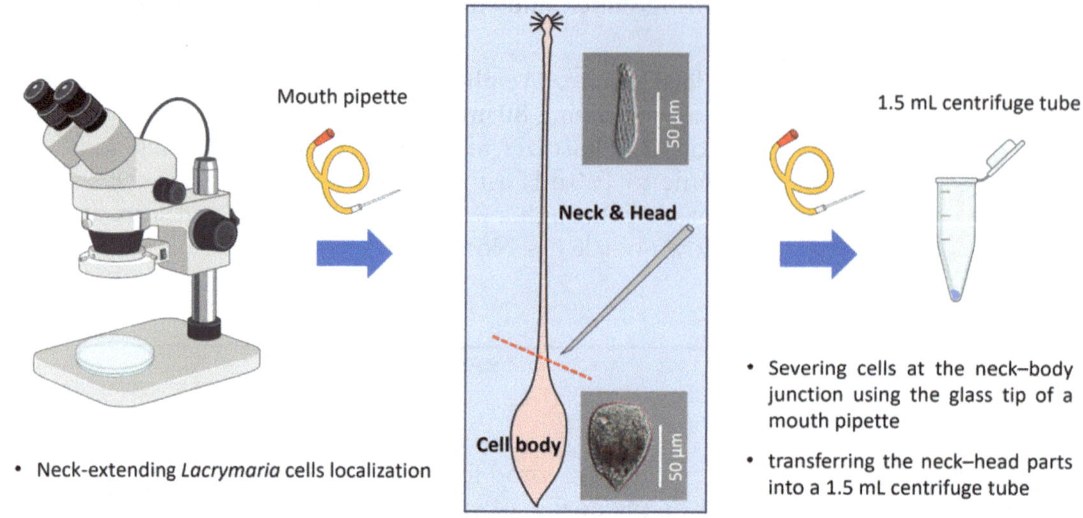


**Fig. 1** Diagram illustrating the process of isolating neck-extending *Lacrymaria* cells. On the left, a stereomicroscope and a mouth pipette are shown. An arrow points to the central panel depicting the neck and head of the cell, with a dashed line indicating the severing point at the neck-body junction. Insets show magnified views of the cell parts, each labeled with a scale bar of 50 µm. Another arrow points to the right, where a 1.5 mL RNase-free centrifuge tube is depicted. Text annotations describe the steps: sever the cells with a mouth piptte and transfer the neck-head parts into the centrifuge tube

9. *(Optional)* Alternative 1: for in-solution protease hydrolysis, reduce the protein extract with 25 mM DTT for 40 min at RT.
10. Alkylate the sample with 50 mM iodoacetamide for 10 min at RT in the dark.
11. Precipitate the proteins by adding five volumes precooled acetone to remove pigment molecules and other small molecules as much as possible.
12. Centrifuge at 6,000 × *g* for 40 min at 4°C. Discard the supernatant.
13. Redissolve the pellet in 50 mM $NH_4HCO_3$ and digest with 1% (wt/wt) trypsin overnight at 37°C.
14. Quench the enzyme reaction by adding 0.1% trifluoroacetic acid.
15. Fractionate the digested peptides using high-pH reverse-phase high-pressure liquid chromatography (*see* **Note 23**).
16. Collect the digested peptides and store at −80°C until processed for MS.
17. *(Optional)* Alternative 2: In-gel protease digestion can be performed. To this end, add 1 mg protein extract to the 5× SDS loading buffer.

18. Denature at 95°C for 5 min before fractionation by NuPAGE gel.
19. Stain gels with Coomassie blue super fast staining solution.
20. Excise gels into about ten pieces, and further cut each band into 1 $mm^3$ fractions.
21. Destain the gel pieces thoroughly with 35% acetonitrile and 50 mM $NH_4HCO_3$.
22. Dehydrate with 75% acetonitrile, followed by disulfide reduction with 25 mM dithiothreitol and alkylation with 50 mM iodoacetamide.
23. Digest proteins overnight at 37°C with 50 mM $NH_4HCO_3$ containing 20 μg/mL sequencing grade modified trypsin (1:100, wt/wt).
24. Transfer the supernatant into a fresh RNase-free centrifuge tube.
25. Sonicate the remaining gels twice with extraction buffer (67% acetonitrile, 5% trifluoroacetic acid).
26. Mix the peptide supernatants and extracts.
27. Desalted for further MS analysis.
28. Dissolve the digested peptides from Sects. 3.1.16 to 3.1.27 in solvent A.
29. Separate the peptides using an online nano-flow EASY-nLC 1200 system with a 75 μm × 15 cm analytical column, and analyze them on a Q Exactive HF-X mass spectrometer.
30. Peptides were eluted using a gradient of solvent B as follows: 4–6% over 1 min, 6–10% over 2 min, 10–18% over 40 min, 18–33% over 10 min, ramped to 90% in 0.5 min, and held at 90% for 6.5 min, all at a constant flow rate of 300 nL/min.
31. The MS was operated in data-dependent acquisition (DDA) mode, with full MS scans acquired over an m/z range of 350–1,800 at a resolution of 60,000, using an automatic gain control (AGC) target of $3 \times 10^6$.
32. The top 20 most intense precursor ions were selected for MS/MS fragmentation using higher-energy collision dissociation (HCD) with normalized collision energy (NCE) of 28%, and the fragments were analyzed at a resolution of 15,000 in the Orbitrap.
33. Dynamic exclusion was set to 25 s, and the isolation width of precursor ions was 1.6 m/z. The maximum injection times were 20 ms for MS and 50 ms for MS/MS. The intensity threshold was set to 5,000. Acquire MS/MS spectra from the above parameters.

34. Analyze all MS/MS spectra by pFind software (version 3.1) against the *Lacrymaria* protein database, combined with the reverse decoy database and common contaminants.
35. Use the open-search algorithm in pFind for database searching. Two missed cleavages are routinely allowed for trypsin. Methylation (K) and dimethylation (K) were set as variable modifications.
36. Set the precursor and fragment ion mass tolerances at 20 ppm. The minimum peptide length was set at six, and the estimated false discovery rate (FDR) threshold for peptides and proteins was a maximum of 1% as parameters to analyze the results.
37. Select the interesting proteins from the results to study their function.

### 3.2 Myosin and Giant Recombinant Protein Production

This protocol is intended to describe the expression and purification of recombinant proteins identified by MS analysis. In this specific case, the cytoskeletal components centrin and myosin are selected as targets. Usually, recombinant proteins are required for antibodies production, when the target proteins are commercially unavailable or when custom antibodies are preferred (*see* **Note 24**).

1. If required for vector construction, select the hydrophilic regions of a C-terminal fragment (1,065–1,395) containing 331 residues and a C-terminal fragment (17,051–17,248) containing 198 residues to generate antibodies against myosin (evm.model.Lac.contig.5314.4) and giant protein (evm.model.Lac.contig.5913.2). The antigen proteins are approximately 40 kDa and 22 kDa, respectively.
2. Codon optimization for *E. coli* expression using an online tool.
3. Clone the codon-optimized sequence into the pET-32a expression vector, which includes a His-tag and approximately 20 kDa Trx- and S-tags.
4. Transfect the resulting plasmid into *E. coli* BL21 strain using the heat shock method.
5. Culture the transformed BL21 clones in liquid LB medium containing ampicillin at 37°C in a shaking incubator for 1 h.
6. To select a single colony, plate the clones on the solid LB medium. Heat the liquid LB medium containing agar in a microwave. When it cooled to about 40°C, add ampicillin to a final concentration of 10 μg/mL and mix well.
7. Pour the medium into sterile Petri dishes and allow it to cool and solidify at RT.

8. Spread the clones onto the solid LB plates containing ampicillin.
9. Incubate plates at 37°C for approximately 12 h until colonies appear.
10. Use a sterile toothpick to transfer a single colony into a conical flask containing liquid LB medium containing ampicillin.
11. Incubate the culture at 37°C with vigorous shaking for 12–15 h to allow bacterial growth.
12. Dilute the culture 1:100 into 10 mL liquid LB medium containing ampicillin.
13. After approximately 2–3 h, the transformed cultures reach the logarithmic phase ($OD_{600}$ ≈0.6).
14. Induce protein expression by adding 0.5 mM IPTG for 4 h at 28°C (*see* **Note 25**).
15. Centrifuge the culture at 5000×*g* for 10 min to pellet the cells, then discard the supernatant.
16. Resuspend the pellet in 500 μL ice-cold 1× PBS in a 1.5 mL RNase-free centrifuge tube to wash the cells (*see* **Note 26**).
17. Centrifuge to pellet the cells and repeat the process twice. After the final wash, transfer 10 μL resuspended cells into labeled tubes for detection.
18. Subject the mixture to sonication on ice using a sonicator equipped with a 10 mm probe, applying cycles of 3 s on and 3 s off for a total of 10 min.
19. Centrifuge at 5,000×*g* for 10 min, then transfer the supernatant to fresh tubes. Save 10 μL insoluble and soluble material for detection, respectively (*see* **Note 27**).
20. Assess protein expression levels by Coomassie blue super fast staining solution.
21. To get more recombinant proteins, select the colony with the highest protein expression level. Perform large-scale culture and induce protein expression as described above, followed by sonication and centrifugation.
22. Place the soluble supernatant on ice or in an ice bath.
23. Protein purification according to the manufacturer's protocol of the His-tag Protein Purification Kit (*see* **Note 28**). To collect the purified His-tagged protein, transfer 1 mL of 50% BeyoGold™ His-tag Purification Resin or other resin to a centrifuge tube, centrifuge at 1,000×*g* for 10 s at 4°C.
24. Discard the storage solution. Add 0.5 mL of the provided non-denaturing lysate buffer and mix well to equilibrate.
25. Centrifuge at 1,000×*g* for 10 s at 4°C and discard the supernatant. Repeat this step one to two times.

26. Add the soluble supernatant material containing the produced recombinant protein.
27. Shake gently at 4°C on a side shaker for 2 h.
28. Transfer the mixture of the sample and resin to the empty affinity column tube provided with the kit.
29. Open the bottom lid and allow the liquid to drain from the column by gravity flow.
30. Wash the column five times with the provided 0.5–1 mL non-denatured washing liquid.
31. Elute the target protein six to ten times with the provided 0.5 mL of non-denaturing eluent.
32. Collect each eluate separately into a different RNase-free centrifuge tube. The collected eluent contains the purified His-tagged protein. The eluent is an imidazole-containing solution. The imidazole concentration and pH of the eluent may vary depending on the specific protein.
33. Transfer 10 μL purified protein into a labeled tube (for later use in SDS-PAGE analysis).
34. NuPAGE Bis-Tris precast polyacrylamide gels are used for electrophoresis, with NuPAGE MES running buffer.
35. Initiate electrophoresis at 80 V for 30 min, then increase to 150 V for an additional 1 h.
36. Turn off the power supply when the dye front reaches the bottom of the gel.
37. Following electrophoresis, carefully separate the gel plates using a spatula. The gel remains on one of the glass plates.
38. Rinse the gel with water and carefully transfer it to a container filled with Western blotting transfer buffer.
39. For the following Western blotting transfer, cut a polyvinylidene difluoride (PVDF) membrane to match the size of the gel and immerse it in methanol.
40. Rinse the PVDF membrane with transfer buffer. Rinse the gel with deionized water to remove traces of running buffer.
41. Transfer proteins from the gel to the PVDF membrane by electroblotting at 220 mA for 1 h.
42. After transfer, block the PVDF membrane with 5% skim milk in 1× PBS containing 0.1% Tween-20 for 1 h at 25°C.
43. Wash the PVDF membrane three times with 1× PBS containing 0.1% Tween-20.
44. Incubate the PVDF membrane overnight at 4°C with the corresponding primary antibodies (myosin, 1:1000 dilution;

giant protein, 1:3000 dilution; positive control: mouse anti-His-tag antibody at a 1:2000 dilution; *see* **Note 29**).

45. On the following day, wash the PVDF membrane three times with 1× PBS containing 0.1 % Tween-20 for 10 min each.
46. Incubate the PVDF membrane with horseradish peroxidase (HRP)-conjugated goat anti-mouse/rabbit IgG secondary antibody (1:5000 dilution) for 2 h at RT.
47. Wash the PVDF membrane three times with 1× PBS containing 0.1% Tween-20 for 10 min each.
48. Place the membrane in 1× PBS containing 0.1% Tween-20.
49. Visualize the chemoluminiscence signal using a fluorescence scanner such as Alpha Innotech FluorChem Q (USA) with dedicated substrate.

### 3.3 Recombinant Centrin-GST and IQ Myosin-His Tag Protein Expression

1. If vector construction is required, select the whole sequence of centrin (evm.model.Lac.contig.4004.2) and the first IQ motif of myosin (evm.model.Lac. contig.5314.4) to express (*see* **Note 30**).
2. Codon optimization for *E. coli* expression using an online tool.
3. Clone the codon-optimized sequences of centrin and the first IQ motif of myosin into the pGEX-4T (bearing GST-tag) and pET-32a (bearing His-tag) expression vectors, respectively.
4. For protein expression, transfect the plasmids into *E. coli* BL21 strain to express the target proteins in vitro, following the previously described method.
5. The protein expression procedure was also carried out as previously described.
6. After sonication, centrifuge at 10,000 ×*g* for 10 min to collect the supernatant, which contains the total protein, including the GST- or His-tagged recombinant proteins.
7. Purify His-myosin recombinant protein as described in Sects. 3.2.23 to 3.2.33.
8. Purify GST-centrin recombinant protein according to the manufacturer's protocol of the GST-tag Protein Purification methods.
9. Transfer Glutathione Sepharose™ 4B to a 1.5 mL RNase-free centrifuge tube.
10. Centrifuge at 500 ×*g* for 5 min, then carefully discard the supernatant.
11. Add 1 mL of cold 1× PBS to wash the Glutathione Sepharose™ 4B.

12. Invert the tube to mix, then sediment the matrix by centrifugation at 500×*g* for 5 min. Discard the supernatant (*see* **Note 31**).
13. Repeat Steps 11 and 12 to perform a total of four washes.
14. Equilibrate a 50% slurry of Glutathione Sepharose™ 4B with 1× PBS (*see* **Note 32**).
15. Add 400 μL Glutathione Sepharose™ 4B to each sonicated supernatant containing either GST-centrin protein or GST protein (*see* **Note 33**).
16. Incubate with gentle agitation at RT for 2 h (*see* **Note 34**).
17. Centrifuge at 1,000×*g* for 5 min, then remove the supernatant (*see* **Note 35**).
18. Wash the matrix with ten bed volumes of 1× PBS (e.g., 2 mL 1× PBS per 400 μL 50% slurry).
19. Centrifuge at 500×*g* for 5 min, then remove the supernatant.
20. Repeat the wash step two times.
21. Keep the beads on ice for later pull-down experiments (*see* **Note 36**).

### 3.4 Pull-Down Experiments to Assess Protein-Protein Interaction

The pull-down experiments protocol can be used to detect protein-protein interaction in vitro, by using a combination of His- and GST-tagged recombinant proteins. The GST-tagged recombinant protein binds to GSH immobilized on the beads. Upon incubation with candidate His-tagged recombinant proteins, any interacting partner is co-purified with the beads and can be subsequently identified. Here, this protocol is used to examine the interaction between the recombinant tagged myosin and centrin.

1. Add the His-myosin protein to the GST-centrin-bound beads, and incubate at 4°C overnight.
2. Centrifuge at 1,000×*g* for 5 min and remove the supernatant.
3. Add 1 mL 1× PBS and vortex briefly.
4. Centrifuge at 1,000×*g* for 1 min to pellet the sepharose beads. Discard the supernatant.
5. Wash the beads with 1× PBS for a total of ten times.
6. Elute the fusion protein by adding 10 μL elution buffer.
7. Resuspend the Sepharose beads and incubate at RT for 5 min.
8. Centrifuge at 1,000×*g* for 1 min, then transfer the supernatant to a fresh tube.
9. Add an equal volume of SDS sample buffer to 20 μL eluate.
10. Boil the mixture at 100°C for 10 min.

11. Detect the interaction between the two proteins using Western blotting with specific antibodies to observe the presence of both proteins in the same sample (*see* **Note 37**).

### *3.5 Use of Cytoskeleton Inhibitors to Assess Cell Treatment*

1. Dissolve 1% agar by heating. Aliquot 500 μL 1% agar into each well of a 24-well culture plate. Allow it to solidify naturally, forming a smooth base layer.
2. Warm the aliquoted LAT-A stock solution to RT in the dark for 1 h (*see* **Note 38**).
3. Dilute the stock solution with sterile lake water to prepare a working solution at a final concentration of 20 μM.
4. Add 500 μL of the working solution to each well of the 24-well plate, with four replicates per treatment group.
5. Control groups include sterile lake water and DMSO solution at matching concentrations, each with four replicates.
6. Select 50 active *Lacrymaria* cells from the same culture batch using a mouth pipette under a stereomicroscope (*see* **Note 20**).
7. Inoculate them into each well, ensuring even distribution across the experimental and control groups.
8. Place the 24-well plate in a dark incubator at 22°C, maintaining standard laboratory culture conditions.
9. Observe the cells at the following time points: 0, 1, 2, 4, 8, 12, 16, 20, and 24 h post-treatment.
10. Use three groups to quantify the number of cells in different states (*see* **Note 39**).
11. One additional group is used for image and video capture, with cells fixed at the corresponding time points for subsequent fluorescence staining experiments (*see* **Note 40**).

### *3.6 Immunofluorescence*

A dedicated figure (Fig. 2) is provided as a workflow guideline for this protocol.

Discrepancies in staining patterns may arise when targeting the same structure with different dye or antibody, or when applying the same dye or antibody to different species. These variations require case-specific evaluation to determine which approach yields results more consistent with biological reality. As illustrated in Fig. 3, although both phalloidin and SiR-Actin target actin (*see* **Note 41**), only the SiR-Actin staining reveals short, rod-shaped actin filaments within *Lacrymaria*.

1. To prepare cells, transfer 100 μL of sterile lake water or corresponding buffer into a clean 8-well glass chamber.

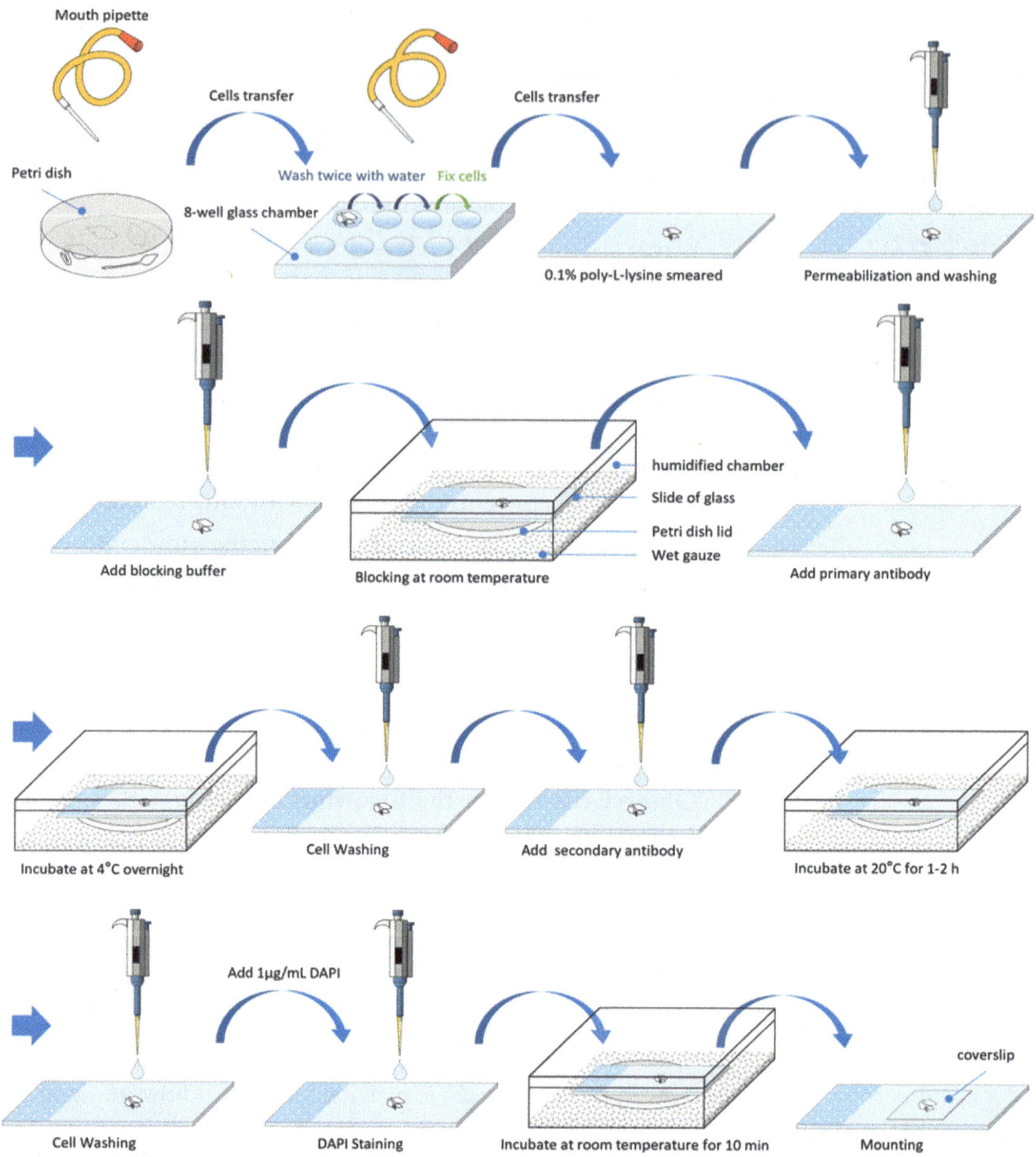


**Fig. 2** Flow chart illustrating a laboratory procedure for cell preparation and immunofluorescence staining. The procedure begins with transferring cells from a petri dish to an 8-well glass chamber using a mouth pipette, followed by washing and fixation. Next, transfer cells to the slide pre-coated with 0.1% poly-L-lysine, permeabilize with 1× PBS containing 0.5% Triton X-100, and wash. Then, block the slide with blocking buffer in a humidified chamber at RT. Incubate with primary antibody at 4°C overnight, wash, and incubate with secondary antibody at 20°C for 1–2 h. Stain the cells with DAPI at RT for 10 min. Finally, mount the silde with a coverslip. Arrows indicate the sequence of steps

2. Transfer the cells into the chamber using a mouth pipette to wash away impurities. Repeat this process two or three times.
3. After thorough washing, transfer the cells into a well containing 50 μL water and wait for fixation.

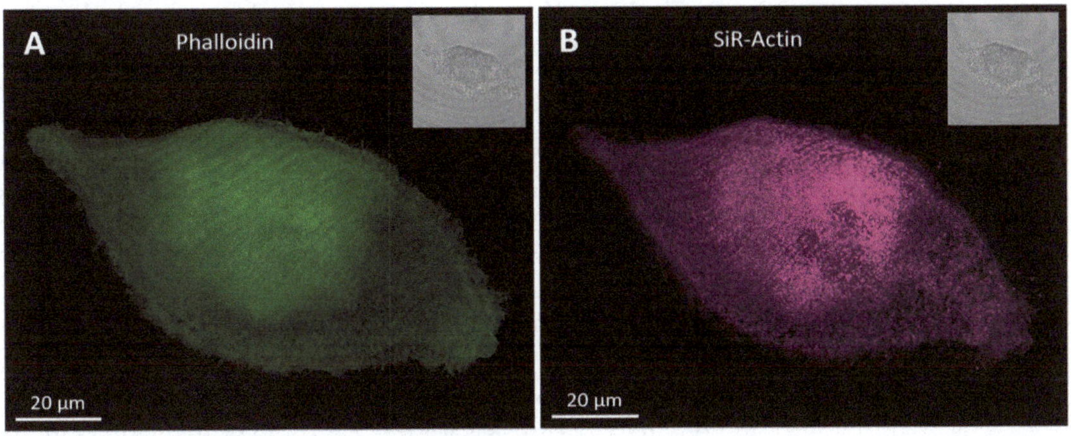


**Fig. 3** Fluorescence microscopy image showing of *Lacrymaria* cells stained with two actin probes. Panel **A** shows a green-stained cell using Alexa Fluor™ 488 Phalloidin, highlighting actin filaments. Panel **B** shows a magenta-stained cell using SiR-Actin, also highlighting actin filaments. Both panels include an inset of a gray-scale image of the same cell. Notably, SiR-Actin distinctly reveals the short, rod-shaped actin structures. Scale bars indicate 20 μm

4. Add 50 μL fixation solution A to the chamber and incubate at 4°C for 2 h (*see* **Note 42**).
5. To promote the fixed-cell adhesion on the slide, apply 10 μL 0.1% poly-L-lysine to a clean glass slide, and spread it evenly using a glass rod.
6. Incubate the slide in an oven at 55°C for 1–2 h to dry and fix the poly-L-lysine coating (*see* **Note 43**).
7. Transfer the fixed cells onto a pre-equilibrated glass slide using a mouth pipette.
8. Remove the excess liquid, and allow the cells to air-dry at RT (*see* **Note 44**).
9. Wash and permeabilize the cells three times using 1× PBS containing 0.5% Triton X-100 for 10 min each to thoroughly remove residual fixative.
10. Incubate the cells in blocking buffer at RT for 30 min to block nonspecific binding sites.
11. For indirect immunofluorescence to enhance the fluorescence signal, a primary antibody and a secondary antibody with a fluorescent dye are needed. For primary antibody incubation (*see* **Note 45**), prepare primary antibody by adding 2 μL antibody to a 1.5 mL RNase-free centrifuge tube containing 18 μ L 1× PBS.
12. Mix gently by pipetting to prepare a 1:10 primary antibody stock solution. Store at −20°C (*see* **Notes 45 and 46**).

13. Before use, add 5 μL primary antibody stock solution to a 1.5 mL RNase-free centrifuge tube containing 45 μL blocking buffer. Mix gently by pipetting to avoid bubble formation. Store at 4°C.
14. Remove excess blocking solution using a mouth pipette. Add 20 μL of the primary antibody working solution to the cells. Incubate the slides in a humidified chamber at 4°C overnight. Wash the cells three times with 1× PBS containing 0.5% Triton X-100 using a mouth pipette for 10 min each time to remove the primary antibody working solution.
15. For secondary antibody incubation (*see* **Note 45**), prepare secondary antibody solutions by adding 2 μL secondary antibody to a 1.5 mL RNase-free centrifuge tube with 18 μL 1× PBS. Mix gently by pipetting to prepare a 1:10 secondary antibody stock solution. Store at −20°C.
16. Before use, add 5 μL secondary antibody stock solution to a 1.5 mL RNase-free centrifuge tube containing 95 μL blocking buffer. Mix gently by pipetting to avoid bubble formation. Store at 4°C.
17. Remove the wash buffer using a mouth pipette, and add 20 μ L secondary antibody working solution. Incubate the slide in a humidified chamber at 20°C for 1–2 h (*see* **Note 45**). If required, coincubate with fluorescent dyes (e.g., Super Tubulin Green, phalloidin) during this step (*see* **Note 47**).
18. Wash the cells three times with 1× PBS containing 0.5% Triton X-100 using a mouth pipette, incubating for 10 min each time to remove unbound secondary antibody.
19. Remove wash buffer using a mouth pipette.
20. For DAPI staining, add 20 μL 1 μg/mL DAPI to cover the cells. Incubate the slide in a dark, humidified chamber at RT for 10 min. Then wash the cells once with 1× PBS.
21. To mount samples, heat the solid anti-fade mounting medium in a 55°C water bath for 15 min, then allow it to equilibrate to RT (*see* **Note 48**).
22. Remove any residual liquid from the slide using absorbent paper.
23. Add 15 μL mounting medium onto the cell area.
24. Gently lower a 22 × 22 mm coverslip onto the sample, using small pieces of modeling clay at each corner to stabilize.
25. Seal the edges of the coverslip with Fixogum to prevent drying.
26. Store the slides at 4°C, protecting from light until image acquisition.

27. Perform imaging using a confocal microscope equipped with a 100× oil immersion objective, such as LEICA STELLARIS 8.
28. Analyse cellular architecture using both Lightning super-resolution mode and STED (stimulated emission depletion) nanoscopy to resolve subcellular structures with high resolution.

### 3.7 Transmission Electron Microscopy (TEM)

1. Culture cells as previously described in Sect. 3.1.1.
2. Collect cells using a mouth pipette and wash two to three times with sterile water to eliminate excess bacteria and impurities.
3. In the final wash step, collect and transferr cells to a 1.5 mL RNase-free centrifuge tube using a mouth pipette under a stereomicroscope (*see* **Note 49**).
4. Add 500 μL of 2.5% glutaraldehyde fixative to the cells, and fix overnight at 4°C (*see* **Note 50**).
5. To wash cells, centrifuge at 200×*g* for 5–10 min to remove the fixative buffer (*see* **Note 51**).
6. Add 500 μL 0.1 M PB to the tube and gently mix to ensure the cells are fully immersed. Incubate for 20 min (*see* **Note 52**).
7. Centrifuge the cells again at 200× *g* for 5 min to discard the supernatant.
8. Repeat steps 6 and 7 to wash the cells thoroughly.
9. After the final wash, resuspend the cell pellet in 100–200 μL 0.1 M PB and store it for subsequent processing.
10. To dehydrate samples, prepare a graded ethanol series by diluting absolute ethanol with water to obtain concentrations of 30%, 50%, 70%, 80%, 90%, and 100%. Also prepare mixtures of 100% ethanol and 100% acetone in volume ratios of 3:1, 1:1, and 1:3.
11. Centrifuge at 200×*g* for 5 min to remove PB.
12. Add 500 μL of 30% ethanol to the centrifuge tube, and gently mix by inversion to ensure the cells are fully immersed.
13. Incubate the cells at RT for 15–20 min, then centrifuge at 200×*g* for 5 min to remove the ethanol.
14. Repeat steps 12 and 13 to dehydrate the cells with 50%, 70%, 80%, and 90% ethanol, followed by two changes of 100% ethanol (*see* **Note 53**).
15. Add 500 μL of the 3:1 ethanol-acetone mixture to fully cover the cells and mix by inverting the tube several times.
16. Incubate the cells at RT for 15–20 min and centrifuge at 200×*g* for 5 min to remove ethanol-acetone mixture.

17. Repeat steps 15 and 16 to dehydrate the cells using 1:1 and 1:3 ethanol-acetone mixtures, followed by two washes with 100% acetone (*see* **Note 54**).
18. To embed the sample, remove the 100% acetone, add 400–500 μL of a 1:1 mixture of 100% acetone and Son812 epoxy resin to the sample. Allow the sample to stand at RT for 3–4 h for initial resin infiltration and structural stabilization (*see* **Note 55**).
19. Centrifuge at 200 ×*g* for 5 min, then carefully remove the supernatant using a pipette (*see* **Note 56**).
20. Add 400–500 μL of a 1:2 mixture of acetone and son812 epoxy resin to the sample. Infiltrate overnight at RT to enhance resin penetration and improve curing within the sample.
21. Centrifuge at 200 ×*g* for 5 min, then carefully remove the supernatant using a pipette.
22. Add 400–500 μL son812 epoxy resin to the sample and incubate at 37°C for 2–3 h to further promote curing and increase embedding stability (*see* **Note 57**).
23. To polymerize the samples, place them in a 37°C oven overnight for precuring.
24. Prepolymerize the samples overnight in an oven at 37°C after thorough resin infiltration (*see* **Note 58**).
25. Transfer the sample to 60°C and incubate for 48 h to complete the curing process (*see* **Note 59**).
26. To section the samples, mount the embedded resin blocks onto a Leica UC7 ultramicrotome.
27. After installing a glass knife, adjust the distance between the knife and the sample block.
28. Perform ultrathin sectioning by setting the appropriate cutting speed and heating current.
29. Transfer the resulting sections carefully onto copper grids coated with support film for further processing.
30. To stain the samples, place the sectioned copper grids in 2% uranyl acetate solution for 5 min in the dark.
31. After staining, rinse gently with deionized water and air-dry (*see* **Note 60**).
32. Repeat the rinsing process three times to thoroughly remove residual stain (*see* **Note 61**).
33. Stain the grids in 3% lead citrate solution for another 5 min, rinse again, and dry at RT.
34. To image the samples, use a suitable transmission electron microscope such as the Hitachi H-7700 transmission electron microscope operated at an accelerating voltage of 80 kV.

### 3.8 Focus Ion Beam Scanning Electron Microscopy (FIB-SEM)

Sample preparations for conventional TEM and FIB-SEM are largely overlapping, with many steps sharing identical procedures and precautions. Therefore, FIB-SEM preparation steps are identical or highly similar to those used in conventional TEM sample preparation, these shared steps will be abbreviated, and emphasis will be placed on the procedures that differ.

1. For FIB-SEM sample preparation, double fixation is required.
2. Rinse the cells two to three times with sterile water.
3. Collect and transfer the cells to a 1.5 mL RNase-free centrifuge tube.
4. Add 500 μL of 2.5% glutaraldehyde fixative to the centrifuge tube, then fix overnight at 4°C.
5. After fixation, wash the cells three times with 0.1 M PB for 20 min to remove excess fixative as discribed above.
6. Transfer the cells to a fresh tube containing 1% OsO4 and fix overnight at 4°C.
7. Wash the cells three times with ultrapure water for 20 min each to remove OsO4.
8. After washing, add 500 μL 3% lead citrate staining solution to the cells and incubate at 37°C in a drying oven for 2–3 h.
9. Centrifuge at 200 × *g* for 5 min, then remove the supernatant using a pipette. Wash three times with ultrapure water for 20 min each.
10. Immerse the cells in an aqueous 2% uranyl acetate solution and stain overnight at RT in the dark (*see* **Note 62**).
11. Centrifuge at 200 × *g* for 5 min, then remove the solution using a pipette. Repeat washing three times with ultrapure water for 20 min each (*see* **Note 63**).
12. To dehydrate, embed, and polymerize samples, follow the same protocol as described for TEM sample preparation from Sects. 3.7.10 to 3.7.26.
13. Section the samples using a Zeiss Crossbeam 350 dual-beam electron microscope. A valid scanning voltage is 2 kV, FIB beam current is 30 kV/3 nA, resolution is 6 nm, slice thickness is 3786 nm, and dwell time per pixel is 533.3 μs.
14. Perform continuous imaging of the cross-sections to obtain a series of consecutive images (Fig. 4A).
15. Process the consecutive images using Dragonfly software (version 2020) to segment and label the relevant ultrastructures (Fig. 4B), resulting in a 3D reconstruction of the target structures.

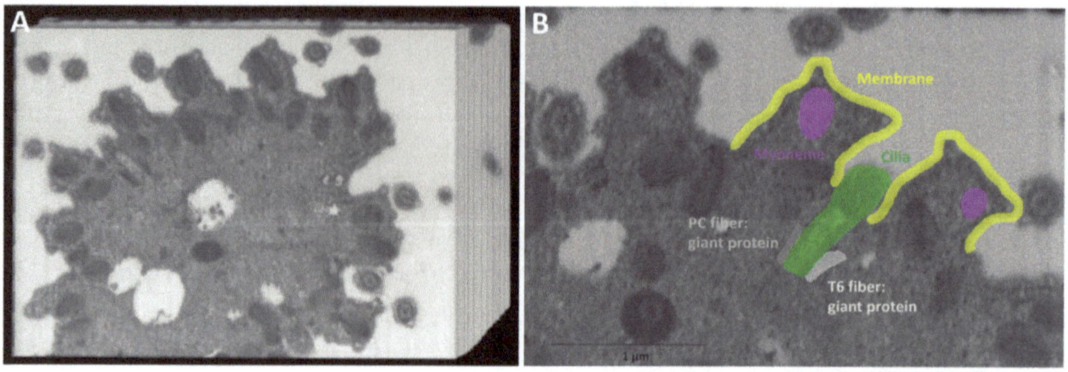


**Fig. 4** Panel figure showing electron microscopy images. Panel **A** displays a series of consecutive cross-sectional images from a specific region of the neck. Panel **B** highlights specific subcellular structures with color annotations: the membrane in yellow, myoneme in purple, and cilia in green. "PC fiber: giant protein" and "T6 fiber: giant protein" are also indicated. Scale bar, 1 μm

## 4 Notes

1. 1% solid agar is dissolved at high temperature and solidifies at RT. Before use, it should be reheated in a microwave and then cooled to approximately 50°C at RT.
2. PMSF is toxic and unstable. It is immediately added to the lysis buffer from the storage solution before use in the fume cupboard.
3. Acetonitrile and formic acid are volatile and irritating, all operations should be performed in a fume hood. The solution should be stored in a sealed container at 4°C, protected from light. Because acetonitrile is volatile, it is recommended to prepare and use it immediately to avoid concentration changes.
4. LB medium with 1% agar is dissolved at high temperature and solidifies at RT. Before use, we should heat it in the microwave and cool it to about 50°C at RT. Then, add the ampicillin to a final concentration of 100 μg/mL.
5. The medium containing IPTG should be stored at 4°C in a dark and used within 1–2 weeks. The IPTG stock can be kept at RT for one month.
6. Tween-20 is a viscous liquid. To aspirate Tween-20 easily, cut the end of a blue pipette tip.
7. In addition to Glutathione Sepharose™ 4B, alternative commercially available resins such as BeyoGold™ GST-tag Purification Resin (Beyotime Biotechnology) can also be used for the purification of GST-fusion proteins. Despite potential differences in purification efficiency, these affinity

matrices demonstrate similar capacities for binding GST-tagged proteins.

8. The solution should be cooled to RT before adjusting the pH, as the pH of the solution varies greatly with temperature.
9. Weigh 0.154 g glutathione (GSH) into 50 mL of 50 mM Tris-HCl, pH 8.0. Shake gently until the powder is completely dissolved. Dispense into 1 mL aliquots, and store at –20°C until needed. Avoid repeated freeze-thaw cycles.
10. When the stock buffer is stored at –20°C, SDS precipitation may occur. The precipitation should be completely dissolved before adding DTT. DTT must be added fresh prior to use, as prolonged storage leads to oxidation and loss of reducing activity. If short-term storage is unavoidable, aliquot the buffer and store at –20°C.
11. Briefly centrifuge the tube before dissolving the powder to ensure that the product is collected at the bottom. It is best to prepare this solution fresh each time. However, it can be stored as aliquots in tightly sealed vials at –20°C. Generally, these will be usable for up to 1 month. Prior to use, the stock solution should be diluted with water to the working concentration.
12. Triton X-100 is a viscous liquid. To aspirate Triton X-100 easily, cut the end of a blue pipette tip.
13. The preparation should be carried out in a fume hood, because paraformaldehyde (PFA) is highly irritating. In addition, the temperature should not be too high during heating to prevent PFA from degrading and becoming ineffective. The solution can be stored at 4°C for short-term use, or at -20°C for long-term storage.
14. DAPI should be protected from light during preparation and use to prevent photodegradation. For use, dilute the stock solution 1:1000 to achieve a working concentration of 1 μg/mL.
15. $OsO_4$ is highly toxic and volatile. Appropriate personal protective equipment must be worn by users. The preparation of 1% $OsO_4$ solution should be carried out strictly under a fume hood.
16. Ethanol is volatile. Consequently, when preparing a graded ethanol series, it is generally recommended to preparefresh and use it immediately.
17. Acetone is a flammable and volatile organic solvent. It should be manipulated in the fume hood to prepare a graded ethanol and acetone mix.

18. Since the reagent may dissolve slowly in water, heat and use a magnetic stirrer to accelerate dissolution. However, avoid excessive temperatures to prevent volatilization or decomposition. In addition, the reagent is radioactive and toxic, and must be handled in a specialized laboratory in a fume hood. When using the reagent, wear full laboratory protective clothing, gloves, and safety goggles to avoid contact with skin and eyes. In case of contact, immediately rinse the affected area with plenty of cold water.
19. For this culture, *Lacrymaria* cells will be more concentrated. Using this method, the density of *Lacrymaria* can reach up to 30 cells/mL in the culture medium. However, this density is still relatively low, so we need to culture a sufficient number of cells for the molecular biology experiments. Because *Lacrymaria* cells are very small and fragile, they are difficult to collect by centrifugation.
20. Due to the collection of the head-neck parts from active cells, the glass micropipette on the mouth pipette serves as a surgical knife to cut cells. In the following methods, due to the limited number of cells, all operations with a low number of cells, including selection of cells, fixation, and washing, should be carried out using a mouth pipette.
21. cOmplete proteinase inhibitor (Roche Diagnostics, Indianapolis, IN, USA) EDTA-free, effectively inhibits serine and cysteine proteases present in bacterial, yeast, plant, and animal cell extracts.
22. The large size of proteins does not allow identifying them by chemical methods. It is necessary to use a protease to digest the protein into small peptide fragments before identification through proteomics.
23. Peptides were first separated with a gradient of 10–100% acetonitrile over 60 min into 60 fractions. Then, the peptides were combined into 30 fractions and dried by vacuum centrifuging.
24. If the primary antibody is not commercialized available and needs to be custom-produced, many companies require approximately six mouths for production, which should be taken into account by researchers. In our particular case, myosin and giant protein primary polyclonal antibodies were produced. Followed by protocol from Sects. 3.2.1 to 3.2.22 to obtain the recombinant proteins, the recombinant proteins were purified by Ni–NTA agarose (Merck KGaA, Darmstadt, Germany), and the protein molecular weight, purity, and concentration were identified by SDS-PAGE. As a short guide for production, 2 mg of purified protein per injection was needed as an antigen for rabbit or mouse antibodies. Freund's

complete adjuvant (FCA) and Freund's incomplete adjuvant (FIA) were used for immunization, Japanese White rabbits received 5 injections and Balb/C mice received 4 injections. Antiserum titers can be tested routinely by enzyme-linked immunosorbent assay (ELISA). An affinity column was used for antibody purification and was made by coupling 1 mg of antigen to CNBr-activated Sepharose 4B (GE). Usually, 10 mL antiserum will be required to purify a large amount of antibody for subsequent studies. Serum can be loaded into the column up to 12 h to maximize recovery of antibodies. Antibodies are eluted with 0.15 M glycine buffer, pH 2.5 [1].

25. The expression level can be controlled by adjusting the IPTG concentration to avoid excessive formation of inclusion bodies. Because low concentration induction tends to promote soluble proteins, while excessively high concentrations may cause nonspecific effects. The condition of protein expression could also be at 37°C for 4 h or at 16°C overnight. Since bacterial growth is slower at 16°C, the induction period is necessarily longer, with overnight induction being the typical maximum.
26. Resuspend the pellet in an appropriate volume of ice-cold 1× PBS, typically 5–10 mL per gram of wet cell pellet, depending on the level of target protein expression.
27. The supernatant can be flash-frozen in liquid nitrogen and stored at −80°C for long-term preservation.
28. His-tag protein purification kit provides the relevant reagents required for the His-labeled protein and the empty column tube of the affinity chromatography column. The relevant reagents include lysis buffer, wash buffer, and eluent buffer. The eluent buffer contains imidazole. The imidazole concentration and pH of the eluent may vary depending on the specific protein.
29. Primary antibodies can be reused at least three times according to our attempt.
30. To verify the interaction between centrin and myosin in vitro, we individually expressed the centrin identified by the highest number of matched peptides in MS analysis and the first IQ motif of myosin, followed by GST pull-down assays.
31. Sepharose 4B should be thoroughly washed with 1× PBS to remove the 20% ethanol storage solution, as residual ethanol may interfere with subsequent procedures.
32. Glutathione Sepharose 4B, equilibrated with 1× PBS, may be stored at 4°C for up to a month.
33. Dilute GST-beads with naked beads appropriately (to adjust the same amount of input for GST and GST fusion protein).

Add the same number of beads to the medium. GST protein was added to the Glutathione Sepharose 4B was the control group.

34. The incubation time can be adjusted or extended based on the binding efficiency.
35. Reserve a small aliquot for SDS-PAGE analysis to measure the binding efficiency to the matrix.
36. Take 20 μL samples for SDS-PAGE gel and Coomassie staining to quantify the amount of bound protein.
37. Keep the beads hydrated throughout the experiment by ensuring the liquid column never fully depletes. In addition, to confirm the consistency of results between the IF and the pull-down assay, we performed Western blotting using an anti-centrin antibody (mouse) and centrin-GST expressed in vitro. This approach can be applied to each protein assayed.
38. LAT-A is a temperature- and light-sensitive natural product, and the solution after preparation is typically stored at −20°C in the dark. When used, warm the solution to RT, which helps prevent the formation of tiny crystals or precipitates due to low temperature. This step ensures that the drug solution is fully dissolved before use and warrants the accuracy of the drug solution in concentration and biological activity. For CB, CK666, and para-Nitroblebbistatin, the treatment procedure was the same as that described for LAT-A, with their final concentrations of 75 μM, 100 μM, and 50 μ M, respectively.
39. To ensure consistency and comparability of the experimental results, cells of similiar viability should be selected from the same batch at comparable times.
40. The phenotypes should be described carefully and quantitatively in preliminary experiments at different conditions, including active, resting, short neck, and globose, which enabled us to identify their real phenotype by inhibitors.
41. Phalloidin is a mushroom-derived toxin commonly conjugated to fluorescent dyes for labeling cytoskeletal F-actin. As a highly selective bicyclic peptide, it binds to and stabilizes F-actin, thereby inhibiting actin filament depolymerization. SiR-actin is a live-cell compatible probe that combines the silicon rhodamine (SiR) fluorophore with the actin-binding natural product jasplakinolide. It enables specific F-actin labeling in live cells with minimal background. Notably, SiR-actin can also be used to stain F-actin in PFA-fixed cells with efficiency comparable to that of phalloidin when following standard phalloidin staining protocols.

42. Fixation methods may vary depending on the cell type and the specific experimental purpose. This information is provided for reference only. We have found that the results improve upon preincubation of the washed cells in 8-well glass chambers containing 200 μL precooled 0.5× PHEM on ice for 1 h to get the extended forms of other ciliates, like *Spirostomum* or *Stentor*. Wash cells with ultrapure water, then add 100 μ L fixation solution B on ice for 15 min. Wash the cells again, and transfer cells to the new chambers containing 100 μL fixation solution B [1, 29]. For *Euplotes*, it is good to first permeabilize in 1% saponin for 1 min and wash once with 1× PHEM. Fix the cells with fixation solution C for 3 min, and wash once with 1× PHEM again. Then, treat the cells with 1× PBS containing 0.5% Triton X-100 for 3 min to enhance membrane permeability. Wash once with 1× PHEM again [30]. This fixation method is unsuitable for *Schmidingera*, as it results in its deformation or lysis due to its high-salt habitat. Fix the cells by using the fixation solution D and incubate at 4°C for 1 h [31].
43. The slides were freshly prepared and used immediately to ensure mounting quality.
44. The cells-containing area is outlined on the backside of the slide with a black marker, Fixogum is applied along the corresponding front-side outline on the front side. After complete drying, this form a hydrophobic barrier that effectively minimizes antibody consumption and prevents diffusion during subsequent immunostaining steps.
45. The use and storage of antibodies (including primary and secondary antibodies) should follow the manufacturer's instructions, they are generally stored at −20°C in the dark. To prevent degradation caused by repeated freeze–thaw cycles, the antibodies should be aliquoted into a 1.5 mL RNase-free centrifuge tubes for storage. All steps involving the secondary antibody should be performed under light-protected conditions.
46. The primary antibody working solution should completely cover the cells (usually 20 μL per sample is sufficient, avoiding the formation of bubbles) to ensure thorough incubation. For colocalization experiment, only antibodies raise in different sources can be used can be mixed and used directly. The role of BSA is to block some nonspecific sites, reducing background, while the role of Triton X-100 is to permeabilizes the cells to facilitate antibodies access to intracellular antigens. After adding the primary antibody, care should be taken to prevent sample drying, which may increase the background signal.

47. The dye can also be used independently, the subsequent steps can then be carried out as usual.
48. Equilibrate the medium to RT in the dark is beneficial for reducing the impact of condensationon imaging. The Abberior solid mount can reduce the "focus drop" caused by the water evaporation during the image acquisition, and also provide some degree of anti-fading protection.
49. Centrifugation may cause cells deformation; it is better to collect them using a mouth pipette. The tube contains as little moisture as possible, as residual liquid may dilute the fixative and influence the fixing quality.
50. The fixation process should be performed overnight at 4°C to prevent cell membrane rupture or structural damage caused by excessive heat. Ensure that the fixative completely covers the sample to avoid incomplete fixation in certain areas.
51. Centrifugation should be performed at low speed and for a short duration to minimize the risk of disrupting cellular structures.
52. Each rinse should last for at least 20 min to ensure thorough removal of residual fixative, thereby preventing glutaraldehyde from interfering with subsequent steps.
53. Discard the previous ethanol solution, and immediately add the next concentration of ethanol promptly to avoid prolonged exposure of the sample to air.
54. Both ethanol and acetone are highly volatile. Minimize exposure time during handling and promptly seal the tube after each liquid exchange.
55. It is recommended to prepare the acetone-resin mixture at least 30 min in advance to minimize bubble formation. When adding the mixture to the tube, the bottom of the tube is gently rinsed with a pipette to ensure thorough contact between the mixture and the cell pellet.
56. It should be carefully performed to prevent the pipette tip from touching the bottom of the tube to disturb the cell pellet.
57. It should be ensured that a stable temperature is maintained during the infiltration steps, particularly during incubation at 37°C. It should be avoided to overheat the sample, which leads to premature resin polymerization.
58. This step initiates gelation of the resin, allowing partial solidification while maintaining sufficient fluidity.
59. This procedure promotes cross-linking of resin molecules by elevated temperature to fully polymerize. This ensures that the samples are stably embedded and suitable for sectioning.

During polymerization, samples should remain undisturbed to prevent mechanical disruption of the polymerization process. Maintain a stable temperature throughout polymerization to prevent the resin from hardening. It is better to sustain the curing time, because incomplete cross-linking may compromise the structural integrity of the resin and affect subsequent sectioning.

60. Use a pipette to dispense an appropriate volume of water gently onto the grid surface, ensuring the water flows gradually across the sample to rinse away excess staining solution. Slightly tilt or gently shake the grid or container to aid the water flow over the sample. Avoid rapid water flow or direct impact on the grid to prevent dislodging the sample.
61. Blot the edge of the grid with nonwoven fabric or clean absorbent paper to remove excess water gently. Do not scratch or disturb the sample to maintain its structural integrity.
62. Care should be taken to keep the samples completely immersed to avoid flotation.
63. Following the rinsing steps, the samples may be temporarily stored in ultrapure water at 4°C until further processing.

## References


1. Zhang J, Qin WW, Hu C et al. (2023) Giant proteins in a giant cell: molecular basis of ultrafast Ca2+-dependent cell contraction. Sci Adv 9:eadd6550
2. Qin WW, Hu C, Gu SY et al. (2024) Dynamic shape-shifting of the single-celled eukaryotic predator *Lacrymaria* via unconventional cytoskeletal components. Curr Biol 34:4869–4883
3. Skillman KM, Diraviyam K, Khan A et al. (2011) Evolutionarily divergent, unstable filamentous actin is essential for gliding motility in Apicomplexan parasites. PLoS Pathog 7(10):e1002280
4. Martinez M, Mageswaran SK, Guerin A et al. (2023) Origin and arrangement of actin filaments for gliding motility in Apicomplexan parasites revealed by cryo-electron tomography. Nat Commun 14: 4800
5. Mast SO (1911) Habits and reactions of the ciliate, *Lacrymaria*. J Anim Behav 1:229–243
6. Pearse AS (1912) General reviews and summaries: recent literature on the behavior of the lower invertebrates. Psychol Bull 9:281
7. Yanase R, Nishigami Y, Ichikawa M et al. (2018) The neck deformation of *Lacrymaria olor* depending upon cell states. J Protistol 51:1–6
8. Coyle SM, Flaum EM, Li H et al. (2019) Coupled active systems encode an emergent hunting behavior in the unicellular predator *Lacrymaria olor*. Curr Biol 29:3838–3850
9. Wan KY (2019) Ciliate biology: the graceful hunt of a shape-shifting predator. Curr Biol 29:R1174–R1176
10. Flaum E, Prakash M (2024) Curved crease origami and topological singularities at a cellular scale enable hyper-extensibility of *Lacrymaria olor*. Science 384:adk5511
11. Nath T, Mathis A, Chen AC et al. (2019) Using DeepLabCut for 3D markerless pose estimation across species and behaviors. Nat Protoc 14:2152–2176
12. Aebersold R, Mann M (2016) Mass-spectrometric exploration of proteome structure and function. Nature 537:347–355
13. Guo T, Steen JA, Mann M (2025) Mass-spectrometry-based proteomics: from single cells to clinical applications. Nature 638:901–911
14. Fenn JB, Mann M, Meng CK et al. (1989) Electrospray ionization for mass spectrometry of large biomolecules. Science 246:64–71
15. Kim SY, Hakoshima T (2019) GST pull-down assay to measure complex formations. Methods Mol Biol 1893:273–280

16. Gnanasekaran P, Pappu HR (2023) Detection of protein-protein interactions using glutathione-s-transferase (GST) pull-down assay echnique. Methods Mol Biol 2690: 111–115
17. Datta A, Kamthan A, Kamthan M (2015) A simple protocol to detect interacting proteins by GST pull down assay coupled with MALDI or LC-MS/MS analysis. Protoc Exch. https://doi.org/10.1038/protex.2015.093
18. Brymora A, Valova VA, Robinson PJ (2004) Protein-protein interactions identified by pull-down experiments and mass spectrometry. Curr Protoc Cell Biol 22:17.5.1–17.5.51
19. Coling D, Kachar B (1998) Principles and application of fluorescence microscopy. Curr Protoc Mol Biol 44:14.10.1–14.10.11
20. Combs CA (2010) Fluorescence microscopy: a concise guide to current imaging methods. Curr Protoc Neurosci Chapter 2:Unit2.1
21. Tam J, Merino D (2015) Stochastic optical reconstruction microscopy (STORM) in comparison with stimulated emission depletion (STED) and other imaging methods. J Neurochem 135:643–658
22. Im K, Mareninov S, Diaz MFP et al. (2019) An introduction to performing immunofluorescence staining. Methods Mol Biol 1897:299–311
23. Winey M, Meehl JB, O'Toole ET et al. (2014) Conventional transmission electron microscopy. Mol Biol Cell 25:319–323
24. Lomholdt WB, Leth Larsen MH, Valencia CN et al. (2024) Interpretability of high-resolution transmission electron microscopy images. Ultramicroscopy 263:113997
25. Richard W (2009) Focused Ion Beam (FIB) combined with SEM and TEM: advanced analytical tools for studies of chemical composition, microstructure and crystal structure in geomaterials on a nanometre scale. Chem Geol 261:217–229
26. Peterson JR, Mitchison TJ (2002) Small molecules, big impact: a history of chemical inhibitors and the cytoskeleton. Chem Biol 9:1275–1285
27. Fujiwara I, Zweifel ME, Courtemanche N et al. (2018) Latrunculin A accelerates actin filament depolymerization in addition to sequestering actin monomers. Curr Biol 28:3183–3192
28. Vondráková Z, Eliášová K, Vágner M (2014) The anti-actin drugs latrunculin and cytochalasin affect the maturation of spruce somatic embryos in different ways. Plant Sci 221-222:90–99
29. Xu LX, Bhamla MS (2019) Biophysical mechanism of ultrafast helical twisting contraction in the giant unicellular ciliate *Spirostomum ambiguum*. bioRxiv https://doi.org/10.1101/854836
30. Lin Q, Fan XP, Sun L et al. (2013) Microtubular organelles in *Euplotes aediculatus* (Ciliophora: Hypotrichida) revealed by fluorescent labeling. J East China Norm Univ (Natl Sci) 5:53–60
31. Li FC, Zhang WY, Luo S et al. (2025) Cell supporting cytoskeletons and phagocytic acquisition of compatible solutes emerge as common strategies for high-salt adaptation in different ciliates. Sci China Life Sci 68:3664–3681

# Chapter 12

# Methods to Study the Role of Cytoskeletal Motor Proteins in the Organization of the Postsynaptic Membrane

**Amanda L. Neisch, Thomas Pengo and Thomas S. Hays**

## Abstract

The organization of cytoskeletal elements and proteins on the postsynaptic side of a synaptic connection is crucial for neuronal function. The postsynaptic compartment receives and processes the signal for proper neuronal communication and signaling. Here, we provide methods to analyze the role of cytoskeletal motors, such as cytoplasmic dynein, in the organization of the postsynaptic membrane. In this chapter, we describe how to measure and quantify the protein abundance at the postsynaptic membrane. Additionally, we describe how to measure the size of ionotropic glutamate receptor (iGluRs) clusters to understand their organization on the postsynaptic side of the junction.

**Keywords** Motor proteins, Dynein, Postsynaptic membrane, Neuromuscular junction, Ionotropic glutamate receptors, Immunofluorescence, Drosophila, Quantification

## 1 Introduction

While the importance of the microtubule cytoskeleton and motor proteins is well established in neurons [1–5], the function of these same cellular components on the postsynaptic side of neuronal connections has been understudied [6, 7]. Electrical impulses from a neuron result in the release of neurotransmitters at the synapse, which are received on the postsynaptic side by receptors. Neurons often synapse onto the dendritic spine of another neuron, but in some instances, neurons synapse onto another tissue type to modulate function. This is true of the neuromuscular junction (NMJ), where an excitatory motor neuron innervates a muscle fiber. Electrical impulses from the neuron to the muscle result in neurotransmitter release and reception by the muscle membrane, which results in depolarization and muscle contraction (reviewed in [8]). The coordination of this signaling relies on the proper organization of components on the postsynaptic side of the junction.

Noa B. Martín-Cófreces et al. (eds.), *The Cytoskeleton*, Methods in Molecular Biology, vol. 3070,
https://doi.org/10.1007/978-1-0716-5515-3_12,

In the *Drosophila*NMJ, type-I motor neurons are glutamatergic, releasing the neurotransmitter glutamate that is received by ionotropic glutamate receptors (iGluRs) in the skeletal muscle cell. *Drosophila* iGluRs are functionally equivalent to mammalian AMPA (α-amino-3-hydroxy-5-methyl-4-isoxazolepropionic acid) receptors of the central nervous system. Comparatively, most glutamatergic synaptic transmission in humans occurs on dendritic spines of other neurons in the CNS. In addition to having equivalent receptors, the structural, cytoskeletal, and membrane-associated components that function to organize and regulate the postsynaptic compartment of the *Drosophila* NMJ are functionally conserved with orthologous components in the dendritic spines of mammals [9–14]. Thus, *Drosophila* is a great model system in which to study the function of the microtubule motor proteins, such as dynein, in regulating the NMJ postsynaptic membrane organization and may advance our understanding of motor protein function in the postsynaptic organization of mammalian dendritic spines.

Here, we describe methods for studying the role of motor proteins, such as dynein, in the organization of the postsynaptic membrane, using *Drosophila*. We outline approaches to identify protein components and complexes that may be altered in abundance on the postsynaptic side of the NMJ following motor protein depletion. In addition, we detail analysis tools used to quantify these changes and determine which components are affected. Finally, we describe methods for assessing the size of glutamate receptor clusters at the NMJ, as changes in cluster size can impact synaptic function [9, 15].

## 2 Materials

1. Incubator capable of holding a constant temperature of 29°C.
2. Standard cornmeal agar food (Bloomington *Drosophila* Stock Center recipe; BDSC): 1.5% inactive yeast (w/v), 0.9% soy flour (w/v), 6.7% yellow cornmeal (w/v), 0.07% light corn syrup (v/v), 0.53% agar (w/v), 59 mM propionic acid in water (*see* **Note 1**).
    a. Mix the yeast and soy flour together, hydrate with water (approximately 7% of the final food volume), and set aside.
    b. Measure out the corn syrup, then set aside.
    c. Measure the cornmeal and agar into a large container and add water (about 12% of the final food volume). Mix until smooth, adding more water if necessary to produce a smooth slurry.
    d. Set aside water (about 9% of the final food volume) and bring the rest to a full boil.

e. Stir the cornmeal and agar mixture continuously and pour the smooth slurry into the rapidly boiling water.
f. If possible, stir the pot at the same time, if possible. If lumps begin to form while pouring, add some of the reserved water to the cornmeal and agar mixture. Mix well before pouring the rest of it into the boiling mixture. Use any remaining water to rinse any residue from the pot.
g. Continuously stirring, add the yeast and soy mixture, followed by the corn syrup.
h. When the mixture begins to bubble at the edges, cook for 10 min with frequent stirring.
i. Turn off the heat and leave to cool to 70°C.
j. Stir in the propionic acid.
k. Dispense the medium.
l. Leave to cool before filling aliquots of approximately 10 mL.

3. Fly strains of choice for genetic crosses (see Table 1).
4. Stereomicroscope.
5. Confocal microscope with imaging capacities.
6. Tungsten wire 0.125 mm diameter to make dissection pins.
7. 9 V battery, to make dissection pins.
8. Dissection pins, six per dissected animal (*see* **Note 2**).
9. Assembling forceps.
10. Dissection forceps (*see* **Note 3**).
11. Dissecting scissors (*see* **Note 4**).
12. Custom-made dissection dishes, as described in [16]; however, these are 5 mm thick with a depression well that is 4 mm deep (*see* **Note 5**).
13. SYLGARD™ 184 Silicone Elastomer Kit: Mix ten parts base with one part of curing agent by weight in a 50 mL conical tube. Mix thoroughly, and then put on a rocker at room temperature for ~1 h. Pour into customized dishes or 9-well glass depressions, leaving enough room in the depression/dish for staining solutions to be added. Let cure overnight at room temperature.
14. Coverslips (24×40 mm, No. 1.5) and slides
15. HL3 saline (hemolymph-like saline): 5 mM Hepes pH 7.2,70 mM NaCl, 5 mM KCl, 20 mM $MgCl_2$, 10 mM $NaHCO_3$, 5 mM trehalose, 115 mM sucrose (*see* **Note 6**).
16. 10× PBS (phosphate-buffered saline): 100 mM $Na_2HPO_4$ pH 7.4, 17.6 mM $NaH_2PO_4{\cdot}H_20$, 1.37 M NaCl. Add

**Table 1**

**Genetic tools to manipulate motor proteins/and or investigate epitope-tagged proteins on the postsynaptic side of the NMJ**

| Genotype | Use | Source/Publication |
|---|---|---|
| *Mef2Gal4.R-3* | Gal4 strain to express UAS-driven strains in all embryonic skeletal muscles | BDSC [20] |
| *G14Gal4/CyoGFP* | Gal4 strain to express UAS-driven strains in all embryonic skeletal muscles | [21] |
| *elavGal80; Mef2Gal4.R-3* | Postsynaptic expression of UAS-driven strains in the somatic muscle while blocking any presynaptic neuronal expression | *elavGal80* [22], *Mef2Gal4.R-3:* BDSC |
| *UAS-dhc64cRNAi*$^{HMS01587}$ | UAS-driven dsRNA to deplete the heavy chain subunit of the dynein complex, Dhc64c | BDSC |
| *UAS-swRNAi*$^{HM05249}$ | UAS-driven dsRNA to deplete the intermediate chain subunit of the dynein complex, sw | BDSC |
| *Mhc-GluRIIB-Myc-3* | Muscle-specific expression of the Myc-tagged GluRIIB subunit | BDSC |
| *Mhc-GluRIIA-Myc-2* | Muscle-specific expression of the Myc-tagged GluRIIA subunit | BDSC |
| *C57Gal4, UAS-DlgS97-GFP* | Postsynaptic, muscle-specific expression of the GFP-tagged DlgS97 isoform | [23] |
| *UAS-Nlg-GFP* | UAS-driven strain for postsynaptic expression/localization of GFP-tagged Neuroligin1 | [10] |
| *UAS-PLC-δ-PH-eGFP-3* | UAS-driven GFP-tagged reporter for $PIP_2$ levels | BDSC |
| *UASp-6xMyc-sktl* | UAS-driven strain for expression/localization of 6xMyc epitope-tagged Skittles, a phosphatidylinositol 4-phosphate 5-kinase (PIP5K) | [24] |
| *w*$^{1118}$ | *white* mutant allele, in an otherwise wild-type strain. Used to outcross Gal4 lines as a control | BDSC |
| *UAS-mCh-RNAi* | Control dsRNA that targets the mCherry sequence | BDSC |

approximately 750 ml of ultrapure water and a stir bar to a 1 L plastic beaker. Weigh 90 g NaCl, 20 g $Na_2HPO_4$, 8.3 g $NaH_2PO_4{\cdot}H_20$ and transfer to the plastic beaker. Stir to dissolve. Bring up the final volume to 1 L in a graduated cylinder.

**Table 2**
**Antibodies to investigate proteins on the postsynaptic side of the NMJ**

| Antibody | Concentration | Use | Source/publication |
|---|---|---|---|
| Mouse α Dlg (4F3-c) | 1:500 | Localization of Dlg, a postsynaptic scaffolding protein | Developmental Studies Hybridoma Bank |
| Mouse α α-Spectrin (3A9-S) | 1:10 | Localization of α-Spectrin, an actin cytoskeleton regulator | Developmental Studies Hybridoma Bank |
| Guinea Pig α β -Spectrin | 1:1000 | Localization of β-Spectrin, an actin cytoskeleton regulator | [15] |
| Rabbit α GluRIIC | 1:3000 | Localization of the GluRIIC subunit of iGluRs | [25] |
| Mouse α Myc (9B11) | 1:2000 | Localization of Myc-epitope tagged iGluR subunits GluRIIA/B | Cell Signaling Technology |
| Rabbit α GFP (TP401) | 1:1000 | Localization of GFP tagged proteins (Nlg1-GFP) and the $PIP_2$ reporter (*PLC-δ-PH-eGFP*) | Torrey Pines Biolabs |
| Rabbit α Nlg1 | 1:500 | Localization of Neuroligin1, a postsynaptic cell-adhesion molecule | [10] |
| Guinea pig α Nlg1 | 1:250 | Localization of Neuroligin1, a postsynaptic cell-adhesion molecule | [26] |
| Mouse α Ten-m (mAB20-s) | 1:10 | Localization of Ten-m, a postsynaptic cell-adhesion molecule | Developmental Studies Hybridoma Bank |
| Mouse α-adducin (1B1-c) | 1:100 | Localization of Adducin, an actin cytoskeleton regulator | Developmental Studies Hybridoma Bank |
| Rabbit α skittles | 1:800 | Localization of Skittles, a PI4P5 Kinase | [27] |
| Goat α HRP Alexa Fluors | 1:1000 | Labels the presynaptic side of NMJs | [28] *See* **Note 8** |
| Alexa Fluor secondaries | 1:500 | For visualization of primary antibodies | *See* **Note 8** |

Filter sterilized and store at room temperature. To make PBS, dilute in ultrapure water.

17. PBT (permeabilization buffer): PBS, 0.1% Triton X-100.
18. PTN (permeabilization/blocking buffer): PBS, 0.1% Triton X-100, 1% normal goat serum.
19. 4% PFA (paraformaldehyde) solution: dilute from 16% paraformaldehyde in 1× PBS.

20. 50% glycerol/PBS solution: PBS, 50% glycerol (v/v). Place on a platform rocker to mix thoroughly before use.
21. Anti-fading mounting media (*see* **Note 7**).
22. Primary antibody of choice to visualize postsynaptic membrane components (see Table 2).
23. Fluorescently conjugated HRP primary antibodies of choice, to label the presynaptic compartment.
24. Secondary antibodies of choice conjugated with Alexa Fluorophores.
25. Nail polish.
26. ImageJ software: https://imagej.net/downloads; https://github.com/imagej/ImageJ

## 3 Methods

In this section, we describe how to measure and quantify changes in protein intensity at the postsynaptic NMJ. The postsynaptic structure of the *Drosophila* NMJ is known as the subsynaptic reticulum (SSR). SSR is an elaboration of the muscle membrane with numerous membrane folds and invaginations that form a complex structure that surrounds the presynaptic motor neuron terminal. Proteins enriched at the postsynaptic membrane (postsynaptic density proteins) are found throughout the SSR; these include scaffolding proteins, cell-adhesion molecules, which link the presynaptic and postsynaptic compartments, membrane phospholipids, actin, and actin regulator proteins [9–14, 17, 18]. In Table 1, genetic tools are listed to deplete dynein complex components from the postsynaptic, skeletal muscle cell, and in addition genetic tools to visualize tagged postsynaptic proteins at the NMJ. Table 2 lists antibodies that can be used to visualize specific postsynaptic membrane proteins, such as postsynaptic cell-adhesion molecules and actin cytoskeletal regulators.

### 3.1 Genetic Crosses, Dissections, and Immunostaining of Samples

1. Cross ~10 virgin females to ~5 males of the appropriate *Drosophila* genotypes at 25°C. Include control crosses of the Gal4 driver to $w^{1118}$ or a control RNAi, such as *UAS-mCherryRNAi*, at the same time.
2. The next day, transfer the adults to a new vial and shift to 29°C, discard the old vial (*see* **Note 9**).
3. Allow flies to lay eggs for 24 h, then remove adults. Optionally, transfer the adults to a new vial for a second 24 h egg lay collection.
4. Select third instar wandering larvae of the appropriate genotype. You may need to select against balancer chromosomes in a stock, using phenotypic (e.g., Tubby (Tb) pupae) or

fluorescently marked balancers (e.g., YFP or GFP). The larvae are typically at the wandering stage ~4 days after the eggs were laid. Use larvae of the same sex for comparisons. Briefly rinse larvae in water.

5. Pin larvae dorsal side up, trachea centered, at both ends using assembly forceps to position the pins (see Fig. 1A, B). The pins are placed at a ~75° angle (see Fig. 1A). Before starting

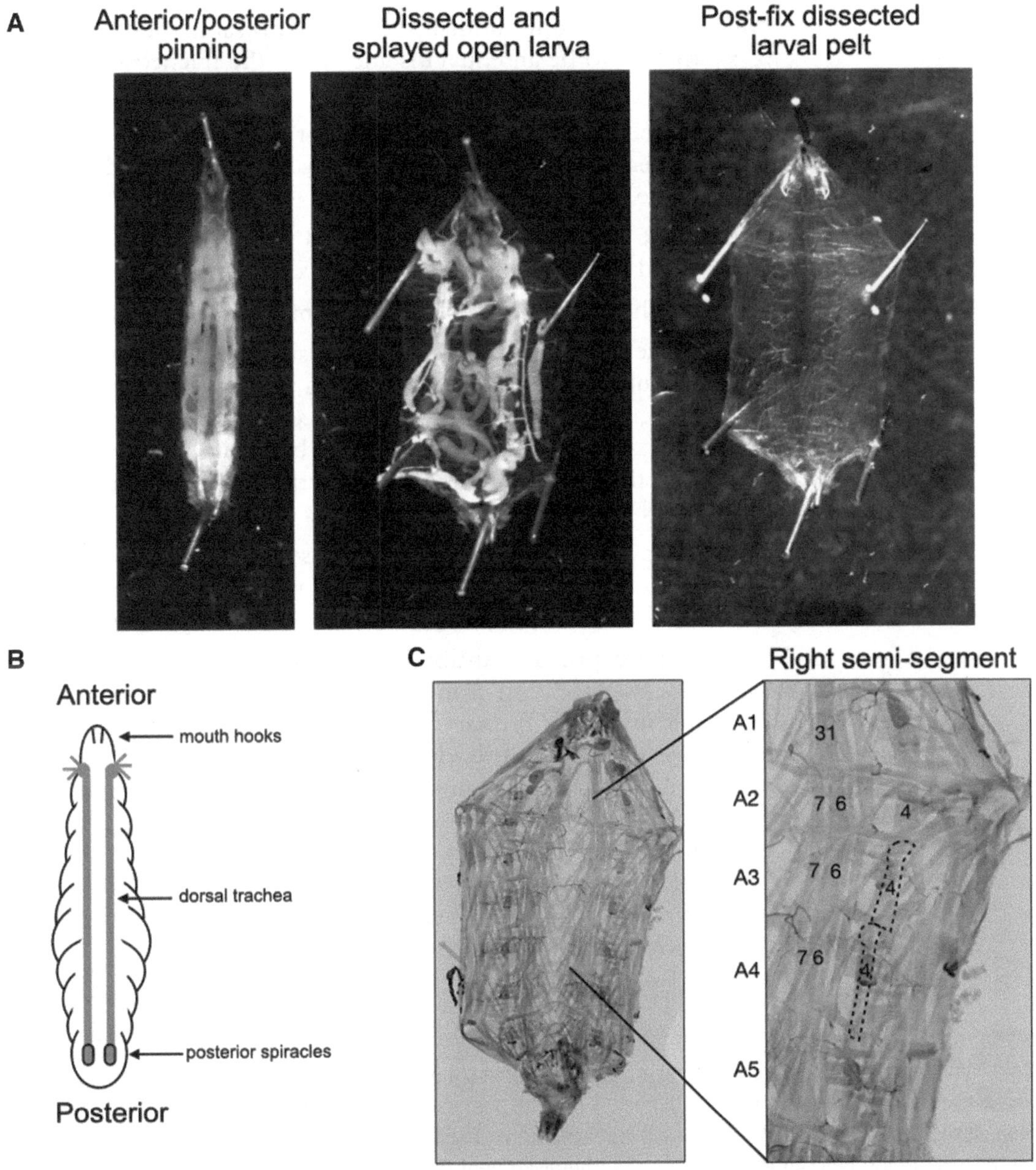


**Fig. 1** **(A)** To dissect larvae for immunostaining, each larva is pinned in a SYLGARD™ dish at the anterior and posterior ends. The larva is then cut open along the dorsal side and pinned to form an elongated hexagon, splaying open the larva. The animal is then fixed, and the fat body, trachea, gut, disks, and brain are removed.

dissections, pin down all the larvae to be dissected. We routinely dissect six animals of each genotype at one time. After all the animals are pinned, fill the dish with HL3 saline to cover the larvae.

6. Make a transverse incision at the posterior dorsal surface just in front of the pin using dissecting scissors. The incision is perpendicular to the anterior/posterior axis and cuts through only the dorsal surface of the animal.
7. Insert the scissors into the posterior transverse incision and gently cut longitudinally along the dorsal midline, centering the cut to be an equal distance from the tracheostomy tubes on either side.
8. Place two pins each at the anterior sides and two each at the posterior sides of the larva to make an elongated hexagon shape (see Fig. 1A–C).
9. Once all the animals are dissected and splayed open as elongated hexagons, add 4% PFA to fix the dissected larvae.
10. Fix the animals in 4% PFA for 20 min at room temperature.
11. While the animals are being fixed, using a #4 or #5 forceps, gently pull on and remove the trachea, the digestive tract, the fat body, and the disks. Using the dissecting scissors, cut the axons near the ventral nerve cord of the brain and remove the brain with the forceps. During the dissection, be careful not to damage the muscle tissue in the process (*see* **Note 10**).
12. After 20 min, remove the fixation solution and rinse the pelts three times in PBS. Replace PBS with PTN to permeabilize and block the tissue for 30 min at room temperature.
13. Prepare primary antibody dilutions in PTN. Remove the blocking solution from the pelts and replace it with ~400–500 μL of the primary antibody solution for each genotype. Use the same antibody solution for genotypes that will be compared. Dissection dishes are then placed in a humidified petri dish (150×15 mm petri dish with a wetted Kimwipe® and placed at 4°C to prevent evaporation of the antibody solution during incubation. Pelts are left pinned in the dish (*see* **Note 11**) to maintain shape and are left in primary antibody solution overnight.

**Fig. 1** (continued) **(B)** A cartoon image of a *Drosophila* larva shows key features used to identify the dorsal side (dorsal trachea), the anterior end (mouth hooks), and the posterior end (posterior spiracles). **(C)** A larva that has been dissected and stained for HRP, which labels neurons. Tissue background fluorescence is used to identify the muscles in the pelt. An image of the right hemisegment of the animal, identifying abdominal segments (A1–A5). Muscles 4, 6, 7, and 31 are indicated. Muscles/NMJ4s of abdominal segments A3 and A4, which are routinely used for imaging and quantification, are outlined with a dotted yellow line

14. The next day, remove the primary antibody solution and wash the pelts 3× in PBT, leaving the solution for 5 min for each wash. Replace PBT with PTN to block the tissue for 1 h at room temperature.
15. Add the appropriate fluorescent secondary antibodies diluted in PTN solution, 500 μL/sample. Incubate for 2 h at room temperature or overnight at 4°C in a humidified petri dish in the dark. Again, use the same antibody solution for genotypes that will be compared.
16. After the 2 h or overnight incubation, wash samples 3× in PBT, leaving the PBT solution for 5 min for each wash.
17. Wash once in PBS, 5 min.
18. Replace the PBS solution with ~500 μL of 50 % glycerol/PBS solution.
19. Remove the pins carefully, using forceps while holding the pelts. Using a dissecting forceps to hold down the pelt, gently remove the pins with a second assembling forceps, being careful not to rip the pelts.
20. Put 3–4 drops of mounting media directly on the slide and spread in a rectangular shape to just under the size of a 22 × 24 mm coverslip in the center of the slide.
21. Lift up the pelts and transfer them into the mounting media. Line the pelts up on the slide so that all are oriented the same way, with the anterior at the top of the slide, and the posterior at the bottom, and with the animals spaced roughly equally on the slide.
22. Place a 22 × 24 mm coverslip on top of the sample, placing it at an angle and gently lowering it to avoid bubbles.
23. Place the slides in the dark at room temperature for 24 h for the mounting media to harden. Seal the coverslips with nail polish and store at 4°C if not imaging immediately.

### 3.2 Imaging and Quantification of Postsynaptic NMJ Protein Abundance

1. Acquire images of the samples on a confocal microscope, using Nyquist sampling.
2. Image NMJs from the same abdominal segments per animal. We image NMJ4 (muscle 4) of abdominal segments 3 and 4 for analysis (*see* **Note 12**).
3. Acquire images of the complete NMJ using a high-resolution objective. We use a 63× Plan Apo NA 1.4 objective and QuantEM 512SC camera (Photometrics), with a pixel size of 0.21 μm × 0.21 μm. Take a Z-stack by setting the Z-stack top and bottom images using the postsynaptic signal. Start the Z-stack just above the NMJ and finish just below the NMJ. We use a step size of 0.24 μm (*see* **Note 13**).

4. To quantify the postsynaptic intensity, images must be acquired with two channels, one staining the presynaptic membrane with HRP, and the other staining a postsynaptic membrane component.
5. Download the HRPtoCh2 analysis script from the GitHub repository (https://github.com/tp81/neisch-dynein-2021) by clicking on HRPtoCh2_membrane_ROI_analysis with Channel selection option Pixel 47 (63× NA 1.4).py. This will take you to the code. Download the raw file.
6. Open ImageJ. Select File > Open and then select code file [HRPtoCh2_membrane_ROI_analysis with Channel selection option Pixel 47 (63× NA 1.4).py]. Once the code is open, at the bottom select either the button "Run" or "Batch." "Batch" will allow you to select a folder with all the images to be analyzed or add individual images to be analyzed all at the same time, while "Run" allows you to select and analyze single images.
7. Once you have selected either "Run" or "Batch," you will need to select the image files to be analyzed in the "inputFile," and where the results files should go in the "outputDir" (Output Directory).
8. Next, assign the channels for HRP (HRP) and the postsynaptic membrane protein (CH2). To determine the channel assignments, open the images in ImageJ and split the channels. The first channel will be assigned as $C=0$ at the end of the file name, whereas the second channel will be assigned as $C==1$, and so on.
9. Once the channels are assigned for the script, click "OK."
10. The image will then open as a Max projection, and there will be a yellow 9.96 μm × 9.96 μm box that is present (see Fig. 2A, *see* **Note 14**). Use the slider at the bottom to change between the channels to open the postsynaptic protein channel.Position the box around one of the terminal boutons (the last boutons of the synaptic terminal). Press "OK."
11. The script will then analyze the Z-stack of images. The background signal will be measured for each channel and subtracted out before creating a mask to define first the HRP membrane signal, and then a mask to define the postsynaptic protein signal (CH2).
12. Three images will open after the script has run through and analyzed an image. One will be a duplication of the Z-stack of the 9.96 μm × 9.96 μm region that was selected. Another called "V" is the region that was defined as membrane, and a third is the ratio image of the postsynaptic membrane signal/HRP. These files are present to allow the researcher to verify what the script is analyzing (see Fig. 2A).

13. A "Results" file (.csv spreadsheet) will be saved to the Output directory. This file contains the measured background signal and the mean membrane signal for both HRP and the postsynaptic membrane protein (CH2), as well as the ratio of the two signals. This allows the data to be expressed in

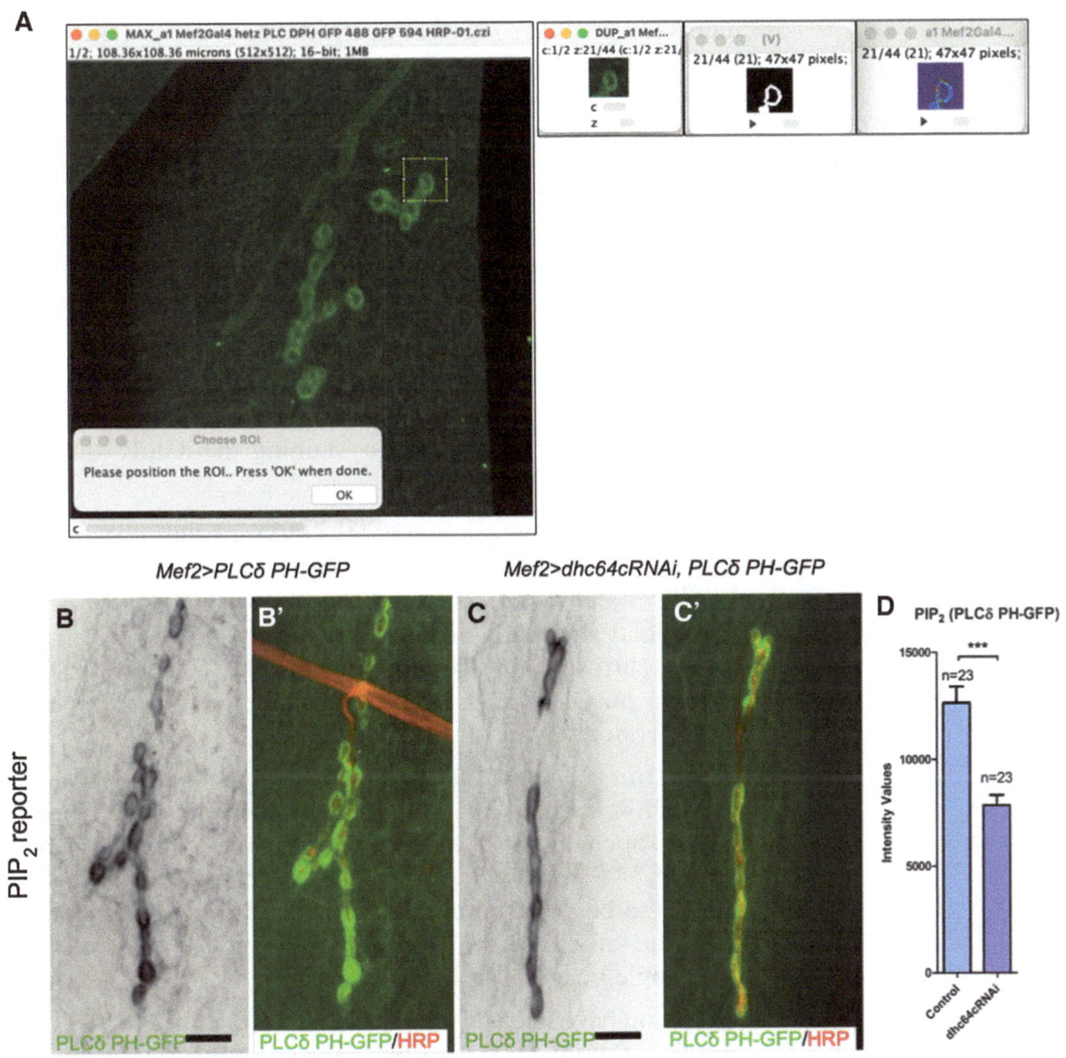


**Fig. 2** (**A**) A maximum projection image is generated by running the script used to measure postsynaptic protein abundance. A 9.96 μm × 9.96 μm box appears to be placed over the terminal bouton to be measured. Once the terminal bouton/region of interest (ROI) is selected and the script is run, a duplicate image of the ROI is generated, an image showing the region defined as membrane in the ROI (V), and a ratio image of the postsynaptic membrane signal/HRP signal is generated. Representative images of the $PIP_2$ reporter (*PLCδ PH-GFP*) and HRP in control animals, *Mef2>PLCδ PH-GFP* (B–B′), and in animals depleted of postsynaptic dynein, *Mef2>dhc64cRNAi, PLCδ PH-GFP* (C–C′). Postsynaptic $PIP_2$ reporter abundance was significant decrease in animals depleted of postsynaptic dynein compared to controls, as measured by the postsynaptic membrane script. Unpaired one-tailed *t*-test, $^{***}P < 0.0001$. (Panels **B–D** are adapted from [15])

two ways, one that is the mean postsynaptic membrane signal and one that is the ratio of postsynaptic membrane signal/HRP. Results can be graphed to compare genotypes (see Fig. 2B–D).

### 3.3 Imaging and Quantification of iGluR Size

In this section, we will describe how to measure, in 3D, the iGluR cluster size in different genetic backgrounds. These volume measurements can be used to determine the radius of each of the clusters for comparison across genotypes (comparable to measurements in [9]). In *Drosophila*, there are five glutamate receptor subunits that make up the heteromeric tetramer receptors. Three of the subunits are essential: GluRIIC, GluRIID, and GluRIIE, and they assemble with either a GluRIIA or GluRIIB subunit to form a tetrameric receptor [19]. Tables 1 and 2 list antibodies and genetic tools to visualize GluRIIC, GluRIIA, and GluRIIB subunits. Enlarged iGluR clusters can result in synaptic transmission and active zone spacing defects [9, 15, 17].

1. To analyze the radii of iGluR spheres, the images must be acquired with two channels, one staining the presynaptic membrane with HRP, and the other staining the iGluR subunit to be analyzed.
2. Image the samples as described above, being sure that the Z-stack starts just above and finishes just below the GluRII subunit staining.
3. Download the GluR analysis script from the GitHub repository (https://github.com/tp81/neisch-dynein-2021) by clicking on GluR_analysis_20190429-1.py. This will take you to the code. Download the raw file.
4. Open ImageJ. Select File>Open and then select the code file (GluR_analysis_20190429-1.py) (*see* **Note 15**). Once the code is open, at the bottom select either the button "Run" or "Batch." "Batch" will allow you to select a folder with all the images to be analyzed or add individual images to be analyzed all at the same time, while "Run" allows you to select and analyze single images.
5. Once you have selected either "Run" or "Batch," you will need to select the image files to be analyzed in the "inputFile" and where the results files should go to "outputDir". Additionally, use the dropdown menu to select the image bit depth based on your acquisition system (8, 12, 14, or 16 bit).
6. Running the script will result in two saved results files for each image analyzed that will go into the Output Directory. One is a.csv spreadsheet of the data from the analysis. All the iGluR clusters (objects) that are identified using the script are above 10 voxels (0.10584 $\mu m^3$). They are individually analyzed and

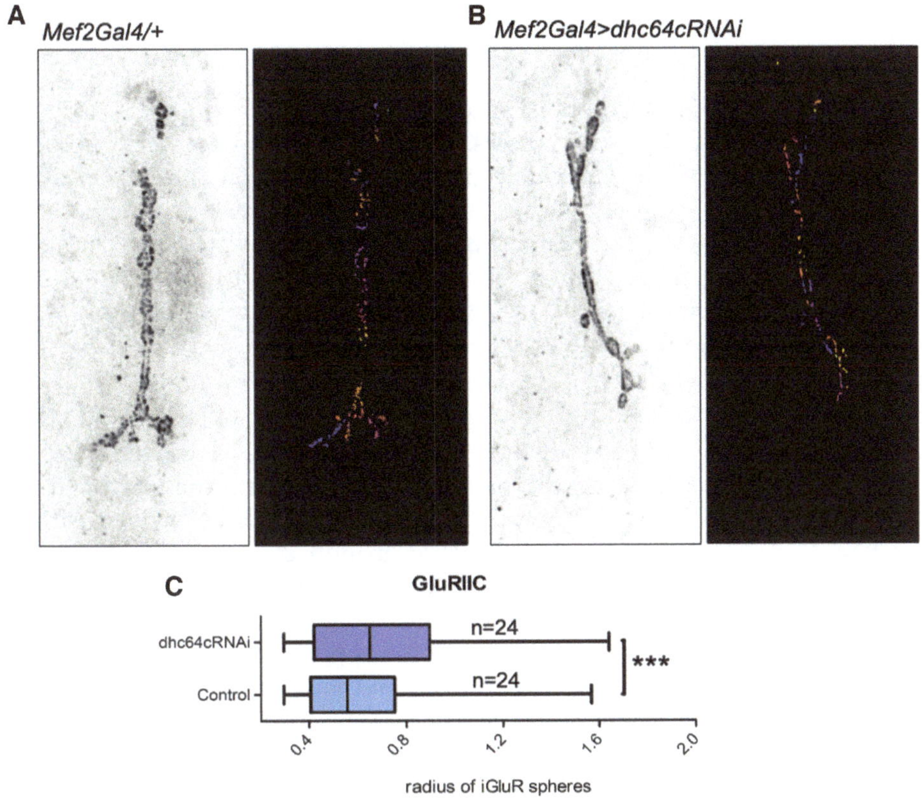


**Fig. 3** (**A, B**) Single sections are shown of NMJs immunostained for the iGluR subunit, GluRIIC. These and other NMJs were analyzed using the GluR analysis script. Next to the raw image is the same section showing the objects that were identified and measured in 3D by the script. Each object that was measured above ten voxels is shown in a different color. Depletion of dynein, by RNAi, results in enlarged iGluR clusters (**B**) compared to controls (**A**). This difference is quantified in (**C**). *n* indicates the number of NMJs analyzed, boxes show the interquartile range, the median is indicated with a line, and whiskers show the complete range. Unpaired one-tailed *t*-test, ***$P < 0.0001$. (Panel **C** is adapted from [15])

put into the spreadsheet. In the spreadsheet, you will find the volume in μm³. From these values, the radius of the equivalent sphere is calculated and given in the spreadsheet as well. The second file that is generated is a TIFF file of the 3D objects that were counted. This allows the researcher to visually look at the iGluR clusters/objects that were counted in through the Z-stack. Each object is individually color-coded (see Fig. 3A, B).

7. After all of the data has been analyzed, the results are then analyzed and graphed to determine if there is a significant difference in iGluR cluster size between control and experimental samples (see Fig. 3C).

## 4 Notes

1. Drosophila media/fly food that is too dry is not suitable for genetic crosses. Even seemingly minor changes to the cooking process can significantly impact the quality of the food, particularly if they result in increased water loss. You may need to increase the amount of water in recipes, if you are cooking small batches in a large kettle, if you do not cover the kettle, if you extend the cooking time, or if the environment is particularly dry.
2. To make pins, as described previously in [16], a 9v battery is connected to the ends of insulated electrical wire that has been stripped at both ends. The positive and negative terminals of the battery are wrapped to one end of the electrical wire. The opposite end of the wire on the positive battery terminal is wrapped around one arm of a #5 forceps. The opposite end of the wire on the negative battery is connected to a loop made at one end of an unfolded steel paper clip. The other end of the paper clip is fashioned into a hook that gets clipped onto the end of the 60 × 15 mm petri dish and is partially submerged in NaOH. Tungsten wire is cut into ~6 mm pieces and held with a Durmont #5 forceps to be dipped into a 60 × 15 mm petri dish containing 5 N NaOH. The ends of the pins are sharpened to be 2–3 mm in length.
3. For dissection forceps, we use Durmont #4 straight forceps, FineScience, 11241-30, or Durmont #5 straight fine forceps (FineScience, 11254-20).
4. Micro dissection scissors are used to cut open the larva. We use Vannas spring scissors, 0.025 × 0.015 mm tips, straight, 8.5 cm long (World Precision Instruments, 500086).
5. As an alternative to custom-made dissection dishes, Borosilicate Glass depression 9-well spot plates can be used to make dissection dishes (Corning, 7220-85).
6. For 1 L of HL3, add approximately 750 mL of ultrapure water and a stirbar to a 1 L plastic beaker. Weigh 4.09 g NaCl, 0.37 g KCl, 4.07 g $MgCl_2$, 0.8 g $NaHCO_3$, 1.89 g trehalose, 39.4 g sucrose, 1.19 g Hepes, and transfer to the plastic beaker. Stir to dissolve, measure pH, and adjust to 7.2 with HCl or NaOH. Bring up the final volume to 1 L in a graduated cylinder. Filter-sterilize the HL3 saline and store at 4°C.
7. We use Prolong Glass mounting media (Invitrogen, P36980).
8. We purchase Alexa fluorophore-conjugated HRP antibodies and Alexa fluorophore-conjugated secondary antibodies from Jackson ImmunoResearch. Concentrations needed for immunostaining may vary by the provider.

9. Crosses set up initially at 29°C do not take well and have fewer progeny. Allowing flies to mate at 25°C for 1-day results in good egg laying at 29°C the next day. For optimal RNAi-mediated mRNA knockdown, we use a temperature of 29°C.

10. While the gut, fat body, disks, and trachea system can be removed prior to fixing, we find it easier to visually see and dissect out the tissue after it has been fixed. This prevents accidental damage to the muscle tissue that can occur. Note that the brain is removed so that the coverslip can lie flat for imaging of the NMJs in the muscle. The brain is large and bulky, which can prevent the coverslip from lying evenly flat on the specimen.

11. Pelts remain flat if the immunofluorescence tissue staining occurs in the dish in which the animals are dissected into to maintain shape. Typically, after fixation, we remove the four pins on the sides and leave the most anterior and posterior pins in the animal throughout the staining procedure. Transferring of pelts to an Eppendorf tube at this point results in curling of the pelts, making them harder to mount and image.

12. Abdominal segments A2–A7 have a stereotypical repeated muscle pattern. To identify specific abdominal segments and muscles, at the anterior of the animal, first find the ventral midline. This is done by finding the herringbone pattern toward the center of the animal. Move either left, to find the left hemisegments, or right, to find the right hemisegments. The most internal muscle near the midline in abdominal segment A1 is muscle 31, which is wide and triangular-shaped. Below muscle 31 in the abdominal segment A2 are muscles 7 and 6. Moving out toward the dorsal midline, muscle 4 is wide and has two thin, angled muscles that meet the top (muscle 5) and bottom (muscle 8) of this muscle.

13. The step size used was determined as the optimal step size for Nyquist sampling within our imaging software.

14. Depending on the objective used and pixel size, you may need to change the number of pixels to define a 10 μm × 10 μm box to measure. To edit the script to change the number of pixels for 10 μm, find the line that says # Parameter and lists the region size as [regionSize=47 # pixels (~10 μm)]. Change the number 47 to the number of pixels that corresponds to ~10 μm in your images. The box size can also be changed if a user wants to measure an area larger than 10 μm × 10 μm.

15. The posted script uses a rolling ball background subtraction of 2. We compared the raw image files to the file that the script saves after analysis, which shows all the objects counted

as iGluR clusters. We felt that the "objects" were representative of what we observed in the raw image as iGluR clusters. The background subtraction can be modified if the "objects" are not representative of the iGluR clusters observed. The background subtraction can be found under #BKD SUB RAD = 5. To change the radius from 2, change the number after rolling=stack. (c2, "Subtract Background...," "rolling= 2 stack"). The script was originally written with a background subtraction of 5.

## Acknowledgments

This work was supported by NIH grants GM044757 to TSH and AG078690 to ALN.

### References


1. Bhattacharyya A, Watson FL, Pomeroy SL, Zhang YZ, Stiles CD, Segal RA (2002) High-resolution imaging demonstrates dynein-based vesicular transport of activated Trk receptors. J Neurobiol 51(4):302–312. https://doi.org/10.1002/neu.10062
2. Okada Y, Yamazaki H, Sekine-Aizawa Y, Hirokawa N (1995) The neuron-specific kinesin superfamily protein KIF1A is a unique monomeric motor for anterograde axonal transport of synaptic vesicle precursors. Cell 81(5):769–780. https://doi.org/10.1016/0092-8674(95)90538-3
3. Klopfenstein DR, Vale RD (2004) The lipid binding pleckstrin homology domain in UNC-104 kinesin is necessary for synaptic vesicle transport in *Caenorhabditis elegans*. Mol Biol Cell 15(8):3729–3739. https://doi.org/10.1091/mbc.e04-04-0326
4. Delcroix JD, Valletta JS, Wu C, Hunt SJ, Kowal AS, Mobley WC (2003) NGF signaling in sensory neurons: evidence that early endosomes carry NGF retrograde signals. Neuron 39(1):69–84. https://doi.org/10.1016/s0896-6273(03)00397-0
5. Maday S, Wallace KE, Holzbaur EL (2012) Autophagosomes initiate distally and mature during transport toward the cell soma in primary neurons. J Cell Biol 196(4):407–417. https://doi.org/10.1083/jcb.201106120
6. Vilmont V, Cadot B, Vezin E, Le Grand F, Gomes ER (2016) Dynein disruption perturbs post-synaptic components and contributes to impaired MuSK clustering at the NMJ: implication in ALS. Sci Rep 6: 27804. https://doi.org/10.1038/srep27804
7. Fan R, Lai KO (2022) Understanding how kinesin motor proteins regulate postsynaptic function in neuron. FEBS J 289(8):2128–2144. https://doi.org/10.1111/febs.16285
8. Rodriguez Cruz PM, Cossins J, Beeson D, Vincent A (2020) The neuromuscular junction in health and disease: molecular mechanisms governing synaptic formation and homeostasis. Front Mol Neurosci 13:610964. https://doi.org/10.3389/fnmol.2020.610964
9. Pielage J, Fetter RD, Davis GW (2006) A postsynaptic spectrin scaffold defines active zone size, spacing, and efficacy at the Drosophila neuromuscular junction. J Cell Biol 175(3):491–503. https://doi.org/10.1083/jcb.200607036
10. Banovic D, Khorramshahi O, Owald D, Wichmann C, Riedt T, Fouquet W, Tian R, Sigrist SJ, Aberle H (2010) Drosophila neuroligin 1 promotes growth and postsynaptic differentiation at glutamatergic neuromuscular junctions. Neuron 66(5):724–738. https://doi.org/10.1016/j.neuron.2010.05.020
11. Wang S, Yang J, Tsai A, Kuca T, Sanny J, Lee J, Dong K, Harden N, Krieger C (2011) Drosophila adducin regulates Dlg phosphorylation and targeting of Dlg to the synapse and epithelial membrane. Dev Biol 357(2):392–403. https://doi.org/10.1016/j.ydbio.2011.07.010
12. Sun M, Xing G, Yuan L, Gan G, Knight D, With SI, He C, Han J, Zeng X, Fang M, Boulianne GL, Xie W (2011) Neuroligin 2 is required for synapse development and function at the Drosophila neuromuscular junction. J Neurosci 31(2):687–699. https://doi.org/10.1523/JNEUROSCI.3854-10.2011

13. Kohsaka H, Takasu E, Nose A (2007) In vivo induction of postsynaptic molecular assembly by the cell adhesion molecule Fasciclin2. J Cell Biol 179(6):1289–1300. https://doi.org/10.1083/jcb.200705154
14. Guan B, Hartmann B, Kho YH, Gorczyca M, Budnik V (1996) The Drosophila tumor suppressor gene, dlg, is involved in structural plasticity at a glutamatergic synapse. Curr Biol 6(6):695–706. https://doi.org/10.1016/s0960-9822(09)00451-5
15. Neisch AL, Pengo T, Avery AW, Li MG, Hays TS (2025) Dynein-driven regulation of postsynaptic membrane architecture and synaptic function. J Cell Sci 138(5). https://doi.org/10.1242/jcs.263844
16. Neisch AL, Avery AW, Machamer JB, Li MG, Hays TS (2016) Methods to identify and analyze gene products involved in neuronal intracellular transport using Drosophila. Methods Cell Biol 131:277–309. https://doi.org/10.1016/bs.mcb.2015.06.015
17. Blunk AD, Akbergenova Y, Cho RW, Lee J, Walldorf U, Xu K, Zhong G, Zhuang X, Littleton JT (2014) Postsynaptic actin regulates active zone spacing and glutamate receptor apposition at the Drosophila neuromuscular junction. Mol Cell Neurosci 61:241–254. https://doi.org/10.1016/j.mcn.2014.07.005
18. Wang SJ, Tsai A, Wang M, Yoo S, Kim HY, Yoo B, Chui V, Kisiel M, Stewart B, Parkhouse W, Harden N, Krieger C (2014) Phospho-regulated Drosophila adducin is a determinant of synaptic plasticity in a complex with Dlg and PIP2 at the larval neuromuscular junction. Biol Open 3(12):1196–1206. https://doi.org/10.1242/bio.20148342
19. Qin G, Schwarz T, Kittel RJ, Schmid A, Rasse TM, Kappei D, Ponimaskin E, Heckmann M, Sigrist SJ (2005) Four different subunits are essential for expressing the synaptic glutamate receptor at neuromuscular junctions of Drosophila. J Neurosci 25(12):3209–3218. https://doi.org/10.1523/JNEUROSCI.4194-04.2005
20. Ranganayakulu G, Schulz RA, Olson EN (1996) Wingless signaling induces nautilus expression in the ventral mesoderm of the Drosophila embryo. Dev Biol 176(1):143–148. https://doi.org/10.1006/dbio.1996.9987
21. Shishido E, Takeichi M, Nose A (1998) Drosophila synapse formation: regulation by transmembrane protein with Leu-rich repeats, CAPRICIOUS. Science 280(5372):2118–2121. https://doi.org/10.1126/science.280.5372.2118
22. Yang CH, Rumpf S, Xiang Y, Gordon MD, Song W, Jan LY, Jan YN (2009) Control of the postmating behavioral switch in Drosophila females by internal sensory neurons. Neuron 61(4):519–526. https://doi.org/10.1016/j.neuron.2008.12.021
23. Bachmann A, Timmer M, Sierralta J, Pietrini G, Gundelfinger ED, Knust E, Thomas U (2004) Cell type-specific recruitment of Drosophila Lin-7 to distinct MAGUK-based protein complexes defines novel roles for Sdt and Dlg-S97. J Cell Sci 117(Pt 10):1899–1909. https://doi.org/10.1242/jcs.01029
24. Jouette J, Guichet A, Claret SB (2019) Dynein-mediated transport and membrane trafficking control PAR3 polarised distribution. Elife 8. https://doi.org/10.7554/eLife.40212
25. Ramos CI, Igiesuorobo O, Wang Q, Serpe M (2015) Neto-mediated intracellular interactions shape postsynaptic composition at the Drosophila neuromuscular junction. PLoS Genet 11(4):e1005191. https://doi.org/10.1371/journal.pgen.1005191
26. Mozer BA, Sandstrom DJ (2012) Drosophila neuroligin 1 regulates synaptic growth and function in response to activity and phosphoinositide-3-kinase. Mol Cell Neurosci 51(3–4):89–100. https://doi.org/10.1016/j.mcn.2012.08.010
27. Claret S, Jouette J, Benoit B, Legent K, Guichet A (2014) PI(4,5)P2 produced by the PI4P5K SKTL controls apical size by tethering PAR-3 in Drosophila epithelial cells. Curr Biol 24(10):1071–1079. https://doi.org/10.1016/j.cub.2014.03.056
28. Jan LY, Jan YN (1982) Antibodies to horseradish peroxidase as specific neuronal markers in Drosophila and in grasshopper embryos. Proc Natl Acad Sci USA 79(8):2700–2704. https://doi.org/10.1073/pnas.79.8.2700

# Chapter 13

# STED Super-Resolution Microscopy for Studying Actin Cytoskeleton Organization

**Marta Sampietro, Elvira Arza, Cristina Scielzo and Valeria R. Caiolfa**


## Abstract

Actin is one component of the intracellular cytoskeleton, forming filaments that are thin, flexible fibers approximately 7 nm in diameter and extending up to several micrometers in length. These filaments organize into higher-order assemblies, creating bundles of three-dimensional networks that behave like semisolid gels. Individual actin filaments (~7 nm) cannot be resolved with conventional light microscopy; however, super-resolution techniques such as STED, PALM, and STORM might increase resolution up to ~20–30 nm, enabling visualization of single filaments and fine structural details within cellular networks. For even higher resolution and ultrastructural detail, electron microscopy remains indispensable. Here, we describe STED procedures for imaging immunostained actin, including critical steps for laser alignment, a method for determining the actual spatial resolution, and criteria for selecting fluorophores and preparing samples.


**Keywords** STED microscopy, Actin, Cytoskeleton, STED microscopy, 3D-STED, Super-resolution microscopy, CLL, Lymphocytes

## 1 Introduction

STED microscopy has become a popular method for obtaining super-resolved images using fluorescent probes [1]. Although commercial STED microscopes offer automatic laser alignment procedures, often there is the need for further optimization, particularly when filaments and three-dimensional protein networks are imaged. STED collects the fluorescence from a field smaller than the diffraction-limited focus, superimposing the excitation focus with the focus of a second laser (the depletion laser) that is designed to have a central region at almost zero intensity [2], forming the so-called "STED doughnut." A strong depletion laser effectively quenches fluorescence through stimulated emission and inhibits the emission of excited fluorophores within the doughnut region.

The optimal resolution and performance of an STED system critically depend on various factors [3], among which is the precise alignment of the excitation and depletion beams. Alignment can be

Noa B. Martín-Cófreces et al. (eds.), *The Cytoskeleton*, Methods in Molecular Biology, vol. 3070, https://doi.org/10.1007/978-1-0716-5515-3_13,

carried out using commercially available gold nanobeads to visualize the laser beam profiles and ensure proper overlap between the excitation and depletion beams [4]. Notably, in 2D-STED, depletion is applied only to the in-focus plane, normally the XY plane, while along the *z*-axis the spatial resolution does not improve, being the same as in a confocal image. In 3D-STED, depletion on the plane and in the *z*-axis are equilibrated to achieve the most isotropic spatial resolution along the three XY and Z dimensions. It is important to measure the actual point spread function (PSF) to define the spatial resolution achieved under the applied conditions, either for 2D or 3D STED.

Here, we use a STED microscope with a simple design that is potentially accessible in many microscopy facilities. However, it is worth noting that the latest generation of super-resolution STED microscopes is designed to deliver sharper images, reduce light-induced damage to samples, and offer versatile imaging options for different specimen types, all while making the technology easier and more intuitive to use. The newest microscopes share the same fundamental principle of stimulated emission depletion but differ in their approach to beam shaping and phototoxicity control. The TauSTED technology leverages fluorescence lifetime to optimize depletion efficiency [5]. In the STED microscope used here, we applied a simpler version of TauSTED, namely the time-gating by which we can empirically refine the precision and contour of the STED doughnut and improve spatial resolution [6, 7]. Other new STED microscopes are based on integrating adaptive illumination modes such as RESCue and DyMIN to minimize light exposure, offer multicolor and 3D imaging capabilities, and versatility for cell biology and tissue studies [8, 9]. In contrast, the MINFLUX solution represents a paradigm shift in super-resolution microscopy, achieving molecular-scale resolution (1–3 nm) by combining STED-like depletion with photon-efficient localization strategies, which enables precise tracking of individual molecules in live cells with minimal photobleaching [10].

The choice of the fluorophore is also crucial, as not all conventional fluorophores used in fluorescence microscopy are suitable for STED imaging, and more restrictions arise for colocalization analyses. The STED beam forces fluorophores to release energy as photons without fluorescence; thus, the dyes must be photostable and compatible with both excitation and STED wavelengths. Several public catalogs from commercial companies and recent publications are available for guiding the fluorophore choice depending on the STED laser lines and colocalization requirements [11, 12]. Latest advances in fluorophore chemistry are significantly improving the performance of STED microscopy, particularly for imaging cytoskeletal structures such as actin filaments and microtubules. Modern STED-compatible dyes are designed for high photostability, minimal phototoxicity, and efficient depletion at commonly used wavelengths (592, 660, and 775 nm). Among the most notable developments are the Abberior STAR series (e.g., STAR 580, STAR 635P), which offer exceptional brightness and stability for multicolor imaging in fixed samples. For live-cell applications, SiR-based probes (Silicon Rhodamine derivatives) are becoming the gold standard, enabling dynamic visualization of actin and tubulin with minimal perturbation. Additionally, fluorogenic probes such as SPY-actin and SPY-tubulin from Spirochrome provide high signal-to-noise ratios by activating fluorescence only upon binding to their target structures, reducing background, and improving image clarity. These dyes allow to resolve individual filaments (~7 nm diameter) and complex cytoskeletal networks at resolutions down to 30 nm, revealing structural details previously inaccessible with confocal microscopy. Their compatibility with advanced STED modes, including 3D STED and FLIM-STED, further expands their utility for studying cytoskeletal dynamics in both fixed and live specimens in a variety of cell types, including cardiovascular, immunological, and oncological model cell lines [13–15].

In this method, using a simple STED microscope, we describe the critical steps required for proper STED laser alignment, the determination of the Point Spread Function (PSF) to define the experimental resolution achievable under sample-specific conditions, and we provide examples of imaging protocols for analyzing the actin cytoskeleton in MEC1 cells—a human chronic lymphocytic leukemia (CLL) cell line [16] as well as in primary CLL patient samples. The procedures are implemented using simple and widely available commercial fluorophores that are fully compatible with STED imaging. These examples will serve as guidelines for applying 2D and 3D STED actin imaging across a wide range of cell types.

## 2 Materials

1. CD19 cells were negatively selected from fresh peripheral blood from patients with CLL. The MEC1 cell line was obtained from Deutsche Sammlung von Mikroorganismen und Zellkulturen GmbH (DSMZ). Cells are maintained in complete culture medium at 37°C and 5% $CO_2$.
2. Complete culture medium: RPMI 1640 medium, 10% fetal bovine serum (FBS; v/v), 15 mg/mL gentamicin.
3. Blocking buffer: PBS, 0.1% BSA (w/v), 10% FBS (v/v).

## 3 Methods

### 3.1 Alignment of the STED Lasers with Gold Nanobeads

This step is gold nanobeads sample preparation (*see* **Note 1**).

1. Rinse a coverslip (No. 1.5, high precision, *see* **Note 2**) thoroughly with Milli-Q water.
2. Immerse in 70% ethanol for 5–10 min to remove contaminants.
3. Dry the coverslip using nitrogen or compressed air.

4. Vortex the gold nanobead suspension (Gold nanoparticles 80–150 nm) to ensure homogeneity.
5. Dilute the nanobead suspension in Milli-Q water to achieve a sparse distribution (e.g., 1:1000 or as needed).
6. Pipette 10–20 μL of the diluted suspension onto the center of the coverslip (22 × 22 mm high-precision (high-quality, #1.5).
7. Allow the droplet to air-dry completely at room temperature (avoid touching the surface).
8. Once dry, mount the coverslip onto a microscope slide or chamber using minimal immersion medium (if required).
9. Ensure the nanobeads are well adhered and the surface is clean.

### 3.2 Imaging Settings

1. Use a STED microscope with an inverted stage (*see* **Note 3**) and appropriate excitation and depletion lasers suitable for the chosen fluorophores in the sample (*see* **Note 4**).
2. Select a high magnification and high NA objective (e.g., Objective 100× N.A1.4 oil) (*see* **Note 5**).
3. Use ultra-pure Immersion oil for a high magnification objective.
4. Activate an excitation line at a wavelength lower than the depletion beam, such as an excitation at 488 nm with depletion at 592 nm (*see* **Note 6**).
5. Use photomultipliers (PMT) detectors for collecting reflection (*see* **Note** 7).
6. To collect reflected signals on two channels, select two emission windows crossing the two laser lines: Reflection of the gold nanobeads (channel 1) and reflection of the STED laser (channel 2).
7. Select image dimensions: a frame size of 512 × 512 pixels is sufficient.
8. Adjust the gain and the excitation laser power to get a neat and bright image.

### 3.3 STED 2D Alignment

1. Set z-depletion at 0%.
2. Set the pinhole value to 1.
3. Find the focus plane at zoom = 1 (Fig. 1A; *see* **Note 8**).
4. Collect with high zoom (e.g., zoom 8) the reflection of the nanobeads and of the STED beam in channels 1 and 2. The reflection of the STED laser on this imaging planc is called the "doughnut" (Fig. 1B, C).
5. Merge the two channels (Fig. 1D) to verify the overlapping of the two reflections: the gold nanobead (round ball) at the center of the STED doughnut mask (Fig. 1E).
6. Activate the autocalibration of the microscope if the reflections are not centered or if the STED doughnut is irregular.

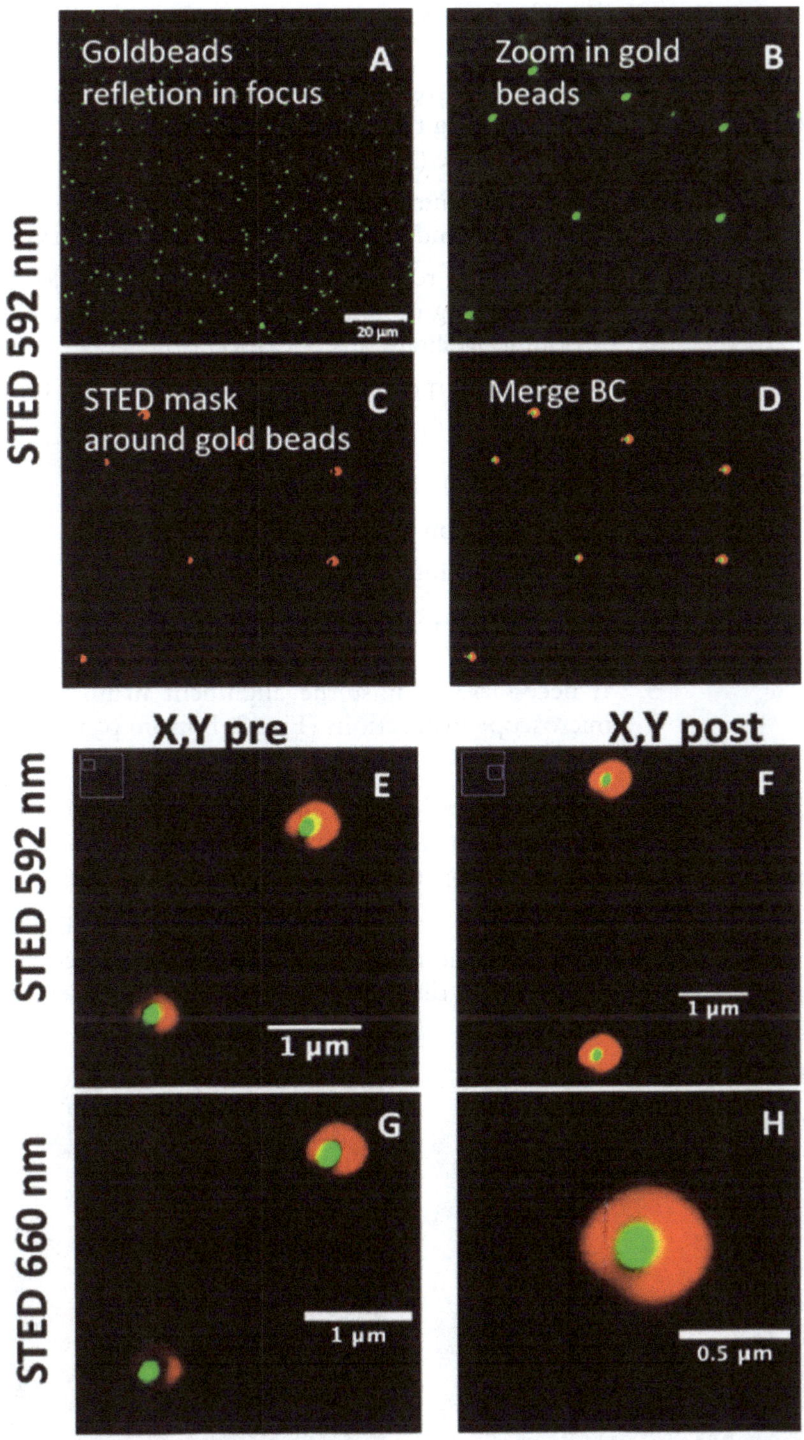


**Fig. 1** Steps of XY STED laser alignment in reflection mode. (**A**) Visualization of gold nanobead reflections. (**B**) Zoom in to find a few isolated beads. (**C**) Reflected STED doughnut mask at 592 nm depletion. (**D**) Merged image of the two reflection channels. (**E**) Prealignment merged image of the STED doughnut-mask and the gold bead reflections. (**F**) Postalignment image displaying the merging of the STED doughnut mask and the gold bead reflection. Calibration at other STED depletion lines should be verified if required for the sample imaging, as shown in (**G**, **H**)

7. Verify that the realignment works correctly in the merged image (Fig. 1F).
8. If necessary, readjust the alignment of the doughnut manually following the microscope instructions or repeat the alignment routine (*see* **Note 9**).
9. Verify the alignment of the STED laser with the gold beads, also at 660 and 775 nm depletion lines (Fig. 1G; *see* **Note 10**)
10. If necessary, readjust the alignment of the doughnut manually following the microscope instructions. Apply a very high zoom to facilitate this operation (Fig. 1H).
11. Verify the perpendicular alignment by setting the image in the XZ mode (Fig. 2).

### 3.4 STED 3D Alignment

1. Set the pinhole full open.
2. Set Z-depletion at 100%.
3. Continue setting as for 2D alignment.
4. Verify the perpendicular reflected image (X, Z) of the bead with respect to the STED mask (Fig. 3, Top panels).
5. If necessary, readjust the alignment manually following the microscope instructions (Fig. 3, bottom panels; *see* **Note 9**).

### 3.5 2D-STED/3D STED on Immunolabeled Actin and Myosin

This step describes how to choose the fluorophores for the staining with antibodies and the critical settings for the procedure. Ensure that antibodies are properly labeled with STED-compatible fluorophores.

1. Check depletion compatibility under the available STED laser lines: Examine the fluorescence emission spectra of the dyes. The tails of their fluorescence emission spectra must cross

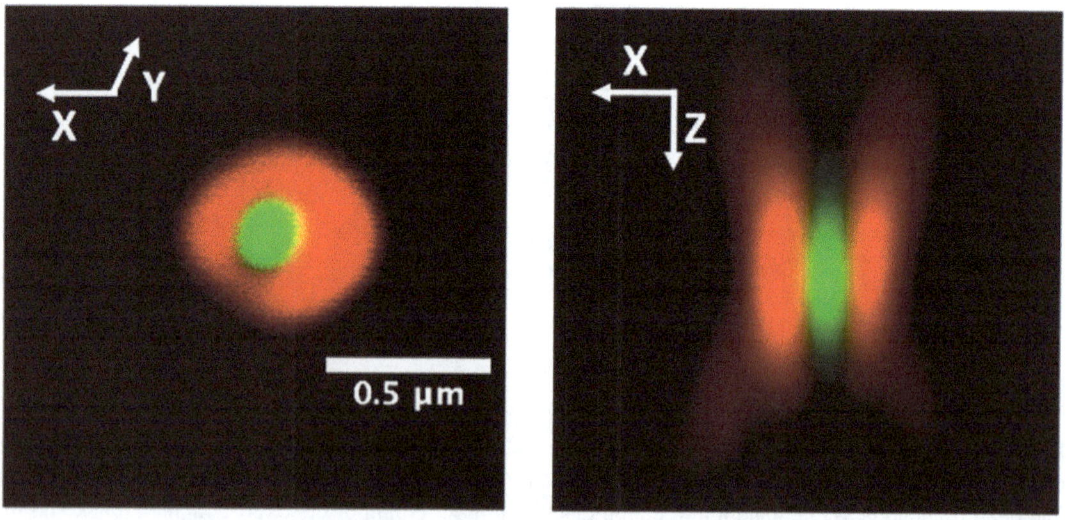


**Fig. 2** Visualization of a gold nanobead reflection and STED mask overlay on the XZ plane after XY STED alignment (right). The Z resolution (left) does not improve under these conditions because Z-depletion is not applied

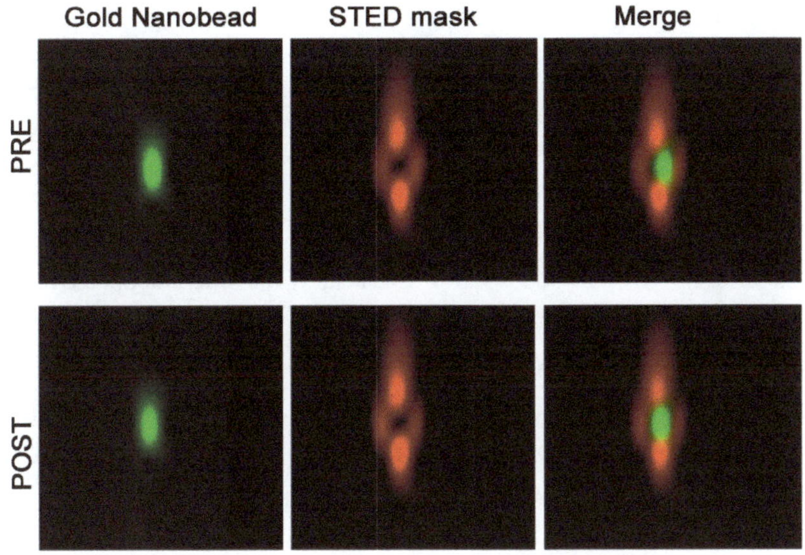


**Fig. 3** STEDalignment with Z-depletion. Prealignment (top panel) and postalignment (bottom panel) of the gold nanobeads reflections. The Z-STED masks and the merged images of the two channels illustrate the recovery of the STED mask centered along the *z*-axis

the STED laser line that is available for depletion. If not, single-beam depletion is not possible.

2. Ensure spectral separation. Examine the fluorescence emission spectra of the fluorophores and check if there is enough spectral separation for collecting their emission without crosstalk (*see* **Note 11**).
3. Verify photostability and brightness. Public commercial catalogs provide lists of dyes suitable for STED imaging and capable of withstanding high depletion power (e.g., Abberior STAR series, ATTO dyes; see examples in Tables 1 and 2).
4. Check that the STED laser line does not unintentionally excite the dyes (namely, Anti-Stokes Excitation). Lower the excitation laser close to zero, open the STED laser, run an image without direct excitation, and verify that the signal is zero.
5. Use antifade mounting media compatible with STED (*see* **Note 12**) as for the fluorescent nanobeads.
6. Validate labeling density and dye performance under STED conditions before planning multicolor imaging.
7. Run STED imaging on single-labeled samples first; optimize immunostaining and STED settings for single color.
8. Compare the optimized conditions for the individual fluorophores before doing double labeling and double color STED.
9. For dual color STED, either in a 2D or 3D setting, select a pair of fluorophores that can be depleted with the same STED laser (to avoid Z-chromatic aberrations and post-processing registration routines [17, 18]).

**Table 1**
**Example of reference fluorophores for single-beam dual-color STED (592 nm depletion)**

| Fluorophore | Excitation (nm) | Emission peak (nm) | 660 nm Depletion compatibility | Notes |
|---|---|---|---|---|
| Alexa Fluor 532 | 532 | ~554 | ✓ Good | Bright, common for green-yellow |
| ATTO 532 | 532 | ~553 | ✓ Good | High photostability |
| Alexa Fluor 555 | 555 | ~565 | ✓ Excellent | Ideal for orange-red |
| ATTO 550 | 550 | ~575 | ✓ Excellent | Robust for STED |
| TMR/TRITC | 555 | ~580 | ✓ Good | Traditional dye |
| Alexa Fluor 568 | 578 | ~603 | ✓ Moderate | Works but is less efficient |
| Abberior STAR 580 | 561 | ~595 | ✓ Good | Optimized for STED |

Source: Leica STED sample prep guides and Abberior dye recommendations

**Table 2**
**Example of reference fluorophores for single-beam dual-color STED (775 nm depletion)**

| Fluorophore | Excitation (nm) | Emission peak (nm) | STED depletion (775 nm) | Notes |
|---|---|---|---|---|
| Abberior STAR 580 | 561 | ~595 | ✓ Efficient | High photostability |
| Alexa Fluor 594 | 561 | ~617 | ✓ Good | Common for actin |
| Abberior STAR 635 P | 640 | ~655 | ✓ Excellent | Ideal for far-red |
| ATTO 647 N | 640 | ~665 | ✓ Excellent | Very bright |
| SiR (Silicon Rhodamine) | 640 | ~674 | ✓ Good | Live-cell compatible |
| CF594 | 561 | ~614 | ✓ Good | Alternative to Alexa 594 |

Source: Leica STED sample prep guides and Abberior dye recommendations

### *3.6 Actin and Myosin Staining*

1. In this example, we used the MEC1 cell line and primary CLL cells from patients and B-cells from healthy donors [19, 20]. We immunolabel actin with Alexa 568 or with phalloidin–Atto 565, and myosin with Alexa 532 (*see* **Note 13**).
2. Seed the cells on coated coverslips (high-quality, #1.5; *see* **Note 2**) to facilitate cell adhesion. In this example, coating is performed with a polyornithine: PBS solution (1:5), incubation for 2 h at 37°C and 5% $CO_2$, and washed with PBS 1×.
3. Seed about $3 \times 10^6$ primary cells in their medium in the coverslip, incubate 2 h in controlled conditions.
4. Remove nonadhered cells and fix with 4% PFA for 15 min in the dark and at room temperature, and wash with PBS.
5. Permeabilize in blocking buffer solution containing 0.3% v/v Triton-X 100, to limit nonspecific antibody binding. 30 min-incubation at RT in the dark.

6. Wash with PBS.
7. Incubate in a humid chamber with primary monoclonal anti-mouse beta actin monoclonal (BA3R) antibody 1:1000 in blocking solution.
8. For double actin/myosin staining, samples are incubated with primary monoclonal anti-mouse beta actin (BA3R) antibody and primary monoclonal myosin light chainEssential light chain (ELC)Regulatory light chain (RLC) 2 antibody, together.
9. Incubate overnight at 4°C in the dark.
10. Wash with PBS.
11. Incubate in a humid chamber with secondary goat-anti-mouse–Alexa-568 (1:500) in blocking solution for single actin staining for 2 h at R/T.
12. For double actin/myosin staining, goat-anti-mouse–Alexa568 and goat-anti-rabbit–Alexa532 in blocking solution are added simultaneously for 2 h at R/T and in the dark.
13. For phalloidin-Atto 565 staining, dye is added at a 1:250 dilution directly for 2 h incubation at room temperature in the dark.
14. Wash with PBS.
15. Mount with ProLong Gold antifade reagents.
16. Incubate at room temperature in the dark for 24 h to allow the mounting medium to dry.
17. Block the coverslip with transparent polish.
18. Store at 4°C.

### 3.7 STED Imaging

The overall procedure is designed to maximize spatial resolution while minimizing photobleaching, autofluorescence, and reflections, keeping the excitation laser power as low as possible. The depletion laser power, image format, zoom, and scanning parameters should be optimized according to sample brightness and fluorophore properties to achieve the highest resolution with minimal photobleaching.

1. Focus on the cells and select a flat, neat field of view.
2. Run a confocal image.
3. For STED 2D, adjust the depletion laser power on the X, Y plane. 100% depletion is expected to give the maximum resolution on this plane.
4. For STED-3D, adjust the depletion laser power on the XY and XZ planes to improve the 3D resolution in the image (Fig. 4).
5. Run the STED image.

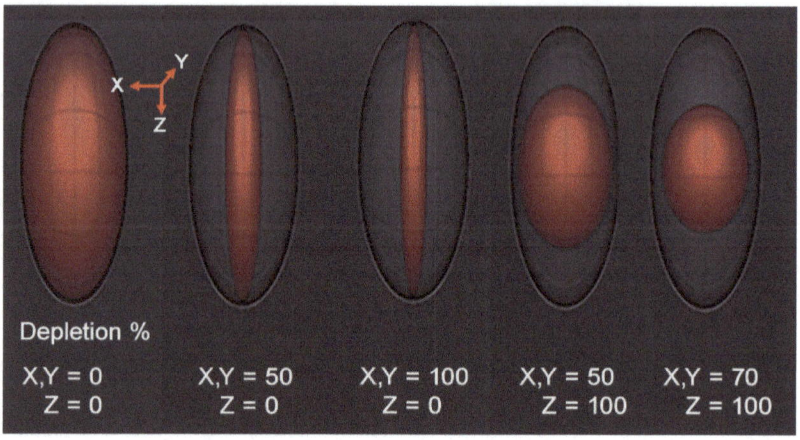


**Fig. 4** PSF as a function of X, Y, Z depletion. Schematic illustration of the point spread function (PSF) volume (shown in red) in a confocal setup and at different X, Y, and Z-STED depletions depending on the STED laser power. Notice that the XY depletion does not improve the PSF along the *z*-axis. In 3D-STED, the asymmetry of the confocal PSF can be mitigated by reducing the XY resolution to improve the axial Z resolution, resulting in a more isotropic volume

6. Compare the STED and confocal images to evaluate the resolution gain.
7. Examine the signal/noise ratio for optimizing the scanning conditions (gain/frame accumulation/line accumulation), while still achieving minimal photobleaching (*see* **Notes 14 and 15**).

Immunolabeled actin is imaged in the MEC1MEC1 cell line and in primary CLL cells. Costaining with myosin is also shown in CLL primary cells (see Fig. 5) according to the setup shown in Table 3.

### *3.8 Experimental resolution of the STED Images: Determination of the PSF. Fluorescent Nanobeads Preparation (see Note 16 and Fig. 4B, C)*

1. Use fluorescent nanobeads (e.g., 20–40 nm diameter, suitable for STED imaging; *see* **Note 17**) labeled with the same fluorophores that are used for immunostaining of the cells, in this example, Alexa532 and Alexa568
2. Dilute the fluorescent nanobeads in distilled water or PBS to a concentration that ensures a sparse distribution (e.g., 1:10,000).
3. Clean coverslips (high-quality, #1.5; *see* **Note 2**) thoroughly (e.g., ethanol wash).
4. Apply a small drop (~10 μL) of the diluted fluorescent nanobead solution onto the coverslip.
5. Let the nanobeads settle for 10–15 min.
6. Rinse gently with distilled water to remove excess beads.
7. Mount the coverslip using the mounting medium (ProLong-Gold antifade reagents) and seal with transparent nail polish or appropriate sealant.

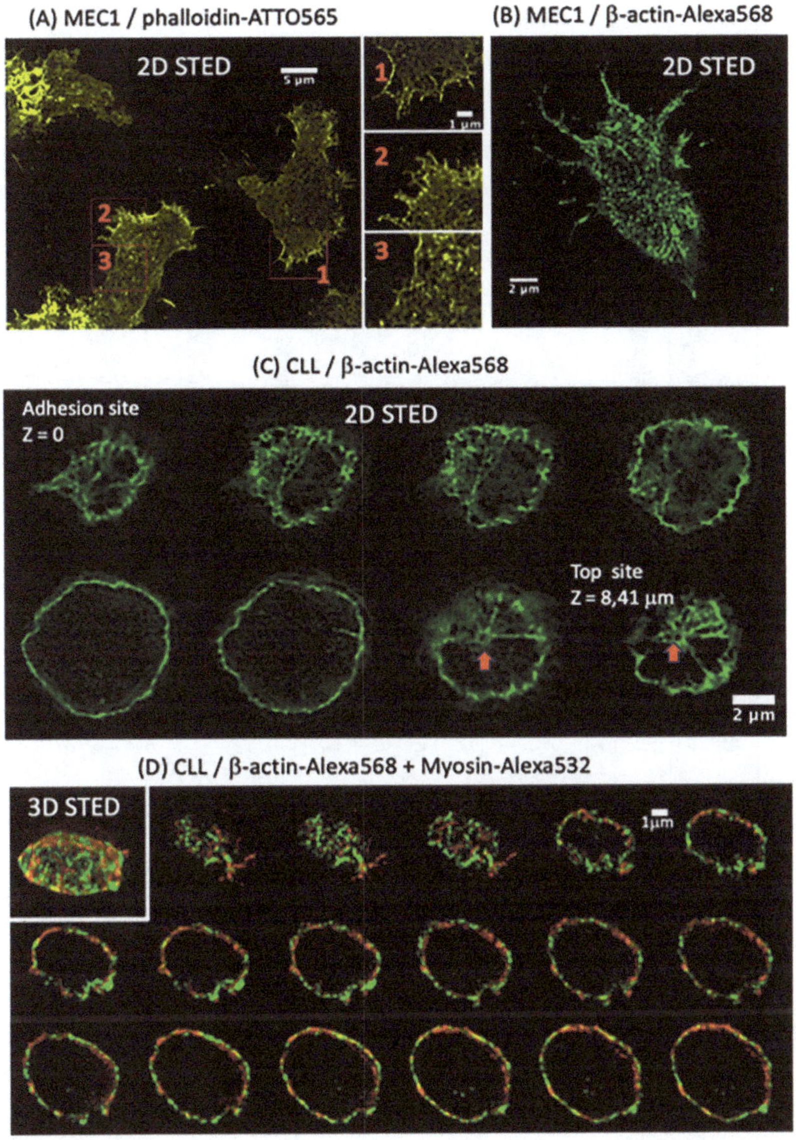


**Fig. 5** Representative examples of 2D-STED β-actin and 3D-STED β-actin/myosin immunostaining in MEC1MEC1 and CLL primary cells. Images were acquired according to the procedures described above. (**A**) 2D-STED image of immunostained cells with primary anti-phalloidin–ATTO565 antibody and (**B**) secondary anti-β-actin–Alexa 568 antibody, revealing similar structural details in MEC1 cells. (**C**) In CLL cells, STED images acquired at increasing distances from the cell adhesion site ($Z=0$) show a filamentous β-actin–Alexa 568 network. The structure forms an interconnected mesh of elongated strands, providing cytoskeletal support. Notably, a small ring-like formation is visible at the center of the network (red arrow, top site), suggesting a localized structural organization. (**D**) 3D-STED stack of an adhesion site in CLL primary cells showing immunostained β-actin (green) and myosin (red). Actin filaments form an extensive network that provides structural support, while myosin appears as discrete filaments and clusters sparsely associated with actin. Images were acquired using STED super-resolution microscopy under optimized conditions for 3D reconstruction

**Table 3**
**2D-STED/3D-STED imaging conditions on immunolabeled-actin and myosin in Fig. 5**

| | MEC1/actin 2D-STED 1 color | | CLL/actin 2D-STED 1 color | CLL/actin/myosin 3D-STED 2 colors | |
|---|---|---|---|---|---|
| **Image procedure** | **Phalloidin–Atto565** | **β-actin–Alexa568** | **β-actin–Alexa568** | **β-actin–Alexa568** | **Myosin–Alexa532** |
| Objective | 100×1.4 | 100×1.4 | 100×1.4 | 100×1.4 | 100×1.4 |
| Excitation (nm) | 563 | 575 | 575 | 575 | 525 |
| Ex laser power (%) | 5 | 10 | 13 | 18 | 15 |
| Em range (nm) | 571–616 | 590–612 | 589–632 | 608–632 | 531–546 |
| Depletion (nm) | 775 | 775 | 660 | 660 | 660 |
| Depletion of laser power (%) | 60 | 30 | 80 | 80 | 100 |
| Scanning | Linear | Linear | Resonant | Resonant | Resonant |
| Scanning direction | Unidirectional | Unidirectional | Unidirectional | Unidirectional | Unidirectional |
| speed (Hz) | 400 | 700 | 8000 | 8000 | 8000 |
| HyD mode | Standard mode | Standard mode | Standard mode | Standard mode | Standard mode |
| Zoom | 2.38 | 2 | 3 | 3,5 | 3,5 |
| Frame size | 1200×1200 | 2048×2048 | 1248×659 | 1024×1024 | 1024×1024 |
| Voxel size (nm) | 41×41×1000 | 28×28×1000 | 31×31×1000 | 32×32×100 | 32×32×100 |
| line average (Y/N) | yes | yes | yes | yes | yes |
| Frame average (Y/N) | no | yes | yes | yes | yes |

8. Allow curing if using hard-setting mounting medium (e.g., overnight/room temp).

### *3.9 Fluorescent Nanobead Imaging Settings (see Note 14)*

1. Switch to STED mode.
2. Locate isolated nanobeads using the confocal mode first.
3. Acquire images of single, well-separated beads; Use optimal imaging settings for excitation and depletion laser powers as for your sample imaging.
4. Adjust pixel size to Nyquist sampling: pixel size $\leq$ PSF/2 (*see* **Note 18**).
5. Set Dwell time and averaging.
6. Set high zoom.
7. Capture multiple beads across the field of view to account for variability (*see* **Note 19**).

### *3.10 Fluorescent Nanobead Data Analysis (see Note 20)*

1. Open the images in ImageJ/Fiji or your preferred analysis software (see Fig. 6).
2. Draw intensity profiles across individual beads along the *x*- and *y*-axes, ensuring that each profile passes through the nanobead center (maximum intensity point) (see the white segments in Fig. 6A)
3. Maintain the profile length constant for all beads so that measurements can be treated as replicates.
4. Repeat the above steps for each fluorophore that is used for immunostaining (see Fig. 6B).
5. Export each intensity profile as gray values into a data table (e.g., Excel spreadsheet) (see Fig. 6C).
6. Normalize each profile between its minimum and maximum values before averaging, to reduce variability across beads.
7. Compute the mean intensity and standard deviation (SD) across all beads for each position along the profile.
8. Fit the mean profile to a Gaussian function, and optionally fit individual bead profiles as well to check consistency
9. Calculate the full width at half maximum (FWHM) of the fitted Gaussian curve using:

   FWHM = 2.35482 × SD. This value represents the PSF.
10. Repeat the process for multiple beads, 1, 2, 3, 4,..., and compute the average FWHM to obtain a robust estimate (Fig. 6C).
11. Compare the average FWHM with the confocal PSF obtained under the same experimental conditions to quantify resolution improvement.

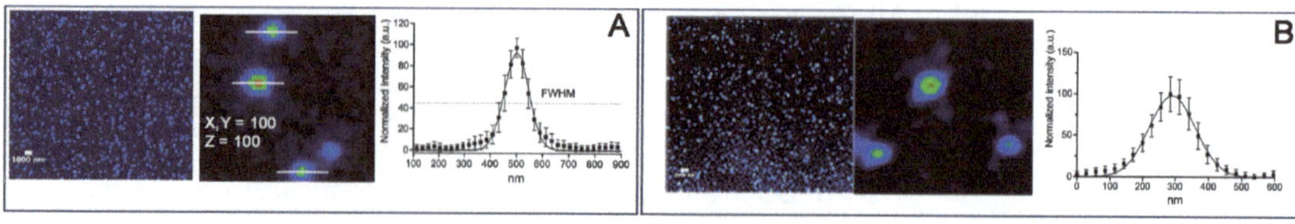


C

Raw Data

| nm | 1 | 2 | 3 | 4 | 5 |
|---|---|---|---|---|---|
| 295 | 0,00 | 0,00 | 0,00 | 0,00 | 0,00 |
| 318 | 0,00 | 0,00 | 0,00 | 0,00 | 0,00 |
| 341 | 0,00 | 0,00 | 0,00 | 0,00 | 0,00 |
| 364 | 0,00 | 0,00 | 0,00 | 0,17 | 0,00 |
| 386 | 0,00 | 0,00 | 1,00 | 5,33 | 2,17 |
| 409 | 0,17 | 0,83 | 23,00 | 28,00 | 18,33 |
| 432 | 9,00 | 18,17 | 82,00 | 73,70 | 64,50 |
| 455 | 42,67 | 70,50 | 117,00 | 142,00 | 128,17 |
| 477 | 86,83 | 122,33 | 165,67 | 205,17 | 188,33 |
| 500 | 112,67 | 186,33 | 170,00 | 229,67 | 214,17 |
| 523 | 111,83 | 166,17 | 125,00 | 213,17 | 180,83 |
| 545 | 80,83 | 88,70 | 75,00 | 161,00 | 130,17 |
| 568 | 21,33 | 28,33 | 18,67 | 109,83 | 94,00 |
| 591 | 0,83 | 4,17 | 1,00 | 80,33 | 49,50 |
| 614 | 0,00 | 0,00 | 0,00 | 43,33 | 11,00 |
| 636 | 0,00 | 0,00 | 0,00 | 9,83 | 0,50 |
| 659 | 0,00 | 0,00 | 0,00 | 0,50 | 0,00 |
| 682 | 0,00 | 0,00 | 0,00 | 0,00 | 0,00 |
| 705 | 0,00 | 0,00 | 0,00 | 0,00 | 0,00 |
| 727 | 0,00 | 0,00 | 0,00 | 0,00 | 0,00 |
| 750 | 0,00 | 0,00 | 0,00 | 0,00 | 0,00 |

Norm Data

| nm | 1 | 2 | 3 | 4 | 5 |
|---|---|---|---|---|---|
| 295 | 0,00 | 0,00 | 0,00 | 0,00 | 0,00 |
| 318 | 0,00 | 0,00 | 0,00 | 0,00 | 0,00 |
| 341 | 0,00 | 0,00 | 0,00 | 0,00 | 0,00 |
| 364 | 0,00 | 0,00 | 0,00 | 0,07 | 0,00 |
| 386 | 0,00 | 0,00 | 0,59 | 2,32 | 1,01 |
| 409 | 0,15 | 0,45 | 13,53 | 12,19 | 8,56 |
| 432 | 7,99 | 9,75 | 48,24 | 32,09 | 30,12 |
| 455 | 37,87 | 37,84 | 68,82 | 61,83 | 59,84 |
| 477 | 77,07 | 65,65 | 97,45 | 89,33 | 87,94 |
| 500 | 100,00 | 100,00 | 100,00 | 100,00 | 100,00 |
| 523 | 99,26 | 89,18 | 73,53 | 92,82 | 84,44 |
| 545 | 71,75 | 47,60 | 44,12 | 70,10 | 60,78 |
| 568 | 18,94 | 15,21 | 10,98 | 47,82 | 43,89 |
| 591 | 0,74 | 2,24 | 0,59 | 34,98 | 23,11 |
| 614 | 0,00 | 0,00 | 0,00 | 18,87 | 5,14 |
| 636 | 0,00 | 0,00 | 0,00 | 4,28 | 0,23 |
| 659 | 0,00 | 0,00 | 0,00 | 0,22 | 0,00 |
| 682 | 0,00 | 0,00 | 0,00 | 0,00 | 0,00 |
| 705 | 0,00 | 0,00 | 0,00 | 0,00 | 0,00 |
| 727 | 0,00 | 0,00 | 0,00 | 0,00 | 0,00 |
| 750 | 0,00 | 0,00 | 0,00 | 0,00 | 0,00 |

| **Gaussian** | |
|---|---|
| Best-fit values | |
| Amplitude | 92,17 |
| Mean | 500 |
| **SD** | **47,29** |
| 95% CI (profile likelihood) | |
| Amplitude | 90,17 to 94,19 |
| Mean | 498,9 to 501,2 |
| SD | 46,01 to 48,60 |
| Goodness of Fit | |
| Degrees of Freedom | 1097 |
| R squared | 0,9 |
| Sum of Squares | 64608 |
| Sy.x | 7,674 |
| Constraints | |
| SD | SD > 0 |
| Number of points | |
| # of X values | 1100 |
| # Y values analyzed | 1100 |

**Fig. 6** PSF calculation. Examples of intensity profile measurements of Alexa 568 nanobeads used to derive the experimental resolution achieved under the STED conditions applied for imaging β-actin–Alexa568 in MEC1 cells (Fig. 5B). (**A**) XY large view image, zoom in to select isolated beads and measure their intensity profiles (white traces), and the Gaussian fit of >100 replicate beads (mean ± SD); (**B**) the same procedure is reproduced on the XZ imaging mode. Note that the spatial resolution is not identical along the lateral and axial imaging planes. In this example, the experimental lateral FWHM is 111 nm, and the axial FWHM is 166 nm. (**C**) Analysis Steps: Raw intensity values of the nanobeads are listed according to their coordinates (nm) along the white trace (left). Each intensity profile is normalized between the minimum value (0%) and the maximum value (100%) of each bead (center). The normalized profiles are then averaged across replicates and fitted with a Gaussian function to determine the standard deviation (SD) of the fit (right). The effective spatial resolution is finally computed as: $FWHM = 2.35482 \cdot SD$

For the example considered here, the β-actin–Alexa568–immunolabeled component in the double-color 3D-STED image (Fig. 5D and Table 3) is detected with an experimental spatial resolution of 111 nm laterally and 166 nm axially (Fig. 6). In comparison, the theoretical FWHM for a confocal acquisition of the same sample, using a 63×/1.40 oil-immersion objective and 575 nm excitation, would be approximately 165 nm lateral and 627 nm axial.

This clearly demonstrates how 3D-STED with balanced X, Y, Z depletion markedly enhances the axial resolution and produces a more isotropic detection volume. Confocal imaging, by contrast, suffers from a much poorer axial resolving power, which significantly hinders the extraction of reliable conclusions in colocalization analyses. Similar limitations are also encountered with 2D double-color STED without Z-depletion, which retains the same confocal uncertainty along the axial dimension (Fig. 4).

### 3.11 Key Points

Spatial resolution in STED imaging can vary significantly depending on several factors, including:

STED depletion laser power, alignment and quality of excitation and STED beams, fluorophore properties, label density, refractive index mismatch and spherical aberration, objective quality and numerical aperture (NA), signal-to-noise ratio, and sample movement and drifts.

1. Align the STED lasers using gold beads before any imaging session. It is critical to ensure precise overlap of excitation and depletion beams.
2. When choosing the fluorophores, prefer far-red, which are more photostable and will reduce photobleaching.
3. Measure the Point Spread Function (PSF) under the experimental conditions as applied to the sample. This is crucial for understanding the spatial resolution of the images and optimizing the choice of the fluorophore, if needed.
4. Use proper fluorescent nanobeads for each detection channel to ensure spatial co-localization across wavelengths.
5. The nanobead dyes for PSF analysis must be the same as for immunostaining.
6. For 3D STED, the PSF must also be determined in the XZ dimension.
7. Apply fine-tune STED power and time-gated detection when possible, to balance resolution and contrast.

## 4 Notes

1. Gold beads are ideal for alignment because they provide strong scattering and do not bleach. They enable the collection of a reflection signal that allows direct assessment of the excitation and STED depletion beam profiles by scanning them through the focal point.
2. This number identifies the thickness of the coverslip. It is mandatory to use the thickness that matches the working distance of the high numerical aperture objective needed for STED imaging.
3. The inverted stage configuration of the STED microscope is not mandatory; however, it is preferred when working with thin samples, such as adherent cells.
4. See Table 1 and 2 for examples of the fluorophores and their combination.
5. Maintain the same objective throughout the entire procedure.

6. Generally, calibration starts at the 592 nm line of the STED laser.
7. Reflection signals are extremely intense; therefore, PMTs are generally preferred at this stage because of their large dynamic range and robustness. Hybrid detectors are more sensitive but also more fragile. New-generation hybrid detectors are equipped with an attenuator that allows their safe use for reflection signals, preventing potential damage.
8. Nanobeads can be difficult to bring into focus. A useful tip is the following: lower the stage until the immersion oil just touches the coverslip, then move slightly back so that the objective is out of the glass. From this position, you can reliably find the focus by moving the Z-knob (or joystick) in a single, consistent direction while watching the live video. This allows us to follow the stage as it approaches the coverslip and capture the in-focus plane more easily.
9. Generally, the STED routine starts with an automatic alignment procedure based on the most blue-shifted STED line. Therefore, any excitation wavelength below that depletion line is suitable for the reflection-based alignment. Some STED microscopes also provide an "expert alignment procedure," which is used when the automatic routine performed at the microscope's reference STED wavelength (typically the 592 nm depletion line) does not achieve optimal alignment, particularly for other depletion wavelengths being tested (e.g., 660 or 775 nm).
10. In addition to the 592 nm line, the other two most common STED lines are at 660 and 775 nm.
11. Emission/excitation spectra of many STED-compatible fluorophores are available in online databases, as the distance between the two points on the curve where the value is half of the maximum height. FLIM-STED microscopes provide an additional tool that is useful for resolving crosstalk of fluorophores based on the difference in fluorophore fluorescence lifetime.
12. Abberior Mount Liquid can be used as a mix of PBS, glycerol, and antifading reagents.
13. Cells were isolated by using the RosetteSep B-lymphocyte enrichment kit (StemCell Technologies, Vancouver, Canada), but any other commercial kit may also be useful and must be determined by researchers. The present study was approved by the Ospedale San Raffaele (OSR) ethics committee under the protocol CLL-BIO. Researchers should ask for relevant authorization from competent authorities following ethical concerns and local laws.

14. Use consistent imaging parameters across measurements.

15. Photobleaching can be a major limitation in STED imaging, especially during 3D acquisitions. When using a basic STED system, photobleaching can be reduced by employing resonant scanning together with a high number of line and frame averages.

16. In STED microscopy, both lateral (horizontal) and axial (vertical) spatial resolution depend on the STED laser power applied to the sample and the selected fluorophores (see Fig. 4A). Resolution of the point spread function (PSF), typically expressed as the full width at half maximum (FWHM), should be quantified under acquisition conditions optimized for the specific sample. This parameter is determined by measuring the intensity of nanobeads labeled with the same fluorophore as the sample and applying consistent imaging procedures. Determining the experimental FWHM value is essential in 3D-STED microscopy to prevent axial distortions and in colocalization STED imaging to accurately define the boundaries for colocalization analysis. Notice that the spatial resolution should not be confused with the voxel size of an image stack, although the two parameters are related. The Voxel Size is the physical size of each 3D pixel (voxel) in the image, determined by the sampling parameters (pixel size in XY and step size in Z). It depends on the acquisition settings, such as zoom, pixel dimensions, and z-step. The Spatial Resolution (expressed as FWHM of the point spread function) is the smallest distance between two points that can still be distinguished as separate in the image. It depends on the optics (objective NA, wavelength), and in STED, on the depletion laser power and fluorophore properties. Key point: A voxel size can be smaller than the spatial resolution (oversampling), which is recommended to satisfy the Nyquist criterion. But voxel size does not improve resolution—it only affects sampling. In 3D STED, depletion in the lateral (XY) and axial (Z) directions is balanced to achieve the most isotropic 3D PSF, minimizing distortions along the axes. The balance is achieved by distributing the depletion laser power differently between lateral and axial directions and, if necessary, by adjusting the pinhole to improve the lateral resolution. The specific approach may vary depending on the type of STED microscope used (Fig. 4A).

17. Ensure fluorescent nanobeads are smaller than the expected resolution limit.

18. The application of the Nyquist criterion helps in optimizing the pixel size as a function of the imaging conditions to ensure that the smallest details are well-defined in the final image and

are not "lost" due to overly coarse sampling. The digital calculator is available at https://svi.nl/Nyquist-Calculator.

19. Consider 3D PSF measurement if axial resolution is relevant by imaging the nanobeads in the X, Z mode.
20. Measure enough nanobeads to obtain a Gaussian fitting with a reasonable statistic (low and random residuals). The standard Gaussian function can be edited in an analysis SW as:

$$f(x) = a\ e^{-\frac{1}{2}\left(\frac{x-\text{Mean}}{\text{SD}}\right)^2}$$

where: $a$ = amplitude$_1$ is the maximum value of the function (peak height), Mean$_1$ is the central value (mean), SD$_1$ is the standard deviation (curve width), $x$ is the independent variable (the diameter of the nanobead). FWHM = 2.35482·SD. FWHM (full width at half maximum) is a measure of the width of a curve, particularly of the Gaussian function, and is defined.

## Acknowledgments

Microscopy was performed at the Microscopy & Dynamic Imaging Unit, CNIC, cofunded by MCIN/AEI /10.13039/501100011033 and FEDER "Una manera de hacer Europa" (#ICTS2018-04-CNIC-16). We thank Veronica Labrador Cantarero and Alfonsa Díaz Torres for their precious help. The CNIC is supported by the Instituto de Salud Carlos III (ISCIII), the Ministerio de Ciencia, Innovación y Universidades (MICIU), and the Pro CNIC Foundation, and is a Severo Ochoa Center of Excellence (grant CEX2020-001041-S funded by MICIU/AEI/10.13039/501100011033). CS received financial support from Associazione Italiana per la Ricerca sul Cancro (AIRC) under IG 2018—ID. 21332. M. Sampietro is supported by an AIRC—Fondazione AIRC per la Ricerca sul Cancro fellowship.

**Author contributions** Conceptualization: VRC and CS, Methodology: MS and EA, Formal analysis: MS, Writing—original draft: VRC, Writing—review & editing: VRC, CS, MS, and EA, and Supervision: VRC.

## References

1. Blom H, Widengren J (2017) Stimulated emission depletion microscopy. Chem Rev 117(11):7377–7427
2. Hell SW, Wichmann J (1994) Breaking the diffraction resolution limit by stimulated emission: stimulated-emission-depletion fluorescence microscopy. Opt Lett 19(11): 780–782
3. Jahr W, Velicky P, Danzl JG (2020) Strategies to maximize performance in STimulated Emission Depletion (STED) nanoscopy of biological specimens. Methods 174:27–41

4. van der Wee EB et al. (2021) 3D test sample for the calibration and quality control of stimulated emission depletion (STED) and confocal microscopes. Commun Biol 4(1):909
5. Akter Y et al. (2025) Combining nanobody labeling with STED microscopy reveals input-specific and layer-specific organization of neocortical synapses. PLoS Biol 23(4):e3002649
6. Vicidomini G et al. (2011) Sharper low-power STED nanoscopy by time gating. Nat Methods 8(7):571–573
7. Lanzano L et al. (2015) Encoding and decoding spatio-temporal information for super-resolution microscopy. Nat Commun 6: 6701
8. Staudt T et al. (2011) Far-field optical nanoscopy with reduced number of state transition cycles. Opt Express 19(6):5644–5657
9. Heine J et al. (2017) Adaptive-illumination STED nanoscopy. Proc Natl Acad Sci USA 114(37):9797–9802
10. Balzarotti F et al. (2017) Nanometer resolution imaging and tracking of fluorescent molecules with minimal photon fluxes. Science 355(6325):606–612
11. Wu Y, Williams RM (2024) The ATTO 565 dye and its applications in microscopy. Molecules 29:17
12. Jeong S, Widengren J, Lee JC (2021) Fluorescent probes for STED optical nanoscopy. Nanomaterials (Basel) 12:1
13. Takagi T et al. (2021) Discovery of an F-actin-binding small molecule serving as a fluorescent probe and a scaffold for functional probes. Sci Adv 7(47):eabg8585
14. Nasufovic V et al. (2025) SiR-XActin: a fluorescent probe for imaging actin dynamics in live cells. Angew Chem Int Ed Engl 64(50):e202509285
15. Zhai R et al. (2023) Small-molecule fluorogenic probes for mitochondrial nanoscale imaging. Chem Soc Rev 52(3):942–972
16. Stacchini A et al. (1999) MEC1 and MEC2: two new cell lines derived from B-chronic lymphocytic leukaemia in prolymphocytoid transformation. Leuk Res 23(2):127–136
17. Hebisch E et al. (2017) A protocol for registration and correction of multicolour STED super-resolution images. J Microsc 267(2):160–175
18. Gonzalez Pisfil M et al. (2022) Stimulated emission depletion microscopy with a single depletion laser using five fluorochromes and fluorescence lifetime phasor separation. Sci Rep 12(1):14027
19. Sampietro M et al. (2023) The nanomechanical properties of CLL cells are linked to the actin cytoskeleton and are a potential target of BTK inhibitors. Hemasphere 7(8):e931
20. Sampietro M et al. (2021) 3D-STED super-resolution microscopy reveals distinct nanoscale organization of the hematopoietic cell-specific Lyn substrate-1 (HS1) in normal and leukemic B cells. Front Cell Dev Biol 9:655773

# Chapter 14

# Immunofluorescence-Based Study of Cargo Transport Mediated by Cytoskeleton-Associated Molecular Complexes

Francesca Finetti and Cosima T. Baldari

## Abstract

The cytoskeleton plays a fundamental role in orchestrating the intracellular cargo transport, providing a dynamic scaffold that ensures the appropriate sorting, movement, and distribution of membrane-bound compartments. Through the coordinated activity of microtubules and actin filaments, cargo vesicles are driven along defined routes to ensure efficient intracellular trafficking. Among the molecular machineries that interact with the microtubule network, the intraflagellar transport (IFT) system has emerged as a key regulator of vesicular dynamics. Initially identified for its essential role in ciliogenesis, IFT is now recognized as a broader trafficking coordinator, mediating intracellular transport in different cellular processes. In T lymphocytes, IFT20, in collaboration with other IFT components, orchestrates the selective targeting of recycling vesicles within the endosomal system, promoting the accurate delivery of key receptors and signaling molecules. This chapter provides a detailed immunofluorescence-based protocol to study the role of the IFT system in vesicular trafficking, with a focus on its involvement in receptor recycling to the immune synapse and the retrograde transport of the mannose-6-phosphate receptor to the trans-Golgi network.

**Keywords** IFT, Cargo transport, Recycling, Retrograde trafficking, Immune synapse, TCR, MPR, Endosome, Immunofluorescence

## 1 Introduction

The dynamic localization of receptors and signaling molecules relies on the interplay between the cytoskeletal networks and the intracellular transport machinery. Microtubules serve as dynamic tracks for the intracellular transport of proteins and organelles, a process mediated by molecular complexes, such as the intraflagellar transport (IFT) complex, and their associated motors, dynein and kinesin. IFT is well known for its role in the bidirectional movement of molecular cargo along the axoneme of cilia and flagella [1]. In addition to its role in ciliary transport, IFT contributes to the trafficking of vesicles within the cytoplasm, ensuring the proper delivery of cargo between cellular compartments [2]. The extraciliary role of IFT is particularly important in specialized cells, such as immune cells, where it regulates

Noa B. Martín-Cófreces et al. (eds.), *The Cytoskeleton*, Methods in Molecular Biology, vol. 3070,
https://doi.org/10.1007/978-1-0716-5515-3_14,

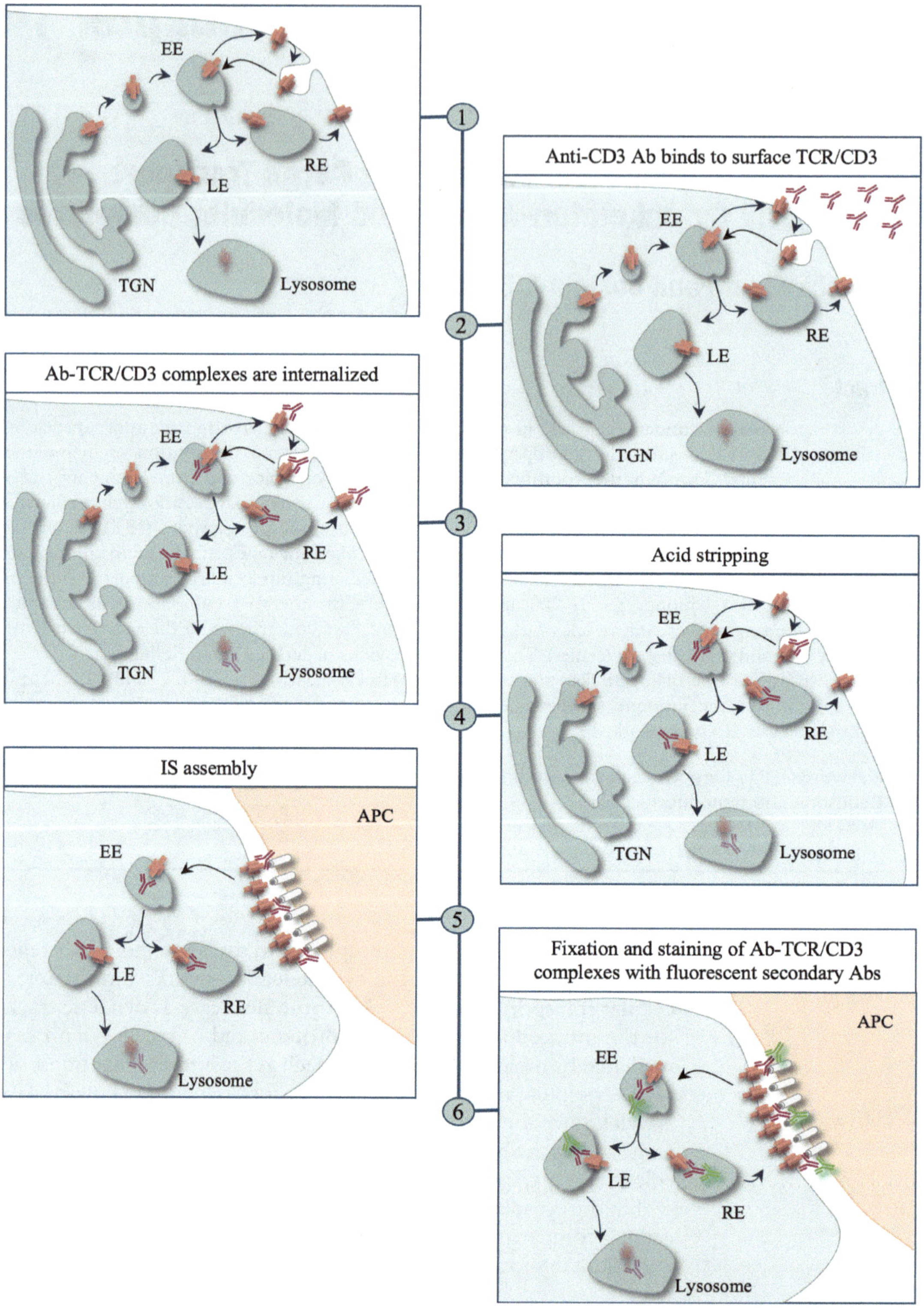


**Fig. 1** Schematic representation of a protocol to specifically track transport of TCR/CD3 complexes to the IS. (1) Under basal conditions, the TCR/CD3 complex, assembled in the endoplasmic reticulum, is transported from the TGN to the plasma membrane. From there, the receptor may undergo endocytosis, reaching early endosomes

the trafficking of signaling molecules associated with endosomes [3–7]. The IFT complex component IFT20 cooperates with other IFT proteins, including IFT88, IFT52, IFT54, and IFT57, to selectively regulate intracellular trafficking in T cells by coordinating a recycling route that is utilized by the T cell antigen receptor (TCR) and the transferrin receptor (TfR), but not by the chemokine receptor CXCR4 [6]. The TCR undergoes continuous recycling between the plasma membrane and the cytoplasm, maintaining a dynamic equilibrium between surface and intracellular pools (Fig. 1). This constitutive recycling acts as a quality control mechanism, ensuring TCR integrity by allowing for the selective degradation of defective complexes. Additionally, it enables the formation of an intracellular TCR pool, which is essential for T cell activation. Indeed, the interaction of TCRs at the plasma membrane with peptide-loaded major histocompatibility complex (pMHC) molecules on the surface of antigen-presenting cells (APCs) triggers the assembly of a specialized structure known as the immunological synapse (IS). IS formation depends on a major reorganization of receptors, signaling proteins and cytoskeletal elements, as well as on the recruitment of the intracellular TCR pool to the IS [5, 8, 9].

To investigate the role of IFT20 in receptor recycling to the IS, we developed an immunofluorescence-based method that enables the visualization of recycling TCR/CD3 complexes both in the cytoplasm and at the cell surface (see Sect. 3.5) [6]. This protocol has proven valuable for studying polarized recycling to the IS, as it distinguishes between surface TCR/CD3 complexes that cluster at the IS and those that recycle from the intracellular pool in T cell–APC conjugates loaded with staphylococcal superantigens [10, 11]. This technique utilizes the anti-CD3 antibody OKT3, known for its ability to bind the TCR-associated CD3 complex, stimulate TCR/CD3-dependent signaling, and promote its internalization and recycling. Receptor recycling analysis involves the antibody (Ab) binding to surface TCR/CD3 complexes and their ligand-dependent internalization. To exclude complexes that have not undergone internalization, we capitalize on the "acid stripping"

**Fig. 1 (Continued)** (EE) before progressing to late endosomes (LE) and lysosomes or, alternatively, returning to the plasma membrane through recycling endosomes (RE). (2) To study TCR/CD3 trafficking to the plasma membrane, T lymphocytes are incubated with saturating concentrations of anti-CD3 (OKT3) Abs, which bind to surface TCR/CD3 complexes. (3) During this incubation, TCR/CD3–Ab complexes are internalized into intracellular vesicular compartments and can be recycled back to the plasma membrane. (4) To ensure that only internalized TCR/CD3–Ab complexes retain their bound antibodies, primary antibodies bound to TCR/CD3 complexes remaining at the plasma membrane are removed by acid stripping. (5) IS assembly is then promoted by incubating T cells with superantigen-loaded APCs. During this process, endosomes containing TCR/CD3–Ab complexes polarize to the synaptic membrane, and a subset of these complexes is recycled back to the IS. (6) Conjugates are fixed, permeabilized, and stained with fluorochrome-conjugated secondary antibodies, which label TCR/CD3–Ab complexes both intracellularly and at the cell surface. Alternatively, if the conjugates are stained without permeabilization, only the recycled TCR/CD3 complexes on the cell surface are visualized (see example in Fig. 2)

step [12, 13] done prior to conjugate formation, which removes Ab-bound TCR/CD3 complexes from the cell membrane. This allows for the specific tracking of internalized complexes, enabling the analysis of their subsequent fate: either recycling to the plasma membrane via recycling endosomes or degradation in lysosomes. Conjugates are formed by mixing acid-stripped T cells and APCs loaded with staphylococcal superantigens. Ab-bound TCR/CD3 complexes can be visualized by immunofluorescence. Staining antigen-specific conjugates using secondary fluorescent antibodies without permeabilization allows for the tracking of surface Ab-bound TCR/CD3 complexes that recycle back to the T-APC cell contact area (Fig. 2). Alternatively, conjugates can be permeabilized and, if necessary, co-stained with antibodies targeting specific endosomal subpopulations, to provide insights into the intracellular localization of TCR/CD3 complexes during recycling (Fig. 2) [6, 10, 11].

IFT20 has also been implicated in lysosomal targeting of hydrolases by orchestrating dynein-dependent retrograde transport of the mannose-6-phosphate receptor (MPR) to the trans-Golgi network

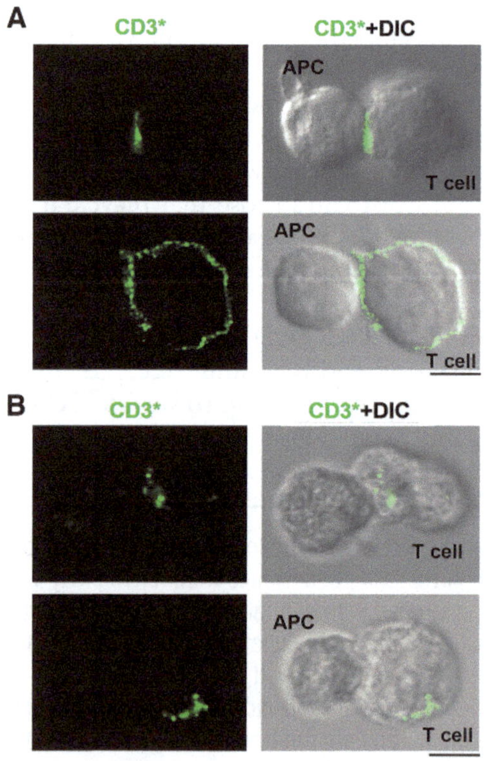


**Fig. 2** Immunofluorescenceanalysis under non-permeabilizing conditions (**A**) or permeabilizing conditions (**B**) of recycled TCR/CD3 complexes (CD3*) in conjugates of Jurkat cells and superantigen-pulsed Raji cells. Each panel shows a representative conjugate harboring either polarized (top) or unpolarized (bottom) TCR/CD3–Ab recycled complexes. Medial optical sections and differential interference contrast (DIC) microscopy images are shown. Scale bar: 5 μm

(TGN) [4]. Lysosomal hydrolases undergo a series of glycosylation events as they transit through the endoplasmic reticulum and the cis-Golgi compartment. Upon reaching the TGN, these enzymes are tagged with a mannose-6-phosphate (M6P) moiety, which facilitates their recognition by MPRs. MPRs bound to M6P-tagged cargo are incorporated into clathrin-coated vesicles that bud from the TGN and are transported to early endosomes before progressing to late endosomes. Progressive acidification of endosomal compartments promotes cargo dissociation, allowing hydrolases to enter lysosomes. Meanwhile, MPRs are retrieved from endosomal membranes and transported back to the TGN via retrograde traffic to enable subsequent rounds of cargo transport (Fig. 3). This recycling process

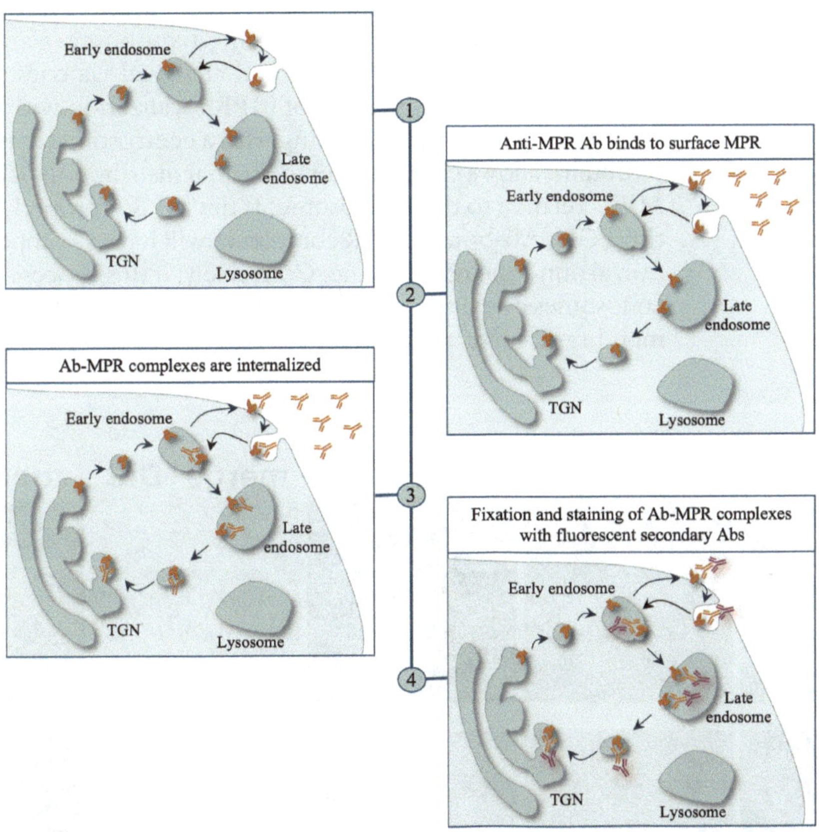


**Fig. 3** Schematic representation of a protocol to specifically track MPR retrograde transport. (1) Under basal conditions, after cargo binding, MPRs are sorted into clathrin-coated vesicles that bud from the TGN and are transported to early endosomes before progressing to late endosomes. The progressive acidification of endosomal compartments promotes the dissociation of MPRs from their cargo, allowing MPRs to be transported back to the TGN via retrograde traffic for subsequent rounds of cargo transport. Additionally, some MPRs follow the default secretory pathway and are exposed at the plasma membrane, where MPRs can bind and internalize extracellular M6P-tagged proteins. (2) To investigate the intracellular trafficking of MPR, T lymphocytes are incubated with saturating concentrations of anti-MPR Abs. (3) During this incubation, MPR–Ab complexes are internalized into intracellular vesicular compartments and undergo retrograde trafficking to the TGN. (4) Finally, cells are fixed, permeabilized, and stained with fluorochrome-conjugated secondary antibodies to label MPR–Ab complexes (see example in Fig. 4)

involves IFT20, the retromer complex, and Rab7, which mediate MPR sorting into tubular endosomal structures that associate with microtubules and facilitate dynein-driven retrograde transport [4, 14–18].

To investigate the dynamics of MPR recycling, we developed an assay based on the pool of MPRs that follow the default secretory route and are exposed at the plasma membrane, where they can bind and internalize extracellular M6P-tagged proteins (see Sect. 3.6) [4]. In this experimental approach, T cells are incubated with saturating concentrations of an anti-MPR antibody, allowing receptor binding and internalization to occur. After incubation, unbound antibodies are removed by washing, and cells are fixed and permeabilized. Internalized MPR–Ab complexes are then labeled with fluorochrome-conjugated secondary antibodies for visualization. To assess retrograde traffic, lymphocytes are costained with the TGN marker TGN38 and analyzed via confocal microscopy, enabling quantification of MPR localization within the TGN (Fig. 4). Alternatively, costaining with a centrosomal marker, such as γ-tubulin, allows evaluation of the spatial distribution of internalized MPRs relative to the centrosome. If the MPR is properly recycled to the TGN, Ab-bound MPR complexes will have a compact distribution around the centrosome. Conversely, if this process is defective, endosomes carrying internalized MPRs will show a dispersed, plasma membrane-proximal pattern.

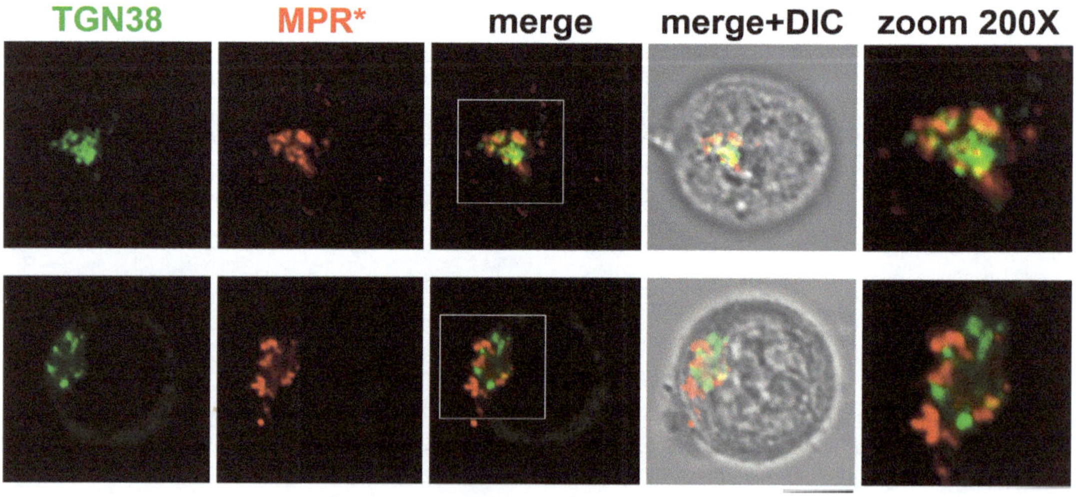


**Fig. 4** Immunofluorescenceanalysis of recycled MPR (MPR*) in Jurkat cells costained with the TGN marker TGN38. The upper images show the retrograde traffic of MPRs to the TGN, as shown by their partial colocalization with TGN38. In contrast, the lower images show impaired MPR recycling, characterized by their failure to reach the TGN and a dispersed, plasma membrane-proximal pattern. Representative medial optical sections and overlays of immunofluorescence and differential interference contrast (DIC) images are shown. The relevant section of the image is shown at a higher magnification (zoom 200×). Scale bar: 5 μm

The integration of these methodologies with RNA interference and gene-targeting technologies specific for IFT proteins (see Sect. 3.1–3.4) or other cytoskeleton-associated molecular complexes offers a robust framework for dissecting the cargo transport process and the regulatory mechanisms underlying this trafficking pathway [4, 6, 19].Furthermore, these approaches can be extended to a broader range of receptors that exploit recycling mechanisms to gain insight into their intracellular trafficking [6, 10, 11].

## 2 Materials

### 2.1 Equipment

1. Electroporator and Amaxa Nucleofector (*see* **Note 1**).
2. Scanning or spinning disk confocal microscope (*see* **Note 2**).
3. Image analysis computer. ImageJ software is used to quantify fluorescence intensity, determine Manders' colocalization coefficient (JACoP plug-in), and distance of $MPR^+$ vesicles from the centrosome.

### 2.2 Cell Purification and Maintenance

1. Raji B cell line (Burkitt's lymphoma) and Jurkat T cell line (acute T cell leukemia, clone E6-1) (*see* **Note 3**).
2. Primary T cells purified from the peripheral blood of healthy donors (*see* **Note 4**).
3. Medium for lymphocyte maintenance: RPMI 1640 supplemented with penicillin (2 mg/mL) and 10% bovine calf serum (BCS) inactivated at 56°C for 30 min.
4. Anti-CD3/CD28- T lymphocytes-coupled Magnetic Beads (*see* **Note 5**).
5. Recombinant human IL-2.

### 2.3 Transfection

1. IFT20-specific esiRNAs (Mission human IFT20 esiRNA, #EHU123981) and a Renilla luciferase control esiRNA (Mission RLUC, Renilla luciferase, esiRNA, #EHURLUC) (Sigma-Aldrich), although alternative siRNA can be used.
2. gRNA targeting GFP (control) or IFT20 (*see* **Note 6**).
3. Opti-MEM medium.

### 2.4 Immunofluorescence-Based Quantification of Receptor Recycling

1. RPMI 1640, 1% bovine serum albumin (BSA) (w/v): Add 5 g of BSA, suitable for cell culture, to 500 mL of RPMI. Filter the solution using a 0.2 μm vacuum filter.
2. Anti-CD3 OKT3 monoclonal antibody (mAb) (*see* **Note** 7).
3. Anti-CI-MPR antibody [4].
4. Stripping solution: 100 mM glycine, 100 mM NaCl, pH 2.5 (*see* **Note 8**).

5. Staphylococcal enterotoxin A (SEA), Staphylococcal enterotoxin B (SEB), and Staphylococcal enterotoxin E (SEE) (*see* **Note 9**).
6. Ten-well Teflon-printed diagnostic microscope slides. The following materials are required for preparation and washing: absolute ethanol, poly-L-lysine, 3MM paper, and a Hellendahl-type dish containing phosphate-buffered saline (PBS).
7. Fixation solution: 8% paraformaldehyde (commercial stock 32%), diluted in PBS and prewarmed to 37°C, and methanol cooled to −20°C.
8. Permeabilization solution: PBS, 1% BSA, and 0.1% Triton X-100 (*see* **Note 10**).
9. Anti-γ-tubulin, anti-TGN38, or anti-dynein antibodies.
10. Secondary fluorochrome-conjugated Alexa Fluor goat antibodies (*see* **Note 11**).
11. Slide mounting medium: PBS, 90% glycerol.
12. Coverslips 24 × 60 mm and conventional nail polish.

## 3 Methods

### 3.1 IFT Protein Silencing

Transfect T cells prior to the experiment to achieve IFT protein knockdown (*see* **Note 12**).

We transfect Jurkat or primary T cells 24 h prior to the experiment using human IFT20-specific esiRNAs or knockout IFT20 expression in T cells using CRISPR–Cas9 gene-editing technology (*see* **Note 13**) following the protocols below.

### 3.2 IFT20 Knockdown in Jurkat Cells

1. Take $5 \times 10^6$ Jurkat cells for each sample, centrifuge at 800 × *g* for 5 min and carefully remove the culture medium.
2. Wash with 1 mL of Opti-MEM® and then discard the supernatant.
3. Resuspend in 800 μL of Opti-MEM® and add 800 ng/$1 \times 10^6$ cells of human IFT20-specific esiRNAs or Renilla luciferase control esiRNA (*see* **Note 14**).
4. Incubate on ice for 5 min.
5. Transfer the sample into an electroporation cuvette and transfect using the BioRad electroporator at 300 mV and 950 μF capacitance.
6. Transfer the transfected sample to a 6-well plate containing 5 mL of prewarmed RPMI 7.5% BCS and incubate in a humidified 37°C 5% $CO_2$ incubator for 24 h (*see* **Note 12**).

### 3.3 IFT20 Knockdown in Primary T Cells

To KD IFT20 in primary T cells (*see* **Note 15**), use the Amaxa nucleofector device and the conditions for T-cell transfection recommended by the manufacturer [4, 5].

1. Add the Nucleofector® supplement to the Nucleofector® Solution: mix 82 µL of Nucleofector® solution and 18 µL of Nucleofector® supplement for each sample.
2. Prepare 6-well plates by filling an appropriate number of wells with 4 mL RPMI 10% BCS and equilibrate plates in a humidified 37°C 5% $CO_2$ incubator for at least 30 min.
3. Take $5 \times 10^6$ T cells and wash 2× with PBS, then discard supernatant completely.
4. Resuspend the cell pellet carefully in 100 µL room temperature Nucleofector® Solution containing siRNA (800 ng of esiRNA/$1 \times 10^6$ cells) and transfer to a nucleofection cuvette.
5. Transfect using the Nucleofector® Program V-024.
6. Add ~500 µL of the pre-equilibrated culture media to the cuvette and gently transfer the sample into a 6-well plate using a pasteur pipette. Incubate the cells in a humidified 37°C 5% $CO_2$ incubator for 24 h (*see* **Note 12**).

### 3.4 IFT20 Knockout in Primary T Cells

Generate T cell blasts from purified human T cells by activating them for 48 h with Dynabead® Human T-Activator CD3/CD28 (ThermoFisher Scientific), or alternative commercially available reagents, following the manufacturer's instructions:

1. Resuspend the Dynabeads in the vial before use by vortexing for at least 30 s.
2. Transfer the required volume of Dynabeads into a tube and wash them by adding 1 mL of PBS 0.1% BCS, then mix thoroughly by vortexing.
3. Remove the buffer by placing the tube on a magnet for 1 min. Then, take the tube off the magnet and resuspend the washed Dynabeads in a volume of RPMI 10% BCS equal to the initial volume of Dynabeads taken from the vial.
4. Add 25 µL of Dynabeads to $2 \times 10^6$ purified T cells resuspended in 2 mL of RPMI 10% BCS supplemented with 100 U/mL of human recombinant IL-2.
5. Seed the mixture into 12-well plates and incubate at 37°C in a humidified incubator with 5% $CO_2$ for 48 h.

   T cell blast can be transfected using the following protocol:
6. Carefully transfer the sample from the 12-well plate into a tube.

7. Place the tube on a magnet for 1 min to separate the beads from the solution. Then, transfer the supernatant containing the cells to a new tube.
8. Wash the sample twice with PBS, then discard the supernatant completely.
9. Transfect $2 \times 10^6$ human primary T cell blasts with Cas9:gRNA ribonucleoprotein (RNP) complexes (Cas9:gRNA ratio 5:3).
10. Prior to transfection, mix 6 μg of gRNA targeting GFP (control) or IFT20 (*see* **Note 6**) with 10 μg of Cas9 and incubate at room temperature for 10–15 min.
11. After transfection, resuspend the cells in 2 mL of RPMI supplemented with 10% BCS and 500 U/mL IL-2, then incubate in a humidified 37°C, 5% $CO_2$ incubator.
12. Perform WB analysis to assess genome editing in the cells at least 72 h post-transfection (*see* **Note 16**), and carry out the assays between days 3 and 7 after transfection (*see* **Note 17**).

### *3.5 Analysis of TCR/CD3 Recycling to the IS by Immunofluorescence*

Set up all samples in duplicate:

1. Count $1 \times 10^5$ T lymphocytes (*see* **Note 18**) for each sample and transfer them into 1.5 mL tubes (*see* **Note 19**). Centrifuge at 16,000 ×*g* for 30 s, carefully remove the culture medium, briefly vortex the samples, and incubate in 50 μL of RPMI supplemented with 1% BSA for 30 min at 37°C. After incubation, centrifuge the samples again at 16,000×*g* for 30 s, then discard the supernatant.
2. Resuspend the samples in 50 μL of RPMI 1% BSA (*see* **Note 20**) with saturating concentrations of OKT3 antibodies (*see* **Note 21**). Incubate the plate at 37°C for 2 h (*see* **Note 22**).
3. In parallel, count $8 \times 10^5$ Raji B lymphocytes per sample, centrifuge at 16,000 ×*g* for 30 s, and resuspend in RPMI without serum to a final concentration of $1 \times 10^6$ cells/200 μL in a 2 mL Eppendorf tube. Add Staphylococcal enterotoxin E (SEE) to achieve a final concentration of 1 μg/mL (*see* **Notes 9 and 23**) and incubate the tube in the cell incubator for 30 min (*see* **Note 24**). Gently mix the suspension every 10 min to prevent cell settling.
4. Prepare the ten-well Teflon-printed diagnostic microscope slides:
    1. wash carefully with ethanol before use;
    2. coat with poly-l-lysine for 15 min at room temperature (see **Note 25**);
    3. discard the solution and carefully rinse the well twice with PBS; and

4. air-dry the slides and maintain in the dark (if prepared in advance, store at 4°C).
5. Wash T lymphocytes with 500 μL of PBS to remove excess primary antibody, centrifuge at 16,000 × *g* for 30 s, and briefly vortex the samples.
6. Add 50 μL of stripping solution (*see* **Note 26**) to the samples and incubate for 30 s at room temperature (*see* **Note 27**).
7. Wash by quickly adding 200 μL of RPMI 1% BSA, and centrifuge the samples at 16,000 × *g* for 30 s. Discard the supernatant.
8. Wash T lymphocytes with 500 μL of RPMI 1% BSA to remove excess of primary antibody, centrifuge at 16,000 × *g* for 30 s, and briefly vortex the samples.
9. Resuspend each sample in 15 μL of RPMI 1% BSA.
10. Centrifuge Raji B cells (used as APCs) at 16,000 × *g* for 30 s, remove the supernatant, wash with 500 μL of PBS to remove excess of superantigens, and carefully resuspend the pellet in 15 μL/sample of RPMI 1% BSA.
11. Transfer 15 μL of the B cell solution to each 1.5 mL tube containing the resuspended T lymphocytes (as described in step 6) and mix by pipetting. Immediately place all samples in a waterbath set to 37°C and incubate for 15 min.
12. Gently pipette and slowly dispense a total volume of 30 μ L of each sample onto the poly-L-lysine-coated diagnostic microscope slides, placing one sample per well (*see* **Note 28**). Protect the slides from accidental hits and allow the samples to adhere for 15 min at room temperature (*see* **Note 29**).
13. Carefully add 30 μL of 8% paraformaldehyde and incubate for 20 min at room temperature in the dark.
14. Wash the wells: immerse the slides in a Hellendahl-type dish filled with PBS for 2 min. Carefully dry the space between the wells using 3MM paper, making sure they do not dry out completely.
15. To analyze intracellular recycled receptors, add 30 μL of permeabilization solution to each well and incubate for 20 min at room temperature (*see* **Notes 10 and 30**). Then, wash the wells following the procedure outlined in step 14 (*see* **Note 31**).
16. Add 15 μL per well of fluorochrome-conjugated Alexa Fluor goat anti-mouse antibody, diluted in PBS (*see* **Note 32**). Keep the slide at room temperature in the dark (*see* **Note 33**).Then, wash the wells following the procedure outlined in Step 14.

17. Add one drop of slide mounting solution per well using a Pasteur pipette, then carefully place the coverslip. Gently press on the coverslip to remove any excess mounting solution and fix it using conventional nail polish.
18. Store the slides at 4°C in the dark.
19. Examine the slides using confocal microscopy with a 60× or higher magnification objective (see example in Fig. 2) (*see* **Note 34**).
20. Analyze T–B cell conjugates using Image J software (*see* **Note 35**).

### 3.6 Visualization of MPR Retrograde Trafficking at the TGN by Immunofluorescence

Set up all samples in duplicate:

1. Count $1 \times 10^5$ T lymphocytes for each sample (*see* **Note 36**) and transfer them to 1.5 mL tubes (*see* **Note 19**). Centrifuge at 16,000 ×*g* for 30 s, carefully remove the culture medium, briefly vortex the samples, and incubate in 50 μL RPMI supplemented with 1% BSA for 30 min at 37°C. After incubation, centrifuge the samples again at 16,000 ×*g* for 30 s, then discard the supernatant.
2. Resuspend the samples in 50 μL RPMI 1% BSA (*see* **Note 20**) with saturating concentrations of anti-MPR antibodies (*see* **Note 37**). Incubate the plate at 37°C for 60 min (Jurkat cells) or 4 h (primary T cells) (*see* **Note 22**).
3. Prepare the 10-well Teflon-printed diagnostic microscope slides (see Sect. 3.5, point 4).
4. Wash T lymphocytes with 500 μL of PBS to remove excess primary antibody, centrifuge at 16,000 ×*g* for 30 s, and briefly vortex the samples. Resuspend each sample in 30 μL of RPMI 1% BSA.
5. Gently pipette and slowly dispense a total volume of 30 μL of each sample onto the poly-L-lysine-coated diagnostic microscope slides, placing one sample per well. Protect the slides from accidental hits and allow the samples to adhere for 15 min at room temperature (*see* **Note 29**).
6. Place the slides in a Hellendahl-type dish containing cold methanol, which simultaneously fixes and permeabilizes the cells, and incubate for 10 min at −20°C.
7. Wash the wells: immerse the slides in a Hellendahl-type dish filled with PBS for 2 min. Carefully dry the space between the wells using 3MM paper, making sure they do not dry out completely.
8. Cell immunostaining (*see* **Note 38**): add 15 μL per well of primary antibody solution obtained by diluting anti-γ

-tubulin, anti-TGN38, or anti-dynein antibodies in PBS 1% BSA. Incubate overnight at 4°C (*see* **Note 33**).

9. Add 15 μL per well of fluorochrome-conjugated Alexa Fluor goat anti-rabbit antibody, diluted in PBS (*see* **Note 32**). Keep the slide at room temperature in the dark (*see* **Note 33**). Then, wash the wells following the procedure outlined in Step 7.
10. Add one drop of slide mounting solution per well using a Pasteur pipette, then carefully place the coverslip. Gently press on the coverslip to remove any excess mounting solution and fix it using conventional nail polish.
11. Store the slides at 4°C in the dark.
12. Examine the slides using a confocal microscope with a 60× or higher magnification objective (see example in Fig. 4) (*see* **Note 34**).
13. Analyze images using ImageJ software. To quantify colocalization between recycled MPR and TGN38 or dynein in median optical sections, use the JACoP plug-in to calculate the Manders' coefficient (*see* **Note 39**). Measure the distance between $MPR^+$ vesicles and the centrosome using ImageJ by drawing a line connecting these structures and recording its length (*see* **Note 40**).

## 4 Notes

1. We use BioRad electoporator for Jurkat cells and Amaxa Nucleofector II system (Lonza) with human T cell nucleofector kit (#VPA-1002, Amaxa Biosystem), although other systems can be used.
2. We use the Zeiss LSM700 confocal microscope with a 63x oil objective or the spinning disk confocal and super-resolution microscope (CSU-W1-SoRA Nikon) with a 60× oil objective.
3. Raji B lymphoblastoid and Jurkat E6-1 lymphoblastic cell lines can be obtained from different sources; we have used the lines available at the ATCC CCL-86 ™ and ATCC TIB-152™, respectively.
4. Several methods are commercially available to purify T cells; in our hands, the RosetteSept™ Human T Cell Enrichment Cocktail (STEMCELL Technologies) followed by Ficoll gradient centrifugation works correctly for this method.

5. We use Dynabead® Human T-Activator CD3/CD28 (ThermoFisher Scientific), but several methods are commercially available.
6. Generate DNA templates for sgRNAs synthesis by PCR amplification using the primers hIFT20 gRNA1 (ttaatacgactcactataggAGTGTAGCCCTGCTTCACCCgttttagagctagaaatagc), hIFT20 gRNA2 (ttaatacgactcactataggCTTCAGCTCTATGGTCTGCTgttttagagctagaaatagc), and a common reverse primer (AGCACCGACTCGGTGCCACT) with pSpCas9(BB)-2A-GFP (Addgene; deposited by Fengh Zhang) as a template. Transcribe sgRNAs in vitro using the HiScribe™ T7 High Yield RNA Synthesis Kit (NEB) and purify them with the RNA Clean & Concentrator™ (Zymo Research) [4]. Store gRNA at −80°C and thaw it on ice before use.
7. Anti-CD3 OKT3 mAb is either commercially available or can be purified from the supernatant of the OKT3 mouse hybridoma (ATCC CRL-8001™). Culture OKT3 hybridoma cells in RPMI supplemented with 7.5% BCS until they reach a concentration of $1 \times 10^6$. Then, purify and titrate OKT3 using the MabTrap purification kit (GE Healthcare), following the manufacturer's instructions. For long-term storage, OKT3 hybridoma cells can be preserved in liquid nitrogen.
8. The stripping solution maintains its functionality for several months when stored at 4°C. It is recommended to periodically check the pH to ensure its stability.
9. SEE is used for studying conjugate formation between Raji B cells and Jurkat T cells. For experiments involving primary T cells, B cells should be preloaded with a mixture of Staphylococcal enterotoxins A (SEA), B (SEB), and E (SEE), each at a final concentration of 1 μg/mL, to maximize coverage of TCR Vβ specificities [5].
10. The permeabilization step ensures cell membrane permeabilization while simultaneously blocking nonspecific binding using 1% BSA (alternatively, 5% goat serum can be used). We use 0.1% Triton X-100 for irreversible permeabilization; alternatively, saponin can be used. However, note that saponin-mediated permeabilization is reversible; hence, it must be included in all subsequent staining steps. Otherwise, fix and permeabilize the cells by placing the slides in a Hellendahl-type dish containing cold methanol and incubating for 10 min at −20°C.
11. Prepare dilutions and carry out incubations of the fluorescently labeled secondary antibody in the dark to protect it from light and preserve its fluorescence.
12. All samples should be analyzed by immunoblotting at various time points after transfection to assess the efficiency of IFT

protein knockdown and determine the optimal timing for the experiment.

13. Similar to human primary T cells, ciliated cells, such as human fibroblasts immortalized with hTERT (ATCC® CRL-4001™), can be knocked out for IFT20 expression by CRISPR–Cas9 gene-editing technology using the Lipofectamine™ CRISPRMAX™ Cas9 Transfection Reagent (Invitrogen) [4]. Alternatively, use other available gene-editing methods to knock out IFT.
14. Alternatively, we generated stable control and IFT20-knockdown Jurkat transfectants using the empty pGP676.13 plasmid or the same construct containing the complementary oligonucleotides corresponding to the coding region of human IFT20 (gatccccGGAAGAGTGCAAAGACTTTatcaagagtAAAGTCTTTGCACTCTTCCtttttggaaa, agcttttccaaaaaGGAAGAGTGCAAAGACTTTactcttgatAAAGTCTTTGCACTCTTCCggg) kindly provided by Gregory J. Pazour [5, 20]. Jurkat cells were transfected by electroporation and subsequently selected using 1 mg/mL G418. Several clones, selected through Western blot analysis, were pooled to generate a stable population of IFT20KD Jurkat cells.
15. Purify T cells from the peripheral blood of healthy donors by using the RosetteSept™ Human T Cell Enrichment Cocktail followed by Ficoll gradient centrifugation. This kit employs tetrameric antibody complexes that recognize non-T cells and glycophorin A on red blood cells (RBCs), thereby enabling the selective depletion of unwanted cells. The antibody cocktail crosslinks these target cells to RBCs, forming rosettes that pellet together with free RBCs during Ficoll centrifugation. Consequently, untouched T cells are recovered at the interface between the plasma and the density gradient medium.
16. IFT20 expression is not completely knocked out, as the transfection efficiency is not 100%, and not all cells will be edited in a homozygous manner.
17. Add fresh complete medium and 100 U/mL IL-2 to the culture and expand the cells. Note that the downregulation of IFT20 expression may decrease as the cells proliferate.
18. Both T cell lines and primary T cells can be utilized for this analysis. By adjusting the time course for internalization and recycling, this method can be applied to a wide range of surface receptors [6]. To ensure reliable and meaningful data, it is essential that both surface receptor expression and internalization rates remain comparable across different cell lines analyzed.

Surface expression of CD3 can be assessed by flow cytometry using the OKT3 antibody, followed by a secondary anti-mouse IgG conjugated to Alexa Fluor 488 (or other fluorochrome detected by your flow cytometer) or a directly labeled OKT3 antibody. For internalization analysis, cells are incubated on ice for 30 min with the OKT3 antibody, washed to remove unbound antibody (defined as time 0), and subsequently transferred to 37°C for 15 or 30 min. At each time point, surface receptor levels are quantified by flow cytometry after staining with the secondary anti-mouse IgG Alexa Fluor 488-conjugated antibody (or other fluorochrome detected by your flow cytometer). Internalization is expressed as a percentage relative to the time 0 value, which is set at 100%.

19. Each T cell sample should be prepared in a 1.5 mL tube to ensure individual processing.
20. To prevent dehydration during prolonged incubation at 37°C, use a minimum of 50 μL of solution to resuspend the samples.
21. Since TCR/CD3 surface expression levels vary between cell lines and primary T cells, each new batch of OKT3 is typically tested on both cell types. To determine the saturating concentration of OKT3, follow this protocol for each new preparation:
    1. Stain T cells by incubating them with serial dilutions of the antibody for 30 min on ice in 1.5 mL tubes;
    2. Wash the cells twice with cold PBS, centrifuge at 16,000 × g for 30 s, and incubate with 15 μL of fluorochrome-labeled secondary antibodies diluted in PBS for 30 min on ice;
    3. Wash twice with cold PBS, centrifuge again at 16,000 × g for 30 s, and resuspend in 200 μL of cold PBS;
    4. Analyze the samples by flow cytometry to determine the primary antibody concentration that achieves saturation of surface binding.
22. This step maximizes internalization of receptor–Ab complexes.
23. Using SEE at a final concentration of 1 μg/mL maximizes the formation of T–B cell conjugates [5].
24. The loading time for B cells with staphylococcal enterotoxins can range from 30 min to 3 h without affecting the efficiency of conjugate formation.
25. Poly-L-lysine, due to its polycationic nature, interacts with the anionic sites on cells, resulting in enhanced adhesive properties. For slide coating, add 30 μL per well of poly-L-lysine,

prepared by diluting the 0.1% (w/v) stock solution 1:10 in $H_2O$.

26. The low pH of the stripping solution removes any residual surface-bound antibodies. This step enables the visualization of TCR/CD3 complexes that are internalized from the surface into the cytoplasm, and subsequently polarize and/or recycle to the IS during conjugate formation.
27. The stripping time can be modified if the cells show signs of damage or if there are residual surface-bound antibodies. The efficacy of stripping can be evaluated by flow cytometry by incubating both stripped and unstripped cells with 15 μL of fluorochrome-labeled secondary antibodies, diluted in PBS, for 30 min on ice.
28. To preserve the integrity of the conjugates during transfer, pipette gently and trim approximately 0.5 cm from the tip of each 200 μL pipette tip, using a separate tip for each sample.
29. Cell adhesion can be assessed by examining the slide under an optical microscope.
30. The permeabilization of T–B cell conjugates allows the secondary antibody to detect both Ab-bound TCR/CD3 complexes that are transported back to the plasma membrane at the IS and those that remain confined within an endosomal compartment, potentially positioned beneath the IS. To specifically monitor only the Ab-bound TCR/CD3 complexes that have successfully recycled to the plasma membrane, this permeabilization step should be omitted (Fig. 2).
31. Introducing an optional staining step with an endosomal marker can be useful for visualizing the vesicular compartment where recycling TCR/CD3 complexes may accumulate. Additionally, centrosome staining may allow for the assessment of its polarization in T–B cell conjugates, thereby helping distinguish between intrinsic trafficking defects in the recycling pathway and those resulting from the centrosome failure to polarize toward the IS.
32. The secondary fluorescently labeled antibodies specifically detect receptor–Ab complexes.
33. Prepare a humidified, covered chamber with damp paper towels to maintain optimal moisture levels and prevent sample drying caused by the small volume in the wells during staining.
34. Images, either median optical section or z-stack, should be acquired by adjusting the detectors to capture an optimal signal while avoiding saturation.
35. The results can be quantified by determining the percentage of conjugates harboring receptor–Ab complexes at the IS relative

to the total number of conjugates. Additionally, the polarization of receptor–Ab complexes to the IS can be quantified in the median optical confocal section as a recruitment index, calculated as the ratio of fluorescence intensity at the synaptic area to the total cell area using ImageJ software [6, 10, 11].

36. To ensure the reliability of retrograde transportanalysis, it is essential to assess both the surface expression and internalization of MPR across the different cell lines tested, as variations in these processes could influence receptor trafficking.
Surface expression of MPR is assessed by flow cytometry using an anti-MPR antibody, followed by a secondary anti-rabbit IgG conjugated to Alexa Fluor 488 (or other fluorochrome detected by your flow cytometer) or a directly labeled anti-MPR antibody. To analyze receptor internalization, incubate the cells on ice for 30 min with the anti-MPR antibody, wash them to remove unbound antibody (defined as time 0), and then incubate the samples at 37°C for 15 or 30 min. At each time point, surface receptor levels are quantified by flow cytometry after staining with the secondary anti-rabbit IgG Alexa Fluor 488-conjugated antibody (or other fluorochrome detected by your flow cytometer). Internalization is expressed as a percentage relative to the time 0 value, which is set at 100%.

37. Analyze the samples by flow cytometry to determine the primary antibody concentration that achieves saturation of surface binding. We use the anti-MPR (cation-independent) antibody (ab32815, Abcam) 1:200. Each new batch of antibody should be tested.

38. To evaluate MPR recycling samples can be stained with:
    1. A centrosome marker, such as γ-tubulin (we use anti-γ -tubulin mouse, #T6557, Sigma Aldrich, 1:200): the quantification of the distance between $MPR^+$ vesicles and γ-tubulin in medial confocal sections allows the determination of whether $MPR^+$ vesicles aggregate correctly around the centrosome or are dispersed at the cell periphery.
    2. A TGN marker, such as TGN38 (we use anti-TGN38 mouse #sc-166224, Santa Cruz, 1:400): the quantification of Manders' co-localization coefficient between $MPR^+$ vesicles and TGN38 in medial confocal sections is used to assess the ability of MPR to recycle to the TGN (Fig. 4).
    3. A marker of the molecular motor dynein (we use anti-dynein mouse #MAB1618, Merck Millipore, 1:500): the quantification of Manders' colocalization coefficient between dynein and $MPR^+$ vesicles in medial confocal sections provides information as to whether MPR correctly couples with dynein for retrograde microtubule transport.

39. Manders' coefficient is a value ranging from 0 (indicating no colocalization) to 1 (indicating complete colocalization) [21].
40. MPR dispersion can be assessed by measuring the distance of each vesicle containing recycled MPR from the centrosome and calculating the mean distance for each cell in single medial confocal sections with any specific software for image analysis [4].

## Acknowledgements

The authors acknowledge the support of the European Commission (ERC_2021_SyG 951329—ATTACK) to CTB and of the University of Siena (PSR-FNF2024_001) to FF.

### References

1. Rosenbaum JL, Witman GB (2002) Intraflagellar transport. Nat Rev Mol Cell Biol 3:813–825. https://doi.org/10.1038/NRM952
2. Finetti F, Onnis A, Baldari CT (2022) IFT20: an eclectic regulator of cellular processes beyond intraflagellar transport. Int J Mol Sci 23. https://doi.org/10.3390/IJMS232012147
3. Finetti F, Cassioli C, Cianfanelli V et al (2021) The intraflagellar transport protein IFT20 recruits ATG16L1 to early endosomes to promote autophagosome formation in T cells. Front Cell Dev Biol 9:1–15. https://doi.org/10.3389/fcell.2021.634003
4. Finetti F, Cassioli C, Cianfanelli V et al (2020) The intraflagellar transport protein IFT20 controls lysosome biogenesis by regulating the post-Golgi transport of acid hydrolases. Cell Death Differ 27:310–328. https://doi.org/10.1038/s41418-019-0357-y
5. Finetti F, Paccani SR, Riparbelli MG et al (2009) Intraflagellar transport is required for polarized recycling of the TCR/CD3 complex to the immune synapse. Nat Cell Biol 11:1332–1339. https://doi.org/10.1038/ncb1977
6. Finetti F, Patrussi L, Masi G et al (2014) Specific recycling receptors are targeted to the immune synapse by the intraflagellar transport system. J Cell Sci 127:1924–1937. https://doi.org/10.1242/jcs.139337
7. Vivar OI, Masi G, Carpier J et al (2016) IFT20 controls LAT recruitment to the immune synapse and T-cell activation in vivo. Proc Natl Acad Sci USA 113:386–391. https://doi.org/10.1073/pnas.1513601113
8. Onnis A, Baldari CT (2019) Orchestration of immunological synapse assembly by vesicular trafficking. Front Cell Dev Biol 7:110. https://doi.org/10.3389/fcell.2019.00110
9. Das V, Nal B, Dujeancourt A et al (2004) Activation-induced polarized recycling targets T cell antigen receptors to the immunological synapse; involvement of SNARE complexes. Immunity 20:577–588. https://doi.org/10.1016/S1074-7613(04)00106-2
10. Onnis A, Finetti F, Patrussi L et al (2015) The small GTPase Rab29 is a common regulator of immune synapse assembly and ciliogenesis. Cell Death Differ 22:1687–1699. https://doi.org/10.1038/cdd.2015.17
11. Finetti F, Patrussi L, Galgano D et al (2015) The small GTPase Rab8 interacts with VAMP-3 to regulate the delivery of recycling T-cell receptors to the immune synapse. J Cell Sci 128:2541–2552. https://doi.org/10.1242/jcs.171652
12. Liu GY, Kulasingam V, Alexander RT et al (2005) Recycling of the membrane-anchored chemokine, CX3CL1. J Biol Chem 280:19858–19866. https://doi.org/10.1074/JBC.M413073200
13. Patel S, Mukovozov I, Robinson LA (2011) Assessment of the recycling of the membrane-bound chemokine, $CX_3CL1$. Methods Mol Biol 748:143–153. https://doi.org/10.1007/978-1-61779-139-0_10
14. Wassmer T, Attar N, Harterink M et al (2009) The retromer coat complex coordinates endosomal sorting and dynein-mediated transport, with carrier recognition by the trans-Golgi

network. Dev Cell 17:110–122. https://doi.org/10.1016/j.devcel.2009.04.016
15. Hong Z, Yang Y, Zhang C et al (2009) The retromer component SNX6 interacts with dynactin p150Glued and mediates endosome-to-TGN transport. Cell Res 19:1334–1349. https://doi.org/10.1038/cr.2009.130
16. Rojas R, Van Vlijmen T, Mardones GA et al (2008) Regulation of retromer recruitment to endosomes by sequential action of Rab5 and Rab7. J Cell Biol 183:513–526. https://doi.org/10.1083/JCB.200804048
17. Arighi CN, Harmell LM, Aguilar RC et al (2004) Role of the mammalian retromer in sorting of the cation-independent mannose 6-phosphate receptor. J Cell Biol 165:123–133. https://doi.org/10.1083/JCB.200312055
18. Seaman MNJ (2004) Cargo-selective endosomal sorting for retrieval to the Golgi requires retromer. J Cell Biol 165:111–122. https://doi.org/10.1083/JCB.200312034
19. Galgano D, Onnis A, Pappalardo E et al (2017) The T cell IFT20 interactome reveals new players in immune synapse assembly. J Cell Sci 130:jcs.200006. https://doi.org/10.1242/jcs.200006
20. Follit JA, Tuft RA, Fogarty KE, Pazour GJ (2006) The intraflagellar transport protein IFT20 is associated with the Golgi complex and is required for cilia assembly. Mol Biol Cell 17:3781–3792. https://doi.org/10.1091/mbc.e06-02-0133
21. Manders EMM, Verbeek FJ, Aten JA (1993) Measurement of co-localization of objects in dual-colour confocal images. J Microsc 169: 375–382. https://doi.org/10.1111/J.1365-2818.1993.TB03313.X

Check for updates

# Chapter 15

# CryoEM Strategies for Elucidating the Closed State Architecture of CCT

**Jorge Gutiérrez-Seijo, Ana Cuervo, Sergio Pipaón, David Gil-Cantero, Carmen Majano, César Santiago, José María Valpuesta and Jorge Cuéllar**

## Abstract

Cryo-electron microscopy (CryoEM) has emerged as a state-of-the-art technique in structural biology, thanks to advances in sample preparation, high-voltage electron microscopy, direct electron detectors, and sophisticated image processing software. In this study, we investigate the eukaryotic chaperonin CCT, a ~1 MDa hetero-oligomeric complex essential for the folding of key substrates such as actin, tubulin, and WD40 family members. While several open-state structures of CCT have been resolved, the highest-resolution reconstructions have been obtained in the closed state. This is primarily due to the increased structural rigidity of the complex when its apical domains become immobilized in this conformation. However, the pseudo-symmetric arrangement of subunits in the closed state poses challenges for subunit identification. Here, we present a workflow for obtaining a high-resolution CryoEM structure of closed CCT, achieved without the use of nanobodies, substrates, or any tools designed to increase CCT asymmetry. Our results demonstrate that subunit assignment and atomic modeling are achievable by exploiting subunit-intrinsic structural features alone.

**Keywords** CCT, Molecular chaperone, Chaperonin, ATPase cycle, ATP-AlFx, Closed conformation, Electron microscopy, Cryo-electron microscopy, Image processing, Three-dimensional reconstruction, Docking and atomic model

## 1 Introduction

In recent years, cryo-electron microscopy (CryoEM) has undergone remarkable technological advancements, enabling the determination of protein and macromolecular complex structures at near-atomic resolution. These improvements have positioned CryoEM as a powerful competitor to traditional structural biology techniques like X-ray crystallography and nuclear magnetic resonance (NMR) spectroscopy, firmly establishing it as a cornerstone method in the field [1, 2]. Improvements in sample preparation methods, the development of new high-voltage electron microscopes, the implementation of cutting-edge direct electron detectors, and significant advances in

Noa B. Martín-Cófreces et al. (eds.), *The Cytoskeleton*, Methods in Molecular Biology, vol. 3070,
https://doi.org/10.1007/978-1-0716-5515-3_15,

image processing software have overcome previous limitations of the technique. Together with its many advantages—such as requiring minimal protein sample, accommodating a wide molecular weight range, and preserving proteins in their native, hydrated state without the need for crystallization—these developments have solidified Cryo-EM as a powerful and versatile tool for high-resolution structural analysis [1, 2].

The key player in this chapter is the eukaryotic chaperonin CCT, also known as the chaperonin containing TCP-1 or TRiC. As a member of the group II chaperonins, CCT is distinguished by its intricate architecture. It forms a double-ring structure, with each ring comprising eight distinct but related subunits that occupy defined positions within the assembly. This arrangement creates a central chamber where protein folding is orchestrated through ATP-dependent conformational changes [3]. CCT plays a pivotal role in cellular proteostasis by facilitating the folding of a diverse yet highly specific set of substrates. Among its clients are the cytoskeletal proteins actin and tubulin, which are obligate substrates that rely entirely on CCT to achieve their native conformations. Beyond these, CCT's substrate repertoire extends to other essential protein families, most notably the WD40-repeat proteins, which represent the largest family of CCT substrates and are involved in a wide range of fundamental cellular processes [4, 5].

Each CCT subunit is divided into three domains: apical, intermediate, and equatorial. The equatorial domain is responsible for establishing inter-subunit interactions and houses the ATP-binding pocket of the protein; the apical domain is involved in the recognition and binding of substrates, and contains a helical protrusion that acts as a built-in lid for the cavity of CCT; and the intermediate domain connects the other two [6]. All CCT subunits possess a variable substrate-binding site and a highly conserved ATP-binding pocket when compared across subunits. However, there is substantial heterogeneity in their substrate and ATP-binding affinities. Notably, only four subunits—CCT1, CCT2, CCT4, and CCT5—bind ATP at physiological concentrations, while the remaining four—CCT3, CCT6, CCT7, and CCT8—recognize substrates more specifically [7].

ATP hydrolysis facilitates the mechanism of aperture and closure of the cavity, and the associated conformational changes. The open CCT can bind and hydrolyze ATP in order to trigger the closure of its cavity through the action of the built-in lid of its subunits. The complex remains in this closed conformation until the inorganic phosphate from the hydrolyzed ATP is released, leaving CCT in an open ADP-bound conformation. Then, ADP is released, and the cycle starts over [8].

CCT poses a challenge for high-resolution structure determination. The flexibility of the apical domains of the subunits prevents the uniform distribution of local resolution across all regions, which can result in poorly resolved regions of the protein appearing in the final reconstruction. Although several open-state CCT structures have been successfully resolved [9–11], the highest-resolution structures reported to date correspond to the closed state of the complex, in which substrates such as actin and tubulin, as well as other clients like G$\beta_5$ from the WD40 protein family, can be observed [12–14].

During the closing process, CCT subunits tend to adopt a more symmetrical arrangement, increasing the difficulty of identifying individual subunits. This can only be achieved by identifying the specific residues of each subunit, or by using alternative approaches, such as subunit-specific nanobodies to overcome the pseudo-symmetric structure of the closed CCT [15].

The methodology described herein (see Fig. 1 for workflow) does not employ nanobodies or other strategies, demonstrating that an atomic-level reconstruction of the closed CCT complex can be achieved by focusing exclusively on subunit-intrinsic structural features.

## 2 Materials

### 2.1 Biological Sample

Purified human CCT from any source, such as Chapters [purification of endogenous CCT from cell cultures] and [purification of the eukaryotic chaperonin CCT from native tissues tissues; Chap. 1].

### 2.2 Buffers and Solutions

1. 5 mM $MgCl_2$ in ultrapure water.
2. 5 mM $AlNO_3$ in ultrapure water.
3. 30 mM NaF in ultrapure water.
4. 5 mM ATP in ultrapure water.
5. Closing Buffer: 25 mM Tris buffer pH 7.3, 100 mM KCl.
6. Fixation solution: 40% ethanol, 10% acetic acid in water.
7. Staining solution: 50% ethanol in water.
8. Coomassie solution: 25% (w/v) Coomassie Brilliant blue, 50% methanol, 10% acetic acid in water.
9. Destaining solution: 40% methanol, 10% acetic acid.
10. 2% Uranyl Acetate solution.

### 2.3 Gels

1. 4%–15% gradient native polyacrylamide gels to assess the quality of the closed conformation induction.

### 2.4 Acquisition of Transmission Electron Microscopy (TEM) and Cryo-Electron Microscopy images

1. Transmission electron microscope.
2. Vitrification device (*see* **Note 1**).
3. Cryo-electron microscope equipped with a direct detector (*see* **Note 1**).

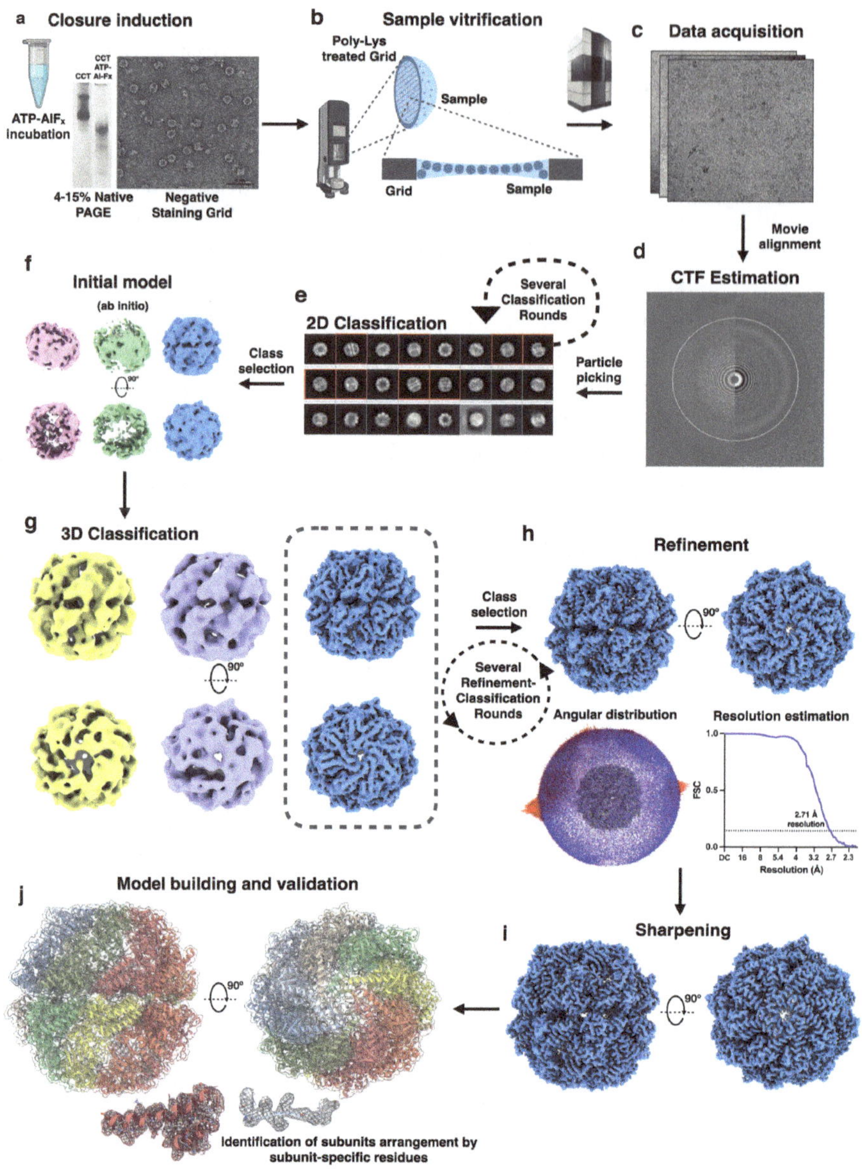


**Fig. 1** CryoEM workflow: from lid closure to atomic model. This figure outlines the single-particle cryo-electron microscopy pipeline. (**a**, **b**) Sample preparation and vitrification are illustrated first: purified CCT oligomers were incubated in the presence of closure stabilizing agents, and their conformational states were checked by negative staining EM and native-PAGE. Then, complexes were applied to the CryoEM grid, vitrified, and loaded into the microscope. (**c**) Acquired images undergo preprocessing including motion correction, (**d**) contrast transfer function (CTF) estimation, and automated particle picking. (**e**) Selected particles were subjected to several reference-free 2D classification, and (**f**) the optimal class averages were used to generate an initial 3D model ab initio. (**g**) Subsequent 3D classifications were used to select particles belonging to a 3D model with improved structural details (squared with dotted lines). (**h**) Final 3D model with refined alignment, angular distribution, and resolution FSC curve. (**i**) Model sharpening showing high-frequency details. (**j**) Model building, showing an example of side chain assignment

### 2.5 Workstation Hardware

The workstation used is usually equipped with at least one GPU device for calculation and 32 CPU threads. In this example, the workstation presented the following specifications. For minimum specification (*see* **Note 2**).

1. Intel Xeon(R) Gold 6230 CPU @ 2.10 GHz. Comprising 2 sockets with 20 cores per socket.
2. 377 GB of Random Access Memory (RAM).
3. 4× Nvidia GPU Tesla T4 with 15 GB of VRAM.
4. 29 TB of HDD local storage.

### 2.6 Workstation Software

1. Scipion v3.7.1—Eugenius.
2. MotionCor 3-1.1.2.
3. gctf 1.18.
4. RELION 5.0.1.
5. cryoSPARC v4.6.2.
6. XmippSrc-3.23.11.0.
7. PHENIX 1.20.1-4487.
8. Coot 1.1.14.

## 3 Methods

### 3.1 Induction of Closed Conformation of CCT

This procedure aims to induce the closed conformation of CCT by mimicking the ATP hydrolysis transition state using the ATP-AlFx analog. Modifications have been made to optimize the formation of fully closed-state conformations based on the protocol described by Knowlton et al. [16]. To induce the ATP-AlFx-driven CCT lid closure reaction, perform the following reaction:

Dilute purified CCT sample to 1 μg/μL in closing buffer.

1. Mix sequentially with the following reagents, maintaining 30 s intervals between additions: 5 mM $MgCl_2$, 5 mM $Al(NO_3)_3$, 30 mM NaF.
2. Immediately after the final addition, add 5 mM ATP to the reaction mix (*see* **Note 3**). CCT complex final concentration should be around 0.7 μg/μL.
3. Incubate the reaction mixture at 4°C for 10–15 min (*see* **Note 4**).
4. Following incubation, proceed directly to vitrification to capture and preserve the closed-state conformation of CCT (Sects. 3.3 and 3.4).

### 3.2 Confirmation of Closed Conformation of CCT by Native Electrophoresis

To confirm the CCT closure confirmation, it is relevant to use different approaches, including controls for direct comparison and confirmation of the conformational change:

1. Run samples in a 4%–15% gradient native polyacrylamide gel. Include a mobility control of CCT untreated with ATP as a reference (see Fig. 1a and **Note 5**).
2. Incubate the gel in Fixation solution for 15 min at room temperature.
3. Incubate in the staining solution for 15 min at room temperature.
4. Stain the gel with Coomassie solution for 30 min at room temperature.
5. Incubate with Destaining solution from 2 h to overnight at room temperature. Change the solution frequently until only proteins are stained with Coomassie.

### 3.3 Confirmation of Closed Conformation of CCT by Transmission Electron Microscopy (TEM)

The confirmation of the induction of the closed conformation of CCT can also be performed through negative staining and observation through TEM:

1. Coat carbon-coated 300 mesh copper grids by glow-discharge at 25 mA for 15 s.
2. Place 5 μL of the sample on a clean piece of hydrophobic material.
3. Place 5 μL of 2% uranyl acetate nearby.
4. Place the carbon side of the grid onto the sample for 1 min at room temperature.
5. Blot gently to dry the grids using Whatman Grade 1 filter paper.
6. Place the carbon side of the grid onto 2% uranyl acetate for 1 min at room temperature.
7. Blot gently the grids using Whatman Grade 1filter paper.
8. Left the grids to air dry before imaging.
9. Proceed to TEM to visualize CCT complexes (Fig. 1a).

### 3.4 Coating of Grids for Cryo-Electron Microscopy Images

Treatment with poly-L-Lys significantly increases the number of side views in large macromolecular complexes [17]. Treat the grids [9] as follows:

1. Expose grids to acetone vapor for 5 min at room temperature.
2. Dry grids for 5 min at room temperature.
3. Place grids on a 20 μL drop of 0.1% w/v poly-Lysine solution and incubate for 90 s.
4. Wash grids by using two consecutive drops of distilled water.

**Table 1**
**Acquisition parameters employed in this processing**

| Microscope | FEI Titan Krios |
|---|---|
| Voltage (kV) | 300 |
| Detector | Gatan K3 |
| Magnification | 81,000× |
| Total dose ($e^{-}$/) | 28, 79 |
| Dose per frame ($e^{-}$/Å$^2$) | 1 |
| Defocus range | −0.8 μm to −2.2μm, 0.2 μm step |
| Pixel Size | 1.072 Å |
| Number of micrographs | 15,400 |
| Initial particles | 2,593,247 |
| Final Particles | 105,562 |
| Resolution (Å) | 2.71 |
| FSC threshold | 0.143 |

5. Blot gently between washes in high-quality filter paper.
6. Store the treated grids before vitrification or use sequentially.

### *3.5 Vitrification of Samples and Image Acquisition*

1. Load 4 μL of CCT at 0.7 μg/μL on a Quantifoil R 2/2 Cu/Rh treated with poly-Lysine.
2. Vitrify samples in a vitrification device with an ethane bath cooled with liquid nitrogen with 2 s of blotting at a −2 "force" value and 95% humidity. See Fig. 1b.
3. Maintain at liquid nitrogen temperature from this point.
4. Acquire images in a microscope equipped with a direct detector camera with a total electron dose of around 30 e/A$^2$, at a pixel size close to 1 Å/px and with a defocus range between −0.8 μm to 2.2 μm. The parameters summarized in Table 1 may require specific adjustments. See Fig. 1c.

### *3.6 Movie Alignment*

Scipion [18] is used here as an image processing framework; for more details in the parameters and other processing frameworks, *see* **Note 6**. For more information about other movie alignment software, *see* **Note 7**.

1. Open the software.
2. Import the movies.
3. Use MotionCor software [19] to align the movie frames into a single image.

4. Frames to ALIGN and SUM: from 1 to 0.
5. Number of patches: 5 for both axes and 0% overlap.
6. Setting the dose filter to yes.
7. Provide the gain reference image if the movies are not gain corrected.
8. Specify whether images need flipping or rotation.

### *3.7 Contrast Transfer Function (CTF) Estimation and Image Selection*

The motion-corrected micrographs serve as input for CTF estimation (Fig. 1d).

1. Use ctffind4 [20], or Gctf [21] software to estimate the CTF with the following parameters in the Scipion prompt: 1024 pixels for "box size," 30–3 Å for "resolution range," 5000–90,000 Å for "defocus search range," 500 Å for "defocus step," tolerated and 1000 for "expected astigmatism," (*see* **Note 8**).
2. Select images based on their CTF fitting with the following parameters: 10 Å for "resolution," 1000 Å for "astigmatism" and 4000–40,000 Å for "defocus" (*see* **Note 9**).
3. To filter images automatically, use Xmipp3 CTF consensus program [18].

### *3.8 Particle Picking*

1. Select a small subset of around 50 images to fine-tune the picking process (Fig. 2a, b). For more information about other particle picking software, *see* **Note 10**.
2. Run Xmipp3 manual-picking Step 1 in Scipion prompt [18].
3. Select particles by clicking on the left mouse.
4. Activate training while correcting the picking mistakes upon reaching at least 200 coordinates for particles.
5. Run the Automatic picking protocol by providing a manual-picking step as input and the set of images to pick the entire dataset.

### *3.9 Particle Extraction*

The RELION particle extraction protocol [22] is a good option to extract CCT particles with a final box size of around 300 Å.

1. Initial 2D classification is commonly performed on down-sampled particles.
2. Use box sizes reduced by a factor of 2–4 to yield a pixel size in the range of 3–4 Å/pixel (around 100-pixel voxel size). This approach significantly speeds up the early stages of data processing.
3. In the RELION particle extraction Scipion prompt [18], provide the input coordinates coming from the picking protocol.
4. Indicate the initial particle box size in pixels (around 300).

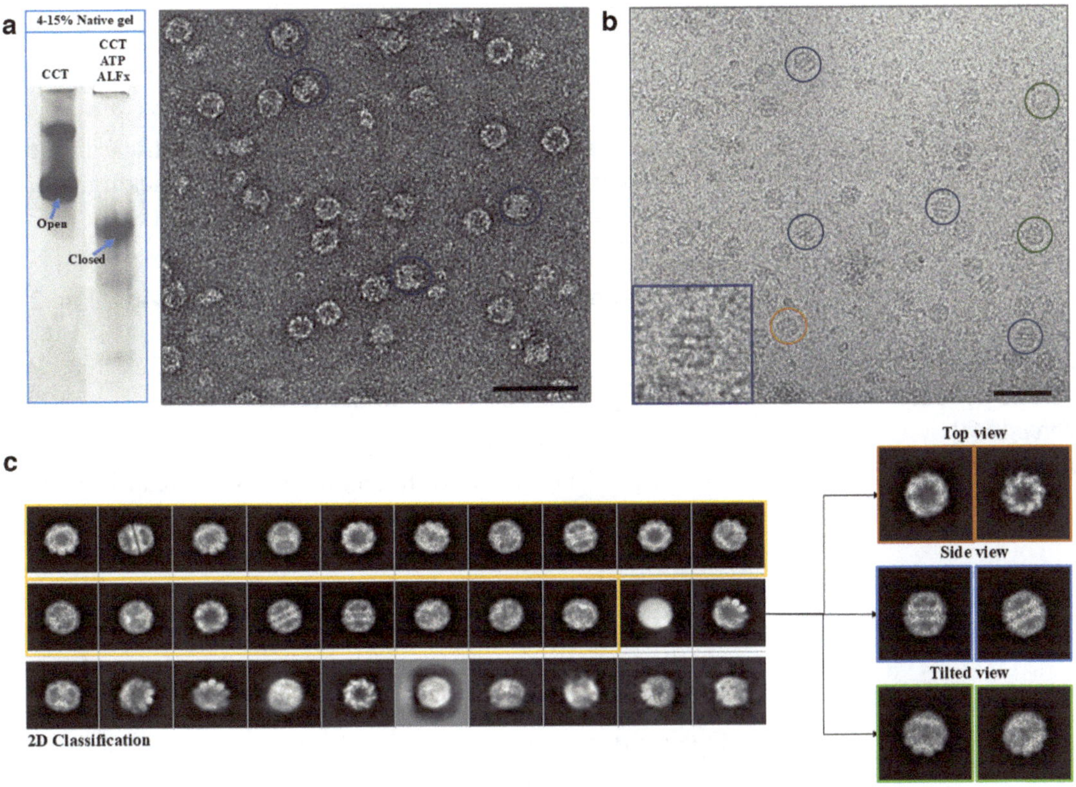


**Fig. 2** CCT lid closure monitoring and cryoEM data processing. (**a**) CCT closing reaction: left panel, 4%–15% gradient native electrophoresis showing different migrations of open and closed CCT conformations; right panel, negative staining analysis of the CCT closing induction with ATP-ALFx. (**b**) CryoEM representative micrographs from CCT closed conformation. The zoomed region shows a characteristic side view of CCT in closed conformation. (**c**) Example of an initial 2D classification. High-quality class averages reveal CCT complexes in various orientations, including side, tilted, and front views. The panel also displays lower-quality 2D averages and classes containing contaminants or incorrectly picked coordinates, demonstrating the range of data quality encountered during the first steps of image processing. The colored circles and squares in panels **a** and **b** highlight particles in top (orange), side (blue), and tilted (green) orientations. Bars, 50 nm

5. Selects rescale particles "Yes."
6. Provide the Re-scaled size for the initial processing (around 100).

### *3.10 2D Classification*

1. Give extracted particles as an input for 2D classification pruning.
2. Run the cryoSPARC software [23] 2D classification program to obtain a rapid 2D classification with the standard parameters and a number of around 200 classes in the initial steps.
3. For class selection, select particles producing high-quality 2D class averages that display secondary structure elements (Figs. 1e and 2c).
4. Open the analysis results window in Scipion prompt from the 2D classification run.

5. Select display particle classes with Scipion.
6. Select on the viewer the classes using left mouse click.
7. Click on particles and mark "ok" to save the selection.
8. Run several times the 2D classification program discarding particles belonging to 2D classes that show wrong picking coordinates, noise, or poorly defined structural elements.

In contrast to the CCT open conformation, the CCT closed form should contain high-resolution details across all domains (Fig. 2).

### 3.11 Optimizing Particle Selection Through 3D Classification

In the example shown, 2 RELION 3D classification [22] steps were performed (Fig. 1g). To obtain similar results:

1. Perform an initial model building using cryoSPARC-initial model [23] before the refinement by using default parameters in the Scipion prompt.
2. Provide a pruned particle set as input.
3. Set the number of classes to 3. Due to the homogeneity of this dataset conformational landscape, no symmetry is needed to obtain a good initial model. The results obtained in this step can be observed in Fig. 1f.
4. Open the "Analyze results" tab and select the model that shows better definition around all the structural domains.
5. Run an initial refinement protocol. This step can be executed using various software packages; the non-uniform refinement protocol available in cryoSPARC [23] is a good candidate for this step. It can be run using default parameters in the Scipion prompt and providing as inputs: the selected particles from the classification and the best initial model from the previous step.
6. 3D classification: Run RELION 3D classification [22] without performing image alignment in order to clean even further the dataset or to remove other CCT conformational states.
7. Use as inputs the 3D model and the particles with the previous alignments coming from the initial refinement step.
8. Run the protocol with default parameters in the RELION prompt from Scipion.
9. Include exception in the setting "Consider previous alignment": "Yes."
10. Include exception in the setting "Is the initial map in absolute gray scale": "No."
11. Include exception in the setting "Initial low-pass filter': "10 A."

12. Include exception in the setting "Perform image alignment": "No."
13. Include exception in the setting "Use GPU acceleration": "No."
14. Open the "Analyze results" tab from the Scipion prompt.
15. Inspect the 3D volumes with your method of choice (slices or Chimera).
16. Select the particles producing the best 3D model by clicking on "Show classification" in Scipion.
17. Select the set with the left mouse click.
18. Click on particles and mark "OK" to save the particles.
19. Repeat this step several times in order to clean the particle set.

### 3.12 3D Refinement

Run additional 3D particlealignment protocols to fine-tune the angular assignment and improve the resolution of the 3D volume. Several protocols might be used for this purpose. In this step, both the RELION auto-refine [22] or non-uniform refinement protocol from cryoSPARC [23] can be used independently or in tandem. An example of a 3D reconstruction, along with the calculated angular distribution and estimated resolution, is shown in Fig. 1h.

1. To use cryoSPARC, run cryoSPARC non-uniform refinement [23] with default parameters using as inputs pruned particles from the 3D classification protocol and their corresponding 3D volume as the initial model.
2. (Optional) To use RELION instead, run auto-refine protocol using the same input particles and 3D volume coming from the 3D classification and set the parameters in Scipion prompt as follows: "Consider previous alignment": "yes"; "Is the initial map in absolute gray scale": "Yes"; "Initial low-pass filter": "10 A"; "Perform image alignment": "Yes"; "Use GPU acceleration": "Yes." For RELION GPU acceleration, choose at least one thread per GPU used, plus an additional thread to control the process.
3. Re-extract the particles to the original pixel-size following the same protocol as in Step 3.9 with a box of around 300 pixels once Nyquist resolution is reached.
4. Assign the previous alignments to the re-extracted particles using the Scipion assign alignment protocol.
5. Resize the 3D volume to the original pixel size using "Crop-resize volume" protocol in Scipion: change the pixel size and the window to the parameters used to extract the particles.
6. Repeat the same alignment steps using non-binned particles and volume. At this stage, change the initial filtering resolution

roughly to the resolution of the previous refinement. When working with particles at their original size, a mask can be used to speed the alignment and decrease the noise. To learn how to create a 3D mask, *see* **Note 11**.

### *3.13 3D Post-Processing: Model Sharpening*

Use a sharpening software to enhance high-resolution features.

1. CryoSPARC gives the sharpened map directly at the end of the refinement; alternatively you can also use DeepEMhancer [24], giving as input the final refined map or RELION post-process protocol providing a previous refinement protocol in the Scipion prompt and a mask. The results can be observed in Figs. 1i and 3a or b, usually the closed maps result in a final volume of higher quality than the open conformation one.
2. To assess the resolution of the different regions of the reconstructed map, perform a local resolution analysis using MonoRes [25] from Xmipp3, providing the final refined map and a mask. The result will generate a map with the local resolution of regions indicated in different colors (see Fig. 3c).

### *3.14 Atomic Model Building*

Once a sharpened volume is obtained, proceed to the next step to build the atomic model within the density through an initial amino acid assignment. This can be achieved using a previously available PDB structure, an AlphaFold prediction [26], or software tools that reconstruct the amino acid chain directly from the electron microscopy map. In this case, the ModelAngelo software, implemented in RELION-5 release [27, 28], provides accurate results (*see* **Note 12**). To obtain good results, it is essential to have enough resolution to clearly identify side chains characteristic of each subunit. Figure 4 depicts well-resolved regions from the apical, intermediate, and equatorial domains that can aid in subunit identification.

To use the RELION-5 ModelAngelo program inside the Scipion interface, proceed as follows:

1. Import into Scipion the FASTA sequence of each CCT subunit and provide them to the program.
2. Provide the sharpened map and a mask that excludes the noisy densities.
3. Perform the "Amino acid chain tracing" with ModelAngelo. This may generate some unconnected amino acid chains, especially in regions where the density is not so well defined (*see* **Note 13**).
4. To manually inspect and join independent chains to ensure that the tracing is well-constructed, open the sharpened map and ModelAngelo output in a modeling software interface, such as Coot [29] or ISOLDE [30]. The model should have

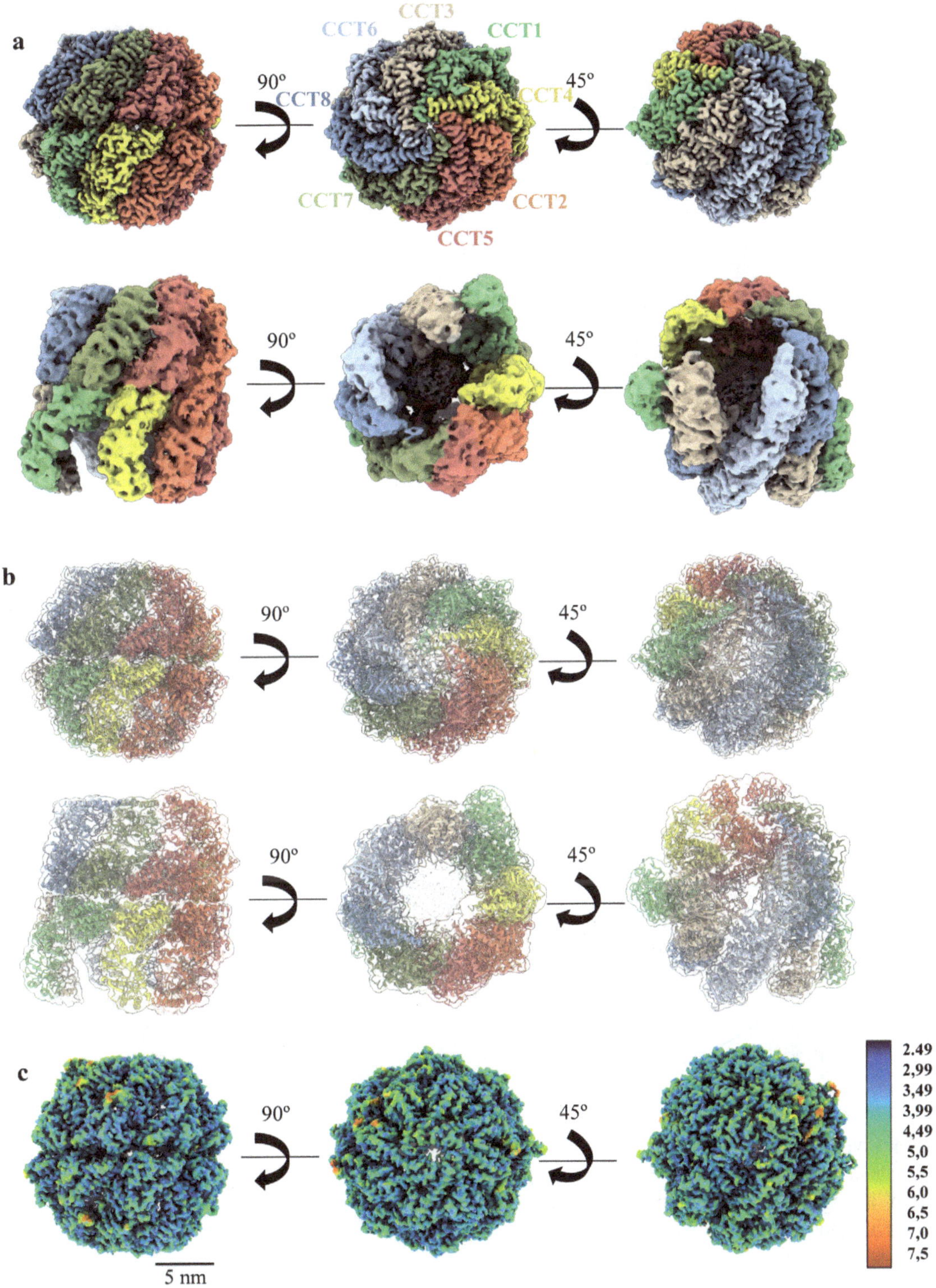


**Fig. 3** CryoEM density map and atomic model of CCT conformations. (a, b) Comparison of the sharpened cryoEM density map after high-frequency enhancement (**a**) and atomic model docking (**b**), between the closed (up, PDB: 9S72) and open conformations of CCT (down, PDB: 6QB8). Each CCT subunit is depicted in a unique color to distinguish individual components within the complex. (c) Local resolution map of the CCT closed conformation illustrating high-resolution regions across the complex. Cooler colors (blue tones) indicate regions of higher resolution, while warmer colors (red tones) represent areas of lower resolution

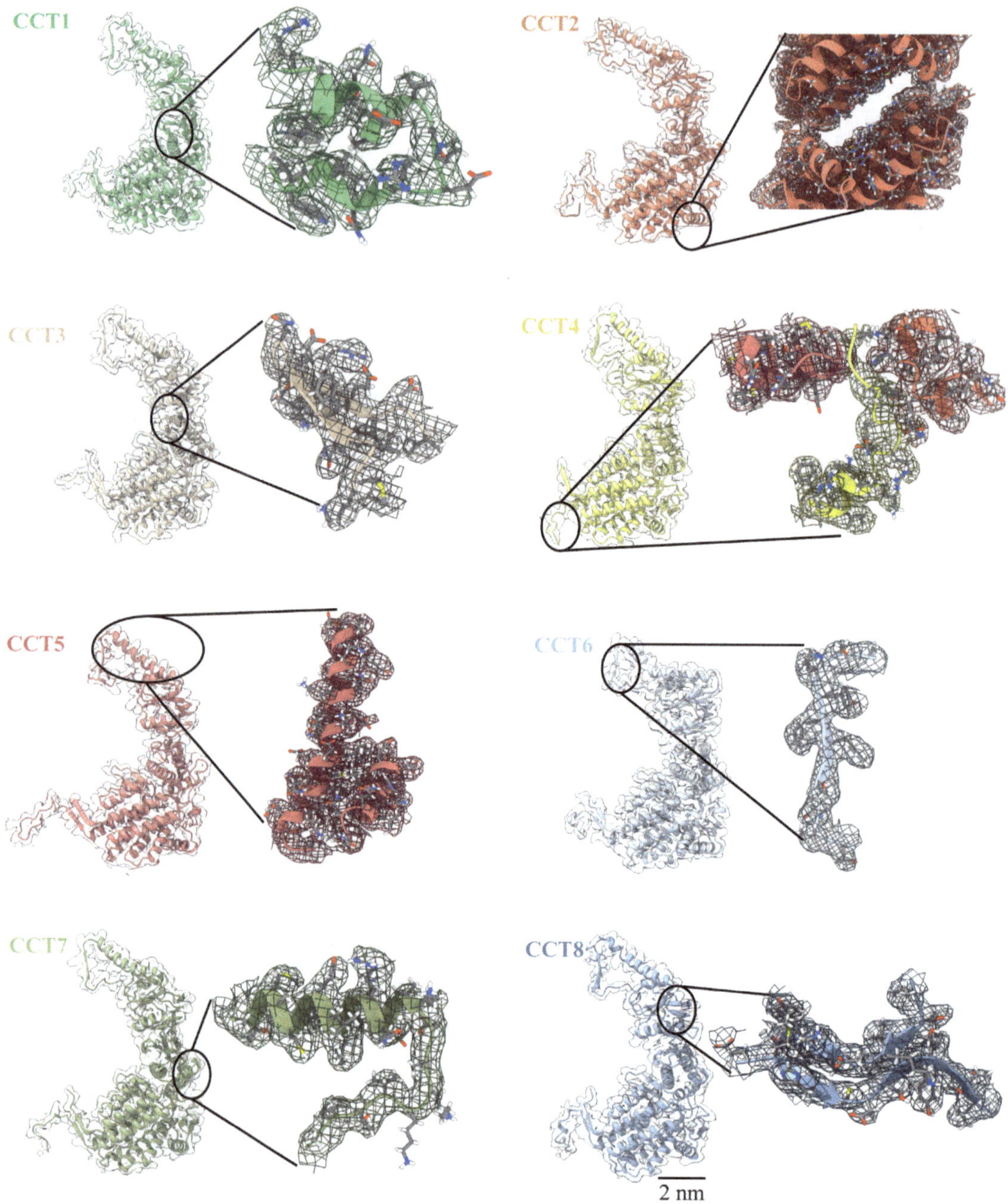


**Fig. 4** CryoEM density maps and atomic models of the regions corresponding to the individual CCT1–CCT8 subunits in the closed conformation of the complex.Each panel displays a density map and the atomic structure of a single CCT subunit. Zoomed-in views highlight the local regions with high-resolution, side chain details used for subunit identification

enough resolution to resolve well-defined secondary structure elements.

5. Adjust the atoms to the map densities using the real space refinement and other modeling tools available in Coot [29]. For more considerations, about modeling tools in Coot, *see* **Note 14**.
6. In the case that there are some densities where the chain was not built previously, select the "Add terminal residue" tool to manually build at this stage.
7. Use the "Mutate and Auto-Fit" tool to configure the correct residue.
8. Open PHENIX [31] real-space refinement prompt to perform an automatic model refinement. There is other software available also for this purpose, like REFMAC5 [32].
9. Provide the atomic structure, the sharpened map indicating the estimated resolution.
10. Define the refinement settings and run the protocol. For more information about Phenix parameters, *see* **Note 15**.
11. After refinement, check the stereochemistry (e.g., bond lengths and angles), Ramachandran plot, and side-chain conformations in the validation tab.
12. Repeat steps in Sects. 3.14.3–3.14.11 iteratively to improve the atomic model.
13. Use the "Validate" tab in Coot [29] to obtain information about all the regions that need to be improved.
14. Select: "Overlaps," "Peptides," "CBeta," and "Rama & Rota Outliers."
15. The final atomic map is obtained as shown in Figs. 1j and 3b.
16. Final structure validations prior to deposition in the database (e.g., Protein Data Bank) can be obtained using the Comprehensive validation module in PHENIX [31].

## 4 Notes

1. Vitrobot Mark IV (Thermo Fisher Scientific) was used to perform vitrification; GP2 Leica could also be used for this task. Images were collected in a 300 kV Titan Krios equipped with a K3 camera, although 200 kV microscopes with a direct detector could also have been used.
2. Minimal workstation requirements for single particle analysis image processing:

a. CPU with 32 or more threads at a minimum clock frequency of 2.5 GHz.
b. RAM capacity of 128–384 = GB.
c. 1–4 NVIDIA GPUs, with 12 GB of VRAM. Consumer gaming models and those from the professional segment are suitable for this setting. AMD and other companies might not work due to most programs relying on the CUDA language.
d. 2 TB NVMe SSD to be used as a scratch drive.
e. 20 TB HDD for local storage.

3. In this work, the closed conformation was stabilized by combining ATP with $AlF_3$, as $AlF_3$ generates a trigonal planar configuration at the γ-phosphate site, thereby preventing ADP release. Alternatively, BeF has also been shown to provide equally effective results [33].
4. This approach aims to fully stabilize the completely closed conformation of CCT, avoiding intermediate closure states, which could prevent achieving the highest possible resolution due to the flexibility of the apical domains. Alternatively, incubation conditions at 37°C for 30 min have also been reported to promote the closed conformation [34], these conditions might be also used to expand the conformational landscape.
5. If you encounter problems obtaining the CCT closed conformation, running a native PAGE could give insights into the structural conformation of your sample. Both 6% and gradient 4%–15% native gels should work, but temperature, time, and amperage should be monitored to maximize band resolution. In our experiments, 4 h at 20 mA worked correctly.
6. Scipion was used in this workflow with the following parameters: pixel size (1.072 Å/px), microscope voltage (300 kV), spherical aberration (2.7 mm), amplitude contrast (0.1), magnification (81,000×), and dose per frame (~1e$^-$/A2). Other frameworks, such as cryoSPARC [23] or doppio, can also be used for image processing. Bear in mind that each framework offers a repertoire of programs and may have different requirements.
7. In this work, we used MotionCor software [19] to align the movie frames. Other software can be used for this task (like RELION's motioncorr [35] or cryoSPARC's patch motion correction [23]). The choice will depend on your framework preferences.
8. While these parameters provide a robust starting point, it is important to note that "box size," "resolution range," and defocused search range" can be fine-tuned depending on the resolution and quality of the acquired images.

9. Once the CTF has been estimated, image selection can be performed based on the quality of the CTF fitting. "Resolution cutoff," "Astigmatism threshold," and "Defocus range" are commonly used for filtering. These thresholds help to exclude images with poor resolutions or extreme defocus values. However, the stringency of these filters can be adjusted depending on the dataset and downstream processing goals.
10. There are several software programs for automatic particle picking, RELION [22], cryoSPARC [23], SPHIRE-crYOLO [36], or Topaz [37]. In our hands, Xmipp [18] provided the best results for CCT.
11. To create a 3D mask, open your 3D volume in ChimeraX and identify the threshold that allows you to observe all the protein domains. In Scipion, use RELION or Xmipp create a 3D mask protocol, providing the 3D model, the threshold, and the low-pass filter of around 15 Å. Usually, 3 pixels of soft-edge and an extended binary mask are also added.
12. Initial amino acid assignment: Alphafold [26] is very powerful when working with small amino acid sequence proteins. However, it is limited to the number of amino acid characters that you can include, to 5000 for the moment, avoiding the complete prediction of larger molecular complexes like CCT. The PDBs belonging to the closed CCT conformation can be used in this particular case.
13. ModelAngelo [28] also generates an output without pruning the amino acid chain. This version can be useful for identifying regions where the chain is missing. Additionally, be aware that poor chain tracing may result from an incorrect hand assignment in the sharpened map.
14. To improve real-space refinement stability, be sure to adjust the restrictions to match the type of secondary structure that you are working with, as well as the chemical restrictions in the "R/RC" tab. To consider information of Hydrogen overlaps, select also the option "Add Hydrogen Atoms" in the "Modeling" tab. Coot 1.1.14 version also offers a variety of refinement options. The "Sphere," "Tandem +3," and "Single Residue" tools are typically sufficient to perform the necessary refinements.
15. In the case where the PDB model fits reasonably well to the map density, select the options "rigid body," "adp" (atomic displacement parameters), "occupancy," and "minimization global refinement" in the "Real Space Refinement" parameter tab. Local grid search-based fit can also be included to resolve side-chain outliers (rotamers or poor map fitting). In the case where the PDB model lies outside the map, the "morphing" option can help to initially adjust the map to the density prior to proceeding with the manual work in Coot [29].

## Acknowledgments

This research was supported by the Spanish Ministry of Science and the Spanish Research Agency through grants PID2022-137175NB-I00 (AEI/FEDER, UE) and AEI/10.13039/501100011033, through the "Severo Ochoa" Programme for Centres of Excellence in R&D (CEX2023-001386-S) to JMV and JC. The support of the grant CaixaResearch Health 2023 (HR23-00837) to JMV and grant P2022/BMD-7232 TomoXliver2 to AC. is also acknowledged. In addition, JC acknowledges the support of a grant from Fundación Ramón Areces. The authors acknowledge the technical support of the cryoEM CNB-CSIC facility.

### References

1. Nogales E, Scheres SH (2015) Cryo-EM: a unique tool for the visualization of macromolecular complexity. Mol Cell 58(4):677–689. https://doi.org/10.1016/j.molcel.2015.02.019
2. Kuhlbrandt W (2014) Biochemistry. The resolution revolution. Science 343(6178):1443–1444. https://doi.org/10.1126/science.1251652
3. Horovitz A, Reingewertz TH, Cuellar J, Valpuesta JM (2022) Chaperonin mechanisms: multiple and (mis)understood. Annu Rev Biophys 51:115–133. https://doi.org/10.1146/annurev-biophys-082521-113418
4. Willison KR (2018) The substrate specificity of eukaryotic cytosolic chaperonin CCT. Philos Trans R Soc Lond B Biol Sci 373(1749). https://doi.org/10.1098/rstb.2017.0192
5. Valpuesta JM, Martin-Benito J, Gomez-Puertas P, Carrascosa JL, Willison KR (2002) Structure and function of a protein folding machine: the eukaryotic cytosolic chaperonin CCT. FEBS Lett 529(1):11–16. https://doi.org/10.1016/s0014-5793(02)03180-0
6. Joachimiak LA, Walzthoeni T, Liu CW, Aebersold R, Frydman J (2014) The structural basis of substrate recognition by the eukaryotic chaperonin TRiC/CCT. Cell 159(5):1042–1055. https://doi.org/10.1016/j.cell.2014.10.042
7. Reissmann S, Joachimiak LA, Chen B, Meyer AS, Nguyen A, Frydman J (2012) A gradient of ATP affinities generates an asymmetric power stroke driving the chaperonin TRIC/CCT folding cycle. Cell Rep 2(4):866–877. https://doi.org/10.1016/j.celrep.2012.08.036
8. Cong Y, Schroder GF, Meyer AS, Jakana J, Ma B, Dougherty MT, Schmid MF, Reissmann S, Levitt M, Ludtke SL, Frydman J, Chiu W (2012) Symmetry-free cryo-EM structures of the chaperonin TRiC along its ATPase-driven conformational cycle. EMBO J 31(3):720–730. https://doi.org/10.1038/emboj.2011.366
9. Cuellar J, Ludlam WG, Tensmeyer NC, Aoba T, Dhavale M, Santiago C, Bueno-Carrasco MT, Mann MJ, Plimpton RL, Makaju A, Franklin S, Willardson BM, Valpuesta JM (2019) Structural and functional analysis of the role of the chaperonin CCT in mTOR complex assembly. Nat Commun 10(1):2865. https://doi.org/10.1038/s41467-019-10781-1
10. Cuellar J, Vallin J, Svanstrom A, Maestro-Lopez M, Bueno-Carrasco MT, Ludlam WG, Willardson BM, Valpuesta JM, Grantham J (2022) The molecular chaperone CCT sequesters gelsolin and protects it from cleavage by caspase-3. J Mol Biol 434(5):167399. https://doi.org/10.1016/j.jmb.2021.167399
11. Jin M, Han W, Liu C, Zang Y, Li J, Wang F, Wang Y, Cong Y (2019) An ensemble of cryo-EM structures of TRiC reveal its conformational landscape and subunit specificity. Proc Natl Acad Sci USA 116(39):19513–19522. https://doi.org/10.1073/pnas.1903976116
12. Kelly JJ, Tranter D, Pardon E, Chi G, Kramer H, Happonen L, Knee KM, Janz JM, Steyaert J, Bulawa C, Paavilainen VO, Huiskonen JT, Yue WW (2022) Snapshots of actin and tubulin folding inside the TRiC chaperonin. Nat Struct Mol Biol 29(5):420–429. https://doi.org/10.1038/s41594-022-00755-1
13. Gestaut D, Zhao Y, Park J, Ma B, Leitner A, Collier M, Pintilie G, Roh SH, Chiu W, Frydman J (2022) Structural visualization of the tubulin folding pathway directed by human chaperonin TRiC/CCT. Cell 185(25):4770–4787 e4720. https://doi.org/10.1016/j.cell.2022.11.014
14. Wang S, Sass MI, Kwon Y, Ludlam WG, Smith TM, Carter EJ, Gladden NE, Riggi M,

Iwasa JH, Willardson BM, Shen PS (2023) Visualizing the chaperone-mediated folding trajectory of the G protein beta5 beta-propeller. Mol Cell 83(21):3852–3868 e3856. https://doi.org/10.1016/j.molcel.2023.09.032

15. Bloch JS, Mukherjee S, Kowal J, Filippova EV, Niederer M, Pardon E, Steyaert J, Kossiakoff AA, Locher KP (2021) Development of a universal nanobody-binding Fab module for fiducial-assisted cryo-EM studies of membrane proteins. Proc Natl Acad Sci USA 118(47). https://doi.org/10.1073/pnas.2115435118
16. Knowlton JJ, Gestaut D, Ma B, Taylor G, Seven AB, Leitner A, Wilson GJ, Shanker S, Yates NA, Prasad BVV, Aebersold R, Chiu W, Frydman J, Dermody TS (2021) Structural and functional dissection of reovirus capsid folding and assembly by the prefoldin-TRiC/CCT chaperone network. Proc Natl Acad Sci USA 118(11). https://doi.org/10.1073/pnas.2018127118
17. Zang Y, Jin M, Wang H, Cui Z, Kong L, Liu C, Cong Y (2016) Staggered ATP binding mechanism of eukaryotic chaperonin TRiC (CCT) revealed through high-resolution cryo-EM. Nat Struct Mol Biol 23(12):1083–1091. https://doi.org/10.1038/nsmb.3309
18. de la Rosa-Trevin JM, Quintana A, Del Cano L, Zaldivar A, Foche I, Gutierrez J, Gomez-Blanco J, Burguet-Castell J, Cuenca-Alba J, Abrishami V, Vargas J, Oton J, Sharov G, Vilas JL, Navas J, Conesa P, Kazemi M, Marabini R, Sorzano CO, Carazo JM (2016) Scipion: a software framework toward integration, reproducibility and validation in 3D electron microscopy. J Struct Biol 195(1):93–99. https://doi.org/10.1016/j.jsb.2016.04.010
19. Zheng SQ, Palovcak E, Armache JP, Verba KA, Cheng Y, Agard DA (2017) MotionCor2: anisotropic correction of beam-induced motion for improved cryo-electron microscopy. Nat Methods 14(4):331–332. https://doi.org/10.1038/nmeth.4193
20. Rohou A, Grigorieff N (2015) CTFFIND4: Fast and accurate defocus estimation from electron micrographs. J Struct Biol 192(2):216–221. https://doi.org/10.1016/j.jsb.2015.08.008
21. Zhang K (2016) Gctf: real-time CTF determination and correction. J Struct Biol 193(1):1–12. https://doi.org/10.1016/j.jsb.2015.11.003
22. Scheres SH (2012) RELION: implementation of a Bayesian approach to cryo-EM structure determination. J Struct Biol 180(3):519–530. https://doi.org/10.1016/j.jsb.2012.09.006
23. Punjani A, Rubinstein JL, Fleet DJ, Brubaker MA (2017) cryoSPARC: algorithms for rapid unsupervised cryo-EM structure determination. Nat Methods 14(3):290–296. https://doi.org/10.1038/nmeth.4169
24. Sanchez-Garcia R, Gomez-Blanco J, Cuervo A, Carazo JM, Sorzano COS, Vargas J (2021) DeepEMhancer: a deep learning solution for cryo-EM volume post-processing. Commun Biol 4(1):874. https://doi.org/10.1038/s42003-021-02399-1
25. Vilas JL, Gomez-Blanco J, Conesa P, Melero R, Miguel de la Rosa-Trevin J, Oton J, Cuenca J, Marabini R, Carazo JM, Vargas J, Sorzano COS (2018) MonoRes: automatic and accurate estimation of local resolution for electron microscopy maps. Structure 26(2):337–344 e334. https://doi.org/10.1016/j.str.2017.12.018
26. Jumper J, Evans R, Pritzel A, Green T, Figurnov M, Ronneberger O, Tunyasuvunakool K, Bates R, Zidek A, Potapenko A, Bridgland A, Meyer C, Kohl SAA, Ballard AJ, Cowie A, Romera-Paredes B, Nikolov S, Jain R, Adler J, Back T, Petersen S, Reiman D, Clancy E, Zielinski M, Steinegger M, Pacholska M, Berghammer T, Bodenstein S, Silver D, Vinyals O, Senior AW, Kavukcuoglu K, Kohli P, Hassabis D (2021) Highly accurate protein structure prediction with AlphaFold. Nature 596(7873):583–589. https://doi.org/10.1038/s41586-021-03819-2
27. Burt A, Toader B, Warshamanage R, von Kugelgen A, Pyle E, Zivanov J, Kimanius D, Bharat TAM, Scheres SHW (2024) An image processing pipeline for electron cryo-tomography in RELION-5. FEBS Open Bio 14(11):1788–1804. https://doi.org/10.1002/2211-5463.13873
28. Jamali K, Kall L, Zhang R, Brown A, Kimanius D, Scheres SHW (2024) Automated model building and protein identification in cryo-EM maps. Nature 628(8007):450–457. https://doi.org/10.1038/s41586-024-07215-4
29. Emsley P, Cowtan K (2004) Coot: model-building tools for molecular graphics. Acta Crystallogr D Biol Crystallogr 60(Pt 12 Pt 1):2126–2132. https://doi.org/10.1107/S0907444904019158
30. Croll TI, Read RJ (2021) Adaptive Cartesian and torsional restraints for interactive model rebuilding. Acta Crystallogr D Struct Biol 77(Pt 4):438–446. https://doi.org/10.1107/S2059798321001145
31. Afonine PV, Poon BK, Read RJ, Sobolev OV, Terwilliger TC, Urzhumtsev A, Adams PD (2018) Real-space refinement in PHENIX for cryo-EM and crystallography. Acta Crystallogr D Struct Biol 74 6(Pt ):531–544. https://doi.org/10.1107/S2059798318006551
32. Kovalevskiy O, Nicholls RA, Long F, Carlon A, Murshudov GN (2018) Overview of refinement procedures within REFMAC5: utilizing data from different sources. Acta Crystallogr

D Struct Biol 74(Pt 3):215–227. https://doi.org/10.1107/S2059798318000979
33. Kalisman N, Schroder GF, Levitt M (2013) The crystal structures of the eukaryotic chaperonin CCT reveal its functional partitioning. Structure 21(4):540–549. https://doi.org/10.1016/j.str.2013.01.017
34. Pereira JH, Ralston CY, Douglas NR, Meyer D, Knee KM, Goulet DR, King JA, Frydman J, Adams PD (2010) Crystal structures of a group II chaperonin reveal the open and closed states associated with the protein folding cycle. J Biol Chem 285(36):27958–27966. https://doi.org/10.1074/jbc.M110.125344
35. Zivanov J, Nakane T, Forsberg BO, Kimanius D, Hagen WJ, Lindahl E, Scheres SH (2018) New tools for automated high-resolution cryo-EM structure determination in RELION-3. Elife 7. https://doi.org/10.7554/eLife.42166
36. Wagner T, Merino F, Stabrin M, Moriya T, Antoni C, Apelbaum A, Hagel P, Sitsel O, Raisch T, Prumbaum D, Quentin D, Roderer D, Tacke S, Siebolds B, Schubert E, Shaikh TR, Lill P, Gatsogiannis C, Raunser S (2019) SPHIRE-crYOLO is a fast and accurate fully automated particle picker for cryo-EM. Commun Biol 2:218. https://doi.org/10.1038/s42003-019-0437-z
37. Bepler T, Morin A, Rapp M, Brasch J, Shapiro L, Noble AJ, Berger B (2019) Positive-unlabeled convolutional neural networks for particle picking in cryo-electron micrographs. Nat Methods 16(11):1153–1160. https://doi.org/10.1038/s41592-019-0575-8

# Chapter 16

# The Open Conformation of CCT: How to Achieve High Resolution Using CryoEM

**Jorge Gutiérrez-Seijo, César Santiago, Sergio Pipaón, David Gil-Cantero, Ana Cuervo, Jorge Cuéllar and José María Valpuesta**


## Abstract

Cryo-electron microscopy (CryoEM) has revolutionized structural biology, establishing itself as a versatile and powerful technique for determining protein structures at near-atomic resolution. Compared to X-ray crystallography and nuclear magnetic resonance (NMR), it enables the study of proteins across a broad molecular weight range in their native state and captures multiple conformational states without requiring crystallization or large amounts of sample. The protocol described herein focuses on the eukaryotic chaperonin CCT (Chaperonin Containing TCP-1), a 1 MDa protein complex whose flexibility and lack of symmetry present challenges for structural determination. Using CryoEM, we achieve a 4 Å resolution 3D reconstruction of CCT in its open conformation, successfully resolving even its highly flexible apical domains.


**Keywords** CCT, Molecular chaperone, Chaperonin, Electron microscopy, Cryo-electron microscopy, Image processing, Three–dimensional reconstruction, Flexibility, Docking and atomic model

## 1 Introduction

Over the past years, CryoEM has revolutionized the field of structural biology, emerging as one of the most powerful and versatile techniques for determining protein and macromolecular complex structures at near-atomic resolution [1, 2]. Compared to other structural methods such as X-ray crystallography or nuclear magnetic resonance (NMR) spectroscopy, CryoEM offers key advantages: it requires only a small amount of protein sample and enables the study of proteins across a broad molecular weight range in their native, hydrated state—without the need for crystallization. This makes it particularly valuable for the study of large macromolecular

The authors "Jorge Gutiérrez-Seijo and César Santiago" are contributed equally to this work and should remain designated as co-first authors.

Noa B. Martín-Cófreces et al. (eds.), *The Cytoskeleton*, Methods in Molecular Biology, vol. 3070,
https://doi.org/10.1007/978-1-0716-5515-3_16,

complexes, membrane proteins, and dynamic structures [1]. Additionally, CryoEM captures multiple conformational states, offering crucial insights into molecular mechanisms and protein dynamics [3–5].With continuous advancements in sample preparation, resolution, and data processing, CryoEM is becoming the preferred technique for determining structures that are challenging to crystallize [1].

The case of study in this manuscript is the eukaryotic chaperonin CCT (Chaperonin Containing TCP-1), a member of the Hsp60 molecular chaperone family. CCT is the most structurally and functionally complex of all chaperonins, uniquely composed of eight distinct subunits. This 1 MDa molecular machine assembles into two stacked oligomeric rings, forming a central cavity for protein folding [6]. Unlike other chaperonins, each ring of CCT consists of eight homologous subunits (CCT1–CCT8), ranging from 55 to 60 kDa, which are encoded by distinct paralogous genes and arranged in a specific order [7–9]. This conspicuous architecture poses significant challenges for structural determination. Its large size hinders analysis by NMR, while the flexibility of its apical domains makes it difficult to obtain high-quality crystals suitable for X-ray crystallography. As a result, only a few CCT structures have been determined using this technique [9–12].

However, with the rise and advancement of CryoEM, the number of resolved CCT structures has increased considerably in recent years. CryoEM has not only enabled the determination of high-resolution CCT structures [13–16] but has also facilitated studies of its interactions with other proteins [14, 15] and provided insights into the folding process of CCT client proteins [16]. However, despite these advancements, resolving the structure of CCT remains a significant challenge. Unlike its prokaryotic counterpart, GroEL [17], which exhibits a clear symmetry axis, CCT is inherently asymmetric, albeit to a limited extent, with each of its eight distinct subunits contributing to its structural complexity. Additionally, the highly flexible apical domains of CCT contain helical protrusions that function as built-in lids for the central cavity [18], further complicating structural analysis by introducing substantial conformational variability.

In this CryoEM protocol, we outline the steps to achieve a high-resolution 3D reconstruction of the eukaryotic chaperonin CCT in its open conformation (see Fig. 1a--i for complete workflow). The obtained map reveals a nearly symmetrical arrangement between the two rings, clearly showing a paired interaction between the subunits. This specific arrangement is closely linked to the functional cycle of CCT, though its details remain unknown [6]. Notably, the resulting volume was of sufficient quality to resolve the apical domains of the CCT subunits, despite their high flexibility, ultimately enabling the building of an accurate atomic model of the protein.

## 2 Materials

### 2.1 Biological Samples

Any endogenous CCT, such as human or bovine complex obtained from any source. Purification is described in precedent Chaps. 1 and 2 of this volume: [Purification of endogenous CCT from cell cultures] and [Purification of the eukaryotic chaperonin CCT from native tissues].

### 2.2 Acquisition of Cryo-Electron Microscopy Images

1. Vitrification Device.
2. Cryo-electron microscope equipped with a direct detector.

### 2.3 Workstation Hardware

The workstation used is usually equipped with at least one GPU device for calculation and 32 CPU threads. In this example, the workstation presented the following specifications. For minimum specification (*see* **Note 1**).

1. Intel Xeon(R) Gold 6230 CPU @ 2.10 GHz. Comprising two sockets with 20 cores per socket.
2. 377 GB of Random Access Memory (RAM).
3. 4× Nvidia GPU Tesla T4 with 15 GB of VRAM.
4. 29 TB of HDD local storage.

### 2.4 Software

1. Scipion v3.7.1—Eugenius.
2. MotionCor 3-1.1.2.
3. gctf 1.18.
4. RELION 5.0.1.
5. cryoSPARC v4.6.2.
6. XmippSrc-3.23.11.0.
7. PHENIX 1.20.1-4487.
8. Coot 0.9.8.96.

## 3 Methods

This chapter will provide insights on the CCT open conformation structural characterization. A general workflow and the main steps are described in Fig. 1.

### 3.1 Pretreatment of Grids

The treatment of grid with poly-L-Lysine has been shown to significantly increase the number of side views in large macromolecular complexes [19]. The grids are treated as in [14].

1. Wash the grids with acetone vapor for 5 min.

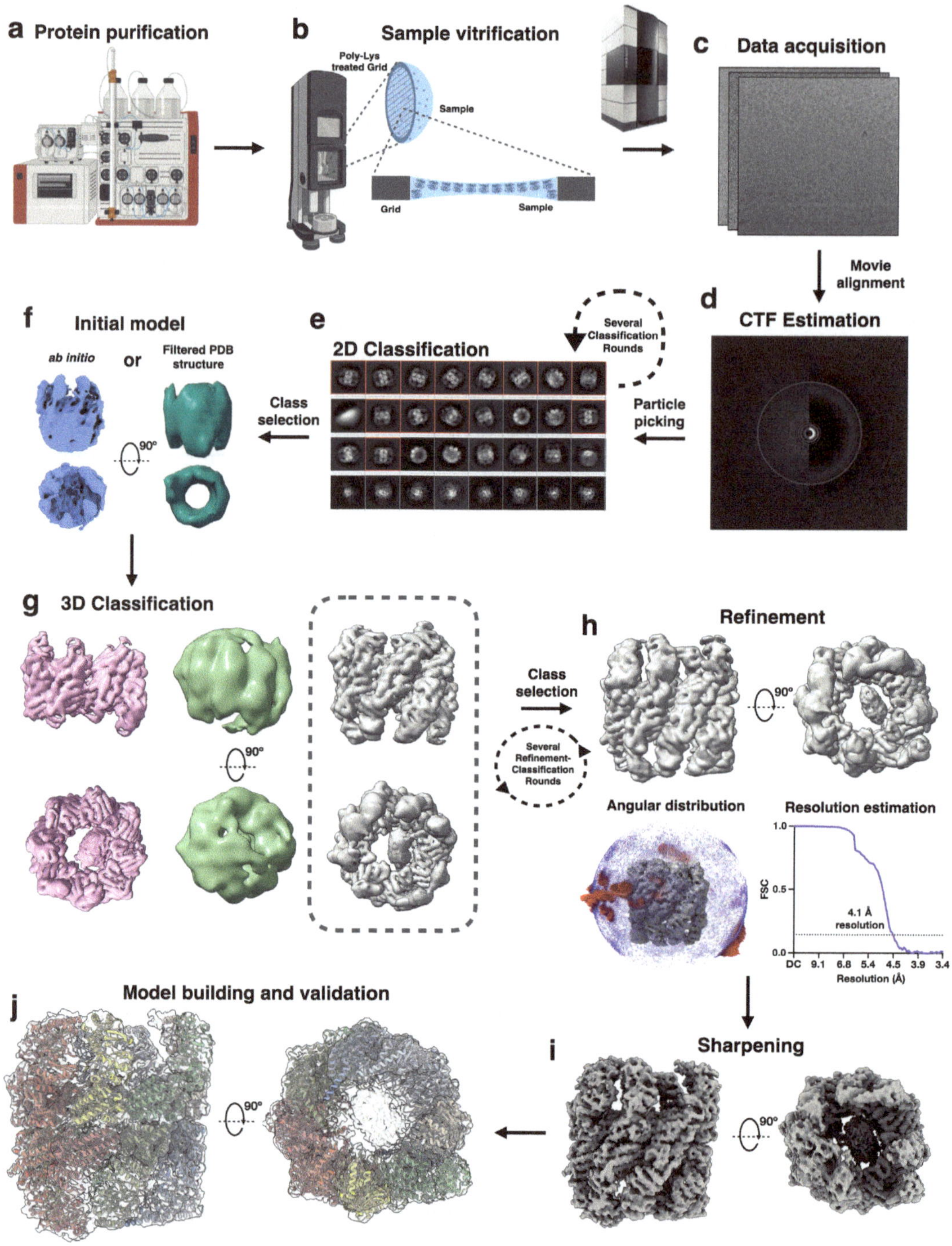


**Fig. 1** CryoEM workflow: from vitrification to atomic model. This figure outlines the single-particle cryoEM pipeline. (**a, b**) Sample purification and vitrification are illustrated first: purified CCT complexes are applied to CryoEM grid, blotted to form thin aqueous films, and flash-frozen in liquid ethane to trap particles in random orientations within vitreous ice. (**c**) Acquired images undergo preprocessing including motion correction, (**d**) contrast transfer function (CTF) estimation and automated particle picking. (**e**) Selected particles are subjected to several reference-free 2D classification, (**f**) with optimal class averages used to generate an initial 3D model

2. Dry for 5 min.
3. Place a drop of 20 µL of 0.1% w/v poly-L-Lysine solution on a clean, hydrophobic surface.
4. Incubate the grids on the poly-L-Lysine for 90 s. The face of the grid meant for sample application should be put in contact with the drop.
5. Wash the grids with two consecutive distilled water droplets.
6. Blot gently between washes with high quality filter paper.
7. Grids can be stored prior to vitrification.

### 3.2 Vitrification and Acquisition of Cryo-Electron Microscopy Images

1. Apply 4 µL of 0.7 µM CCT to the poly-L-lysine-treated side of a Quantifoil Cu/Rh grid.
2. Vitrify in a vitrification device. For a Vitrobot Mark IV (Thermo Fisher Scientific), the used values are 2 s of blotting at a −2 "force" value and 95% humidity. See Fig. 1b.
3. Store grid in liquid nitrogen.
4. Acquire images in a microscope equipped with a direct detector camera with a total electron dose of around 40 $e^{-}/A^{2}$, at a pixel size close to 1 Å/px and with a defocus range between 0.5–2.5 µm. The exact parameters of the acquisition, whose results are shown in the figures, are summarized in Table 1. See Fig. 1c.

### 3.3 Movie Alignment

The values described in this chapter are indicated in brackets. Users can use them to practice, but they are to be adapted to their particular image dataset. To align your dataset:

1. Create a new Scipion project (*see* **Note 2**) [20].
2. Import the movies (dataset): specify the path location. If movies are not corrected for gain, provide the gain reference image and specify whether it needs flipping or rotation.
3. Specify the pixel size (0.84 Å/px).
4. Specify the microscope voltage (300 kV).
5. Specify the spherical aberration (2.7 mm).
6. Specify the amplitude contrast (0.1).
7. Specify the magnification (105,000x).
8. Specify the dose per frame (~1 $e^{-}/A^{2}$).

**Fig. 1** Continued Cab initio. **(g)** Subsequent rounds of 3D classifications using these particles and the initial model serve to filter particles with the highest resolution structural details. **(h)** Their alignment are refined, and angular distribution and resolution are evaluated to assess the quality of the final 3D reconstruction. **(i)** Finally, high-frequency details are enhanced, **(j)** culminating in the building of a refined atomic model

**Table 1**
**Acquisition parameters employed in (14)**

| Microscope | FEI Titan Krios |
|---|---|
| Voltage (kV) | 300 |
| Detector | Gatan K2 |
| Mode | Superesolution |
| Magnification | 105,000 x |
| Total Dose ($e^-/Å^2$) | 36 |
| Dose per frame ($e^-/Å^2$) | 1 |
| Defocus range | −1 μm - −2 μm, 0.2 μm step |
| Pixel Size (Å/px) | 1.37 |
| Number of images | 5576 |
| Initial particles | 1,769,600 |
| Final Particles | 452,000 |
| Resolution (Å) | 3.97 |
| FSC threshold criteria | 0.143 |

9. Use MotionCorr software (*see* **Note 3**) [21] to perform the movie motion correction.
10. Use default values for the number of "Frames to ALIGN and SUM."
11. Select the "Number of patches": 5 for both axis and 0% overlap.
12. Use default parameter for the "Dose filter". For other Movie correction parameters, *see* **Note 4**.
13. Proceed to estimate Contrast Transfer Function (CTF).

### *3.4 CTF Estimation and Selection*

The aligned images are used as an input to estimate the CTF. Different programs can be utilized for this. In the example, described in this chapter, Gctf [22] was used.

1. Open Gctf.
2. Select the aligned movies as input.
3. Indicate "Box size" parameter (1024 pixels).
4. Indicate "Resolution range" parameter (30–3 Å).
5. Indicate "Defocus search range" parameter (5000–90,000 Å).
6. Indicate "Defocus step" parameter (500 Å).
7. Indicate "Expected" (tolerated) and "Astigmatism" parameters (1000). *See* **Note 5** and Fig. 1d.

8. To filter images automatically, open Xmipp3 CTF consensus program (*see* **Note 6**).
9. Import the estimated ctfs.
10. Adjust Defocus range of 4000–40,000 Å.
11. Adjust Astigmatism at 1000 Å.
12. Adjust Resolution at 10 Å (*see* **Note 7**).

### 3.5 Particle Picking and Extraction

1. Select high-quality images for picking.
2. Create a small subset of 30–50 images to fine-tune the picking process.
3. Use Scipion pwem-subset protocol [20] to automatically make a random subset of images with different defocus ranges.
4. Alternatively, higher defocus images can be selected to facilitate initial steps of the manual picking (see Fig. 2a for an example of an optimal image for particle picking).
5. Use Xmipp3 manual-picking Step 1 for picking (*see* **Notes 8 and 9**) [23].
6. Select a small subset of around 2000 coordinates from 10 or more images. See Fig. 2a for an example of different CCT views in the image.
7. Click on "activate training" on a fully picked image.

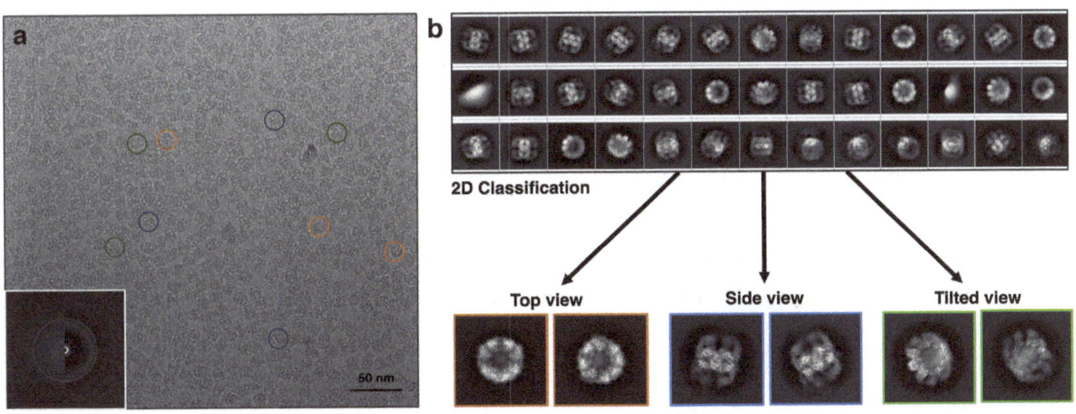


**Fig. 2** CryoEM data processing and 2D classification. (**a**) Representative image obtained from the electron microscope, with their corresponding contrast transfer function (CTF) patterns shown as inset. The colored circles highlight particles in top (orange), side (blue), and tilted (green) orientations, corresponding to the views presented in panel (**b**). (**b**) Results of 2D classification. High-quality class averages reveal CCT complexes in various orientations, including side, tilted, and front views. The panel also displays lower quality 2D averages and classes containing contaminants or incorrectly picked coordinates, illustrating the range of data quality encountered during processing

8. Visualize the following images to assess the automatic picking performance.
9. Improve picking by correcting coordinates: erase bad pickings and include those particles that were not selected.
10. Stop the training once the automatic picking is working correctly.
11. Click on "save coordinates."
12. Perform automated particle selection on the entire dataset using "Xmipp3 auto-picking" (Step 2).
13. Under "Micrographs to pick," select "Other."
14. Provide the motion-corrected dataset as input.
15. Once picking is complete, proceed to extract the particles.
16. Use Relion particle extraction protocol [24]. *See* **Note 10** for other particle extraction considerations.
17. Select the coordinates from "Xmipp3 auto-picking" (Step 2), the aligned images and the estimated CTFs.
18. Use an original box size of 360 pixels (300 Å), normalization applied, and contrast inverted (Fig. 2b) (*see* **Note 11**).

### 3.6 2D Classification and Particle Selection

Several programs can be used to perform 2D classification. To speed up the process, use cryoSPARC2 software [25]:

1. Use particles from Sect. 3.5.
2. Select 100 classes (note that each class average should have at least 500 particles).
3. Indicate a "Resolution limit" to around 8 Å in the initial steps. This parameter can be progressively decreased.
4. Indicate a "Radial mask" of 234 Å.
5. Run the classification.
6. Open the "Analyse results" window from the 2D classification run.
7. Use "Display particles classes" with Scipion.
8. Perform "class selection" with particles producing high-quality 2D class averages (*see* **Note 12**).
9. Click the particles button and save the selection.

To obtain good quality particles, 2D classification and class selection should be carried several times, discarding classes containing poorly picked particles or low-quality images (Fig. 2b). For more information on other considerations in 2D class selection, *see* **Note 13**.

### 3.7 3D Classification: Optimization of Particle Selection

1. A well-defined initial 3D model is essential for further refinement steps; therefore, an initial model building is required. Use cryoSPARC2-initial model [25] to perform this task.
2. Provide the particles from Sect.3.6.
3. Set the number of classes to 3 and run the protocol (*see* **Note 14**). See Fig. 1f.
4. Save the model that best resembles a barrel-like complex.
5. Perform 3D classification (*see* **Note 15**) with Relion 3D classification [24] steps.
6. Set the parameter "Consider previous alignment": "no."
7. Use the previous selected initial model and the particles from Sect. 3.6.
8. Adjust the parameter "Initial volume filtering" to 30–50 Å.
9. Adjust the parameter "Number of classes" from 3–5 classes.
10. Adjust the parameter "Iterations" to 25 (*see* **Note 16** for further considerations).
11. Set the parameter "Perform image alignment: "yes."
12. Select particles belonging to 3D volumes where the apical domains are better defined. CCT is an example of anisotropy due to its flexible apical domains (see Fig. 1g).
13. Alternatively, perform cryoSPARC 3D classification (*see* **Note 17**) [25] with three or four initial models with different apical domain qualities.

### 3.8 Refinement and Postprocessing

The next step for 3D reconstruction is fine-tuning the angular assignment to achieve the best possible resolution. Therefore, a 3D refinement step using the particles selected during the classification process improves the resolution of the 3D volume. Several protocols can be used for this purpose. The present protocol uses Relion auto-refine protocol [24].

1. Use the input particles and 3D volume from the 3D classification.
2. Set the parameter "Consider previous alignment": "yes."
3. Set the parameter "Is the initial map in absolute grey scale": "Yes". If classification has been done with other software that RELION or Xmipp, select "No."
4. Set the parameter "Initial low-pass filter': "30 Å."
5. Set the parameter "Perform image alignment': "Yes."
6. Set the parameter "Use GPU acceleration": "Yes."
7. For RELION, GPU acceleration chose at least one thread per GPU used plus an additional thread to control the process.

8. Use 7.5° sampling as a starting point for the refinement.
9. Repeat the same steps using non-binned particles (*see* **Note 18**) and volume once Nyquist resolution is reached. An example of a 3D reconstruction, along with the calculated angular distribution and estimated resolution, is shown in Fig. 1h.
10. Alternatively, use the "non-uniform refinement" protocol from cryoSPARC2 [26] with the same particle input.
11. Next step is required to refine the flexible apical domain regions of the CCT volume. Relion "auto-refine" combined with "angular sampling for local searches" performs a local refinement with masks [24], while maintaining previous alignment. CryoSPARC Local refinement [25] also uses masks (*see* **Note 19**). There are other alternatives for resolving flexible regions, such as new programs that use neural networks to recover flexible regions (*see* **Note 20**).
12. Perform a model sharpening using the RELION post-process protocol to enhance high-resolution details of the map (see Figs. 1i and 3a). For this, use your desired volume and a mask created from that volume (as described in *see* **Note 19**).Alternatively, sharpening can be achieved with tools like DeepEMhancer [27]. CryoSPARC refinement also provides a sharpened model in its output.
13. To evaluate the resolution of your map, perform a local MonoRes analysis [28] from Xmipp3 to obtain a local resolution map (see Fig. 3b).

### 3.9 Model Building

1. Load the best sharpened map on ChimeraX [29] and the atomic model of CCT in its open conformation (PDB: 6QB8).
2. Align the atomic model to the map using the "Fit in map" tool.
3. Save map in ccp4 format and the aligned model as pdb, so that it is later compatible with Coot. When saving, make sure that you save the relative position of the atomic model to the map.
4. Subject the model to a flexible fitting of the individual subunits with IMODFIT [30].
5. Launch Coot [31] and open both, the aligned model and map (from Step 8).
6. Select all the restrictions in the R/RC tab to improve refinement stability. If computational power is a limitation, work with previously isolated subunits from IMODFIT instead of the entire complex (*see* **Note 21**).

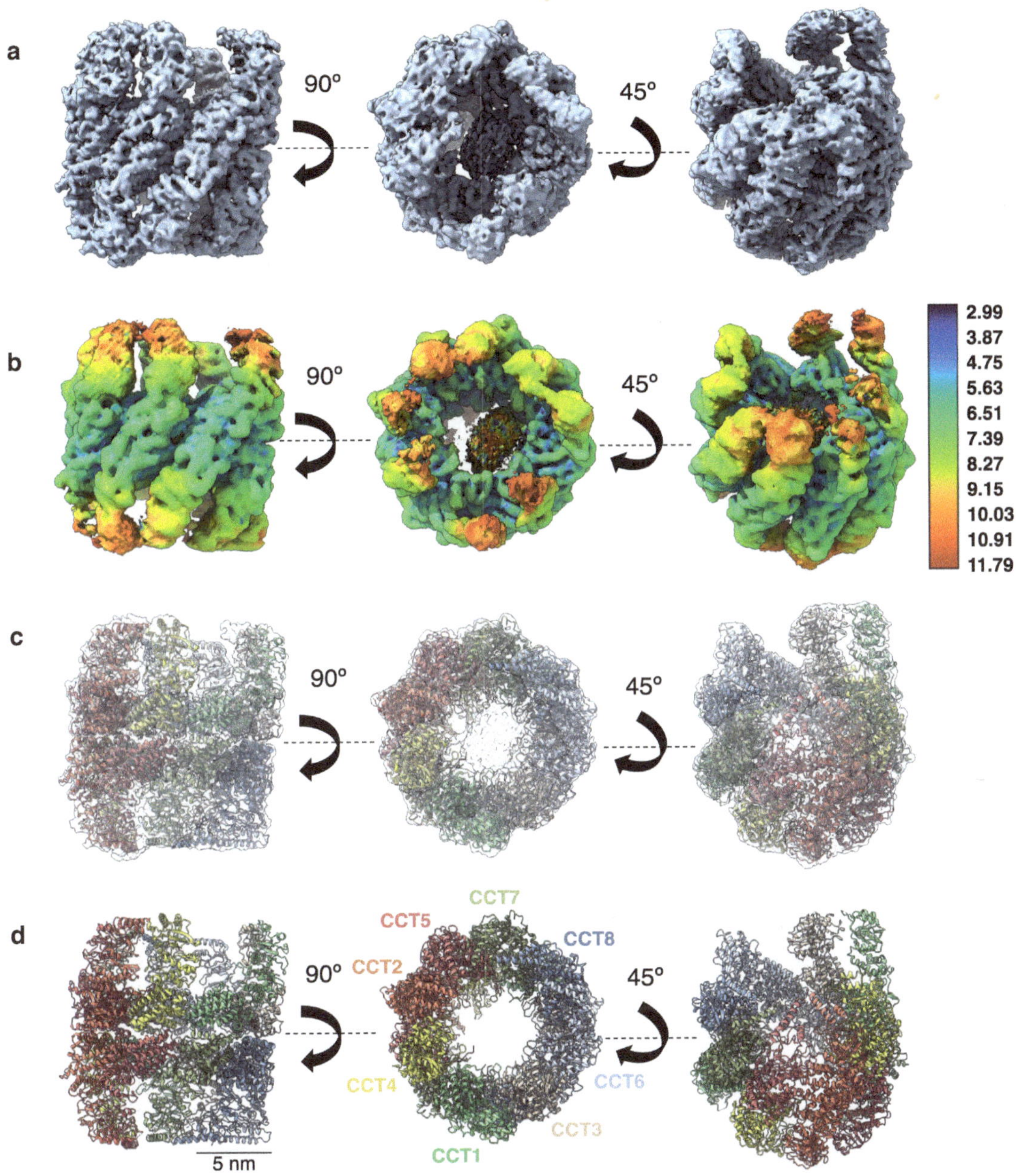


**Fig. 3** CryoEM map refinement and atomic model building. (**a**) Sharpened cryoEM density map after high-frequency enhancement, revealing fine structural details of the CCT complex. (**b**) Local resolution map of the CCT structure. Cooler colors indicate regions of higher resolution, while warmer colors represent areas of lower resolution, illustrating the variability in map quality across the complex. (**c**) Rigid-body docking of the initial atomic model into the refined cryoEM density map. Each CCT subunit is depicted in a unique color to distinguish individual components within the complex. (**d**) Final atomic structure derived from the cryoEM map following coordinate refinement and model building. The color scheme for CCT subunits matches the one used in panel (**c**), facilitating subunit identification and structural comparisons

7. Adjust the atoms to the map by using the "Real space refinement" tool in Coot. Adjust the restrictions to match the type of secondary structure, such as alpha helix or beta sheet, as appropriate in the R/RC tab.
8. Save the new coordinates in Coot [31] once the model has been adjusted.
9. Further refine the structure using real-space refinement in PHENIX [32]. Select the options "rigid body," "adp" (atomic displacement parameters), "occupancy," and "minimization global refinement." Include local grid search-based fit to resolve side-chain outliers (rotamers or poor map fitting).
10. Check the refinement results in the validation tab of PHENIX. The structure should be validated by stereochemistry (e.g., bond lengths, angles), Ramachandran plot, and side-chain conformations.
11. Repeat Steps 3 and 5 iteratively. Continue refining with real-space refinement in Coot [31] and PHENIX [32]. Ensure that the Ramachandran angles and other parameters (like sidechain fitting) are optimized. A correctly adjusted structure should be visible, with the PDB well-aligned to the map (see Fig. 3c).
12. Perform final validation using the "phenix.validation_cryoem" module in PHENIX [32].
13. The structure is ready for deposition in a database (e.g., Protein Data Bank) once the structure is validated and refined. The final structure (see Fig. 3d) can now be prepared for publication or submission to a repository.

## 4 Notes

1. Minimal workstation requirements for single particle analysis image processing:
   a. CPU with 32 or more threads at a minimum clock frequency of 2.5 GHz.
   b. RAM capacity of 128–384 GB.
   c. 1–4 NVIDIA GPUs, with 12 GB of VRAM. Consumer gaming models as well as those from the professional segment are suitable for this setting. AMD and other companies might not work due to most programs relying on CUDA language.
   d. 2 TB NVMe SSD to be used as scratch drive.
   e. 20 TB HDD for local storage.

2. Other frameworks such as cryoSPARC [25] or Doppio can be used for particle analysis. Bear in mind each framework offers a repertoire of programs and may have different requirements.
3. Both RELION's motioncorr [33] and cryoSPARC's [25] patch motion corrections are suitable for motion correction. Both are commonly used, but the choice depends on your framework preferences.
4. Other movie alignment parameters. In some cases, images from super-resolution mode may be binned twice to save disk space. Unweighted images are usually not needed unless they help during the particle picking step. If images are provided in EER format, the EER fractionation parameter should be set to the number of EER fractions used during acquisition.
5. These values may need adjustment based on dataset-specific acquisition parameters. For faster processing, CTF estimation can be computed with a 512-pixel box size, though this may yield slightly less accurate values. If a GPU is unavailable, alternative programs such as CTFFind4 [34] can be used for CPU-based calculations.
6. Images should be inspected to discard those of poor quality based on their CTF fitting, astigmatism, or resolution.
7. These values can be adjusted depending on dataset characteristics.
8. So far, in most cases, Xmipp particle picking protocol [23] has been the most effective for CCT particles selection. However, other software tools can also be used for this step.
9. Particle picking alternatives. There are several alternatives for particle picking, including automated protocols such as RELION [24], cryoSPARC [25], SPHIRE-CrYOLO [35], or Topaz [36]. Some of these methods require 2D classification models as input. All of them are widely used and have demonstrated high efficiency in particle picking
10. In this protocol, Relion particle extraction [24] was used, but Xmipp particle extraction [23] is also a valid alternative. Usually, to speed up initial 2D classification cleaning steps particle box size is resized 2–4× times to a pixel size of around 3 or 4 Å/px.
11. Other particle extraction considerations: To improve particle alignment and allow proper CTF correction, particle Box size must be at least 20%–50% larger than the long side of the particle. The value must also be an even number if RELION is used in any step.
12. Good 2D averages should display secondary structure elements, especially in the central domain where alignment is most accurate (Fig. 2b). Flexibility in the apical domains tends

to yield blurry densities in these areas even in well-aligned classes (Fig. 2b, bottom).

13. Other considerations in 2D classes selection. Although RELION and Xmipp have their own 2D classification protocols, cryoSPARC [25] is generally faster, allowing for quicker processing of large datasets. To speed up 2D classification, it is normally carried out using 2–4 times downsampling boxes (around 40–60 px box size). It has also been observed experimentally that a big amount of side views yields deformed CCT 3D volumes, usually best results are obtained discarding all axial views and most of tilted views and trying to enrich the final dataset with as many side views as possible.
14. Initial model: In some cases, it is difficult to obtain a good initial model. Other protocols like Relion initial model or EMAN initial [37] model could also be used. If working with symmetric structures (like CCT-closed conformation), providing the symmetry information (D8) can help generate better initial models. However, for the open conformation, where apical domains present different orientations not symmetrically related, usually this is not a good option. Alternatively, an already solved open CCT PDB structure (6QB8 or 8SFE), converted into a model and filtered to 50 Å resolution, can be used as a starting point.
15. It is crucial to prune the dataset especially in the apical domains by classifying particles into different 3D volumes. Different programs can be used for this purpose.
16. 3D classification: If the particle number is not too high, and the task is not so much time consuming, Relion [24] 3D classification with alignment may be performed. Select "yes" in the "perform alignment" parameter. Start using 7.5° as initial angular sampling. Then duplicate the protocol, continue from the previous run and select lower angles. Maintain the default parameters for the rest and do not perform local angular search. Use GPU and scratch directory if available to speed up the processing. If number of particles extracted may be too high for your computing setup, consider splitting the dataset into smaller subsets for independent classification.
17. CryoSPARC [25] 3D classification is the protocol for classifying without alignment. If both classification and alignment are needed, heterogeneous refinement from cryoSPARC should be employed, requiring as many initial volumes as classes.
18. Further details on unbinned particle extraction: For unbinned particles, extract the coordinates of particles using pwem-extract coordinates. Reperform particle extraction with the desired box size to get unbinned particles.

19. Mask creation: To generate a mask, first, it is necessary to save the region(s) of interest without the rest of the particle; this task can be performed with Chimera [29]. Afterwards, import the new map and use a protocol to create a mask (in Scipion you can use xmipp-create 3D mask [23] or Relion-create a mask [24]. Set the desired threshold and apply smooth borders (3 pixels).
20. In recent years, a substantial number of programs focused on flexibility have been released. They are divided in two categories, those based on deformation grids, and those based on intensive classifications. Some of them are 3DFlex [38], 3D variability [5], hetSIREN [3], or Dynamight [4].
21. In this case, the PDB and the map are very similar. When this is not the case due to different conformations or other reasons, it is recommendable to run first a Phenix refinement [32]. Try the morphing option, especially when the difference is bigger. Afterwards, proceed with coot [31] and the rest of the protocol as described.

## Acknowledgments

This research was supported by the Spanish Ministry of Science and the Spanish Research Agency through grants PID2022-137175NB-I00 (AEI/FEDER, UE) and AEI/10.13039/501100011033, through the "Severo Ochoa" Programme for Centres of Excellence in R&D (CEX2023-001386-S) to J.M.V. and J.C. The support of the grant CaixaResearch Health 2023 (HR23-00837) to J.M.V. is also acknowledged. In addition, J.C. acknowledges the support of a grant from Fundación Ramón Areces. The authors acknowledge the technical support of the cryoEM CNB-CSIC facility.

### References

1. Nogales E, Scheres SH (2015) Cryo-EM: a unique tool for the visualization of macromolecular complexity. Mol Cell 58(4):677–689. https://doi.org/10.1016/j.molcel.2015.02.019
2. Kuhlbrandt W (2014) Biochemistry. The resolution revolution. Science 343(6178):1443–1444. https://doi.org/10.1126/science.1251652
3. Herreros D, Mata CP, Noddings C, Irene D, Krieger J, Agard DA, Tsai MD, Sorzano COS, Carazo JM (2025) Real-space heterogeneous reconstruction, refinement, and disentanglement of CryoEM conformational states with HetSIREN. Nat Commun 16(1):3751. https://doi.org/10.1038/s41467-025-59135-0
4. Schwab J, Kimanius D, Burt A, Dendooven T, Scheres SHW (2024) DynaMight: estimating molecular motions with improved reconstruction from cryo-EM images. Nat Methods 21(10):1855–1862. https://doi.org/10.1038/s41592-024-02377-5
5. Punjani A, Fleet DJ (2021) 3D variability analysis: resolving continuous flexibility and discrete heterogeneity from single particle cryo-EM. J Struct Biol 213(2):107702. https://doi.org/10.1016/j.jsb.2021.107702

6. Bueno-Carrasco M, Cuéllar J (2018) Mechanism and function of the eukaryotic chaperonin CCT. eLS 1–9
7. Pipaon S, Gragera M, Bueno-Carrasco MT, Garcia-Bernalt Diego J, Cantero M, Cuellar J, Fernandez-Fernandez MR, Valpuesta JM (2021) Chaperonins: nanocarriers with biotechnological applications. Nanomaterials (Basel) 11(2):503. https://doi.org/10.3390/nano11020503
8. Kalisman N, Adams CM, Levitt M (2012) Subunit order of eukaryotic TRiC/CCT chaperonin by cross-linking, mass spectrometry, and combinatorial homology modeling. Proc Natl Acad Sci USA 109(8):2884–2889. https://doi.org/10.1073/pnas.1119472109
9. Leitner A, Joachimiak LA, Bracher A, Monkemeyer L, Walzthoeni T, Chen B, Pechmann S, Holmes S, Cong Y, Ma B, Ludtke S, Chiu W, Hartl FU, Aebersold R, Frydman J (2012) The molecular architecture of the eukaryotic chaperonin TRiC/CCT. Structure 20(5):814–825. https://doi.org/10.1016/j.str.2012.03.007
10. Dekker C, Roe SM, McCormack EA, Beuron F, Pearl LH, Willison KR (2011) The crystal structure of yeast CCT reveals intrinsic asymmetry of eukaryotic cytosolic chaperonins. EMBO J 30(15):3078–3090. https://doi.org/10.1038/emboj.2011.208
11. Munoz IG, Yebenes H, Zhou M, Mesa P, Serna M, Park AY, Bragado-Nilsson E, Beloso A, de Carcer G, Malumbres M, Robinson CV, Valpuesta JM, Montoya G (2011) Crystal structure of the open conformation of the mammalian chaperonin CCT in complex with tubulin. Nat Struct Mol Biol 18(1):14–19. https://doi.org/10.1038/nsmb.1971
12. Kalisman N, Schroder GF, Levitt M (2013) The crystal structures of the eukaryotic chaperonin CCT reveal its functional partitioning. Structure 21(4):540–549. https://doi.org/10.1016/j.str.2013.01.017
13. Knowlton JJ, Gestaut D, Ma B, Taylor G, Seven AB, Leitner A, Wilson GJ, Shanker S, Yates NA, Prasad BVV, Aebersold R, Chiu W, Frydman J, Dermody TS (2021) Structural and functional dissection of reovirus capsid folding and assembly by the prefoldin-TRiC/CCT chaperone network. Proc Natl Acad Sci USA 118(11):e2018127118. https://doi.org/10.1073/pnas.2018127118
14. Cuellar J, Ludlam WG, Tensmeyer NC, Aoba T, Dhavale M, Santiago C, Bueno-Carrasco MT, Mann MJ, Plimpton RL, Makaju A, Franklin S, Willardson BM, Valpuesta JM (2019) Structural and functional analysis of the role of the chaperonin CCT in mTOR complex assembly. Nat Commun 10(1):2865. https://doi.org/10.1038/s41467-019-10781-1
15. Cuellar J, Vallin J, Svanstrom A, Maestro-Lopez M, Bueno-Carrasco MT, Ludlam WG, Willardson BM, Valpuesta JM, Grantham J (2022) The molecular chaperone CCT sequesters gelsolin and protects it from ceavage by Caspase-3. J Mol Biol 434(5):167399. https://doi.org/10.1016/j.jmb.2021.167399
16. Gestaut D, Zhao Y, Park J, Ma B, Leitner A, Collier M, Pintilie G, Roh SH, Chiu W, Frydman J (2022) Structural visualization of the tubulin folding pathway directed by human chaperonin TRiC/CCT. Cell 185(25):4770–4787.e4720. https://doi.org/10.1016/j.cell.2022.11.014
17. Tang YC, Chang HC, Roeben A, Wischnewski D, Wischnewski N, Kerner MJ, Hartl FU, Hayer-Hartl M (2006) Structural features of the GroEL-GroES nano-cage required for rapid folding of encapsulated protein. Cell 125(5):903–914. https://doi.org/10.1016/j.cell.2006.04.027
18. Valpuesta JM, Martin-Benito J, Gomez-Puertas P, Carrascosa JL, Willison KR (2002) Structure and function of a protein folding machine: the eukaryotic cytosolic chaperonin CCT. FEBS Lett 529(1):11–16. https://doi.org/10.1016/s0014-5793(02)03180-0
19. Zang Y, Jin M, Wang H, Cui Z, Kong L, Liu C, Cong Y (2016) Staggered ATP binding mechanism of eukaryotic chaperonin TRiC (CCT) revealed through high-resolution cryo-EM. Nat Struct Mol Biol 23(12):1083–1091. https://doi.org/10.1038/nsmb.3309
20. de la Rosa-Trevin JM, Quintana A, Del Cano L, Zaldivar A, Foche I, Gutierrez J, Gomez-Blanco J, Burguet-Castell J, Cuenca-Alba J, Abrishami V, Vargas J, Oton J, Sharov G, Vilas JL, Navas J, Conesa P, Kazemi M, Marabini R, Sorzano CO, Carazo JM (2016) Scipion: a software framework toward integration, reproducibility and validation in 3D electron microscopy. J Struct Biol 195(1):93–99. https://doi.org/10.1016/j.jsb.2016.04.010
21. Zheng SQ, Palovcak E, Armache JP, Verba KA, Cheng Y, Agard DA (2017) MotionCor2: anisotropic correction of beam-induced motion for improved cryo-electron microscopy. Nat Methods 14(4):331–332. https://doi.org/10.1038/nmeth.4193
22. Zhang K (2016) Gctf: real-time CTF determination and correction. J Struct Biol 193(1):1–12. https://doi.org/10.1016/j.jsb.2015.11.003
23. de la Rosa-Trevin JM, Oton J, Marabini R, Zaldivar A, Vargas J, Carazo JM, Sorzano CO (2013) Xmipp 3.0: an improved software suite for image processing in electron microscopy.

J Struct Biol 184(2):321–328. https://doi.org/10.1016/j.jsb.2013.09.015
24. Scheres SH (2012) RELION: implementation of a Bayesian approach to cryo-EM structure determination. J Struct Biol 180(3):519–530. https://doi.org/10.1016/j.jsb.2012.09.006
25. Punjani A, Rubinstein JL, Fleet DJ, Brubaker MA (2017) cryoSPARC: algorithms for rapid unsupervised cryo-EM structure determination. Nat Methods 14(3):290–296. https://doi.org/10.1038/nmeth.4169
26. Punjani A, Zhang H, Fleet DJ (2020) Non-uniform refinement: adaptive regularization improves single-particle cryo-EM reconstruction. Nat Methods 17(12):1214–1221. https://doi.org/10.1038/s41592-020-00990-8
27. Sanchez-Garcia R, Gomez-Blanco J, Cuervo A, Carazo JM, Sorzano COS, Vargas J (2021) DeepEMhancer: a deep learning solution for cryo-EM volume post-processing. Commun Biol 4(1):874. https://doi.org/10.1038/s42003-021-02399-1
28. Vilas JL, Gomez-Blanco J, Conesa P, Melero R, Miguel de la Rosa-Trevin J, Oton J, Cuenca J, Marabini R, Carazo JM, Vargas J, Sorzano COS (2018) MonoRes: automatic and accurate estimation of local resolution for electron microscopy maps. Structure 26(2):337–344 e334. https://doi.org/10.1016/j.str.2017.12.018
29. Goddard TD, Huang CC, Meng EC, Pettersen EF, Couch GS, Morris JH, Ferrin TE (2018) UCSF ChimeraX: meeting modern challenges in visualization and analysis. Protein Sci 27(1):14–25. https://doi.org/10.1002/pro.3235
30. Lopez-Blanco JR, Chacon P (2013) iMODFIT: efficient and robust flexible fitting based on vibrational analysis in internal coordinates. J Struct Biol 184(2):261–270. https://doi.org/10.1016/j.jsb.2013.08.010
31. Emsley P, Cowtan K (2004) Coot: model-building tools for molecular graphics. Acta Crystallogr D Biol Crystallogr 60(Pt 12 Pt 1):2126–2132. https://doi.org/10.1107/S0907444904019158
32. Afonine PV, Poon BK, Read RJ, Sobolev OV, Terwilliger TC, Urzhumtsev A, Adams PD (2018) Real-space refinement in PHENIX for cryo-EM and crystallography. Acta Crystallogr D Struct Biol 74(Pt 6):531–544. https://doi.org/10.1107/S2059798318006551
33. Zivanov J, Nakane T, Forsberg BO, Kimanius D, Hagen WJ, Lindahl E, Scheres SH (2018) New tools for automated high-resolution cryo-EM structure determination in RELION-3. Elife 7:e42166. https://doi.org/10.7554/eLife.42166
34. Rohou A, Grigorieff N (2015) CTFFIND4: fast and accurate defocus estimation from electron micrographs. J Struct Biol 192(2):216–221. https://doi.org/10.1016/j.jsb.2015.08.008
35. Wagner T, Merino F, Stabrin M, Moriya T, Antoni C, Apelbaum A, Hagel P, Sitsel O, Raisch T, Prumbaum D, Quentin D, Roderer D, Tacke S, Siebolds B, Schubert E, Shaikh TR, Lill P, Gatsogiannis C, Raunser S (2019) SPHIRE-crYOLO is a fast and accurate fully automated particle picker for cryo-EM. Commun Biol 2:218. https://doi.org/10.1038/s42003-019-0437-z
36. Bepler T, Morin A, Rapp M, Brasch J, Shapiro L, Noble AJ, Berger B (2019) Positive-unlabeled convolutional neural networks for particle picking in cryo-electron micrographs. Nat Methods 16(11):1153–1160. https://doi.org/10.1038/s41592-019-0575-8
37. Bell JM, Chen M, Baldwin PR, Ludtke SJ (2016) High resolution single particle refinement in EMAN2.1. Methods 100:25–34. https://doi.org/10.1016/j.ymeth.2016.02.018
38. Punjani A, Fleet DJ (2023) 3DFlex: determining structure and motion of flexible proteins from cryo-EM. Nat Methods 20(6):860–870. https://doi.org/10.1038/s41592-023-01853-8

Chapter 17

# Methods to Study CCT Expression and Impact on Cytoskeleton

**Camila Scagnetti, Hugo Yébenes, Silvia Requena and Noa B. Martín-Cófreces**


## Abstract

Chaperones are proteins whose function is to intervene in the correct folding of newly synthesized polypeptides in the ribosome, thus providing their native conformation. CCT belongs to the Hsp60 family of chaperonins and is specifically a Type II subgroup. Here, we describe methods for determining and manipulating the CCT protein and RNA levels, as well as CCT distribution in cells, using complementary biochemical and imaging approaches. Protocols are provided for samples from primary human cells and established cell lines, including electrophoretic separation and immunoblotting. In parallel, procedures for immunofluorescence staining and microscopy are outlined, with emphasis on fixation, permeabilization, and antibody optimization. Critical parameters, appropriate controls, and troubleshooting considerations are discussed to facilitate robust and reproducible detection of CCT across different cellular systems.


**Keywords** Chaperonine, CCT, siRNA, Western blot, Immunofluorescence, Fluorescence microscopy, Jurkat, CD4+ T lymphocytes

## 1 Introduction

The correct folding of proteins is essential to maintain cell viability and homeostasis. In the crowded environment inside the cell, protein folding remains a challenge that is tackled with molecular chaperones. Chaperones aid nascent polypeptide chains to achieve their native and final conformation [1]. Among molecular chaperones, one of the most complex is CCT (also known as TRiC, TCP-1 ring complex). The CCT complex is ubiquitous in all eukaryotes as a 16-subunit oligomer, which is organized in two identical toroidal rings, each formed by eight different, albeit similar, subunits stacked back to back. In yeast, subunits are designated Cct1p–Cct8p, often—using Greek characters (α, β, γ, δ, ε, η, θ, and ζ); in humans, subunits are numbered CCT1–8. Each CCT subunit contains three specialized domains: equatorial (ATP binding and hydrolysis), apical (substrate

Noa B. Martín-Cófreces et al. (eds.), *The Cytoskeleton*, Methods in Molecular Biology, vol. 3070,
https://doi.org/10.1007/978-1-0716-5515-3_17,

binding), and intermediate (hinge). ATP hydrolysis is critical for CCT function, as it drives a conformational cycle that switches between open and closed conformation, where unfolded protein substrates become encapsulated, and their folding is achieved in part through specific interactions with the inner surfaces of each ring cavity [2]. The inner cavity surface of the complex is quite heterogeneous, with one side of the inner ring being more electropositive and the other more electronegative. This heterogeneity is essential to the specificity of CCT towards different substrates [3].

Inside the cell, CCT is an abundant and widely distributed protein complex, mainly found in the cytoplasm but also present in the cell nucleus [4]. Although initially characterized as a chaperone with few specific substrates, tubulin and actin among them, its protein substrate repertoire was progressively increased, with some studies estimating that it contributes to the folding of about 10% of cell proteins [1]. Thus, the CCT function is required in many essential cellular processes. Tubulin and actin are the building blocks of microtubules and actin microfilaments, respectively. They are obligate CCT clients to acquire their native conformation in the crowded intracellular environment, and this particularity led to an early characterization of CCT involvement in cytoskeletal dynamics [5, 6]. Afterwards, many other proteins were characterized as CCT substrates, and its role in other cellular processes became patent. Among those are cell division, centrosome and basal body dynamics, autophagy, immune synapse formation, and transcription and translation [7]. In this context, CCT is also able to fold some of the protein subunits in mTOR (mammalian target of rapamycin) complex 1 and 2 [8], leading to a potential role in metabolism. This has been observed in T lymphocytes recently, where CCT is required to allow the centrosome reorientation towards synaptic contacts with antigen-presenting cells, regulating the mitochondrial network organization and functional responses [9]. Indeed, low CCT expression promotes changes in tubulin post-translational modifications [9, 10], leading to differential regulation of kinesin motor activity, organelle distribution, and extracellular vesicle production in cells, which show a senescent phenotype [10].

Given the complexity and importance of the CCT complex in a wide range of biological processes, tracing its expression levels and cellular location is a relevant field of research. Using fluorescence-based microscopy allows the study of CCT in a wide range of cell samples and specimens. The present method details the use of specific siRNA to silence the expression of CCT subunits, and molecular biology techniques to examine their expression levels. A combination of these techniques facilitates the study of CCT impact on cell cytoskeleton during physiological processes, such as T lymphocyte activation.

## 2 Materials

### 2.1 Cells

1. Jurkat E6.1: original immortalized Jurkat cell line from NIH derived from a childhood T acute lymphoblastic leukemia (*see* **Note 1**).
2. Primary CD4+ T cell lymphocytes isolated from healthy human donors (*see* **Note 2**).

### 2.2 Reagents and Media

1. Heat-inactivated fetal bovine (hFBS): warm the FBS for 30 min at 56°C in a water bath and allow cooling at R/T. Aliquot and store at −20°C until used.
2. Complete cell culture medium: RPMI 1640, 25 mM HEPES, 10% hFBS.
3. Incomplete cell culture medium: RPMI 1640, 25 mM HEPES.
4. Hank's Balanced Salt Solution (HBSS): 0.3371 mM $Na_2HPO_4$ pH 6.9, 136.893 mM NaCl, 5.366 mM KCl, 1.261 mM $CaCl_2$, 0.4057 mM MgSO, 0.4919 mM $MgCl_2$, 0.4409 mM $KH_2PO_4$, 5.551 mM D-glucose, 4.166 mM $NaHCO_3$.
5. 0.4 cm electroporation cuvettes (4 mm gap size).
6. Transfection medium: Opti-MEM I.
7. P6 and P24 sterile cell culture plates: prewarm 30 min before use with the corresponding media.
8. siRNAs for silencing: scramble and against CCT subunits (see Table 1 and **Note 3**).

**Table 1**
**RNA oligonucleotide sequence for silencing**

| Gene | Forward | Reverse |
|---|---|---|
| *TCP1* | UGGUAAACCUCGAGACAACdTdT | GUUGUCUCGAGGUUUACCAdTdT |
| *CCT2* | CCAUUGGAGACUUGGUAAAdTdT | UUUACCAAGUCUCCAAUGGdTdT |
| *CCT4* | AGGUGGUGCUCCAGAAAUAdTdT | UAUUUCUGGAGCACCACCUdTdT |
| *CCT5* | CCAGACAGGUGAAGGAGAUdTdT | AUCUCCUUCACCUGUCUGGdTdT |
| *Ctrl 1* | CGUACGCGGAAUACUUCGAdTdT | UCGAAGUAUUCCGCGUACGdTdT |
| *Ctrl 2* | UUCUCCGAACGUGUCACGdTdT | CGUGACACGUUCGGAGAAdTdT |

*TCP1* T-complex 1 or chaperonin containing TCP1 subunit 1, *CCT2* chaperonin containing TCP1 subunit 2, *CCT4* chaperonin containing TCP1 subunit 4, *CCT5* chaperonin containing TCP1 subunit 5, *Ctrl 1* scramble oligonucleotide 1, *Ctrl 2* scramble oligonucleotide 2, *dT* overhang DNA chemistry. Homo sapiens sequences

9. Lymphocyte separation medium: any lymphocyte separation medium with a density of 1.077 g/mL.
10. Saline solution: 9 mg/mL NaCl in distilled water. Filter through 0.22 μm sterile filters before use. Non-pyrogenic solutions are recommended.
11. Ethylenediaminetetraacetic acid (EDTA) solution: 0.5 M EDTA, pH 8. Filter through 0.22 μm sterile filters before use.
12. Human CD4+ T Cell biotinylated-antibody cocktail coupled to streptavidine-coated magnetic beads and a magnet for negative isolation from any source.
13. Lymphocyte isolation medium: HBSS, 2% BSA, 1 mM EDTA. Sterile.
14. 0.2 cm plastic cuvettes for nucleofection.
15. Any home-made or commercial kit for RNA isolation.
16. Any home-made or commercial kit for qPCR.
17. Primers for housekeeping and target genes for qPCR (Table 2).

**Table 2**
**Primer sequences for qPCR (5′–3′ ends)**

| Gene | Forward Primer | Reverse Primer |
|---|---|---|
| *TCP1* | GGAAACGATCCGCTCCCAAA | CCAAGCCAACTGGACCAAGA |
| *CCT2* | AGAGGTGATTCTGCGTGTGG | ACAGCACGTGGGAATGCTTA |
| *CCT3* | TGGAAACATCAATGCTGCCAAG | GCCTCCCATTGGGTCCAAAA |
| *CCT4* | AGTATTACATCCAGCAGCCAGAA | GGTGCCATCTCCTGCTTCTA |
| *CCT5* | AGTTAGCCAAGAGGCGGATAAG | GACTTCGGTCATAGTCTGGATGG |
| *CCT6A* | GGCAATGTGCTGCTTCACG | ATCATCCTGGGCTGTTGCTA |
| *CCT6B* | CTGTCAATATATGCGCCGCC | CACCTGCACCAGAAACAAGC |
| *CCT7* | GGCGGGCCGCTGAATAA | TGGGAGCTATCAGTCCCCTC |
| *CCT8* | CAGATGCTCAAGGAGGGAGC | TGTACGAGTGGTTTGGGCAAG |
| *ACTB* | CCAGAGGCGTCAAGGGATAG | CCAACCGCGAGAAGATGA |
| *RPL30* | TTCTCGCTAACAACTGCCCA | GCCACTGTAGTGATGGACACC |
| *RPL31* | ATGGGAACTCCAGATGTGCG | CACACGGATTCGGTATGGCA |
| *RPL37* | GGAACTGGTCGAATGAGGCA | CAGCTGCCCTCTTGGGTTTA |

*TCP1* T-complex 1 or chaperonin containing TCP1 subunit 1, *CCT2* chaperonin containing TCP1 subunit 2, *CCT3* chaperonin containing TCP1 subunit 3, *CCT4* chaperonin containing TCP1 subunit 4, CCT5 chaperonin containing TCP1 subunit 5, CCT6A chaperonin containing TCP1 subunit 6A, CCT6B chaperonin containing TCP1 subunit 6B, CCT7 chaperonin containing TCP1 subunit 7, CCT8 chaperonin containing TCP1 subunit 8, ACTB actin beta, *Rpl30* ribosomal protein L30, *RPL31* ribosomal protein L31, RPL37 ribosomal protein L37. Homo sapiens sequences

18. Lysis buffer: 20 mM Tris-HCl, pH 7.5, 1% NP-40 substitute, 0.2% Triton X-100, 150 mM NaCl, 2 mM EDTA, 1.5 mM $MgCl_2$, phosphatase and protease inhibitors.
19. 5× Laemmli buffer: 312.5 mM Tris-HCl pH 6.8, 10% SDS, 50% glycerol, 0.01% bromophenol blue, 25% $\beta$-mercaptoethanol (or 1.25 M DTT), added fresh.
20. Pre-immune serum solution: 1:200. Sera from different sources are required (mouse, rabbit, goat).
21. Antibodies and probes for immunofluorescence recognizing CCT subunits (2, 3, 6, 7, and 8), mitochondria such as Tom20, tubulin, microfilaments (phalloidins), nucleus (DAPI, Hoesch).
22. Secondary antibodies conjugated to corresponding fluorochromes allow visualization of different proteins and cell compartments in the same samples.
23. Mounting medium containing anti-fading agents with the correct RI for objectives.
24. 2× PHEM: 120 mM Pipes pH 6.9, 60 mM Hepes, 20 mM EDTA, 4 mM $MgCl_2$. Use ultrapure water (deionized $H_2O$).
25. Immunofluorescence fixation buffer (FB): PHEM, 4% paraformaldehyde, 0.12 M sucrose. Use ultrapure water (deionized $H_2O$).
26. Immunofluorescence blocking buffer (BB): PHEM, 3% bovine serum albumin (BSA), 50 μg/mL human gamma globulin, 0.2% $NaN_3$. Use deionized $H_2O$.
27. Immunofluorescence blocking and permeation buffer (BPB): BB, 0.2% Triton Tx-100.
28. Immunofluorescence permeation buffer (PB): FB:BB (ratio 1:1), 0.2 % Triton Tx-100.
29. 10× TBS (Tris-buffered saline): 200 mM Tris-HCl pH 7.4, 1.5 M NaCl. Use deionized $H_2O$. Filter through a 0.22 μm sterile filter before use.
30. Wash buffer: TBS, 0.2% Tween-20
31. Blocking buffer: TBS, 0.2% Tween-20, 5% BSA.
32. Native lysis buffer: 0.1 M MES pH 6.7, 1 mM EGTA, 1 mM $MgCl_2$, 10 mM guanosine 5′-triphosphate (GTP), inhibitors of proteases and phosphatases.
33. Native running buffer CCT: 0.1 M MES pH 6.7, 1 mM $MgCl_2$, 1 mM EGTA.
34. Native running buffer tubulin: 0.1 M MES pH 6.7, 1 mM $MgCl_2$, 1 mM EGTA, 1 mM GTP.

35. 4× native loading buffer: 50 mM MES, pH 6.7, 30% glycerol (v/v) or 25% sucrose (w/v). 0.025% bromophenol (w/v) can be added for visualization.
36. Native PAGE: 0.1 M MES pH 6.7, 4%–6% acrylamide/bisacrylamide (30%:0.8%), 1 mM $MgCl_2$, 1 mM EGTA, 1 mM GTP, 0.1% APS (ammonium persulfate), 0.1% TEMED (v/v).
37. Native transfer buffer: 24.8 mM Tris(hydroxymethyl)aminomethane, 0.192 M Glycine.

### 2.3 Equipment

1. Hematological analyzer or counter chamber.
2. Electroporator.
3. Nucleofector.
4. Vacuum pump.
5. Class II Microbiological Safety Cabinet with vertical laminar flow.
6. Laser scanning confocal or epifluorescence optical microscope.
7. Benchtop centrifuge with swing-out rotor and buckets (*see* **Note 4**).
8. Conventional table-top microcentrifuge.
9. Ultracentrifuge with swing-out rotor and buckets (*see* **Note 5**).
10. Thermal cycler.
11. PCR real-time system.

## 3 Methods

### 3.1 Silencing of CCT in Non-Adherent Jurkat Cells Through Electroporation

1. Prewarm incomplete cell culture medium at 37°C in P6 plates, 4 mL per well.
2. Add half of the percentage of the supplementation serum hFBS, 5% (200 μL).
3. Prewarm the electroporation cuvettes.
4. Count the cells; use $10^7$ cells per point.
5. Collect cells by centrifugation at 193 × *g*, 5 min. Discard the supernatant.
6. Add 3 mL of cold HBSS (4°C).
7. Centrifuge at 250 × *g* (*see* **Note 4**), 2 min, 4°C. Decant the supernatant.
8. Add 3 mL of cold Opti-MEM I (4°C).

9. Centrifuge at 193 × *g*, 2 min, 4°C. Decant the supernatant.
10. Resuspend the pellet in 0.4 mL of cold Opti-MEM I per electroporation.
11. Mix the corresponding siRNAs with cells in the cuvettes.
12. Electroporate at 240 V, 975 μF (*see* **Note 6**).
13. Retrieve the cells (*see* **Note** 7) and plate them in the pre-warmed plates.
14. Maintain the culture at 37°C for 2–3 h.
15. Add hFBS to the cell medium until reaching 10 % hFBS as the working concentration.
16. Culture cells O/N.
17. Discard debris and dead cells 24 h post-transfection by performing Sects. 3.3 or 3.4.

### 3.2 Human CD4+ T Lymphocyte Isolation and Silencing of CCT Through Nucleofection

Prior to cell silencing, CD4+ T lymphocytes or T cells should be isolated from the peripheral blood of healthy donors. Several sources can be used, such as complete blood or buffy coat preparations, which include mostly immune cells from blood. In buffy coats, platelets and erythrocytes are also present and should be separated prior silencing of cells. This section describes a method to silence T lymphocytes directly after specific isolation of CD4 T cells.

1. If complete blood is used, dilute (1:2) with saline solution.
2. If Buffy coat preparations are used, dilute (1:3) with saline solution.
3. Prepare a gradient by adding the lymphocyte separation medium at the bottom of centrifuge tubes and gently adding the cell mix on top in a ratio of (1:2).
4. Centrifuge at 278 × *g* for 30 min with no break.
5. Collect the cells from the interphase; these are peripheral blood mononuclear cells (PBMCs).
6. Wash with saline solution five to six times (*see* **Note 8**).
7. Purify the CD4+ T cells by negative selection with magnetic beads (*see* **Note 9**).
8. Prewarm incomplete cell culture medium at 37°C in plates.
9. Prewarm the nucleofection solution and the cuvettes.
10. Count and use $5 \times 10^6$ cells per nucleofection. Use a 1.5 mL tube per reaction.
11. Spin the cells in a microfuge and discard supernatant (*see* **Note 10**).
12. Wash twice with cold HBSS to eliminate FBS.

13. Resuspend the cells with 100 μL of transfection mix per sample (*see* **Note 11**) and transfer to prewarmed cuvettes.
14. Nucleofect immediately (*see* **Note 12**).
15. Retrieve live cells with a micropipette (*see* **Note 13**).
16. Plate cells in prewarmed plates at a concentration of $4 \times 10^6$ cells/mL.
17. Maintain cells at 37°C for 2–3 h.
18. Add 5% of hFBS to reach the percentage in the complete cell culture medium. Supplement with 5 ng/mL of recombinant human IL-7 to improve viability.
19. Discard debris and dead cells 24 h post-transfection (Sects. 3.3 or 3.4).

### 3.3 Isolation of Live Cells Through Gradients of Density

1. Retrieve electroporated cells by centrifugation at $193 \times g$ for 5 min at R/T. Discard the supernatant.
2. Resuspend the cells in 1 mL of complete cell culture medium.
3. Gently, add 1 mL of separation medium to the bottom of the cell culture tube (*see* **Note 14**).
4. Spin at $360 \times g$ for 6 min with no break (*see* **Note 4**).
5. Retrieve live cells from the interphase between the cell culture medium and the lymphocyte separation medium (*see* **Note 15**) in a new tube.
6. Add 5 mL of HBSS or saline medium (*see* **Note 16**).
7. Centrifuge at $123 \times g$ for 5 min at R/T.
8. Discard the supernatant and resuspend the cells in complete medium.
9. Count the cells.
10. Culture the cells at a minimal concentration of $10^6$ cells/mL.

### 3.4 Isolation of Live Cells Through Slow Centrifugation

1. This is an alternative method to Sect. 3.3 (*see* **Note 17**).
2. Retrieve the electroporated cells from the culture plates gently in a tube.
3. Centrifuge cells at 30–55 $\times g$ for 10 min at R/T, no breaks (*see* **Note 18**).
4. Discard the supernatant very carefully, without touching the pellet (*see* **Note 19**).
5. Resuspend the pellet in complete medium and count.
6. Leave in culture at a concentration of $1 \times 10^6$ cells/mL.

### 3.5 Determination of CCT Expression Through RT-qPCR

1. Use $1.5 \times 10^6$ cells for RNA isolation (*see* **Note 20**).
2. Centrifuge cells and discard the supernatant completely.

3. Resuspend the cell pellet with any suitable reagent that extracts and maintains the integrity of total RNA from cells and tissues (*see* **Note 21**).
4. Isolation of RNA can be performed with any of the currently available commercial kits (*see* **Note 22**).
5. Use 500 ng of mRNA to retrotranscribe to first-strand cDNA in a 20 μL reaction volume (*see* **Note 23**).
6. Run the retro-transcription reaction in a benchtop thermal cycler: 5 min at 25°C, 70 min at 42°C, and 15 min at 70°C for enzyme inactivation (*see* **Note 24**).
7. Use 1 ng/μL of cDNA for a 10 μL qPCR reaction mix (*see* **Note 25**). Gene expression quantification requires the amplification of housekeeping and target genes for further normalization. See Table 2 for primer sequences.
8. Perform cDNA amplification by a first step for polymerase activation for 10 min at 95°C, 40 cycles of 10 s at 95°C, and 1 min at 60°C, followed by 10 s at 95°C prior to the melting curve being finished (*see* **Note 26**).
9. Analyze data with dedicated software for normalization; 2−Δ ΔCT method is recommended (*see* **Note 27**).

### 3.6 Determination of CCT Protein Through SDS-PAGE and Western Blot

1. Count the cells and use 0.5–2 x $10^6$ cells per sample in a 1.5 mL Eppendorf tube.
2. Spin down the samples (*see* **Note 28**).
3. Remove the supernatant carefully to avoid pellet resuspension.
4. Store at −80°C or process immediately.
5. If processed immediately, incubate the cell lysate with 50 μL of lysis buffer per million of cells on ice for at least 30 min.
6. Spin down the samples in a microcentrifuge at 1000 ×*g* for 10 min at 4°C to remove cell debris and nuclei.
7. Retrieve the supernatant carefully to avoid pellet resuspension and transfer it to a new 1.5 mL Eppendorf tube.
8. Store the sample at −20°C or proceed to determine total protein concentration through any assay (*see* **Note 29**).
9. Adjust the sample volume with Laemmli buffer 5× to 1–1.5× final concentration, ensuring equal protein amount in all samples to load the same volume.
10. Incubate samples at 85°C in a thermal block for 10 min.
11. Spin down the samples in a microcentrifuge at 10,000 ×*g* for 1 min.
12. Load samples into SDS-PAGE (*see* **Note 30**).
13. Run at 90 V until proteins are resolved.

14. Transfer to nitrocellulose or PVDF membranes.
15. Block membranes for 30 min at R/T with blocking buffer (*see* **Note 31**).
16. Blot membranes with primary antibodies diluted in blocking buffer for 1–2 h at R/T (*see* **Notes 32 and 33**).
17. Wash four to five times with the wash buffer.
18. Blot membranes with peroxidase-labeled secondary antibodies diluted in blocking buffer for 30–60 min at R/T (*see* **Note 34**).
19. Wash four to five times with the wash buffer.
20. Acquire the chemiluminescence signal from antibodies with any commercial substrate and imaging system.
21. Perform the densitometric analysis of bands with dedicated software.
22. Normalize against loading control.

### 3.7 Expression of CCT Through Native PAGE and Native Blot

1. For native-PAGE assays of protein complexes, use native lysis buffer (*see* **Note 35**).
2. Lyse cells through a 2 mL dounce homogenizer (20 strokes; tight pestle) at R/T.
3. Centrifuge in a minicentrifuge at 14,000 × *g* at 25°C for 30 min.
4. Use supernatant and discard pellets.
5. Determine protein content in samples.
6. Use 5× native loading buffer to obtain samples with equal amounts of protein.
7. Load 20 μg of protein per sample in 4% or 6% native-PAGE (*see* **Note 30**).
8. Run gels for 2 h at 80 V in native running buffer.
9. Transfer to nitrocellulose of PVDF membranes for 2 h at 400 mA in native transfer buffer.
10. Block and probe membranes against corresponding antibodies as in Sect. 3.6 (*see* **Notes 31–34**).

### 3.8 Immunofluorescence: Intracellular Staining

This protocol is described for non-adherent cells, but it can also be used with adherent cells. These cells require previous culture on the coverslip.

1. Prepare a wet chamber (e.g., square P500 plate) with filter paper to fit the plate and wet with ultrapure water (deionized $H_2O$).

2. Cover the wet paper with a hydrophobic, rigid film such as Parafilm M.
3. Add drops of 25–40 μL of poly-L-Lysine solution (50 μg/mL).
4. Place the coverslips over the drops to expand throughout their bottom surface. Incubate at 4°C O/N or 37°C for 1–2 h (*see* **Note 36**).
5. Wash the coverslips by immersion in deionized $H_2O$ and place them back in the chamber with the poly-L-Lysine side facing upwards (*see* **Note 37**).
6. Count the cells. Use $1.0–1.5 \times 10^5$ cells of each line per coverslip.
7. Resuspend the cells in complete medium to a total of 80 μL per coverslip.
8. Add the cells on the coverslips.
9. Fix the cells by adding one volume of FB to the coverslip (about 80 μL) for 10 min at R/T.
10. Remove the FB and wash the crystals with TBS.
11. Add 80 μL of BB and block for a minimum of 30 min (*see* **Note 38**).
12. Permeate cell membranes with 80 μL PB for 5 min at R/T.
13. Wash with TBS for 2 min.
14. Block with BPB at R/T 30 min, 80 μL.
15. Prepare primary antibodies in BB at concentrations of about 1–10 μg/mL.
16. Add 25 μL per coverslip and incubate 1 h at 4°C.
17. Wash with TBS for 2 min, five times.
18. Prepare secondary antibodies in BPB and spin at maximal speed in a minifuge for 2 min (*see* **Notes 39–41**).
19. Add 25–40 μL per coverslip and incubate 1 h at 4°C.
20. Block the samples with 80 μL of corresponding preimmune serum solution, 20 min at R/T.
21. Wash with TBS for 2 min, five times.
22. Repeat Steps 15–21 as many times as primary antibodies are needed in staining (*see* **Note 42**).
23. Wash the microscope slides with neutral detergent and let them dry with ethanol at 70% (*see* **Note 43**).
24. Rinse shortly with deionized $H_2O$ (×2) to remove salt remains.

25. Carefully drain the coverslips on paper without dragging the cells.
26. Add carefully a 8 μL drop of mounting medium onto the coverslips.
27. Place carefully the coverslips onto the microscope slide (*see* **Note 44**).
28. Let dry at R/T in darkness O/N.

### 3.9 Imaging Under a Fluorescence Microscope. Confocal Acquisition

A wide range of microscopes can be used for image acquisition, such as epifluorescence or confocal microscopes, depending on the kind of image and information desired (*see* **Note 45**). Here, we describe a general method for image acquisition in microscopes equipped with detectors. There are also microscopes equipped with cameras, such as cMOS.

1. Place the microscope slide; coverslips face to objective.
2. Use 100× magnification or 63× with zoom due to the small size of cells.
3. Add a drop of proper imaging medium for the immersion objective used (*see* **Note 46**).
4. Focus on the specimen of interest.
5. Use the corresponding excitation beams and detectors for signal acquisition of the different fluorochromes present in the sample (*see* **Note 47**).
6. Adjust excitation beam intensity until the observed difference between the background and the specific signal for the protein or molecule of interest is maximal (*see* **Note 48**).
7. Modulate the intensity of the beam light to obtain the maximal dynamic range for the signal detected. If the signal is low, some systems allow the accumulation of images acquired at the same locations (*see* **Note 49**).
8. Final adjustment of the acquisition settings will depend on your sample and your optical system (*see* **Note 50**).
9. Acquire Z-stacks that include all the volume of your sample to obtain all the information in the specimen (*see* **Note 51**).
10. Analyze the intensity and localization of each component visualized to determine the biology of your cells with dedicated software (*see* **Note 52**).
11. Use 3D reconstructions and maximal projections to visualize whole cells (*see* **Note 53**).
12. Analyze single images from your Z-stack to observe protein localization and structure of cytoskeletons (see Fig. 1 and see **Note 54**).

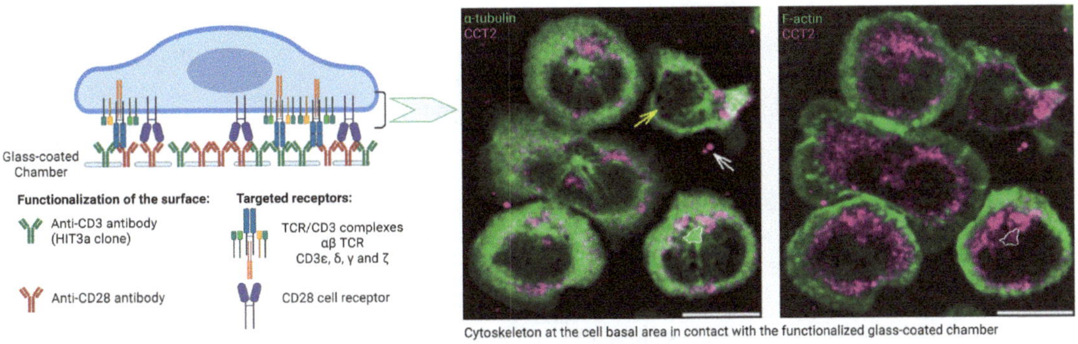


**Fig. 1** CCT2 localization in immune synapses formed by CD4 T cells. Left, Scheme showing the microscope setup for the experiment. Right, images showing CCT2 (magenta) and α-tubulin (green; left image) or filaments of actin (green, right image). CCT2 is enriched in the area of soluble α-tubulin (left image), near the interface with the external, actin-based lamella (right image). CCT is also present at the centrosome (white-surrounded area); radial microtubules (green) are observed in the actin-free area. Yellow arrow, vesicular components surrounded by cytoskeletal components, such as multivesicular bodies, at the immunological synapse [9, 10]. White arrows, extracellular particles enriched in CCT [10]. Acquisition settings used in Confocal STELLARIS: 1024 × 1024 logical size, 400 Hz scanning speed, 0.3 μm Z-step size, 30 steps. Laser settings used: excitation for tubulin (FITC) detection was 1.00% for 492 nm (white laser) and emission detection was in the [498–563] nm range; excitation for filaments of actin (ATTO647N-conjugated phalloidin) detection was 2.40 % for 648 nm (white laser) and emission detection was in the [653–778] nm range; excitation for CCT2 (Alexa-546) detection was 3.00% for 552 nm (white laser) and emission detection was in the [557–597] nm range

## 4 Notes

1. Use 10–15 × $10^6$ cells per sample.
2. Use 5 × $10^6$ per sample.
3. To silence the CCT complex, use of specific oligonucleotides against subunits CCT1, CCT2, CCT4, and CCT5 (ratio 1:1:1:1) is valid. The concentration of the mix shall be 2–2.5 μ M maximum to avoid excessive cell death. Stocks of 100 μM for siRNAs are recommended. The siRNA aliquots should be long-term stored at −80°C. Keep in ice (4°C) until use upon thaw.
4. The speed indicated for the benchtop centrifuge for cell culture and transfection methodologies described here corresponds to a swing-out rotor of 13 and 17 cm radius, including the swinging buckets.
5. For the protocols described herein, an ultracentrifuge equipped with a swing-out bucket for large tubes (about 36 mL capacity) and able to operate at 120,000 × $g$ is required.
6. These electroporation conditions are optimized for a BioRad II device and Jurkat cells but can be determined for other cell types and devices. If samples are not electroporated

immediately (less than 30 min), keep them in ice or refrigerator (4°C). The cuvettes should be completely dry when placed into the device.

7. Approximately half of the cells are lost in the electroporation process and appear as a white foam.

8. Several washes are needed in order to completely drain the platelets. Usually, a first centrifugation at 193 × *g* for 15 min followed by four supplementary centrifugations at 123 × *g*, 10 min are required.

9. EasySep™ Human CD4+ T negative Cell Isolation Kit was used, but any other commercial kit or a specific cocktail of biotinylated antibodies and streptavidin-coated beads can be used to this end.

10. It is relevant to perform each nucleofection in a unique tube to perform the steps as quickly as possible.

11. Mix: 2.5 μL of RNA in 100 μL of Opti-MEM I. Do not keep the cells in the nucleofection solution for more than 2 min before nucleofection.

12. Usually, T lymphocytes require the program U-04 Nucleofector I-Amaxa. Specific conditions must be used for other cell types.

13. Dead cells appear as a white mass in the cell mix. Try to leave it inside the cuvette to avoid toxicity in your cell culture. Plate immediately after the cells upon nucleofection.

14. Two visible phases are distinguished: HBSS or medium being the upper phase and lymphocyte separation medium the lower one. The proportion between these must be such that the HBSS or medium phase is equal to or greater than the lymphocyte separation medium (1:1, 2:1, etc.).

15. Retrieve the live cells either from the upper phase of the gradient or the whole supernatant; include both phases if the number of cells is small and they are not clearly observed as a white cloud floating over the lymphocyte separation medium. Do not recover the cell pellet at the bottom of the tube, since it is composed mainly of dead cells.

16. Wash twice if both phases of the supernatant are collected to retrieve the cells, since the mix will contain lymphocyte separation medium, which can have a deleterious effect.

17. Occasionally, the processing of cells through gradients results in discarding a significant amount of living, valid transduced cells. This is mostly observed when electroporation yields a small number of cells, which can prevent further processing and experimentation. e.g., by using the slow centrifugation method.

18. The low speed of the centrifugation used here makes the cell pellet at the bottom fairly attached.
19. Avoid discarding the supernatant by flipping the tube; instead, use a vacuum pump to gently absorb the supernatant and prevent cell pellet detachment from the bottom.
20. We regularly use $1.5 \times 10^6$ human CD4+ T cells isolated from PBMCs to further avoid the saturation of the spin column's membrane during RNA extraction. The cell size and other conditions, such as cell activation or the use of stable cell lines, influence the amount of mRNA present in the cell. To avoid the collapse of elements for RNA isolation, these factors should be considered when choosing the RNA extraction method.
21. We use TRIsure™ reagent, but any commercial or homemade reagent is suitable.
22. We regularly use the RNAeasy Micro Kit (QIAGEN), following a modified version of the Quick-Start protocol. Briefly, 150 μL of chloroform is added to TRIsure-treated sample defrosting, vortexed for 15–30 s and incubated for 3 min. This mix is centrifuged at full speed in a conventional tabletop microcentrifuge for 15 min at 4°C, resulting in a two-phase separated volume. The aqueous upper phase is carefully collected to avoid TRIsure contamination and mixed by pipetting with one volume of 70% ethanol. Then, the sample is transferred to a 2 mL spin column (provided in the RNAeasy Micro Kit) and centrifuged for 15 s at $\geq 8000 \times g$. Next steps are performed as indicated in the Quick-Start protocol. Incubation with DNase enzyme should be done to avoid contamination and unspecific-product amplifications. Finally, isolated RNA is eluted in 14 μL of nuclease-free water and collected in a 1.5 mL RNAse-free collection tube. If RNA concentration is expected to be higher than 500 ng/μL, we recommend doubling the volume. For samples with expected low RNA concentration, we recommend incubating the RNA-binding membrane with nuclease-free water for 1–2 min before final elution. Elution buffers are provided in commercial extraction kits and could be used instead of nuclease-free water. For long-period storage, isolated RNA is preferentially stored at −80°C. Otherwise, −20°C is recommended until use.
23. We use the GoScript™ Reverse Transcription System from Promega, but any commercial system may be used. Try to obtain isolated RNA as concentrated as possible. In a 20 μL reaction mix, 1 μL GoScript™ Reverse Transcriptase allows the conversion of up to 5 μg of total RNA to cDNA. Before PCR starts, DNA is incubated for 5 min at 70°C with 1 μL of

random primer. The final concentration of $MgCl_2$ in the master mix is 3 mM. cDNA could be stored at 4°C for short-term usage or maintained frozen for long-term storage.

24. We used the Primer Designing tool from the National Center for Biotechnology Information (NCBI) to design the primers. The length of the amplicon was shorter than 300 base pairs, and the sequences included intron-spanning regions. The oligonucleotides were synthesized in a 20 nm scale, dry, and desalted format. The final concentration of primers was 0.3 μM. The expression of *RPL30*, *RPL31*, and *RPL37* genes was quantified to normalize the expression of genes encoding every subunit of the CCT. Those genes were selected based on their stability throughout the human lifespan [11]. As a substrate and major client of CCT, *ACTB* expression might be informative and relevant for comparison with the expression of CCT genes. Before running the experiment, it is recommended to make serial dilutions to ensure the efficiency of the oligonucleotides to further apply the $2^{-\Delta\Delta CT}$ method.
25. We regularly use the GoTaq® qPCR Master Mix system from Promega and target-specific primers, but any commercial method is suitable.
26. We use the CFX384 Touch™ Real-Time PCR detection system (Bio-Rad), but any commercial device and accompanying software is suitable. We recommend using the software accompanying the machine for analysis (*see* **Note 27**).
27. We use the 3.1 version of the CFX Manager software (Bio-Rad), but any commercial device and accompanying software is suitable (*see* **Note 26**).
28. If using a minifuge, 4500 rpm for 5 min at 4°C is enough to recover cells. Do not use maximal speed.
29. To determine protein concentration, any assay can be used, such as Lowry, Bradford, or preferably, BCA (bicinchoninic acid assay) if a low amount of protein is expected.
30. The use of a Hamilton syringe to load the gel is recommended, to load exact volumes and to avoid increasing sample volume by mixing with the electrophoresis buffer.
31. Membranes can be alternatively blocked O/N at 4°C; avoid drying out.
32. Membranes can be alternatively incubated with primary antibodies O/N at 4°C, if required. Avoid drying out. If a low primary antibody amount is available, the incubation can be performed in a wet chamber, over parafilm, and the membrane can be placed facedown with as little as 1 mL of antibody solution.

33. Antibody concentration will depend on the affinity of the antibody and the amount of antigen in the sample. Anti-CCT antibodies are commercially available and usually work at 1 μ g/mL. Anti-tubulins are usually used at very low concentrations to avoid overload of chemiluminescent detectors.
34. HRP-conjugated secondary antibody concentration must be as small as possible to avoid bleaching of membranes when detecting chemiluminescence. Typical dilutions are 1:10,000 to 1:20,000 from most commercially available antibodies. To quantify tubulins, use 1:100,000 dilutions.
35. For native electrophoresis of CCT and tubulins, do not use 4°C during the procedure.
36. For long-term storage, close the wet chamber with Parafilm M and freeze at −20°C.
37. Cells adhere to the poly-L-Lysine side; during the procedure, maintain the coverslips continuously hydrated, covering them with up to 100 μL HBSS or TBS. In particular, between incubations with antibody solutions. Close the plate to avoid evaporation.
38. Fixed cells onto coverslips treated with blocking buffer can be stored in P24 plates at 4°C. The cells should face upwards and be completely covered with about 500 μL of filtered TBS to avoid dryness. Microtubules may undergo changes due to low temperatures.
39. Fluorochrome-conjugated secondary antibody concentration must be as small as possible to avoid photobleaching of fluorochromes and possible cross-talk between channels during image acquisition. Typical dilutions are 1:500 to 1:1000 from most commercially available antibodies.
40. Fluorochrome-conjugated primary antibodies do not need secondary antibodies or serum blocking.
41. Spin all solutions of fluorochrome-conjugated antibodies (primary or secondary antibodies) before use to avoid protein aggregates that may yield high background objects during image acquisition.
42. Organize stain to allow labeling of proteins or compartments of interest with different fluorochromes, securing separation of excitation and emission signals during image acquisition. This will depend on the available microscopes, e.g., antibodies anti-CCT combined with secondary antibodies conjugated to Alexa-546; outer mitochondrial membrane detected with anti-Tom20 combined with secondary antibodies conjugated to Alexa-594; microtubules detected with anti-tubulin antibody conjugated to FITC; microfilaments of actin with

phalloidin-conjugated to ATTO647N; nucleus detected with DAPI or Hoesch (U.V. emission).

43. Microscope slides can contain traces of detergents and other components from manufacturing that may increase unspecific signal and background in samples during acquisition.

44. Place the coverslips slowly, at an oblique angle to reduce the amount of bubbles that form in the medium.

45. When using an epifluorescence microscope, image acquisition is faster than in other devices. An automated Z-motor for stack acquisition may be required to apply ulterior deconvolution to avoid out-of-focus light. Be aware that specific characteristics about Z-step size, pixel size, objective, refractive index of the medium, etc., may be required to apply deconvolution algorithms to the obtained images. Some commercial brands are currently provided with proprietary deconvolution algorithms. There are also specific software dedicated to this kind of post-processing, either freely (NIH Image J) or commercially available (Huygens software). By using a laser scanning confocal microscope (LSCM), greater resolution can be obtained for deeper and more complex analyses.

46. Immersion objectives usually perform better to observe specimens, due to increased numerical aperture (NA) and resolution for the same magnification, avoiding light refraction, scattering, and reflection due to the lack of an air gap between the specimen and the lens.

47. The initial choice of fluorochromes is relevant; selection of molecules with no or low cross-talk at their emission spectra will prevent unspecific detection during acquisition. If several fluorochromes with similar emission spectra are used in the antibody panel, first test whether the different excitation and emission detection conditions allow correct separation, or whether there is bleeding of one of the channels to the other channels used. Prepare samples including only one of the stainings (one fluorochrome) and test them to observe whether a specific signal can be detected by using the acquisition methods of the other fluorochromes. In the case of microscopes with spectral detection, use the corresponding excitation beam for this particular fluorochrome; if available, use the live acquisition mode and a full wavelength range for detection. Then, modify the range to observe the signal detected in the wavelength range corresponding to the other fluorochromes that will be combined in your final preparation. Signals must be near the noise level for specificity. When probing all the stainings together, try to excite the fluorochromes with your method, modify the excitation beam intensity, and observe if the level of the signal detected for the

other fluorochromes changes. If this is the case, a quick fix is to try different fluorochrome combinations.

48. When a very low or no signal is detected, and the visualization system is performing properly, going back to the bench and adjusting the settings of primary and secondary antibodies is the best option in most scenarios. The use of high-intensity beam light for illumination of the sample usually results in rapid photobleaching and damage to the sample.

49. Images should be acquired with equal characteristics of staining (kind and concentration of primary and secondary antibodies conjugated to fluorochromes), objectives (magnification, NA), immersion media if required, pixel size, excitation intensity, and emission detectors to allow comparison and quantification of characteristics to establish statistical differences between samples (*see* **Note 50**).

50. Hybrid detectors are available in a number of microscopes, which allow photon counting modes for image acquisition, improving signal-to-noise ratio, compared to photomultipliers. The formers allow modulating intensity and offset, but hybrid detectors are recommended for quantitative analysis of the dataset. The use of photon counting mode for hybrid detectors found in some microscopes improves the detection of protein and cell components in samples. To prevent the detection of different fluorochromes by the same detector, the selection of fluorochromes with different emission wavelengths is required (see an example in Fig. 1 and *see* **Note 47**).To observe if there is cross-detection in a channel, first test whether the excitation of each fluorochrome is detected in the other channels by using single-stained samples. Modify the laser intensity and capture images. Use these conditions for full-stained samples. Do not use "autoscale mode" for observation of the signal when adjusting the laser level. The maximum signal should not be at the upper limit of the full dynamic range to avoid loss of information in samples with potentially increased signal. Overexposure of samples leads to photobleaching, preventing further imaging of this sample. Observe if the maximum signal of the colors in autoscale mode decreases with acquisition, indicating photobleaching. A quick fix involves lowering the laser intensity. If the signal is too low, some optical devices allow accumulating scans.

51. Single plane images inform about the localization of cell components with the maximal resolution of different optical systems (2D); in general, about 200 nm of resolution limit. Z-stacks with confocal allow acquiring the whole cell volume, but the limit resolution in Z is about 500–700 nm for confocal microscopes, which can be lowered by several methods,

including STED (stimulated emission depletion) microscopy or by applying algorithms (*see* **Note 53**).

52. The combined visualization of different cell components, such as cytoskeleton components or organelles, allows for localizing the protein of interest in the cell. Differences in localization may be related to biological function. This kind of information can be quantified with specific software, allowing characterization of cells in different scenarios.
53. Be aware that volumetric or 3D reconstructions of confocal or epifluorescence Z-stacks will always show distorted elements due to the limit of resolution imposed by the optic system, mainly due to light diffraction in 3D (point spread function; PSF). In addition, epifluorescence shows decreased resolution in 2D due to out-of-focus light, which is improved in confocal microscopes thanks to the use of a pinhole.
54. A combination of silencing and immunofluorescence will serve to observe differences in the cytoskeletons and metabolism due to changes in CCT, as in [8–10].

## Acknowledgments

This study was supported by grants from the Madrid Regional Government (S2022/BMD-7209-INTEGRAMUNE-CM), Obra Social Fundación la Caixa (LCF/PR/HR23/52430018), and MCIN/AEI/10.13039/501100011033 (PID2022–141895OB-I00) to NBMC. We are thankful for the microscopy experimentation and analysis conducted at the Videomicroscopy Facility of the IIS Princesa (Madrid, Spain), cofunded by IFEQ21/00085 and IFCS22/00014 from ISCIII and FEDER.

## References

1. Sontag EM, Samant RS, Frydman J (2017) Mechanisms and functions of spatial protein quality control. Annu Rev Biochem 86:97–122. https://doi.org/10.1146/annurev-biochem-060815-014616
2. Jin M, Zang Y, Wang H, Cong Y (2025) The conformational landscape of TRiC ring-opening and its underlying stepwise mechanism revealed by cryo-EM. QRB Discov 6:e7. https://doi.org/10.1017/qrd.2024.17
3. Shen PS, Willardson BM (2025) Protein folding by the CCT/TRiC chaperone complex. Curr Opin Struct Biol 91:102999. https://doi.org/10.1016/j.sbi.2025.102999
4. Gvozdenov Z, Peng AYT, Biswas A, Barcutean Z, Gestaut D, Frydman J, Struhl K, Freeman BC (2024) TRiC/CCT chaperonin governs RNA polymerase II activity in the nucleus to support RNA homeostasis. bioXriv. https://doi.org/10.1101/2024.09.26.615188
5. Llorca O, McCormack EA, Hynes G, Grantham J, Cordell J, Carrascosa JL, Willison KR, Fernandez JJ, Valpuesta JM (1999) Eukaryotic type II chaperonin CCT interacts with actin through specific subunits. Nature 402:693–696. https://doi.org/10.1038/45294
6. Llorca O, Martín-Benito J, Ritco-Vonsovici M, Grantham J, Hynes GM, Willison KR, Carrascosa JL, Valpuesta JM (2000) Eukaryotic chaperonin CCT stabilizes actin and tubulin folding intermediates in open quasi-native

conformations. EMBO J 19(22):5971–5979. https://doi.org/10.1093/emboj/19.22.5971
7. Zeng C, Han S, Pan Y, Huang Z, Zhang B, Zhang B (2024) Revisiting the chaperonin T-complex protein-1 ring complex in human health and disease: a proteostasis modulator and beyond. Clin Transl Med 14(2):e1592. https://doi.org/10.1002/ctm2.1592
8. Cuéllar J, Ludlam WG, Tensmeyer NC, Aoba T, Dhavale M, Santiago C, Bueno-Carrasco MT, Mann MJ, Plimpton RL, Makaju A, Franklin S, Willardson BM, Valpuesta JM (2019) Structural and functional analysis of the role of the chaperonin CCT in mTOR complex assembly. Nat Commun 10: 2865. https://doi.org/10.1038/s41467-019-10781-1
9. Martin-Cofreces NB, Chichon FJ, Calvo E, Torralba D, Bustos-Moran E, Dosil SG, Rojas-Gomez A, Bonzon-Kulichenko E, Lopez JA, Otón J, Sorrentino A, Zabala JC, Vernos I, Vazquez J, Valpuesta JM, Sanchez-Madrid F (2020) The chaperonin CCT controls T cell receptor–driven 3D configuration of centrioles. Sci Adv 6:eabb7242. https://doi.org/10.1126/sciadv.abb7242
10. Rojas-Gómez A, Dosil SG, Chichón FJ, Fernández-Gallego N, Ferrarini A, Calvo E, Calzada-Fraile D, Requena S, Otón J, Serrano A, Tarifa R, Arroyo M, Sorrentino A, Pereiro E, Vázquez J, Valpuesta JM, Sánchez-Madrid F, Martín-Cófreces NB (2023) Chaperonin CCT controls extracellular vesicle production and cell metabolism through kinesin dynamics. J Extracell Vesicle 12:12333. https://doi.org/10.1002/jev2.12333
11. McLoughlin KJ, Pedrini E, MacMahon M, Guduric-Fuchs J, Medina RJ (2019) Selection of a real-time PCR housekeeping gene panel in human endothelial colony forming cells for cellular senescence studies. Front Med (Lausanne) 6:33. https://doi.org/10.3389/fmed.2019.00033

# Chapter 18

# Methods to Study Arginylation of Tubulin


## Brittany MacTaggart and Anna Kashina



## Abstract

Arginylation is a posttranslational modification in which arginine is transferred from a tRNA molecule onto a protein by arginyltransferase 1 (ATE1). Previous studies have found that *Ate1* knockout causes cytoskeletal defects at both the cellular and organismal levels, but the protein targets of arginylation that mediate these effects remain poorly defined. Some of the *Ate1* knockout phenotypes have been linked to β -actin arginylation, but more recently, the role of tubulin arginylation has become an active area of study. Here, we describe methods for studying the arginylation of tubulin both in vitro and from cell/tissue extracts.



**Keywords** Tubulin, Arginylation, Proteomics, Arginyltransferase, Posttranslational modification


## 1 Introduction

Arginylation is a posttranslational modification (PTM) mediated by arginyltransferase 1 (ATE1), in which arginine is transferred from a tRNA molecule onto N-terminal or internal aspartate or glutamate residues within a protein or peptide substrate (reviewed in [1]). In mice, ATE1 is essential for development [2–5] and facilitates the normal function of different physiological systems in adults [6–8]. However, despite the apparent importance of arginylation in both developmental progression and disease prevention, the specific protein targets that mediate these effects are poorly understood, largely due to the technical difficulty of arginylation identification.

Global screens for arginylation have shown that many cytoskeletal proteins are targets of ATE1 [9, 10]. Accordingly, *Ate1* deletion results in cytoskeletal defects at the cellular and organismal levels [3–5, 8, 11–13]. The most well-studied example of cytoskeletal regulation by arginylation is β-actin, which is N-terminally arginylated at the cell leading edge to promote actin polymerization and cell migration [11, 14–16]. However, β-actin is not the only arginylated cytoskeleton protein. Isoforms of α- and β-tubulin, which are known


Noa B. Martín-Cófreces et al. (eds.), *The Cytoskeleton*, Methods in Molecular Biology, vol. 3070,
https://doi.org/10.1007/978-1-0716-5515-3_18,

to be regulated by PTM [17], have also been identified as targets of arginylation [9, 10, 18].

In our recent work, we have identified arginylation of α-tubulin at Glu77 in mouse embryonic fibroblasts (MEFs) and functionally studied this modification. We found that this arginylation promotes microtubule dynamics by limiting the binding of the stabilizing protein MAP1S to microtubules [19]. This discovery adds arginylation to the list of PTMs that functionally regulate tubulin and microtubules.

Additional tubulin arginylation sites have been identified in various screens for arginylation using different tissues, including whole mouse embryos [9], mouse brain [10], mouse platelets [18], and human brain and iPSCs [20]. None of these sites has been validated or functionally studied yet. It appears likely that these additional arginylated sites may play new roles in microtubule regulation.

Here, we describe methods for studying the arginylation of tubulin, including identifying arginylation sites on endogenous tubulin from cell extracts [21, 22] and assaying tubulin arginylation in vitro [1]. In the protocol below, cultured cells are used as a source of microtubules, and mass spectrometry is performed on excised protein bands from SDS-PAGE. Alternatively, microtubules can also be obtained by purification from native tissues as described, e.g., in [23]. Mass spectrometry can be performed in solution on purified microtubules or tubulin preparations as described in [24, 25].

## 2 Materials

1. Cell lines for culturing (*see* **Note 1**)
2. Tissue culture media (specific to the cell type used)
3. Phosphate-buffered saline (PBS): 10 mM $Na_2HPO_4$, pH 7.4, 137 mM NaCl, 2.7 mM KCl, 1.8 mM $KH_2PO_4$
4. MME Buffer: 100 mM MES Free Acid pH 6.8, 1 mM $MgCl_2$, 1 mM EGTA
5. Protease inhibitor cocktail
6. Dithiothreitol (DTT)
7. Ultracentrifuge (such as Beckman TL-100)
8. Taxol
9. GTP
10. Sucrose
11. NaCl
12. SDS sample buffer
13. Reagents for SDS-PAGE/Bis–Tris electrophoresis

14. Coomassie Blue
15. 4× Assay Buffer: 200 mM HEPES pH 7.5, 100 mM KCl, 60 mM $MgCl_2$, 0.4 mM DTT
16. [$^{14}$C]-Arg (Moravek) (*see* **Note 2**)
17. RRS (*see* **Note 3**)
18. ATE1 (*see* **Note 3**)
19. $tRNA^{Arg}$ (*see* **Note 3**)
20. Tubulin (*see* **Note 4**)
21. ATP
22. SDS sample buffer
23. AutoFluor
24. Gel dryer
25. X-ray film

## 3 Methods

### *3.1 Identification of Arginylation Sites on Tubulin Purified From Cell Extracts*

1. Grow cultured cells in ten 10 cm dishes (*see* **Note 5**) to 80%–90% confluence (see Fig. 1).
2. Wash each plate once with 10 mL of PBS.
3. Collect the cells from each plate by scraping into 1 mL of PBS.
4. Pool the cells in PBS suspension into a 15 mL conical tube (*see* **Note 6**), and spin down at 1000 rpm for 5 min in a tabletop centrifuge (*see* **Note 7**). Discard the supernatant and collect the cell pellet. If needed, the cell pellet at this step can be frozen and stored before proceeding to Step 5 (*see* **Note 8**).
5. Resuspend the cell pellet in 1.5× volume of MME Buffer supplemented with 1× protease inhibitor cocktail and 1 mM DTT.
6. Transfer the suspension to a 1.5 mL microcentrifuge tube.
7. Place the tube on ice and sonicate for 3 s on, 3 s off for 20 times (*see* **Note 9**).
8. Transfer the crude lysate into ultracentrifuge tubes and centrifuge at 120,000×*g* for 1 h at 4°C.
9. Transfer the supernatant (SI) to a new microcentrifuge tube. Discard the cellular debris pellet (PI).
10. Add 20 μM Taxol and 1 mM GTP to the supernatant and incubate at 37°C for 30 min in a heat block to polymerize microtubules.

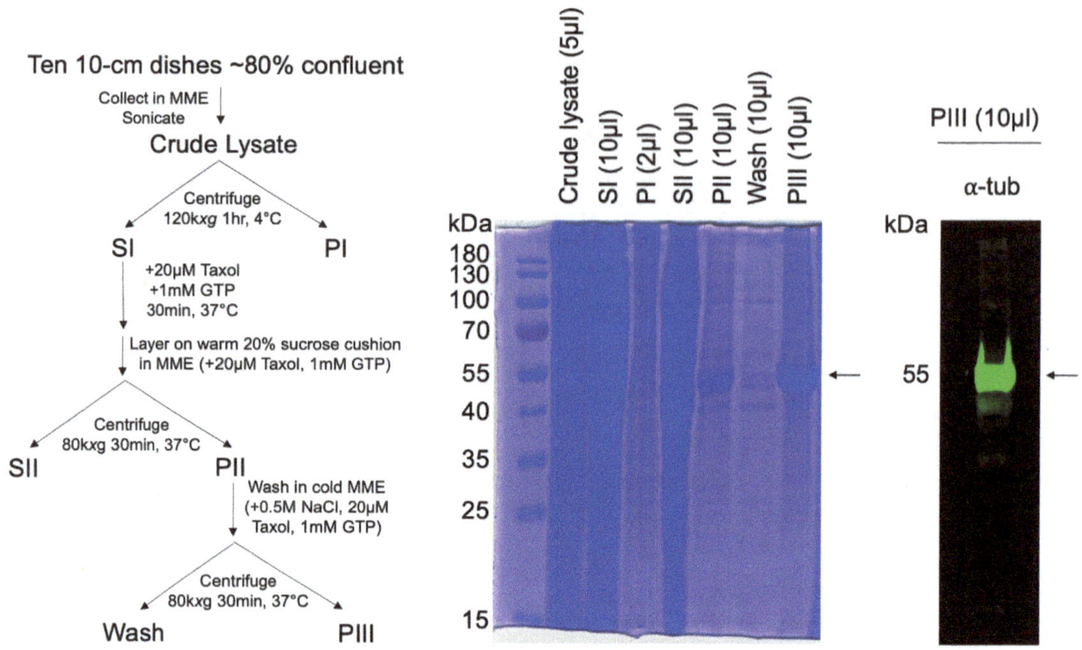


**Fig. 1** Isolation of tubulin from cell extracts for analysis by mass spectrometry. Left, a diagram of the workflow for preparing tubulin from cell extracts. Center, an SDS-PAGE gel stained with Coomassie Blue showing the purification steps. Right, a Western blot on the final fraction (PIII) probed for α-tubulin (green). Arrows represent the tubulin band

11. Prepare a 20% sucrose solution in MME buffer supplemented with 20 μM Taxol and 1 mM GTP. Pre-warm this solution to 37°C.
12. Transfer 300 μL of the pre-warmed sucrose solution into an ultracentrifuge tube, and carefully layer the reaction mixture over the cushion.
13. Centrifuge at 80,000 × *g* for 30 min at 37°C to pellet the polymerized microtubules.
14. Sequentially remove and discard the supernatant (SII) and sucrose cushion. Collect the microtubule pellet (PII).
15. Wash the microtubule pellet (PII) in 500 μL of cold MME Buffer supplemented with 0.5 M NaCl, 20 μM Taxol, and 1 mM GTP.
16. Centrifuge at 80,000 × *g* for 30 min at 37°C. Remove and discard the supernatant (Wash) (*see* **Note 10**). Collect the washed microtubule pellet (PIII).
17. Resuspend the microtubule pellet (PIII) in 200 μL of MME buffer. Add an equal volume of 2× SDS sample buffer, run on a 4%–12% Bis–Tris protein gel, and stain with Coomassie prior to mass spectrometry.

18. Excise the band of interest (~50 kDa) for mass spectrometry (*see* **Note 11**).
19. Search the mass spectrometry data with full-tryptic or semi-tryptic specificity against an appropriate proteome database (e.g., Swiss-Prot mouse proteome database for MEFs) and a common contaminant database using MaxQuant with the protein and peptide false discovery rates set to 1%.
20. To identify arginylation sites, search for modifications with a mass addition of 156.1011 (R), 170.1168 (R+methyl), and 184.1324 (R+dimethyl) on the protein N-terminus, aspartate, and glutamate residues (*see* **Note 12**).

### 3.2 In Vitro Tubulin Arginylation Assay

1. On ice, mix a 100 μL reaction containing 1× Assay Buffer, 12.5 μM [$^{14}$C]-Arg, 40 μM tRNA$^{Arg}$, 2 μM RRS, 1 μM ATE1, and 8 μM tubulin.
2. Carry out the reaction in a heat block at 37°C for 4 h (*see* **Note 13**).
3. Stop the reaction by adding an equal volume of 2× SDS sample buffer and heating to 95°C for 5 min.

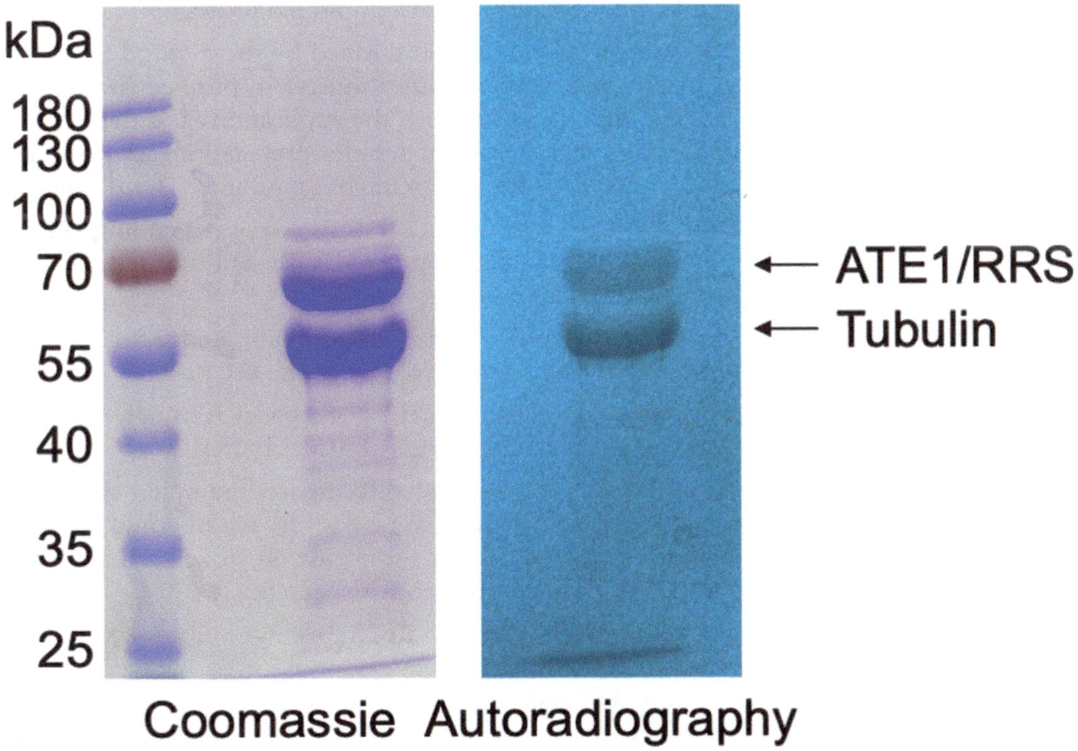


**Fig. 2** Assaying in vitro arginylation of tubulin. Coomassie gel (left) and autoradiography film (right) of an in vitro arginylation reaction performed using porcine brain tubulin

4. Run the samples on a 1 mm 12% SDS-PAGE gel and stain with Coomassie Blue.
5. Incubate the gels with AutoFluor for 1 h on a rocker.
6. Dry the gels at 80°C for 1 h.
7. Expose the dried gels to film at −80°C for 10–14 days before developing (see Fig. 2 and see **Note 14**).

## 4 Notes

1. When possible, it is recommended to collect samples from *Ate1* knockout cells (or tissues) as a negative control to ensure that all identified arginylation sites are genuine ATE1-dependent targets (as opposed to mass ambiguities that may fall below the mass spectrometry resolution).
2. [$^{14}$C]-Arg is used for detecting arginylation by autoradiography, but other types of labeled Arg could also be used (e.g., [$^{13}$C,$^{15}$N]-Arg can be used to identify arginylation by mass spectrometry).
3. Recombinant ATE1, RRS, and tRNA$^{Arg}$ expressed in *E. coli* can be purified according to published protocols [1].
4. In addition to the tubulin purified by the method described above in this protocol, other sources of purified tubulin can also be used, depending on the applications and the goals of the experiment. Different tubulin preparations are commercially available, such as porcine brain tubulin (Cytoskeleton).
5. The number of dishes needed may vary depending on cell type. Ten 10 cm dishes provide sufficient starting material when using MEFs.
6. Pre-mark the conical tube with a 0.1 mL and 0.2 mL line to more easily estimate the volume of the cell pellet.
7. For ten 10 cm dishes of MEFs, the pellet is usually ~200 μL, which is dissolved in 300 μL of MME buffer.
8. Although it is recommended to use fresh samples, cell pellets obtained in Step 4 can also be flash frozen in liquid nitrogen and stored at −80°C for future use. If frozen, rapidly thaw the cell pellets and proceed with Step 5.
9. Starting with the crude lysate (following sonication), the progress of the purification can be monitored by keeping 10 μ L samples at each step (crude lysate, SI, PI, SII, PII, Wash, and PIII) to run them on an SDS-PAGE gel, and stain with Coomassie Blue and/or analyze by Western blotting for α -tubulin (Abcam ab7291, 1:2000) (Fig. 1). The major

~55 kDa protein band corresponding to tubulin (sometimes visible as a doublet for a- and b-tubulin) should be visible in cell extracts and exclusively enriched in PII and PIII fractions.

10. The Wash fraction at this stage contains the microtubule-associated proteins (MAPs), which are eluted off the microtubules by 0.5 M NaCl in the wash buffer. These full wash fractions can also be saved, and TCA precipitated to be assessed by mass spectrometry or Western blot for protein composition and potential arginylation.
11. Procedures for mass spectrometry may vary by facility. Typically, proteins are reduced with Tris(2-carboxyethyl) phosphine, alkylated with iodoacetamide, digested with trypsin, and analyzed using a 90 min LC–MS/MS run on a Thermo Q Exactive Plus mass spectrometer.
12. The identification of any arginylated peptides should be confirmed by using standard unmodified and arginylated peptides.
13. To obtain a time course and look for the dynamics of arginylation over time, take a 10 μL sample of the arginylation reaction at different time points (e.g., 0, 10, 20, 30, 60, 120, and 240 min), mix with an equal volume of 2× SDS sample buffer, and heat to 95°C for 5 min for the subsequent analysis by SDS-PAGE and autoradiography.
14. The exact exposure time depends on the strength of the signal, which can vary between different batches of proteins and reagents. Since $^{14}C$ is an extremely long-lived radioactive isotope, gel can be reexposed to film multiple times as needed to develop the optimal signal.

## References


1. Kashina AS (2023) Protein Arginylation. Springer US, New York, NY
2. Kwon YT, Kashina AS, Davydov IV et al (2002) An essential role of N-terminal arginylation in cardiovascular development. Science 297:96. https://doi.org/10.1126/science.1069531
3. Kurosaka S, Leu NA, Zhang F et al (2010) Arginylation-dependent neural crest cell migration is essential for mouse development. PLoS Genet 6:e1000878. https://doi.org/10.1371/journal.pgen.1000878
4. Wang J, Pavlyk I, Vedula P et al (2017) Arginyltransferase ATE1 is targeted to the neuronal growth cones and regulates neurite outgrowth during brain development. Dev Biol 430:41–51. https://doi.org/10.1016/j.ydbio.2017.08.027
5. Rai R, Wong CCL, Xu T et al (2008) Arginyltransferase regulates alpha cardiac actin function, myofibril formation and contractility during heart development. Development 135:3881–3889. https://doi.org/10.1242/dev.022723
6. Rai R, Zhang F, Colavita K et al (2016) Arginyltransferase suppresses cell tumorigenic potential and inversely correlates with metastases in human cancers. Oncogene 35:4058–4068. https://doi.org/10.1038/onc.2015.473
7. Wang J, Han X, Leu NA et al (2017) Protein arginylation targets alpha synuclein, facilitates normal brain health, and prevents neurodegeneration. Sci Rep 7:1–14. https://doi.org/10.1038/s41598-017-11713-z

8. Kurosaka S, Leu NA, Pavlov I et al (2012) Arginylation regulates myofibrils to maintain heart function and prevent dilated cardiomyopathy. J Mol Cell Cardiol 53:333–341. https://doi.org/10.1016/j.yjmcc.2012.05.007
9. Wong CCL, Xu T, Rai R et al (2007) Global analysis of posttranslational protein arginylation. PLoS Biol 5:e258. https://doi.org/10.1371/journal.pbio.0050258
10. MacTaggart B, Shimogawa M, Lougee M et al (2023) Global analysis of post-translational side-chain arginylation using pan-arginylation antibodies. Mol Cell Proteomics 22:100664. https://doi.org/10.1016/j.mcpro.2023.100664
11. Karakozova M, Kozak M, Wong CCL, et al. (2006) Arginylation of beta-actin regulates actin cytoskeleton and cell motility. Science 313:192–196. https://doi.org/10.1126/science.1129344
12. Zhang F, Saha S, Kashina A (2012) Arginylation-dependent regulation of a proteolytic product of talin is essential for cell-cell adhesion. J Cell Biol 197:819–836. https://doi.org/10.1083/jcb.201112129
13. Batsios P, Ishikawa-Ankerhold HC, Roth H et al (2019) Ate1-mediated posttranslational arginylation affects substrate adhesion and cell migration in *Dictyostelium discoideum*. Mol Biol Cell 30:453–466. https://doi.org/10.1091/mbc.E18-02-0132
14. Saha S, Mundia MM, Zhang F et al (2010) Arginylation regulates intracellular actin polymer level by modulating actin properties and binding of capping and severing proteins. Mol Biol Cell 21:1350–1361. https://doi.org/10.1091/mbc.e09-09-0829
15. Pavlyk I, Leu NA, Vedula P et al (2018) Rapid and dynamic arginylation of the leading edge β-actin is required for cell migration. Traffic 19:263–272. https://doi.org/10.1111/tra.12551
16. Varland S, Vandekerckhove J, Drazic A (2019) Actin post-translational modifications: the cinderella of cytoskeletal control. Trends Biochem Sci 44:502–516. https://doi.org/10.1016/j.tibs.2018.11.010
17. Roll-Mecak A (2020) The tubulin code in microtubule dynamics and information encoding. Dev Cell 54:7–20. https://doi.org/10.1016/j.devcel.2020.06.008
18. Wang J, Han X, Wong CCL et al (2014) Arginyltransferase ATE1 catalyzes midchain arginylation of proteins at side chain carboxylates in vivo. Chem Biol 21:331–337. https://doi.org/10.1016/j.chembiol.2013.12.017
19. MacTaggart B, Wang J, Tang H-Y, Kashina A (2025) Arginylation of α-tubulin at E77 regulates microtubule dynamics via MAP1S. J Cell Biol 224. https://doi.org/10.1083/jcb.202406099
20. Lin Z, Xie Y, Gongora J et al (2024) An unbiased proteomic platform for activity-based arginylation profiling. bioRxiv Prepr Serv Biol https://doi.org/10.1101/2024.06.01.596974
21. Miller LM, Xiao H, Burd B et al (2010) Methods in tubulin proteomics. In: Methods in cell biology. pp 105–126
22. Liu N, Xiong Y, Ren Y et al (2015) Proteomic profiling and functional characterization of multiple post-translational modifications of tubulin. J Proteome Res https://doi.org/10.1021/acs.jproteome.5b00308
23. Gell C, Friel CT, Borgonovo B et al (2011) Purification of tubulin from porcine brain. Methods Mol Biol 777:15–28. https://doi.org/10.1007/978-1-61779-252-6_2
24. Kashina AS, Yates JR (2015) Identification of arginylated proteins by mass spectrometry. Methods Mol Biol 1337:93–104. https://doi.org/10.1007/978-1-4939-2935-1_13
25. Wang J, Yates JR, Kashina A (2019) Biochemical analysis of protein arginylation. Methods Enzymol 626:89–113. https://doi.org/10.1016/bs.mie.2019.07.028

# Chapter 19

# Posttranslational Dynamics of the Subcellular Proteome by Mass Spectrometry-Based Proteomics

Ana Martínez-Val, Jose Manuel Rodríguez, Jesper V. Olsen and Jesús Vázquez

## Abstract

Localization of proteins to different cell compartments is a major posttranslational regulatory mechanism that eukaryotic cells have evolved to coordinate protein homeostasis and responses to stimuli. Subcellular fractionation allows the separation of distinct protein populations from cellular compartments such as the cytosol, cytoskeleton, membrane, endoplasmic reticulum, or nucleolus. This fractionation can be followed by mass spectrometry-based proteomics to provide insights into spatiotemporal regulation or cellular protein networks. Here, we describe a high-throughput workflow based on sequential cell lysis, which enables the profiling of subcellular proteome architecture and the detection of changes in protein localization at global and posttranslational levels.

**Keywords** Subcellular fractionation, Posttranslational regulation, Mass spectrometry, Cell signalling, Proteomics

## 1 Introduction

Cell homeostasis is tightly controlled by a vast network of proteins, whose function is controlled by a myriad of mechanisms, spanning from translation rate control to posttranslational modifications. Posttranslational protein regulation comprehends mechanisms to modulate protein abundance (i.e., targeted degradation) or function, such as chemical modifications like phosphorylation or acetylation, or by spatial localization of proteins. The spatial distribution of proteins within the cell is crucial for their normal function, and also for other mechanisms such as signal transduction. In fact, some proteins are known to perform different functions depending on their subcellular localization, a mechanism known as protein moonlighting [1].

There are multiple strategies to study the subcellular localization of proteins. Traditionally, molecular biology-based imaging techniques have provided great insight into the subcellular proteome

Noa B. Martín-Cófreces et al. (eds.), *The Cytoskeleton*, Methods in Molecular Biology, vol. 3070,
https://doi.org/10.1007/978-1-0716-5515-3_19,

landscape [2]. However, these techniques can only monitor a limited number of proteins per assay. Recent techniques, such as proximity-labeling, have been developed to provide very detailed localization insights at the organelle level [3]. However, these strategies are limited in throughput and proteome depth, and therefore, are not suitable to analyze the full proteome landscape or complex experimental designs with multiple replicates and/or biological conditions. Subcellular protein fractionation combined with mass spectrometry (MS)-based proteomics offers a powerful platform for such analysis. Several strategies have been developed for this purpose. These include LOPIT-DC [4] and SubCellBarcode [5], which have been used to map over 8000 proteins in different cell lines. However, they need complex protocols that require differential centrifugation as well as extensive processing and long analysis time.

In this work, we present a detailed protocol to perform subcellular fractionation followed by MSproteomics using data independent acquisition (DIA) as shown by Martinez-Val et al. [6], which allows for high-throughput analysis of the spatiotemporal regulation of the proteome. Our protocol includes guidelines to perform the subsequent quantitative analysis using DIA-NN [7] and iSanXoT [8–10].

This sequential subcellular fractionation protocol, combined with DIA, enables rapid, sensitive, and reproducible mapping of the subcellular proteome. As a result of this protocol, each cell is fractionated into six subcellular compartments, enriched in proteins from definite subcellular neighborhoods. Fraction 1 is enriched in soluble cytosolic proteins, fraction 2 is enriched in non-soluble cytosolic components and protein complexes, fraction 3 contains proteins from the lumen of membrane organelles, fraction 4 comprises membrane-bound proteins, fraction 5 is mostly comprised of nuclear and chromatin-associated proteins, and fraction 6 contains mainly nucleolar proteins (Fig. 1). Cytoskeleton proteins distribute mainly between fractions 1 and 2.

## 2 Materials

### 2.1 Solutions for Subcellular Fractionation

1. Buffer A: 30 mM HEPES, pH 7.4, 15 mM NaCl, 2 mM $MgCl_2$, 1 mM EDTA. To prepare 200 ml, mix 6 ml of 1 M HEPES, pH 7.4, 600 µl of 5 M NaCl, 1600 µl of 250 mM $MgCl_2$, and 400 µl of 500 mM EDTA. Complete to 200 ml with milliQ water. It can be stored up to 1 year at 4°C.
2. Buffer AS: 30 mM HEPES, pH 7.4; 15 mM NaCl, 2 mM $MgCl_2$, 1 mM EDTA, 350 mM sucrose. To prepare 200 ml, add 23.96 g of sucrose to a measuring cylinder. Add 6 ml of 1 M HEPES, pH 7.4, 600 µl of 5 M NaCl, 1600 µl of 250 mM

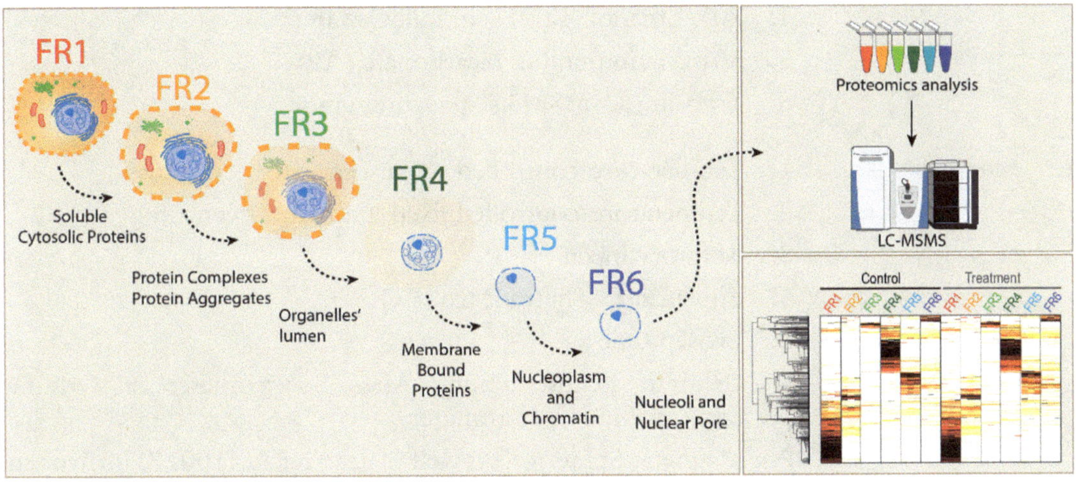


**Fig. 1** Schematic representation of the subcellular fractionation protocol. Fractionation was performed by sequential lysis of the cell to obtain six fractions, followed by LC–MSMS proteomicanalysis to derive the protein abundance profiles across fractions

$MgCl_2$, and 400 µl of 500 mM EDTA. Complete to 200 ml with milliQ water. It can be stored up to 1 year at 4°C.

3. Buffer AG: 30 mM HEPES, pH 7.4, 15 mM NaCl, 2 mM MgCl2, 1 mM EDTA, 20% glycerol. To prepare 200 ml of buffer AG, mix 6 ml of 1 M HEPES pH 7.4, 600 µl of 5 M NaCl, 1600 µl of 250 mM MgCl2, 400 µl of 500 mM EDTA, and 47 ml of 85% glycerol. Complete to 200 ml with milliQ water. It can be stored up to 1 year at 4°C.
4. 0.15% digitonin: take 180 µl of 5% digitonin solution and resuspend it in 5820 ml of milliQ water. Prepare aliquots of 600 µl, and store at −20°C.
5. 10% N-Dodecyl-β-D-maltoside (DDM). Store at 4°C.
6. 10% Tween: mix Tween®20 prepared with milliQ water in a ratio 1:9 (v/v). Store at 4°C.
7. 10% SDS: mix 20% SDS prepared with milliQ water in a ratio of 1:1 (v/v). Store at room temperature.
8. 5 mM sodium chloride.
9. 1.4 mM sodium chloride.
10. Benzonase nuclease 250 units/µl.

### 2.2 Solutions and Reagents for Protein Aggregation Capture (PAC)-Based Digestion

1. Denaturing buffer: 5% SDS, 83 mM Tris (2-Carboxyethyl) phosphine (TCEP), 167 mM iodo-acetamide (IAM).
2. MagResyn® Hydroxyl beads
3. 95% Acetonitrile (ACN) in milliQ water.

4. 70% Ethanol (EtOH) in milliQ water
5. 50 mM ammonium bicarbonate (ABC)
6. Trypsin LC–MSMS sequencing grade

### 2.3 Equipment

1. Temperature-controlled swing-out rotor centrifuge
2. Temperature-controlled fixed-angle rotor centrifuge
3. Rotatory mixer
4. Eppendorf® ThermoMixer® C
5. Evosep One
6. Orbitrap Eclipse Tribrid Mass Spectrometer or Orbitrap Exploris Mass Spectrometer
7. Magnetic Stand for 96 well plates (ref.AM10027, invitrogen) or magnetic rack for eppendorf tubes

### 2.4 Software

1. DIA-NN version 1.8.1 or superior
2. R v 4.3.1.
3. iSanXoT v 2.0.0 or superior

## 3 Methods

### 3.1 Subcellular Fractionation

1. Prepare subcellular fractionation buffers fresh and keep them at 4°C throughout the whole protocol (*see* **Note 1**). Add phosphatase and protease inhibitors to each buffer A, AS, and AG. For 10 ml of buffer: 20 μl of 500 mM TCEP, 20 μl of 500 mM Sodium fluoride, 100 μl of β-glycerol-phosphate, 100 μl of sodium orthovanadate, 1 tablet of cOmplete, Mini, EDTA-free Protease Inhibitor Cocktail (*see* **Note 2**).
2. Prior to starting the protocol, thaw an aliquot of 0.15% digitonin. Once thawed, heat at 95°C for 5 min. Cool down to room temperature before proceeding.
3. Collect the cells. The ideal starting amount of cells is between 10 and 20 million cells, or equivalent to a P15 culture plate. If cells are adherent, trypsinize them. Wash the cell pellet twice with 5 ml of cold PBS. After the second wash, pellet the cells, discard the supernatant, and proceed immediately to the next step. Every subsequent step until Step 9 must be performed at 4°, or on ice.
4. Prepare fraction 1:
   i. Resuspend the cell pellet in 540 μl buffer AS.
   ii. Add 60 μl of 0.15% digitonin.

iii. Rotate for 30 min in a cold room at 4°C (*see* **Note 3**).

iv. Spin down, 3 min, 500×*g* in a swing-out rotor centrifuge at 4°C.

v. Recover supernatant to a clean tube and label "Fraction 1."

vi. Keep on ice.

vii. Wash the cell pellet twice with 1 ml of buffer AS.

viii. Each time, resuspend the cell pellet thoroughly but gently by slowly pipetting up and down.

ix. Spin down, 3 min, 500×*g* in a swing-out rotor centrifuge at 4°C and discard supernatants.

x. Proceed with the pellet to prepare fraction 2.

5. Prepare fraction 2:
   i. Resuspend the cell pellet in 540 μl buffer AS.
   ii. Add 60 μl of 1.4 M NaCl.
   iii. Rotate for 15 min in a cold room at 4°C.
   iv. Spin down, 3 min, 500×*g* in a swing-out rotor centrifuge at 4°C.
   v. Recover the supernatant to a clean tube and label "Fraction 2." Keep on ice.
   vi. Wash the cell pellet twice with 1 ml of buffer AS. Each time, resuspend the cell pellet thoroughly but gently by slowly pipetting up and down.
   vii. Spin down, 3 min, 500×*g* in a swing-out rotor centrifuge at 4°C and discard supernatants.
   viii. Proceed with the pellet to prepare fraction 3.

6. Prepare fraction 3:
   i. Resuspend the cell pellet in 570 μl buffer AS.
   ii. Add 30 μl of 10% Tween-20.
   iii. Rotate for 15 min in a cold room at 4°C.
   iv. Spin down, 3 min, 500×*g* in a swing-out rotor centrifuge at 4°C.
   v. Recover supernatant to a clean tube and label "Fraction 3." Keep on ice.
   vi. Wash the cell pellet twice with 1 ml of AS wash.
   vii. Each time, resuspend the cell pellet thoroughly but gently by slowly pipetting up and down.

viii. Spin down, 3 min, 500 × *g* in a swing-out rotor centrifuge at 4°C and discard the supernatants.

ix. Proceed with the pellet to prepare fraction 4.

7. Prepare fraction 4:

   i. Resuspend the cell pellet in 540 μl buffer AG.

   ii. Add 60 μl of 10% DDM.

   iii. Rotate for 15 min in a cold room at 4°C.

   iv. Spin down, 3 min, 500 × *g* in a swing-out rotor centrifuge at 4°C.

   v. Recover supernatant to a clean tube and label "Fraction 4." Keep on ice.

   vi. Wash the cell pellet twice with 500 μl of buffer AG. Each time, resuspend the cell pellet thoroughly but gently by slowly pipetting up and down.

   vii. Spin down, 3 min, 500 × *g* in a swing-out rotor centrifuge at 4°C and discard the supernatants.

   viii. Proceed with the pellet to prepare fraction 5.

8. Prepare fraction 5:

   i. Resuspend the cell pellet in 540 μl buffer A

   ii. Add 1 μl benzonase and 60 μl of 5 M NaCl.

   iii. Rotate for 15 min in a cold room at 4°C.

   iv. Spin down, 3 min, 2000 × *g* in a fixed-angle rotor centrifuge at 4°C.

   v. Recover supernatant to a clean tube and label "Fraction 5." Keep on ice.

   vi. Wash the cell pellet once with 500 μl of buffer AG. Each time, resuspend the cell pellet thoroughly but gently by slowly pipetting up and down.

   vii. Spin down, 3 min, 2000 × *g* in a fixed-angle rotor centrifuge at 4°C and discard supernatants.

   viii. Proceed with the pellet to prepare fraction 6.

9. Prepare fraction 6:

   i. Resuspend the cell pellet in 522 μl Buffer A.

   ii. Add 60 μl of 1.4 M NaCl and 18 μl of 10% SDS.

   iii. Boil for 10 min at 95°C. Cool down to room temperature.

   iv. Label as "Fraction 6."

10. Spin all fractions in a fixed-angle rotor centrifuge for 5 min at 24,000 × *g* at 4°C (*see* **Note 4**).

11. Transfer fractions to clean tubes and store at −80°C or proceed to Step 3.2 (*see* **Note 5**).

### 3.2 Protein Aggregation Capture-Based digestion

Prepare the denaturing buffer as indicated in the Sect. 2.2.

1. To denature, reduce, and alkylate the samples, add 38 µl of denaturing buffer to each one of the samples so the final concentration of each sample is 0.3% SDS, 5 mM TCEP, and 10 mM IAM. Keep samples protected from light.
2. Incubate samples for 10 min at 95°C while mixing 1200 rpm and protected from light.
3. Store at −80°C or proceed to the following step.
4. For the protein digestion, add 300 µl of each sample to a low-bind eppendorf tube or a well from a low-bind 1 mL 96-well plate. Store the remaining 300 µl at −80°C.
5. Add 700 µl of 100% acetonitrile to each well.
6. Add 50 µl hydroxyl beads to each well (*see***Note 6**).
7. Mix with a multichannel pipette 15 times.
8. Incubate for 10 min.
9. Mix with a multichannel pipette 15 times.
10. Incubate for 10 min.
11. Place the plate on top of the magnet (*see* **Note 7**). Wait 30 s to ensure that all beads have attached to the magnet. Aspirate supernatant. Without removing the plate from the magnet, proceed to perform the washes.
12. Add 900 µl of 95% ACN.
13. Aspirate supernatant.
14. Repeat wash with 95% ACN for a total of three washes.
15. Add 900 µl of 70% EtOH.
16. Aspirate supernatant.
17. Repeat wash with 70% EtOH for a total of two washes.
18. Prepare digestion buffer by mixing 50 mM ABC with trypsin. If trypsin is lyophilized, resuspend it in 50 mM acetic acid. For a full plate (96 samples), use 100 µg of trypsin. Resuspend trypsin in 100 µl of acetic acid and add to 30 ml to 50 mM ABC.
19. Add 300 µl of digestion buffer to each sample, make sure the beads are softened, and mix evenly during shaking.
20. Place the cover lid on top.
21. Incubate for 12 h on a Thermomixer® using the following program: loops of 45 s of mixing at 1200 rpm followed by 6 min without agitation. Keep at 37°C for 12 h, and turn to 4°C afterward.
22. Centrifuge the plates at 400 × *g* for 10 s.

23. Place the sample plate on top of a magnet, and transfer samples to a new plate. Add 10 µl of 25% TFA to each sample and store them at −20°C or proceed to Step 3.3.

### 3.3 Preparation of Evotips for LC–MSMS (see Note 8)

1. Add 20 µl of 0.1% formic acid in acetonitrile on top of each Evotip.
2. Centrifuge for 1 min at 800*g* on a swing-out rotor centrifuge.
3. Place the Evotip on top of a reservoir containing 2-propanol.
4. Add 20 µl of 0.1% formic acid in LC–MSMS water on top of the Evotip.
5. Centrifuge for 1 min at 800*g* on a swing-out rotor centrifuge.
6. Add 20 µl of the acidified sample on top of the Evotip.
7. Centrifuge for 1 min at 800*g* on a swing-out rotor centrifuge.
8. Add 20 µl of 0.1% formic acid in LC–MSMS water on top of the Evotip.
9. Centrifuge for 1 min at 800*g* on a swing-out rotor centrifuge.
10. Add 200 µl of 0.1% formic acid in LC–MSMS water on top of the Evotip.
11. Centrifuge for 10 s at 800*g* on a swing-out rotor centrifuge.
12. Add of 0.1% formic acid in LC–MSMS water at the bottom of the Evotip box to cover the bottom of the tips.
13. Proceed to LC–MSMS analysis or store tips at 4°C.

### 3.4 LC–MSMS Analysis Using DIA

The acquisition parameters for the method should be adapted to the LC–MS instruments available at the laboratory. To maximize the proteome coverage when doing this analysis, we recommend employing a data independent acquisition (DIA) strategy. The following parameters are recommended for an Evosep One interfaced with a Thermo Orbitrap Eclipse.

1. LC–MSMS analysis: separate samples using an Endurance Evosep column (15 cm–150 µm-C18 1.6 µm) interfaced with the Orbitrap Eclipse Tribrid Mass Spectrometer (Thermo Scientific) using an EasySpray Ion Source. Set spray voltage to 2.1 kV, funnel RF level at 40, and heated capillary temperature at 310°C.
2. The recommended gradient to achieve the best compromise between coverage and throughput is 30 samples per day (SPD), which is equivalent to 44 min of chromatographic separation (*see* **Note 9**).

### 3.5 Protein Identification and Extraction of Subcellular Profiles

1. For protein identification and quantification using DIA-NN, download DIA-NN from https://github.com/vdemichev/DiaNN (*see* **Note 10**).

2. Load your raw files in the input window.
3. Click on "Add FASTA" and load your database of interest.
4. Under "Precursor ion generation," click on "FASTA digest for library-free search/library generation" and "Deep learning-based spectra, RTs and IMs prediction."
5. Select "Trypsin/P" in "Protease". Choose up to 2 missed cleavages, 1 maximum number of variable modifications.
6. Click on "N-term M excision," "C carbamidomethylation," and "Ox (M)."
7. Adjust the precursor m/z range and fragment ion m/z range depending on your MS instrument acquisition settings.
8. In "Main output," indicate the folder where you want to store the results.
9. In "Algorithm", leave the default parameters except for "Cross-run normalization," which should be set to "Off."
10. Click on "Run."
11. Once the DIA-NN search is finished, go to the results folder (indicated in Step "8"). The tables "report.pr_matrix.tsv" and "report.pg_matrix.tsv" contain, respectively, the quantitativedata at precursor and protein level that can be used for subsequent analysis (*see* **Note 11**).
12. In case there is an abundance bias between replicates, it is required to normalize protein abundance across samples. To do so, transform the data to logarithm in base 2 and perform sample normalization using the "normalizecyclicloess" function from the "limma" package in R (Fig. 2).
13. Using any table-processing software (i.e., Excel, Perseus, R), import the table generated in the previous step into the software of your choice to calculate the scale intensity of each protein per fraction following the calculation shown in the example in Fig. 3A. As a result, each protein can be depicted as a scaled barplot (Fig. 3B).
14. Transform the previously log2-transformed data to raw values (*see* **Note 12**).
15. Replace NA values with 0.
16. Calculate the sum of all the raw intensities across fractions per protein.
17. Divide the intensity of each fraction by the sum.
18. Using organelle markers as a validation of fractionation protocol: one possible strategy to validate the fractionation protocol is to download well-known organelle markers and evaluate their fractionation profile (*see* **Note 13**).
19. Download "pRoloc" library from Bioconductor [11].

```
#1. Install limma, if not already installed
if (!require("BiocManager", quietly = TRUE))
  install.packages("BiocManager")
BiocManager::install("limma")
#2. Load limma library
library(limma)
#3. Indicate your working directory: the folder where your DIA-NN results are stored.
#If working in Windows Operative System, remember to edit your folder path to change "\" to "\\"
setwd("D:\\TEMP\\DIANN")
#4. Read protein quantification results from DIA-NN
input<-read.table("report.pg_matrix.tsv", head=T, sep="\t", quote="")
#5. Log2-transform the data. Indicate the columns that contain the quantitative data between brackets.
#In this examples, are columns 6 to 23.
#This example correspond to 3 replicates of 6 subcellular fractions.
input[,6:23]<-log2(input[,6:23])
#6. Plot the intensity distribution to evalute if furhter normalization is required.
boxplot(input[,6:23])
#7. Normalize the data and plot it again to evaluate that the normalization has corrected the sample bias.
#Important: normalization must be performed separately for each fraction.
# Fraction 1
input[,6:8]<-normalizeCyclicLoess(input[,6:8])
# Fraction 2
input[,9:11]<-normalizeCyclicLoess(input[,9:11])
# Fraction 3
input[,12:14]<-normalizeCyclicLoess(input[,12:14])
# Fraction 4
input[,15:17]<-normalizeCyclicLoess(input[,15:17])
# Fraction 5
input[,18:20]<-normalizeCyclicLoess(input[,18:20])
# Fraction 6
input[,21:23]<-normalizeCyclicLoess(input[,21:23])
boxplot(input[,6:23])
#8. Export the normalized dataset.
write.table(input, "report.pg_matrix_log2_normalized.tsv", sep="\t", row.names = F)
```

**Fig. 2** Example of code to perform sample-to-sample normalization using the loess function from the limma package in R. In this example, there are three experimental replicates

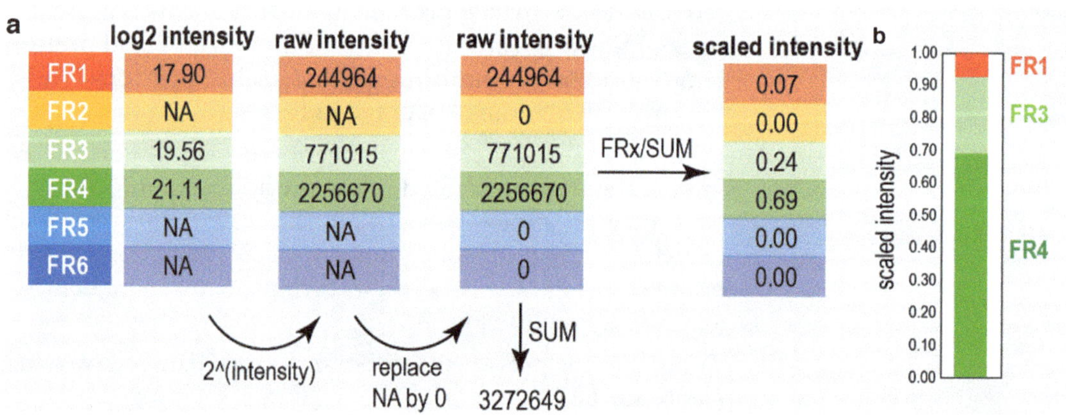


**Fig. 3** Schematic representation of the mathematical transformation required to scale the intensities across fractions. (**a**) Mathematical transformation to scale the intensities across fractions to 1 in order to obtain the abundance profile of each protein, as shown in (**b**)

20. Extract the markers from your organism of interest using the function "pRolocmarkers" and indicating the organism. For humans, use "hsap." As a result, you will obtain a relationship of known markers of 12 subcellular compartments and the corresponding Uniprot Identifier (Fig. 4).

```
if (!require("BiocManager", quietly = TRUE))
  install.packages("BiocManager")
BiocManager::install("pRoloc")
library(pRoloc)
#For Homo sapiens: "hsap"
hs_markers<-as.data.frame(pRolocmarkers("hsap"))
head(hs_markers)

                pRolocmarkers("hsap")
P08865          40S Ribosome
P0CW22          40S Ribosome
P15880          40S Ribosome
P22090          40S Ribosome
P23396          40S Ribosome
P25398          40S Ribosome

table(pRolocmarkers("hsap"))

 40S Ribosome        60S Ribosome   actin cytoskeleton              Cytosol Endoplasmic Reticulum      Golgi Apparatus
           33                  46                   45                   77                    92                   34
     Lysosome        Mitochondria              Nucleus           PEROXISOME             plasma mem           Proteasome
           42                 134                  116                   27                   190                   36
```

**Fig. 4** Example of the code and its output used to extract the organelle markers for the human proteome contained in the pRoloc R package

21. Map the markers to the data obtained in Step 3.5.17, and plot the protein profile of the different subcellular compartments (*see* **Note 14**).

#### *3.6 Differential Regulation Analysis by Subcellular Fraction*

1. Prepare the data: the quantity matrices at precursor level from DIANN's report (see Step 3.5) can be directly used by iSanXoT for quantification and statistical analysis (*see* **Note 15**).
2. Download iSanXoT. The multiple releases are available in the "release" section, which can be found at the following link: https://github.com/CNIC-Proteomics/iSanXoT/releases. The iSanXoT Windows distribution is packaged in an NSIS Launcher (exe file) (*see* **Notes 16** and **17**).
3. Download and uncompress the iSanXoT workflow template ("WPP_DIA") from: https://raw.githubusercontent.com/CNIC-Proteomics/iSanXoT/master/docs/templates/WPP_DIA.zip (*see* **Note 18**).
4. Run iSanXoT: from the Start menu, choose "Programs > iSanXoT" or alternatively double-click the iSanXoT desktop icon (*see* **Note 19**).
5. Create a project by choosing "New Project" from the "Project" menu.
6. Provide a name of your choice for the project folder and indicate a path to locate this folder, then click the "Submit" button.
7. Import the WPP_DIA workflow template using the "Import Workflow" submenu in the "Project" menu. You need to provide the root folder.
8. At the "Input files" tab, go to "Provide the Identification/Quantification file (ID-q)", click on "Choose identification

file" and select the "report.pr_matrix.tsv" file obtained in Step 3.5.11.

9. The task tables for the required modules in this DIA workflow have been filled automatically through the imported workflow. The filled modules are: "RELS_CREATOR" in the "Relation tables" tab; "LEVEL_CREATOR," "LEVEL_CALIBRATOR," and "INTEGRATE" in the "Basic modules" tab; and "REPORT" in the "Reports" tab (*see* **Note 20**).
10. The "REPORT" module generates the resulting reports from the analysis.
11. Click on "Save and Run" to execute the iSanXoT workflow. Repeat for each subcellular fraction.
12. The iSanXoT reports are saved in the "reps" folder within your project folder. The "Npep2prot_Quantprot_FR1.tsv" file compiles data at the protein level across the given samples for the comparison to be studied (e.g., Control vs Treatment, Table 1) describes the content of the different columns provided in the report.
13. The FDR column contains the statistical significance of the comparison (Control vs Treatment). Therefore, ranking the protein by this value would shortlist the proteins that are differentially regulated, which are those than deviated from the normal distribution. However, the values from Xinf and Z columns can also be used to perform analysis outside of iSanXoT, such as limma moderated *t*-tests.

**Table 1**
**The table includes the content of the different columns of the "Control vs Treatment" example provided in the report**

| Reported variables | Description |
|---|---|
| *n* | Number of precursors per peptide and peptides per protein. |
| Xinf | Log2 ratios of proteins from all samples |
| Xsup | The grand mean |
| Vinf | Statistical weights |
| Z | Standardized log2 protein ratios |
| FDR | False discovery rate at the protein level |

## 4 Notes

1. Prepare stocks of solutions to facilitate the protocol: 500 mM Ethylenediaminetetraacetic acid disodium salt (EDTA) and store at room temperature. 500 mM Tris (2-Carboxyethyl) phosphine (TCEP), 500 mM sodium fluoride (NaF), 100 mM beta-glycerol phosphate, 100 mM of sodium orthovanadate and store at −20°C.
2. The use of protease inhibitors that contain EDTA can interfere with the benzonase activity, affecting the outcome of nuclear fractions.
3. Samples must be incubated using a rotatory wheel to ensure that the pellet is properly mixed during incubation.
4. It is critical to use Eppendorf tubes that can withstand centrifugal forces of up to 24,000*g*.
5. It is preferred to use protein-low-binding tubes to minimize losses.
6. Hydroxyl beads must be resuspended in 70% acetonitrile prior to adding them to the sample.
7. The cover lid to be used during the incubation should not contain glue that can be transferred to the samples and contaminate them.
8. This protocol is presented to be performed in combination with LC–MSMS analysis on an Evosep One coupled to a Thermo Orbitrap Eclipse. Since the LC analysis is, therefore, performed on an Evosep One, the samples need to be prepared on Evotips for further analysis.
9. One example of MS acquisition method for an Orbitrap Eclipse: Full MS resolution: 120,000 at m/z 200; full MS AGC target: 300%; full MS maximum injection time (IT): 45 ms; DIA MS2 scan AGC target: 1000%; DIA windows: 50 windows of 12 Th scanning from 400 to 100 m/z with an overlap of 1 Da; MS2 resolution: 30,000 at m/z 200; DIA MS2 scan maximum IT: 54 ms; Normalized collision energy (NCE): 30%.
10. Be aware that DIA-NN supports only Sciex.wiff, Bruker.d, Thermo.raw,.mzML and.dia formats. For.wiff and.raw support or any extra requirements, refer to https://github.com/vdemichev/DiaNN?tab=readme-ov-file#raw-data-formats.
11. In order to evaluate the quality of the results, one strategy to control quality is to measure the completeness of the quantification between replicates in the same fraction. We recommend discarding identifications that are not consistently identified across at least 75% of the replicates.

12. In order to properly calculate the scaled intensity, data must be transformed again from logarithm to raw values.
13. Organelle markers can be obtained from resources different from the ones presented here, such as Gene Ontology Cellular Components, or using localization inference algorithms such as DeepLoc [12].
14. Alternatively, subcellular fractionation success can be performed by extracting the main protein clusters per fraction in the global analysis, and performing an enrichment analysis for cellular compartments to identify which subcellular compartments are overrepresented in each fraction.
15. However, before we proceed, we recommend shortening the column names corresponding to the different samples that have been analyzed to avoid showing the full path to each file as column headers.
16. The installation instructions refer to a Windows operating system. However, iSanXoT is compatible with macOS High Sierra (10.13.6) and Ubuntu 20.04 (×64). For more details on installation on these operating systems, please refer to https://cnic-proteomics.github.io/iSanXoT/#_Installation.
17. To download the exe Launcher, click on: iSanXoT_Launcher_2.0.0.win32-x64.exe. Once downloaded, double-click the Launcher file, and the Installer window will appear. Follow the on-screen instructions to complete the installation. During the installation process, you can choose the iSanXoT installation folder. Select the desired folder where you want iSanXoT to be installed. Once the installation has been completed, click on "Finish," and you are ready to run iSanXoT
18. This folder contains all the information required to perform a complete DIA quantification analysis using iSanXoT for the provided samples.
19. The first time iSanXoT is run, several libraries must be installed; therefore, a window will appear displaying a progress bar, informing you about the percentage of packages that have been installed. These packages contain the libraries required by iSanXoT's backend and are installed during the initial launch of the application.
20. Figure 5A–C shows a display of the task tables contained in the workflow used in this protocol. This example is set for evaluating one fraction, to repeat the experiment with other fractions, the only table that requires to the modified in the one in "Basic Modules" and "LEVEL CREATOR" (Fig. 5B). For example, if analyzing Fraction 2, the names for "Ratio numerator column" and "Ratio denominator column(s)"

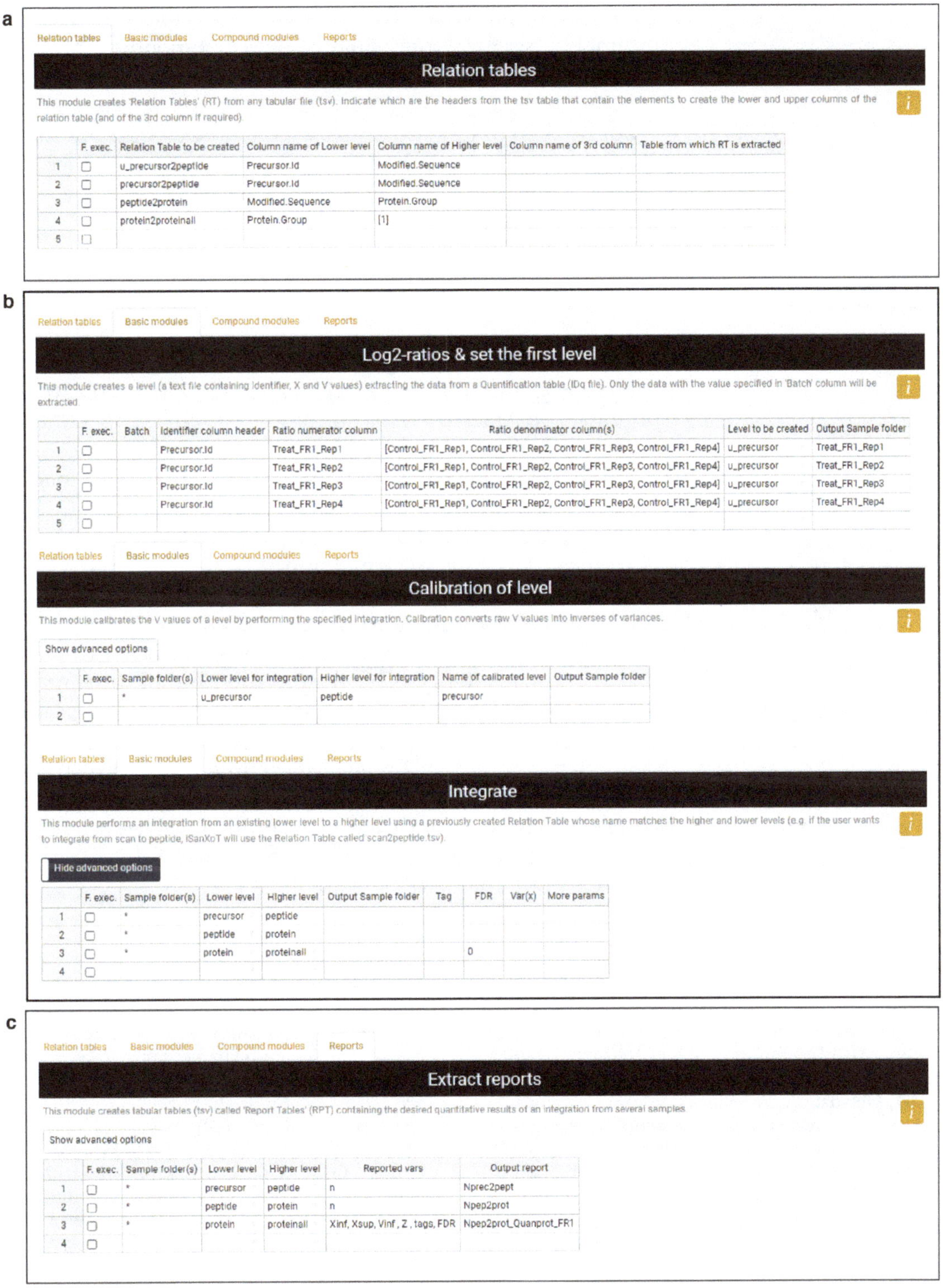


**Fig. 5** Snapshot of iSanXoT task tables from the workflow used in this protocol. (**a**) "RELS CREATOR" in the "Relation tables" tab; (**b**) "LEVEL_CREATOR," "LEVEL_CALIBRATOR," and "INTEGRATE" in the "Basic modules" tab; and (**c**) "REPORT" in the "Reports" tab

must be changed to refer to the column names for fraction 2 samples. Similarly, the name of the corresponding "Output sample folder" must be updated, as we all the name of the "Output report" (Fig. 5C). The remaining task tables should not be altered. It is also possible to analyze all fractions simultaneously, by adding the other comparisons for the other fractions at the "LEVEL CREATOR" task table.

## Acknowledgments

AM-V work is funded by the grant "Ayudas de Atracción de Talento Investigador César Nombela 2023" (2023-T1/SAL-GL-28990, Comunidad de Madrid). This study was supported by competitive grants PID2021-122348NB-I00 funded by MICIU/AEI/10.13039/501100011033 and by "ERDF A way of making Europe," PLEC2022-009298, PLEC2022-009235, and EQC2021-007053-P funded by MICIU/AEI/10.13039/501100011033 and by "European Union NextGenerationEU/ PRTR," and S2022/BMD-7333-CM (INMUNOVAR-CM) funded by Comunidad de Madrid. The project leading to these results has received funding from "la Caixa" Foundation under the project code LCF/PR/HR22/52420019. The CNIC is supported by the Instituto de Salud Carlos III (ISCIII), the Ministerio de Ciencia, Innovación Y Universidades (MICIU), and the Pro CNIC Foundation, and is a Severo Ochoa Center of Excellence (grant CEX2020-p01041-S funded by MICIU/AEI/10.13039/501100011033). Work at The Novo Nordisk Foundation Center for Protein Research (CPR) is funded in part by a donation from the Novo Nordisk Foundation (NNF14CC0001 and NNF24SA0098829)

## References

1. Singh N, Bhalla N (2020) Moonlighting proteins. Annu Rev Genet 54:265–285. https://doi.org/10.1146/annurev-genet-030620-102906
2. Thul PJ, Akesson L, Wiking M, Mahdessian D, Geladaki A, Ait Blal H, Alm T, Asplund A, Bjork L, Breckels LM, Backstrom A, Danielsson F, Fagerberg L, Fall J, Gatto L, Gnann C, Hober S, Hjelmare M, Johansson F, Lee S, Lindskog C, Mulder J, Mulvey CM, Nilsson P, Oksvold P, Rockberg J, Schutten R, Schwenk JM, Sivertsson A, Sjostedt E, Skogs M, Stadler C, Sullivan DP, Tegel H, Winsnes C, Zhang C, Zwahlen M, Mardinoglu A, Ponten F, von Feilitzen K, Lilley KS, Uhlen M, Lundberg E (2017) A subcellular map of the human proteome. Science 356(6340). https://doi.org/10.1126/science.aal3321
3. Dionne U, Gingras AC (2022) Proximity-dependent biotinylation approaches to explore the dynamic compartmentalized proteome. Front Mol Biosci 9:852911. https://doi.org/10.3389/fmolb.2022.852911
4. Geladaki A, Kocevar Britovsek N, Breckels LM, Smith TS, Vennard OL, Mulvey CM, Crook OM, Gatto L, Lilley KS (2019) Combining LOPIT with differential ultracentrifugation for high-resolution spatial proteomics. Nat Commun 10(1):331. https://doi.org/10.1038/s41467-018-08191-w
5. Orre LM, Vesterlund M, Pan Y, Arslan T, Zhu Y, Fernandez Woodbridge A, Frings O, Fredlund E, Lehtio J (2019) SubCellBarCode: proteome-wide mapping of protein localization and relocalization. Mol Cell

73(1):166–182. https://doi.org/10.1016/j.molcel.2018.11.035

6. Martinez-Val A, Bekker-Jensen DB, Steigerwald S, Koenig C, Ostergaard O, Mehta A, Tran T, Sikorski K, Torres-Vega E, Kwasniewicz E, Brynjolfsdottir SH, Frankel LB, Kjobsted R, Krogh N, Lundby A, Bekker-Jensen S, Lund-Johansen F, Olsen JV (2021) Spatial-proteomics reveals phospho-signaling dynamics at subcellular resolution. Nat Commun 12(1):7113. https://doi.org/10.1038/s41467-021-27398-y
7. Demichev V, Messner CB, Vernardis SI, Lilley KS, Ralser M (2020) DIA-NN: neural networks and interference correction enable deep proteome coverage in high throughput. Nat Methods 17(1):41–44. https://doi.org/10.1038/s41592-019-0638-x
8. Rodriguez JM, Jorge I, Martinez-Val A, Barrero-Rodriguez R, Magni R, Nunez E, Laguillo A, Devesa CA, Lopez JA, Camafeita E, Vazquez J (2024) iSanXoT: a standalone application for the integrative analysis of mass spectrometry-based quantitative proteomics data. Comput Struct Biotechnol J 23:452–459. https://doi.org/10.1016/j.csbj.2023.12.034
9. Trevisan-Herraz M, Bagwan N, Garcia-Marques F, Rodriguez JM, Jorge I, Ezkurdia I, Bonzon-Kulichenko E, Vazquez J (2019) SanXoT: a modular and versatile package for the quantitative analysis of high-throughput proteomics experiments. Bioinformatics 35(9):1594–1596. https://doi.org/10.1093/bioinformatics/bty815
10. Navarro P, Trevisan-Herraz M, Bonzon-Kulichenko E, Nunez E, Martinez-Acedo P, Perez-Hernandez D, Jorge I, Mesa R, Calvo E, Carrascal M, Hernaez ML, Garcia F, Barcena JA, Ashman K, Abian J, Gil C, Redondo JM, Vazquez J (2014) General statistical framework for quantitative proteomics by stable isotope labeling. J Proteome Res 13(3):1234–1247. https://doi.org/10.1021/pr4006958
11. Gatto L, Breckels LM, Wieczorek S, Burger T, Lilley KS (2014) Mass-spectrometry-based spatial proteomics data analysis using pRoloc and pRolocdata. Bioinformatics 30(9):1322–1324. https://doi.org/10.1093/bioinformatics/btu013
12. Thumuluri V, Almagro Armenteros JJ, Johansen AR, Nielsen H, Winther O (2022) DeepLoc 2.0: multi-label subcellular localization prediction using protein language models. Nucleic Acids Res 50(W1):W228–W234. https://doi.org/10.1093/nar/gkac278

Check for updates

# Chapter 20

# Computational and Statistical Framework for Quantitative Proteomics Analysis


## Ismail Kirrout and Nuria Montes



## Abstract

Mass spectrometry–based quantitative proteomics usually produces large datasets that require exhaustive analysis to extract underlying biological information. This chapter presents a step-by-step pipeline for the statistical and computational analysis of such data, oriented and generalizable to any mass spectrometry–derived proteomic dataset. These steps include: (i) data preprocessing and quality control, (ii) identification of differentially abundant proteins through statistical modeling, (iii) functional enrichment analyses, including Over Representation Analysis (ORA) and Gene Set Enrichment Analysis (GSEA) to integrate proteomic changes in the context of biological processes and pathways, and, finally, (iv) interactome (protein–protein interaction) construction and visualization to situate proteomic alterations within signaling networks. Throughout, reproducible off-the-shelf R code and practical guidance for each step are provided, facilitating an end-to-end analysis from raw proteomic data to the biological interpretation, illustrated with visualization examples and best-practice recommendations. The complete script and necessary files are freely available at https://github.com/UMBB-IIS-Princesa/Quantitative-Proteomics-Pipeline



**Keywords** Proteomics-analysis, Proteomics, Computational


## 1 Introduction

Proteomics has become an essential field for understanding cellular mechanisms, as it directly measures the functional molecules determining the phenotype. Modern mass spectrometry (MS)–based proteomics allows identification and quantification of thousands of proteins, enabling comprehensive studies of how proteomes change in complex conditions (diseases, treatments, etc.). Unlike transcriptomic data, proteomics provides direct insight into protein abundance, post-transcriptional regulation, and protein-level control mechanisms, which often correlate more closely with phenotypic outcomes [1]. However, analyzing quantitative proteomics data poses unique challenges. MS-based measurements have high dynamic range and technical variability [2], and missing values are common, especially


Noa B. Martín-Cófreces et al. (eds.), *The Cytoskeleton*, Methods in Molecular Biology, vol. 3070,
https://doi.org/10.1007/978-1-0716-5515-3_20,

for low-abundance proteins that may fall below the detection limit [3, 4]. Moreover, these datasets typically have limited replicates (due to experimental cost and effort), which complicates statistical analysis. The scale of MS-based proteomics data also poses substantial analytical challenges, as quantitative proteomics experiments routinely produce matrices with thousands of features, making downstream analysis and biological interpretation non-trivial [5].

To address these challenges, specialized computational tools are required. The general workflow when tackling these datasets involve: (i) Data preprocessing, including quality control checks, handling missing observations and normalizing raw protein/peptide intensities; (ii) statistical analysis, to identify differentially expressed (or differentially abundant) proteins between conditions; and (iii) downstream functional analysis, where the identified proteins are examined for enriched biological functions, pathway, and interactions, to help derive the underlaying biology. Many methods originally developed for other high-throughput data (like transcriptomics) have been adapted to proteomics, but new algorithms continue to emerge to handle proteomic-specific data structures. In this chapter, a computational pipeline fully developed in the R language is detailed, which integrates dedicated packages to perform each of the required steps to go from raw proteomic data to biological sense. The complete analysis pipeline, including scripts and necessary files, is freely available at GitHub (https://github.com/UMBB-IIS-Princesa/Quantitative-Proteomics-Pipeline/) and the Zenodo database (https://doi.org/10.5281/zenodo.18849537) [6].

## 2 Materials

### 2.1 In Silico Resources: Hardware

1. Equipment: A standard desktop or laptop computer with any 64-bit operating system (Windows, macOS, or Linux) and at least 8 GB of RAM; 16+ GB is recommended for large datasets (*see* **Note 1**).
2. Storage: Sufficient disk space to store raw data and required files (*see* **Note 2**).
3. Internet connectivity: Many R packages fetch annotation data (e.g., Gene Ontology terms, KEGG pathway maps) from online resources (*see* **Note 3**).

### 2.2 In Silico Resources: Software

1. R environment: R (version 4.5+ recommended) and an Integrated Development Environment (IDE) such as RStudio for convenience. All analyses in this chapter are performed in R v. 4.5.1. Installation instructions for both R and R studio are available at https://posit.co/download/rstudio-desktop/ (Accessed February 2026).

2. Bioconductor and CRAN packages: The following R packages are required to follow the steps and execute the code portrayed in this chapter (install via "BiocManager::install()" or "install.packages()" as appropriate).
3. clusterProfiler v. 4.16. Installation and documentation accessible at: https://bioconductor.org/packages/release/bioc/html/clusterProfiler.html (Accessed February 2026).
4. colorspace v. 2.1-2. Installation and documentation accessible at: https://cran.r-project.org/web/packages/colorspace/index.html (Accessed February 2026).
5. ComplexHeatmap v. 2.24.1. Installation and documentation accessible at: https://www.bioconductor.org/packages/devel/bioc/html/ComplexHeatmap.html (Accessed February 2026).
6. DEqMS v. 1.26.0. Installation and documentation accessible at: https://www.bioconductor.org/packages/release/bioc/html/DEqMS.html (Accessed February 2026).
7. dplyr v. 1.2.0. Installation and documentation accessible at: https://cran.r-project.org/web/packages/dplyr/index.html (Accessed February 2026).
8. EnhancedVolcano v. 1.26.0. Installation and documentation accessible at: https://bioconductor.org/packages/release/bioc/html/EnhancedVolcano.html (Accessed February 2026).
9. enrichplot v. 1.28.4. Installation and documentation accessible at: https://bioconductor.org/packages/release/bioc/html/enrichplot.html (Accessed February 2026).
10. ggraph v. 2.2.2. Installation and documentation accessible at: https://cran.r-project.org/web/packages/ggraph/index.html (Accessed February 2026).
11. ggplot2 v. 4.0.2. Installation and documentation accessible at: https://cran.r-project.org/web/packages/ggplot2/index.html (Accessed February 2026).
12. grid v. 4..5.1. Base package and is part of the R distribution.
13. igraph v. 2.2.1. Installation and documentation accessible at: https://cran.r-project.org/web/packages/igraph/index.html (Accessed February 2026).
14. limma v. 3.64.3. Installation and documentation accessible at: https://bioconductor.org/packages/release/bioc/html/limma.html (Accessed February 2026).
15. msImpute v. 1.18.0. Installation and documentation accessible at: https://www.bioconductor.org/packages/release/bioc/html/msImpute.html (Accessed February 2026).

16. org.Hs.eg.db v. 3.21.0. Installation and documentation accessible at: https://bioconductor.org/packages/release/data/annotation/html/org.Hs.eg.db.html (Accessed February 2026).
17. pathview v. 1.48.0. Installation and documentation accessible at: https://bioconductor.org/packages/release/bioc/html/pathview.html (Accessed February 2026).
18. pheatmap v. 1.0.13. Installation and documentation accessible at: https://cran.r-project.org/web/packages/pheatmap/index.html (Accessed February 2026).
19. PRONE v. 1.2.2. Installation and documentation accessible at: https://bioconductor.org/packages//release/bioc/html/PRONE.html (Accessed February 2026).
20. RCy3 v. 2.28.1. Installation and documentation accessible at: https://www.bioconductor.org/packages/release/bioc/html/RCy3.html (Accessed February 2026).
21. readr v. 2.1.6. Installation and documentation accessible at: https://cran.r-project.org/web/packages/readr/index.html (Accessed February 2026).
22. rrvgo v. 1.20.0. Installation and documentation accessible at: https://www.bioconductor.org/packages/release/bioc/html/rrvgo.html (Accessed February 2026).
23. scales v. 1.4.0. Installation and documentation accessible at: https://cran.r-project.org/web/packages/scales/index.html (Accessed February 2026).
24. scatterpie v. 0.2.6. Installation and documentation accessible at:https://cran.r-project.org/web/packages/scatterpie/index.html (Accessed February 2026).
25. shadowtext v. 0.1.6. Installation and documentation accessible at: https://cran.r-project.org/web/packages/shadowtext/index.html (Accessed February 2026).
26. STRINGdb v. 2.20.0. Installation and documentation accessible at: https://www.bioconductor.org/packages/release/bioc/html/STRINGdb.html (Accessed February 2026).
27. stringr v. 1.6.0. Installation and documentation accessible at: https://cran.r-project.org/web/packages/stringr/index.html (Accessed February 2026).
28. tidyr v. 1.3.2. Installation and documentation accessible at: https://cran.r-project.org/web/packages/tidyr/index.html (Accessed February 2026).
29. viridis v. 0.6.5. Installation and documentation accessible at: https://cran.r-project.org/web/packages/viridis/index.html (Accessed February 2026).

**2.3 Public Databases**

1. Protein/gene identifier databases (i.e., UniProt) [7] (*see* **Note 4**).
2. Gene Ontology (GO) [8] (*see* **Note 5**).
3. KEGG Pathway database [9] (*see* **Note 6**).
4. STRING Protein–Protein Interaction (PPI) database [10] (*see* **Note 7**).

**2.4 Experimental Data**

1. Protein quantification table: a tabular dataset (e.g., CSV, TSV, or TXT) containing quantitative proteomic data. A minimal setup should consist of rows corresponding to features (peptides, proteins, or genes) and columns to samples or conditions (*see* **Note 8**).
2. Sample metadata: a design matrix containing information about each sample in the quantification table. At minimum, a column specifying the sample name in the table and a column indicating to which experimental condition or group they belong to (e.g., control vs treatment, different time points; *see* **Note 9**).

## 3 Methods

**3.1 Selection of Analysis: MS Techniques and Quantification Data Outputs**

Different MS-based proteomic techniques and labeling strategies can influence the data structure and subsequent analysis. It is important to understand the source of your quantitative data before choosing analysis steps:

1. Data-Dependent Acquisition (DDA): DDA is commonly used for label-free quantification, where the spectrometer stochastically selects a subset of peptide ions for fragmentation (MS/MS) in each cycle [11]. This often results in "Missing Not At Random" (MNAR) values (*see* **Note 10**), where a peptide is identified in some biological replicates but falls below the detection threshold in others. Quantification typically relies on MS1 peak intensities integrated over time, or alternatively, spectral counting (the frequency of MS/MS events), though the latter is less sensitive for subtle fold-change detection.
2. "Data-Independent Acquisition" (DIA): DIA methods systematically fragment all ions within predefined m/z windows, providing a more consistent identification across samples, reducing missing values and improving reproducibility [11]. Quantification in DIA is primarily performed using MS2 chromatogram intensities (fragment-level extraction), yielding a near-complete data matrix across samples, which simplifies many downstream statistical workflows.

3. Isobaric labeling (TMT, iTRAQ): these techniques involve the chemical labeling of peptides from different samples, which are then pooled and analyzed in a single run [12]. Quantification is based on the intensities of reporter ions released during fragmentation. Within a single run data completeness is nearly 100%. However, missingness can occur across different runs (batch effects). The data output is often in the form of peptide- or protein-level ratios or intensities per channel. This data usually require normalization to correct for labeling efficiency and other biases.
4. Bottom-up vs. top-down Proteomics: most quantitative pipelines assume a bottom-up approach, where proteins are enzymatically digested into peptides. This requires aggregating peptide-level data into protein groups afterward [13]. However, in top-down, intact proteins are analyzed yielding protein-level measurements directly.
5. Quantification values: Regardless of the acquisition method, raw intensity values are generally log-normally distributed. For most parametric statistical tests (e.g., *t*-tests, ANOVA), data must be log2-transformed to achieve normality and stabilize variance. Conversely, count-based data (spectral counts) follow discrete distributions (e.g., Poisson or Negative Binomial) and should be handled with specialized count-based models rather than standard linear statistics.

The initial MS acquisition strategy and quantification method will determine how to handle missing data and normalization. For instance, isobaric labeling (TMT) datasets, such as the one in [14] and shown here, typically exhibit high data completeness within a single multiplexed run, and require rigorous normalization across channels and batches to account for labeling efficiency and inter-batch variation. The workflows described in the following sections for preprocessing, normalization, and imputation are specific to this data structure. While the mathematical principles may apply across other formats, it is the responsibility of the reader to identify the characteristics of their own acquisition method. This chapter assumes the reader has already performed the initial protein inference and quantification (e.g., using MaxQuant–Andromeda or similar software) to reach the required tabular protein-level dataset used as the starting point of this analysis. The MaxQuant output files for the example study presented here are publicly available via the PRoteomics IDEntifications Database (PRIDE) [15] under the identifier PXD030883.

### 3.2 Quality Control and Preprocessing

To ensure the suitability of the raw quantification table for posterior analysis, the first step is the assessment of data quality (quality control) and removal of confounding factors that may mask the biological signal (preprocessing). This encompasses (i) filtering out low-quality features, (ii) normalizing and checking for batch effects across samples, and (iii) handling missing values.

1. Open R through the terminal, the R console, or your preferred IDE (i.e., RStudio).
2. Install required libraries if not already done:

```
# Install CRAN packages
install.packages(c("colorspace", "dplyr", "ggraph",
"ggplot2", "igraph", "pheatmap", "readr", "scales",
"scatterpie", "shadowtext", "stringr", "tidyr",
"viridis"))
# Install BiocManager packages
if (!require("BiocManager", quietly = TRUE))
install.packages("BiocManager")
BiocManager::install(version = "3.22")
BiocManager::install(c("clusterProfiler",
"ComplexHeatmap", "DEqMS", "EnhancedVolcano",
"enrichplot", "limma", "msImpute", "org.Hs.eg.db",
"pathview", "PRONE", "RCy3", "rrvgo", "STRINGdb"))
```

3. Load the libraries for the analysis:

```
# Data handeling
handeling <- c("readr", "tidyr", "dplyr",
"stringr")
lapply(handeling, library, character.only = TRUE)
# Visualization
visualization <- c("ggplot2", "grid
lapply(visualization, library, character.only =
TRUE)
# Quality Control and Preprocessing
QC_preprocessing <- c("PRONE", "msImpute")
lapply(QC_preprocessing, library, character.only =
TRUE)
# Differential expression
dea<- c("limma", "DEqMS")
lapply(dea, library, character.only = TRUE)
# Enrichment
ea<- c("clusterProfiler", "org.Hs.eg.db",
"rrvgo","enrichplot")
lapply(ea, library, character.only = TRUE)
# Pathway
library(pathview)
# Network analysis
na<- c("STRINGdb", "igraph", "ggraph",
"scatterpie", "RCy3")
lapply(na, library, character.only = TRUE)
```

4. Import the protein quantification file into R and inspect the data (*see* **Note 11**):

```
MQ_data <- read_tsv("proteinGroups.txt")
dim(MQ_data)
str(MQ_data)
```

5. Load the experiment metadata into R, for metadata creation (*see* **Note 12**). The metadata file for this chapter was generated based on the data's original manuscript [14]:

```
md <- read_tsv("metadata.txt")
dim(md)
str(md)
```

6. Filter columns: "MaxQuant" output data has many columns that are unused in the downstream analysis. Essential columns are "Majority protein IDs" (protein/protein group identifiers) and quantification columns. Quality control columns include "Only identified by site", "Reverse" and "Potential contaminant." Other important columns are "Gene names" and "Peptide counts (razor+unique)" (number of peptides used for identification) (*see* **Note 13**):

```
data <- MQ_data %>%
dplyr::select(
# Identifiers
'Majority protein IDs'
'Gene names'
# Quantification
matches("^Reporter intensity
corrected.*Pat_tbc_pool_")
# QC / filtering
'Peptide counts (razor+unique)'
'Only identified by site'
'Reverse'
'Potential contaminant')
```

7. Trim identifiers. Ensure that rows are uniquely identified by protein ID or gene ID (*see* **Note 14**):

```
data <- data %>%
dplyr::mutate(
# Representative protein ID
protein_id = str_split_fixed('Majority protein IDs',
";", 2)[, 1]
# Representative gene symbol
gene = str_split_fixed('Gene names', ";", 2)[, 1]
# Representative peptide count
n_peptides = as.integer(
str_split_fixed('Peptide counts (razor+unique)',
";", 2)[, 1]) )%>%
# Drop the original columns to avoid ambiguity
```

```
dplyr::select(-'Majority protein IDs', -'Gene
names', -'Peptide counts (razor+unique)')
anyDuplicated(data$protein_id) # Should output 0 if
there are no duplicates
```

8. Eliminate proteins only identified by site, reverse and those labeled as possible contaminants (*see* **Note 15**):

```
data_qc <- data %>%
# Remove contaminants and reverse hits
dplyr::filter(
is.na('Only identified by site') | 'Only identified
by site'!= "+"
is.na('Reverse') | 'Reverse'!= "+"
is.na('Potential contaminant') | 'Potential
contaminant'!= "+") %>%
# (Optional) Remove proteins identified only by 1
peptide
dplyr::filter(n_peptides >= 2) %>%
# Drop used columns
dplyr::select(-'Only identified by site',
-'Reverse', -'Potential contaminant')
```

9. Record zeros as Not Available (NA) values (*see* **Note 16**):

```
data_qc <- data_qc %>% mutate(across(starts_
with("Reporter intensity corrected"), ~ na_if(.,
0)))
```

10. Remove features detected in very few samples or samples containing excessive missing values (*see* **Note 17**):

```
# Select quantification columns
sample_cols <- grep("^Reporter intensity
corrected", names(data_qc))
# Drop features with to many NAs
na_counts <- rowSums(is.na(data_qc[, sample_cols]))
proteins_to_keep <- na_counts < (0.3 *
length(sample_cols))
data_qc <- data_qc[proteins_to_keep,]
# Drop samples with to many NAs
data_qc <- data_qc %>%
dplyr::select(
-all_of(sample_cols)
all_of(sample_cols[colMeans(is.na(dplyr::select(.,
all_of(sample_cols)))) < 0.3]))
# Filter the metadata to keep same samples
md_qc <- md %>%
dplyr::filter(Column %in% intersect(md$Column,
colnames(data_qc))) %>%
dplyr::mutate(Column = factor(Column, levels =
colnames(data
dplyr::arrange(Column) %>%
```

```
dplyr::mutate(Column = as.character(Column))
```

11. Normalize before comparing samples to remove systematic biases. The package "PRONE" [16] implements several normalization and batch effect control methods, allowing for assessment and comparison between them. This package revolves around the "SummarizedExperiment" object. To create this object, an "expression matrix" as well as the corresponding "metadata" object are needed:

```
se <- load_data(data_qc, # Expression matrix
md_qc, # Metadata
protein_column = "protein_id"
gene_column = "gene"
ref_samples = md[md$Condition == "ref",]$Column
batch_column = "Batch"
condition_column = "Condition"
label_column = "Label")
```

12. Apply normalization. Available normalization methods can be displayed by the "get_normalization_methods()" function and several normalization methods can be applied at once [16] (*see* **Note 18**):

```
se_norm <- normalize_se(se, c("Median" #
Simple sample shifting,"Quantile"# Sample-to-
reference,"LoessF" # Sample-to-reference,"VSN" #
Variance stabilizing,"RobNorm" # Model-based))
```

13. Assess the intensity distributions of normalized data. One common way is boxplots (*see* **Note 19**). In addition to qualitative assessments, intragroup variability can be quantitatively evaluated using the "pooled median absolute deviation" (PMAD). PMAD provides a measure of within-condition variability, which should lower as the normalization is performed (see Fig. 1):

```
plot_boxplots(se_norm,
 ain = NULL,
 color_by = NULL,
 label_by = NULL,
 ncol = 3,
 facet_norm = TRUE) +
ggplot2::theme_bw() +
ggplot2::theme(legend.position = "bottom", legend.
direction = "horizontal")
plot_intragroup_PMAD(se_norm
ain = NULL
condition = NULL
diff = TRUE) +
ggplot2::ylab("Intragroup variation compared to
log2") +
```

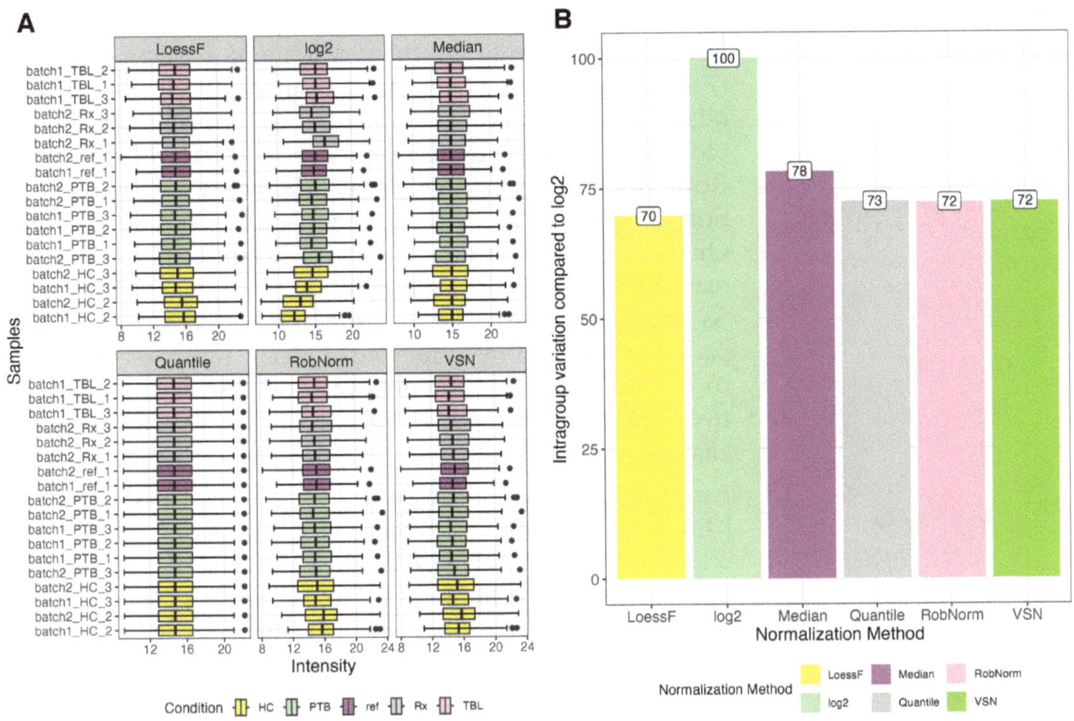


**Fig. 1** Evaluation of normalization methods. (**A**) Boxplots of intensity distributions across samples after applying different normalization methods (LoessF, only log2 transformation, Median, Quantile, RobNorm, and VSN). Samples are colored by condition (HC, PTB, ref, Rx, TBL). (**B**) Bar plot showing intragroup variation assessed using pooled median absolute deviation (PMAD). Values represent the percentage of remaining variation within biological groups after normalization, compared to the log2-transformed data (set to 100%)

```
ggplot2::theme_bw() +
ggplot2::theme(legend.position = "bottom", legend.
direction = "horizontal")
```

14. Interpret the results: In this example, results demonstrate that normalization improved the consistency of protein intensity distributions compared with the log2 transformation alone. In Fig. 1A, the log2 transformation shows noticeable variability in medians and interquartile ranges across samples of the same condition, indicating residual technical bias. In contrast, LoessF, Median, RobNorm, and VSN normalization improved alignment of sample distributions, reducing between-condition dispersion. Quantile normalization produced nearly identical distributions across samples, reflecting a strong distribution enforcement. In Fig. 1B, intragroup variation is quantified using PMAD and expressed relative to the log2-transformed data (100%). This representation illustrates how all normalization methods reduced within-group dispersion. LoessF showed the strongest reduction (70%), followed

by RobNorm and VSN (72%), and Quantile (73%), whereas Median normalization achieved a more modest reduction (78%). These results indicate that although all tested methods reduce technical variability, LoessF provides the greatest reduction in robust dispersion in this dataset, while VSN and RobNorm appear to offer a balance between variability reduction and preservation of biological differences. In contrast, Quantile normalization may be overly aggressive, potentially masking biologically meaningful variation.

As the RobNorm normalization method achieved a comparable reduction in intragroup PMAD and it was intended to handle common proteomics issues [16], it is a conservative, stable choice, and therefore can be selected for further analysis.

15. Assess batch effect. If the experiment is performed in batches [different MS runs, different days, operators, etc.], batch effects can confound biological differences. Common signs of batch effects include clustering of samples by run instead of experimental condition, which can be checked via Principal Component Analysis (PCA) (see Fig. 2 and PCA visualization code in **Note 20**).

```
# Select the desired assay
abund <- as.matrix(assay(se_norm, "RobNorm"))
abund <- abund[complete.cases(abund),]
# Calculate PCs
pca_res <- prcomp(t(abund), center = TRUE, scale. =
TRUE)
# Extract PCA results
scores <- as.data.frame(pca_res$x)
scores$Column <- rownames(scores)
scores <- merge(scores, md, by = "Column", all.x =
TRUE)
```

16. Interpret the results: the PCA reveals that in the log2-transformed data (Fig. 2A), PC1 explains a large proportion of total variance (63.2%), and separation appears greatly driven by batch effects, as samples cluster according to batch rather than by biological condition. Following RobNorm normalization (B), the variance explained by PC1 decreases (28.5%), and samples cluster more distinctly by biological condition, indicating effective reduction of technical variability, however the effect of the batch is still permanent.

17. Control batch effect by using either IRS or "limma::removeBatchEffect" (in PRONE: "limBE") methods on top of normalization. Use the combination pattern "on" to add the batch effect correction on top of the normalization method.

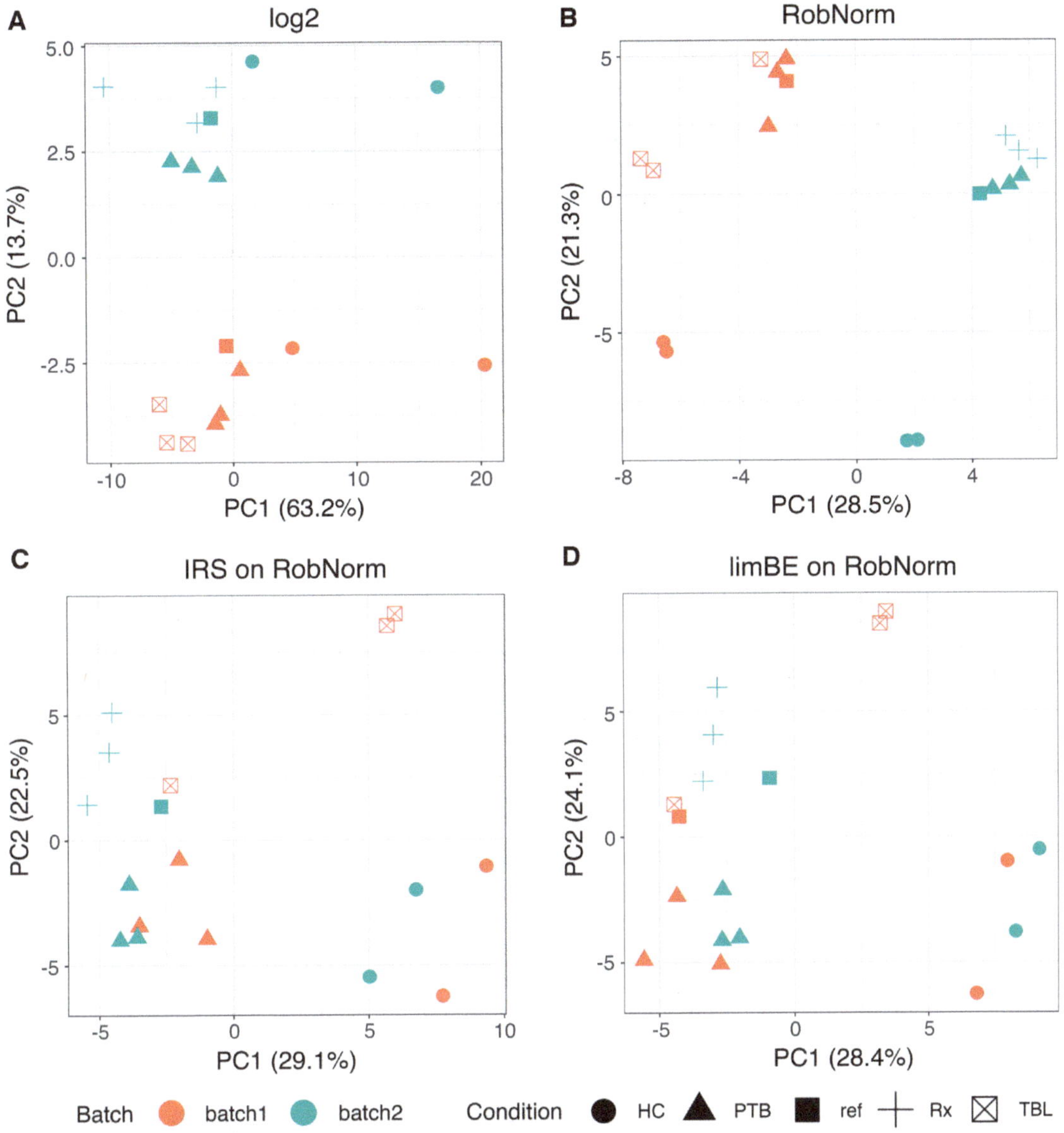


**Fig. 2** Principal component analysis (PCA) of protein intensity data under different normalization and batch-correction strategies. PCA plots showing sample clustering on (**A**) log2-transformed data, (**B**) RobNorm normalized data, (**C**) IRS applied on RobNorm-normalized data, and (**D**) limma batch effect removal (limBE) applied on RobNorm-normalized data. Samples are colored by batch (batch1 in red, batch2 in blue) and shaped by biological condition (HC, PTB, ref, Rx, TBL). The percentage of variance explained by PC1 and PC2 is indicated on each axis

This will remove variation associated with batch while preserving variation due to experimental conditions.

```
se_norm <- normalize_se(se, c("RobNorm", "IRS_on_
RobNorm", "limBE_on_RobNorm"), combination_pattern
= "_on_")
```

18. Interpret the results: after applying IRS or limBE on normalized data (Fig. 2C, D) batch separation is visibly reduced, and condition-specific clustering specially in HC and PTB appear substantially improved compared to RobNorm alone.

19. Extract the normalized quantification matrix from the "SummarizedExperiment" object once the desired method has been selected:

```
data_norm <- assay(se_norm, "limBE_on_RobNorm")
# Rename the columns with sample label
label_map <- setNames(md$Label, md$Column)
new_labels <- label_map[colnames(data_norm)]
colnames(data_norm) <- new_labels
data_norm <- as.matrix(data_norm)
# Recover feature ids
rownames(data_norm) <- rowData(se_norm)$Protein.IDs
```

20. Assess replicate consistency by calculating the correlation of samples (see Fig. 3).

```
# Calculate correlations
cor_mat <- cor(data _norm, use = "pairwise.
complete.obs", method = "pearson")
# Plot correlation matrix
pheatmap::pheatmap(
cor_mat
clustering_distance_rows = "correlation"
clustering_distance_cols = "correlation")
```

21. Interpret the results: the correlation heatmap in Fig. 3 demonstrates strong within-condition similarity and clear clustering. Samples of the same condition exhibit high pairwise correlations, indicating consistent protein expression profiles across replicates. Hierarchical clustering groups conditions into distinct clusters, preserving biological structure after normalization. It is relevant to observe if samples from different batches cluster together within their respective conditions, suggesting effective mitigation of batch-driven effects.

22. Define the experimental factors for downstream missingness diagnostics and group-aware imputation:

```
md_qc<-as.data.frame(md_qc)
batch <- factor(md_qc[, "Batch"])
condition <- factor(md_qc[, "Condition"])
```

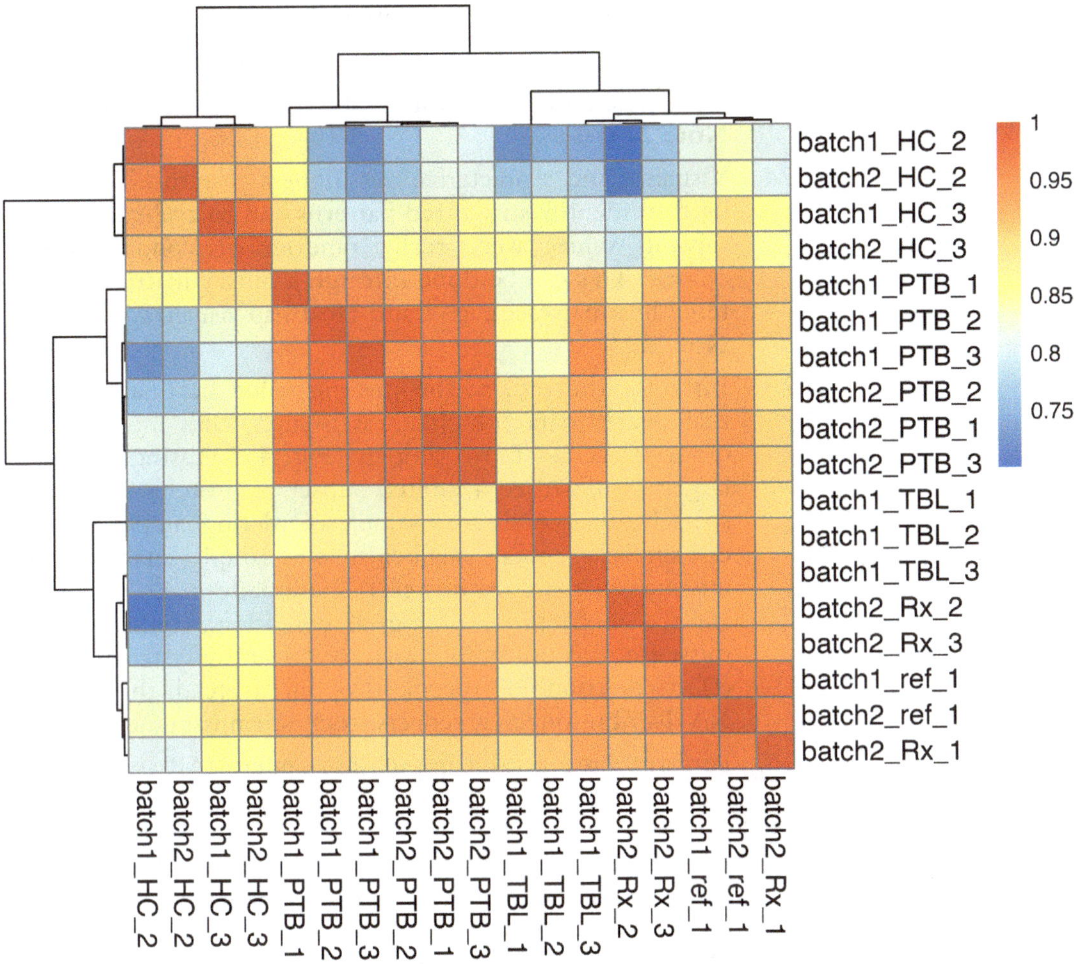


**Fig. 3** Sample correlation heatmap following normalization. Hierarchical clustering heatmap displaying pairwise Pearson correlation coefficients between all samples after normalization. Color intensity represents correlation strength (blue = lower correlation, red = higher correlation). Dendrograms indicate hierarchical clustering of samples based on correlation distance, illustrating similarity patterns across biological conditions and batches

23. Select features for imputation by applying "msImpute" package's [17] feature selection. This function uses the entropy-based missingness (EBM) criterion. EBM summarizes the distribution of observed values across experimental groups:

```
hdp <- selectFeatures(data_norm, method = "ebm",
group = condition)
```

Console output:

```
> No NaN EBMs detected. Peptides are missing evenly
across samples.
```

```
> Switchted to 'hvp' method as final msImpute
features
```

This indicates that the observed missing values are condition-associated and most likely appear at random (*see* **Note 21**).

24. Visualize and characterize missingness patterns. This step is used to identify structured patterns and to evaluate whether missing values were really random or condition/batch related. This can be done through a binary matrix and clustered heatmap (see Fig. 4 and Heatmap visualization code in **Note 22**).

25. Interpret the results: the missing value heatmap indicates that overall data completeness is high, with most proteins observed across most samples (Fig. 4). Missing values are largely confined to a limited subset of proteins and appear sparsely distributed across samples. Ordering samples by dropout similarity reveals small clusters of samples sharing similar missingness patterns; however, these patterns do not strictly align with batch or biological condition, confirming that missingness is not predominantly driven by systematic batch effects or condition-specific dropout. Instead, the observed NA distribution likely reflects stochastic missingness.

26. Perform missing-value imputation under MAR assumption:

```
msImpute_mar <- msImpute(data_norm, method = "v2")
# method = "v2" is for MAR assumptions
```

27. Evaluate the imputation impact to assess whether imputation preserved plausible data structure and variability. Coefficient-of-variation (CV) plots can be generated for each matrix to inspect changes in intensity-dependent variability and detect potential over-smoothing or variance inflation after imputation (see Fig. 5):

```
# Sort the plots in the display
par(mfrow=c(1,2))
# Generate the CV plots
pcv <- plotCV2(data_norm, main = "data")
pcv <- plotCV2(msImpute_mar, main = "msImpute v2")
```

28. Interpret the results: The mean–variance relationship is largely preserved after imputation, with variance decreasing as mean intensity increases. The overall distribution of points and the fitted mean–variance trend remain comparable to the original data, indicating that this approach did not distort the global variance structure. This informs that the imputation strategy will not introduce pronounced artificial variance inflation or

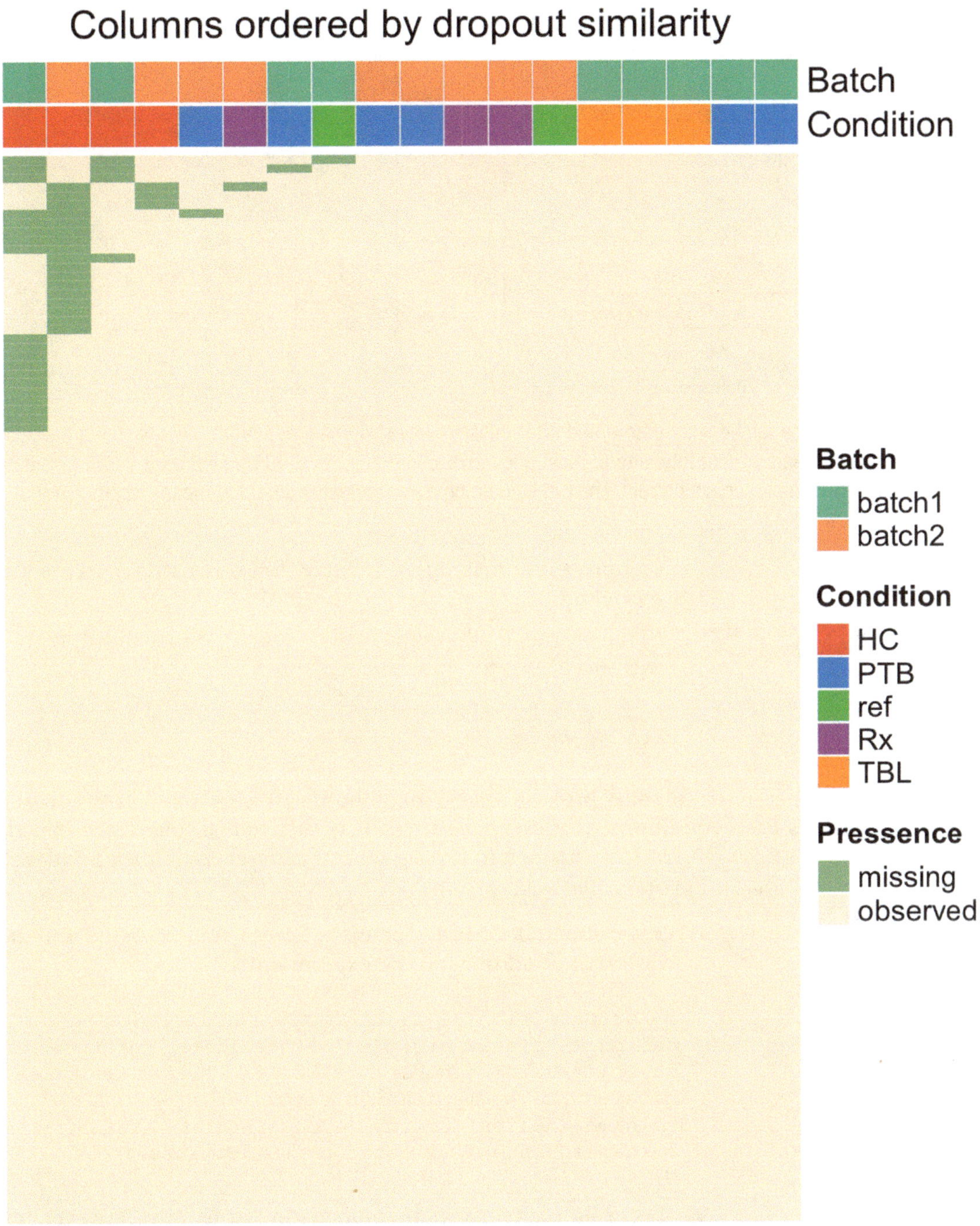


**Fig. 4** Missing value patterns across samples ordered by dropout similarity. Heatmap showing the presence (beige) or absence (green) of protein intensity values across all samples. Columns (samples) are ordered by similarity of missingness patterns (dropout similarity), and rows represent proteins. Annotation bars above the heatmap indicate batch (batch1, batch2) and biological condition (HC, PTB, ref, Rx, TBL)

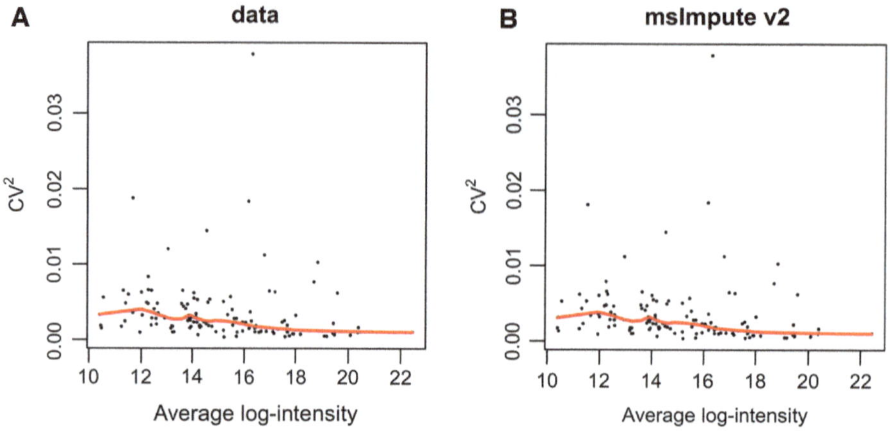


**Fig. 5** Mean–variance relationship before and after missing value imputation. Scatter plots showing the relationship between mean protein intensity (x-axis) and variance (y-axis) for (left) original data and (right) data imputed. Each point represents a protein. The red smoothing curve indicates the fitted mean–variance trend

compression, making it suitable for downstream statistical modeling.

29. Rename the final preprocessed data (a normalized matrix with no NAs) (for convenience).

```
data <- msImpute_mar
```

### 3.3 Differential Expression Analysis

To identify proteins that change significantly between experimental conditions, a differential expression or differential abundance analysis (DAA) is performed to test significant differences in the abundance of proteins between groups.

1. Experimental design: Set up a design matrix based on the different conditions in the experiment.

```
md_qc$Condition <- factor(md_qc$Condition)
design <- model.matrix(~ 0 + Condition, data =
md_qc) # Batch can be modeled if not already taken
care in the normalization step
colnames(design) <- sub("^Condition", "",
colnames(design)) # nicer names (optional)
```

2. For a simple two-group comparison (Control vs Treatment), a *t*-test could be used for each protein. However, proteomic-data often have few replicates, which makes *t*-tests unreliable as variance estimates are noisy. To improve power, a common choice is using the “limma” [18] package, which employs an empirical Bayes moderation to shrink variance estimates towards a pooled mean. limma treats the protein abundance

data similarly to how it treats gene expression data, so first fit a linear model for each protein.

```
fit <- lmFit(data, design)
```

3. For multiple conditions, define contrasts with the following code to perform all possible comparisons in the same analysis:

```
# Get condition combinations
conds <- colnames(design)
pairs <- combn(conds, 2)
# Create contrasts
contrast_names <- apply(pairs, 2, \(x) paste0(x[1],
"_vs_", x[2])) # B_vs_A
contrast_formulas <- apply(pairs, 2, \(x)
paste0(x[1], " - ", x[2])) # B - A
Cmat <- makeContrasts(contrasts = contrast_
formulas, levels = design)
colnames(Cmat) <- contrast_names
```

4. Fit the contrast to the fitted linear model. This step re-parameterizes the model to directly estimate the log2 fold-changes corresponding to each specified comparison.

```
fit2 <- contrasts.fit(fit, Cmat)
fit2 <- eBayes(fit2)
```

5. Extract limma's output table to show the log2 fold change, average expression, moderated *t*-statistic, raw *P*, and adjusted *P* (Benjamini–Hochberg) of proteins using the function "topTable." The false discovery rate (FDR) control is essential as several proteins are tested simultaneously. Typically, an FDR threshold (i.e., 0.05 or 0.01) is used to define significant hits. In the case of multiple comparisons, the desired table should be specified in the "coef" argument of the function.

```
res <- topTable(fit2, coef = "HC_vs_PTB" #
Comparison of interest, adjust.method="BH", sort.
by="P", number=Inf)
```

If multiple comparisons are tested, extract the count of significant features with this code:

```
sig_counts <- data.frame(
Contrast = colnames(fit2),
N_sig = sapply(colnames(fit2), function(cn) {
tt <- topTable(fit2, coef = cn, number = Inf, sort.
by = "none")
sum(tt$adj.P.Val < 0.05 & abs(tt$logFC) > 0, na.rm
= TRUE)
}))
```

6. Incorporate peptide number (*see* **Note 23**). "DEqMS" [19] is a method that extends limma by modeling variance through peptide count. DEqMS can be applied when the quantification software provides the number of peptides corresponding to each identified protein. To employ this method, add a column with the number of peptides for each feature to limma's "fit2" object:

```
# Get peptide column
pep_count <-setNames(data_qc$n_peptides,
data_qc$protein_id)
# Add the peptide count column
fit2$count <- pep_count[rownames(fit2$coefficients)]
```

7. Model the peptide number by fitting the DEqMS model:

```
fit3 <- spectraCounteBayes(fit2)
```

8. Apply the "outputResult" function to format the "fit3" results, obtaining the adjusted *P* for peptide count ("sca.P.Value"):

```
res_deqms <- outputResult(fit3, coef_col =
"HC_vs_PTB")
```

9. Recover the gene names for a clearer visualization of the results:

```
res_deqms$gene <- rowData(se_norm)$Gene.Names[
match(rownames(res_deqms), rowData(se_
norm)$Protein.IDs)]
```

10. Plot –log10(*P*) vs log2 fold-change for all proteins to obtain a Volcano Plot. Significant proteins (above a *P* and fold-change threshold) are highlighted. This gives a global view of magnitude vs significance (see Fig. 6A and the Volcano plot code in **Note 24**).
11. Plot a heatmap of significant proteins to illustrate the expression pattern of top hits across samples, and to ensure clear separation between conditions. This is often used to confirm a coherent behavior for DE proteins (see Fig. 6B and the Heatmap visualization code in **Note 25**).
12. Interpret the results: The differential expression analysis identified a subset of proteins significantly altered (up- and down-expressed) between HC and PTB. The volcano plot (Fig. 6A) shows several proteins exhibiting both statistically significant changes and substantial fold differences, including increased abundance of APOE, VWF, APOC4–APOC2, and APOC2 in HC, and decreased levels of proteins such as HP, ITGB1, ICAM2, and ITGB3. The distribution of significant proteins on both sides of the fold-change axis indicates shifts

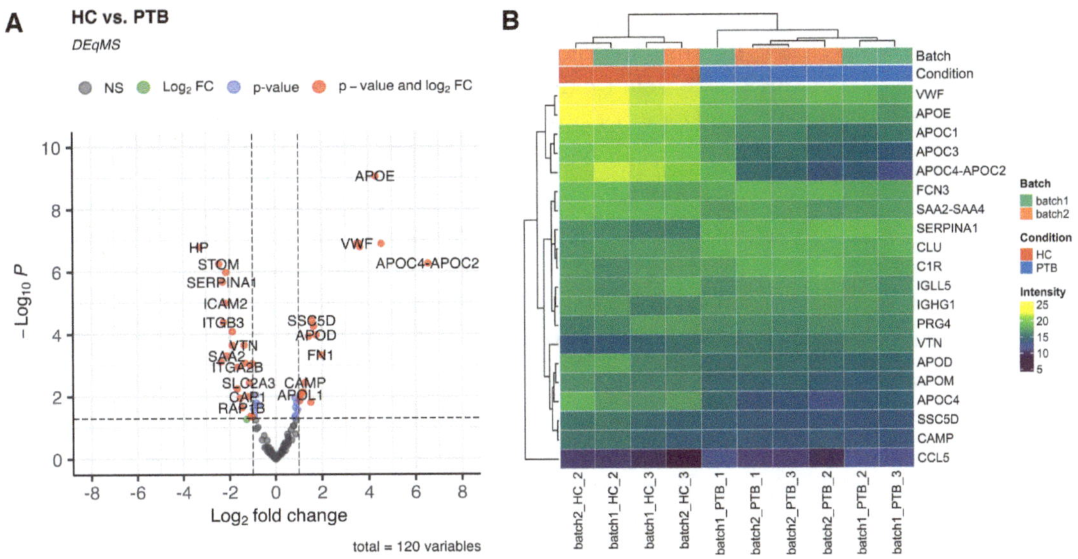


**Fig. 6** Differential protein expression between HC and PTB. (**A**) Volcano plot of the results from the differential expression analysis performed using DEqMS for the conditions HC vs PTB. The x-axis represents log2 fold change (HC relative to PTB), and the y-axis shows $-\log_{10}$ adjusted *P*. Dashed vertical lines indicate fold-change thresholds ($\pm 1$), and the horizontal dashed line indicates the statistical significance threshold (FDR-adjusted $P = 0.05$). Points are colored according to significance criteria: non-significant (gray), significant by fold change only (green), significant by *P* only (blue), and significant by both *P* and fold change (red). Significantly regulated proteins are labeled. A total of 120 proteins were tested. (**B**) Heatmap of the top 20 differentially expressed proteins ($|\log_2 FC| > 1$, ranked by adjusted *P*) identified in panel (**A**). Rows represent proteins and columns represent individual samples, annotated by batch and biological condition. Color intensity reflects normalized protein abundance (log-scale). Hierarchical clustering was performed on both samples and proteins to visualize expression patterns across conditions

in multiple functional protein groups. The heatmap (Fig. 6B) confirms these findings at the sample level, demonstrating clear separation between HC and PTB samples based on the expression patterns of the identified proteins. Additionally, replicates cluster according to biological condition rather than batch, supporting the robustness of the normalization method. These results indicate significant proteomic differences between HC and PTB, supporting the decision of downstream pathway and functional enrichment analyses.

### 3.4 Over Representation Analysis

Once the list of proteins differentially present between biological conditions has been obtained, an enrichment analysis is performed to ascertain the biological functions or processes that are affected by their changes. "Over Representation Analysis" (ORA) or "Gene Ontology" (GO) enrichment analysis are popular methods to find over-represented functional categories in a list of proteins. GO encompasses three ontologies: Biological Process (BP), Molecular

Function (MF), and Cellular Component (CC). In proteomics studies, BP enrichment is commonly used to understand the pathway or processes that are modulated, but MF and CC can provide insights into the types of activities or subcellular localizations involved.

1. Prepare the gene list. The input for GO enrichment is a list of proteins or genes of interest (i.e., the set of proteins found to be significantly altered between experimental conditions). The proteins can be separated into up- vs down-expressed or combined as a unique list if direction can be overlooked. A proper *background* set of genes for enrichment analysis is required (*see* **Note 26**).

```
res_sig <- res_deqms %>% dplyr::filter(sca.adj.pval
<= 0.05, abs(logFC) >= 1)
all_proteins <- unique(data_qc$gene)
sig_proteins <- unique(res_sig$gene)
```

2. Run "enrichGO" function from clusterProfiler [20] (*see* **Note 27**). The resulting go_enrich is an object containing a table of GO terms with enrichment statistics. A pvalueCutoff = 0.05 can be set to only keep terms with FDR corrected $P < 0.05$ (*see* **Note 28**).

```
go_enrich <- enrichGO(gene = sig_proteins,
universe = all_proteins,
OrgDb = org.Hs.eg.db,
keyType = "SYMBOL",
ont = "BP",
pAdjustMethod = "BH",
pvalueCutoff = 0.35,
qvalueCutoff = 0.33)
# Inspect the first couple of terms
head(go_enrich@
result[,c("ID","Description","GeneRatio","BgRatio",
"p.adjust")])
# Create dataframe
enrich_data <- as.data.frame(go_enrich)
```

3. Visualize the enrichment in results with plotting functions from "clusterProfiler" such as "dotplot" and "cnetplot" (see Fig. 7).

```
dotplot(go_enrich, x = "RichFactor", showCategory =
10)
cnetplot(go_enrich, showCategory = 2, node_label =
"none", size_item=0.6, size_category=1.2, color_
category="#97c497", color_item="#c4c4c4") +
geom_cnet_label(node_label = "category", size = 5)
+
geom_cnet_label(node_label = "item", size = 4)
```

4. Interpret the results. Functional enrichment analysis of differentially expressed proteins revealed overrepresentation of biological processes related to receptor-mediated endocytosis, lipid transport, and transmembrane transport. The bubble plot (Fig. 7A) shows "receptor-mediated endocytosis" exhibiting the highest RichFactor (ratio of differentially expressed proteins mapped to a GO term relative to the total number of proteins annotated to that term) and lowest adjusted *P* among the enriched terms, suggesting strong enrichment driven by a substantial proportion of the identified proteins. Additional enriched processes such as "lipid transport" and "lipid localization" indicate coordinated alterations in lipid handling and membrane-associated functions. The gene–concept network (Fig. 7B) further highlights the interconnected nature of these pathways, with proteins like APOE, APOC1, APOC3, ITGB1, ITGB3, CD36, and VTN contributing to multiple enriched terms. The overlap of proteins across morphogenesis- and endocytosis-related processes suggests coordinated regulation

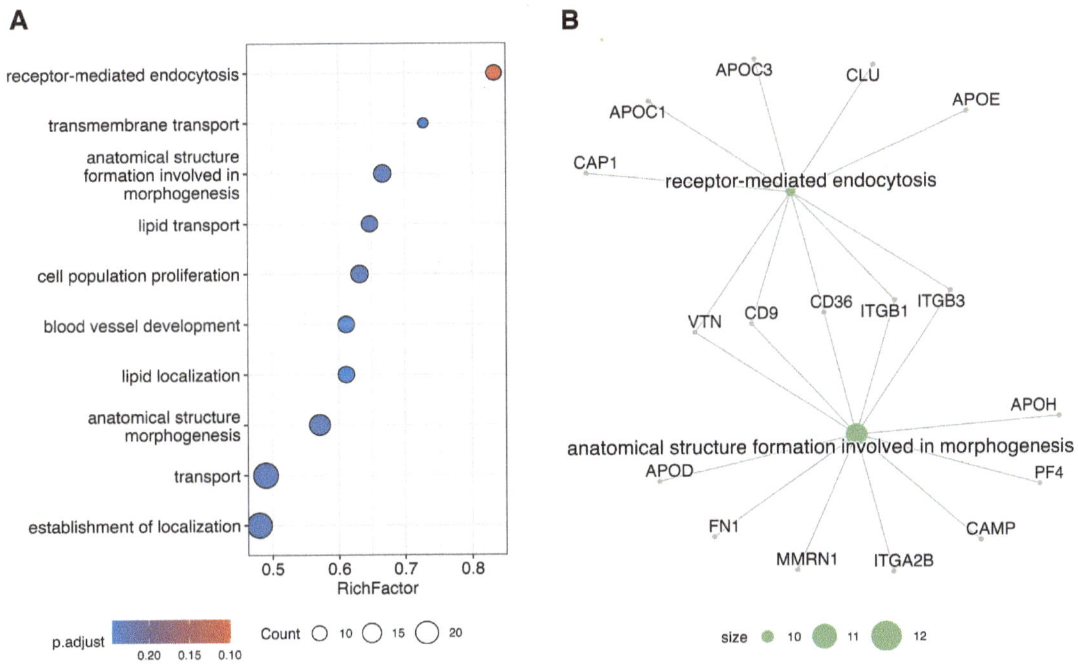


**Fig. 7** Gene Ontology (GO) enrichment analysis of differentially expressed proteins. (**A**) Bubble plot of top ten enriched biological process GO terms. The x-axis represents the RichFactor (ratio of differentially expressed proteins mapped to a GO term relative to the total number of proteins annotated to that term). Bubble size indicates the number of proteins (Count) associated with each term, and color represents adjusted *P* (p.adjust). (**B**) Gene–term network illustrating the relationship between two of the enriched GO terms and their associated proteins. Green nodes represent GO terms (node size proportional to gene count), and grey nodes represent proteins contributing to each term. Edges indicate protein–term associations

of extracellular matrix remodeling, integrin signaling, and lipid-associated transport mechanisms.

5. Reduce redundancy via RRVGO (*see* **Note 29**). "rrvgo" [21] offers the capability of simplifying and clustering GO terms under a more significant or representative "parent term." To this end, calculate a similarity matrix:

```
# Define variables
orgdb="org.Hs.eg.db"
ont="BP"
keytype = "SYMBOL"
# Calculate similarity matrix
simMatrix <- calculateSimMatrix(
enrich_data
orgdb
keytype = "SYMBOL"
ont = "BP"
method = "Rel"
semdata = GOSemSim::godata(annoDb = orgdb, ont =
ont, keytype = keytype))
```

6. Set scoring metrics to determine the most informative terms. Usual choices are enrichment scores or $-\log_{10}$ of the corrected *P*:

```
# Calculate ranking metric
enrich_data$logp <- (-log10(enrich_data$p.adjust))
enrichment_scores <- setNames(enrich_data$logp,
enrich_data$ID)
```

7. Determine the parent term:

```
# Select and group under the parent term
reducedTerms <- reduceSimMatrix(simMatrix
enrichment_scores
threshold=0.8, # Similarity threshold
orgdb="org.Hs.eg.db")
```

8. Visualize RRVGO results once the reduction has been performed to assess the clustering of terms (see Fig. 8).

```
reduced_heatmap <- heatmapPlot(simMatrix,
reducedTerms,
annotateParent=TRUE,
annotationLabel="parentTerm",
fontsize=6)
```

9. Interpret the results: The rrvigo semantic similarity heatmap reveals structured clustering of enriched GO terms into functionally coherent groups (Fig. 8). Highly similar terms cluster

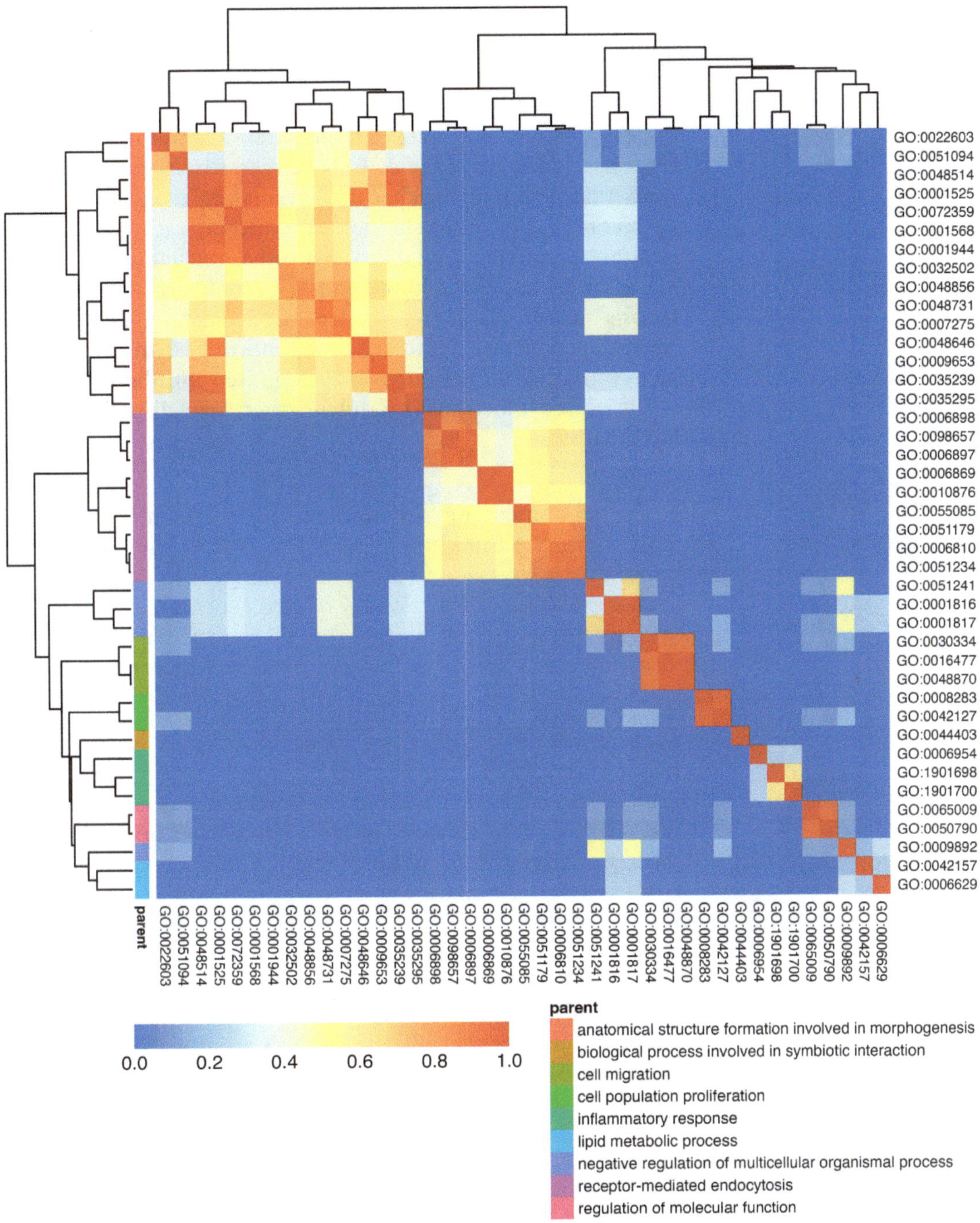


**Fig. 8** Semantic similarity heatmap of enriched Gene Ontology (GO) terms. Heatmap showing pairwise semantic similarity between significantly enriched GO biological process terms summarized using rrvigo. Color intensity represents semantic similarity scores (blue = low similarity, red = high similarity). Hierarchical clustering was performed to group-related GO terms. The colored annotation bar indicates higher-level parent categories assigned by rrvigo

together (red blocks along the diagonal), reflecting redundancy among closely related biological processes. Prominent clusters correspond to processes associated with "anatomical structure formation and morphogenesis," "receptor-mediated endocytosis," "lipid metabolic processes," and "inflammatory response." The presence of well-defined similarity blocks suggests that multiple enriched GO terms represent overlapping or hierarchically related biological concepts rather than independent processes. Notably, "anatomical structure formation involved in morphogenesis" and "receptor-mediated endocytosis" form distinct great separated clusters. Overall, the semantic reduction and clustering confirm that the differential proteomic changes are driven by a limited number of biologically related functional modules rather than dispersed, unrelated pathways.

### 3.5 Gene Set Enrichment Analysis

While the enrichment above considered a *discrete list* of significant proteins, Gene Set Enrichment Analysis (GSEA) uses *all proteins with their continuous values* (often fold changes) to detect coordinated changes in sets of functionally related genes, even if individual proteins are not statistically significant. This can reveal more subtle, but coordinated, shifts in pathway or processes that a strict cutoff might miss.

1. Prepare a Ranked List. For GSEA, all proteins are ranked by a metric reflecting their internal change. A simple choice is to use the signed statistic from the differential analysis, for example, the log2 fold change (ordered from most upregulated to most downregulated). When $\log_2$FC alone, a protein with a small but very significant change can rank lower (towards the middle of the list) than other proteins with a large but noisy change. This can be avoided by using $\log_{10}(P)$ multiplied by the sign of the FC (LPS) as the ranking metric. The "clusterProfiler" GSEA functions expect a named numeric vector, where names are gene IDs and values are the ranking metric, sorted in decreasing order. For the dataset used herein, create a ranked list by log2 fold change:

```
# Get the desired columns from the DEA result
gene_abundance <- res_deqms %>%
dplyr::select(logFC) %>%
dplyr::arrange(desc(logFC)) %>%
tidyr::drop_na()
# Create the named vector
geneList <- setNames(
gene_abundance$logFC
rownames(gene_abundance))
```

2. Run GSEA on the KEGG pathwaydatabase (*see* **Note 30**):

```
gsea_kegg <- gseKEGG(geneList = geneList,
organism = "hsa",
keyType = "uniprot",
pAdjustMethod = "BH",
pvalueCutoff = 1, # Cutoff set to one to keep all
results
verbose = FALSE)
```

3. Visualize GSEA results. With the "clusterProfiler" tool, plot the GSEA running enrichment score for a specific set of pathways/terms (*see* **Note 31**) (see Fig. 9).

```
gsea_kegg2 <- arrange(gsea_kegg, desc(abs(NES)))
# Plot top 3 gene sets
```

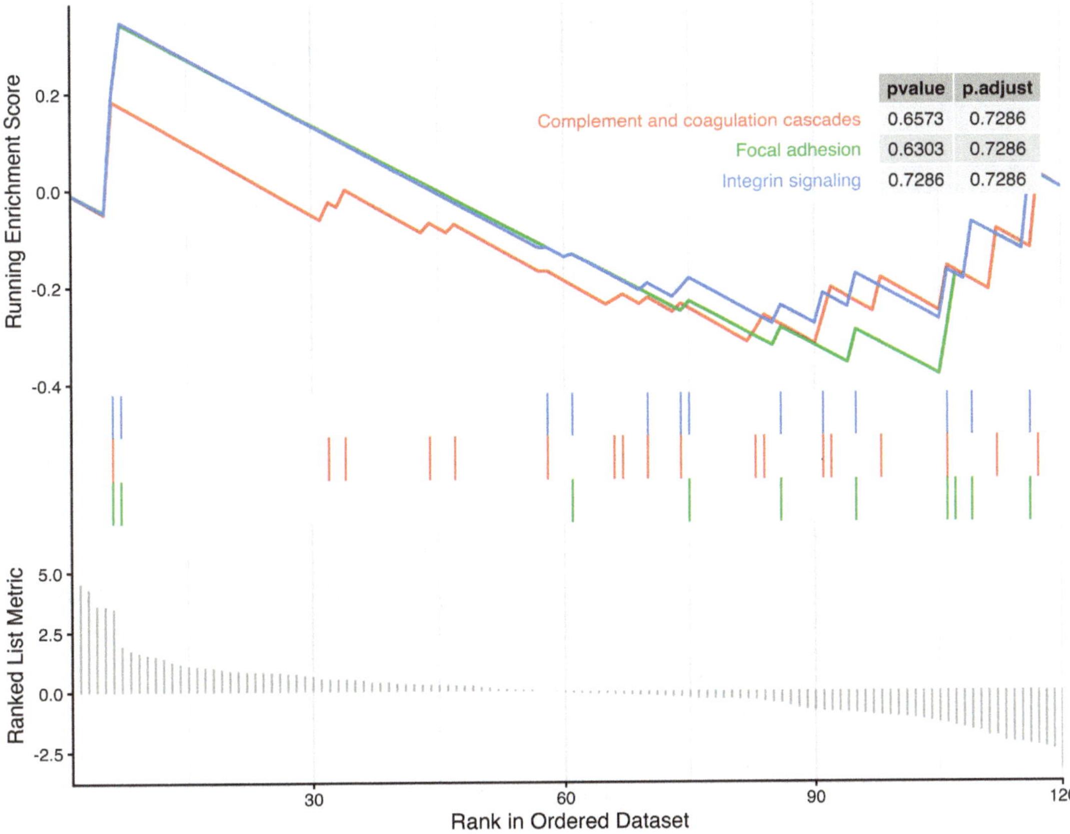


**Fig. 9** Gene set enrichment analysis (GSEA) of selected pathways. Top panel represents the running enrichment score (RES) plots for three pathways: Complement and coagulation cascades (red), Focal adhesion (green), and Integrin signaling (blue). The x-axis represents the rank of proteins in the ordered dataset based on differential expression statistics, and the y-axis shows the running enrichment score. Vertical tick marks in the middle panel indicate the positions of pathway-associated proteins within the ranked list. The lower panel displays the ranked list metric used for ordering proteins. Nominal *P* (pvalue) and adjusted *P* (p.adjust) for each pathway are indicated in the inset

```
gseaplot2(gsea_kegg2, 1:3, pvalue_table=T, base_
size=14)
```

4. Interpret the results. In this case, the GSEA did not identify significant enrichment for the examined pathways (Fig. 9). Although "Complement and coagulation cascades," "Focal adhesion," and "Integrin signaling" show modest fluctuations in running enrichment scores across the ranked protein list, none of the pathways reaches statistical significance. The enrichment curves display small positive deviations near the top of the ranked list followed by gradual declines, indicating that pathway-associated proteins are not consistently concentrated at either extreme of differential expression. The dispersed distribution of pathway member proteins across the ranked dataset, as shown by the tick marks, further supports the absence of coordinated up- or downregulation within these pathways. Overall, these results suggest that while individual proteins within these pathways may be differentially expressed, the pathways as collective functional units do not exhibit significant enrichment in this comparison.

### 3.6 Pathway Visualization

After identifying key pathways or processes in the enrichment analyses, a useful next step is visualizing the data in the context of the enriched pathways. Pathway visualization tools map experimental results onto known biochemical or signaling pathways, allowing intuitive interpretation of how protein abundances change in a network of interactions or reactions.

1. Visualize the pathway using the KEGG pathway maps included in the "Pathview" package [22]. Pathview retrieves pathway diagrams from KEGG and overlays the data (e.g., fold changes or expression values) on the genes or proteins in the pathway. To employ Pathview functions, a named numeric vector of gene data is required, as in the GSEA analysis (*see* **Note 32**).

```
gene_abundance <- res_deqms %>%
dplyr::select(gene, logFC) %>%
dplyr::arrange(desc(logFC)) %>%
tidyr::drop_na()
geneList <- setNames(
gene_abundance$logFC
gene_abundance$gene)
pv_out <- pathview(gene.data = geneList
gene.idtype = "SYMBOL"
pathway.id = "hsa04518", # pathway of interest from
the GSEA analysis
species = "hsa"
out.suffix = "test1"
limit = list(gene = 6)
both.dirs = list(gene = TRUE)
```

```
low = list(gene = "#2166ac")
mid = list(gene = "#f7f7f7")
high = list(gene = "#b2182b"))
```

Upon running this code in R, the pathview utility will download the KEGG pathway diagram for the selected pathway,

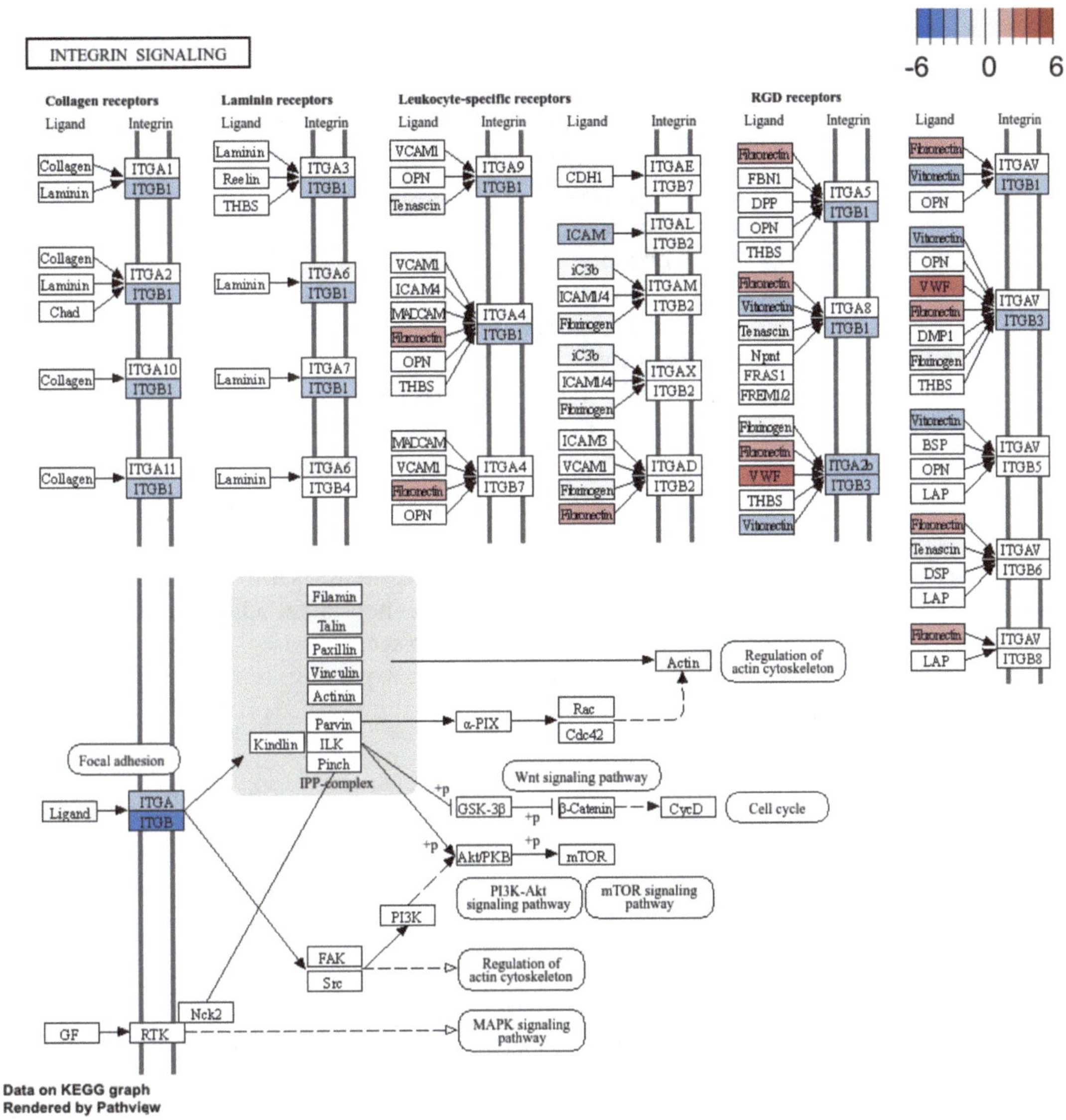


**Fig. 10** Integrin signaling pathway with quantitative colors.KEGG integrin signaling pathway rendered using Pathview with log2 fold change values mapped onto pathway components. Colored nodes represent proteins detected in the dataset, with red indicating upregulation and blue indicating downregulation (scale shown at top). The diagram illustrates ligand–integrin interactions (collagen, laminin, fibronectin, vitronectin, VWF, OPN), integrin heterodimers (e.g., ITGA/ITGB combinations), and downstream signaling modules including focal adhesion, PI3K–Akt, mTOR, MAPK, Wnt signaling, and regulation of the actin cytoskeleton

map the geneList values onto the diagram, and save the output as an image file (PNG) in the working directory, named with the pathway and out.suffix (Fig. 10).

2. Interpret the results. Mapping differential protein expression onto the KEGG integrin signaling pathway reveals coordinated alterations within extracellular matrix and integrin interactions and associated signaling modules (Fig. 10). Several integrin subunits, particularly ITGB1 and multiple ITGA partners, show differential regulation, suggesting remodeling of integrin heterodimer composition. Ligands such as VWF and fibronectin exhibit altered abundance, supporting changes in extracellular matrix engagement and adhesion dynamics. Downstream adaptor and cytoskeletal regulatory components appear comparatively less affected, indicating that the primary changes are concentrated at the receptor–ligand interface rather than broadly across all intracellular signaling branches. The observed pattern suggests modulation of cell–matrix interactions and focal adhesion processes, potentially influencing actin cytoskeleton organization and signaling through PI3K–Akt and MAPK pathways. Overall, the pathway overlay supports enrichment results pointing to altered integrin-mediated adhesion and extracellular matrix signaling as a key functional feature of the proteomic differences observed in this comparison.

### 3.7 Interactome Building

The final step in many proteomics analyses is to place the results into the context of protein interaction networks, building an interactome of the proteins that changed, to see how they connect physically or functionally. This can highlight hub proteins or modules and can help generate hypotheses about signaling cascades or protein complexes affected by the experiment.

1. Retrieve Protein–Protein Interactions (PPI) from the STRING database (version 11.5; *see* **Note 33**). The resulting data frame contains one row per interaction, including interacting protein pairs and their corresponding scores.

```
string_db <- STRINGdb$new(version="11.5", species=
9606, score_threshold=700) # Threshod to filter
only top confidence interactions and reduce network
cluttering
```

2. Transform protein identifier mapping to STRING internal IDs using the database mapping function.

```
map_df <- data.frame(UNIPROT=rownames(res_sig))
mapped <- string_db$map(map_df, "UNIPROT",
removeUnmappedRows = TRUE)
hits <- unique(mapped$STRING_id)
```

3. Extract interaction edges. All pairwise interactions among the mapped proteins are retrieved from STRING.

```
edges <- string_db$get_interactions(hits) #
Retrieve the interaction network for these hits
```

4. To use interaction confidence as a quantitative edge attribute, rescale the STRING combined score to a 0–1 range and store it as an edge weight, as a representation of interaction strength.

```
edges$weight <-edges$combined_score / 1000
```

5. Construct the Network by converting the interaction table into an undirected graph using the "igraph" [23] package. In this representation, "Nodes" correspond to proteins; "Edges" represent protein–protein interactions; "Edge weights" reflect STRING interaction confidence.

```
g <- graph_from_data_frame(
d = unique(edges[, c("from","to","weight")]),
directed = FALSE)
```

6. Integrate the quantitative proteomics results to contextualize the network with experimental results by mapping the quantitative attributes from the differential expression analysis onto the network nodes.

```
# Get variables to add as atributes
node_attr <- res_sig %>%
tibble::rownames_to_column("UNIPROT") %>%
transmute(
UNIPROT
gene = gene
logFC = logFC
fdr = adj.P.Val
neglog10fdr = -log10(adj.P.Val))
node_attr_string <- mapped %>%
left_join(node_attr, by = "UNIPROT")
# Add attributes
m <- match(V(g)$name, node_attr_string$STRING_id)
V(g)$UNIPROT <- node_attr_string$UNIPROT[m]
V(g)$gene <- node_attr_string$gene[m]
V(g)$logFC <- node_attr_string$logFC[m]
V(g)$fdr <- node_attr_string$fdr[m]
V(g)$neglog10fdr <- node_attr_string$neglog10fdr[m]
V(g)$label <- V(g)$gene
```

7. Calculate Network topology metrics (*see* **Note 34**).

```
# Topology metrics
V(g)$degree <- igraph::degree(g, mode="all")
V(g)$betweenness <- betweenness(g, directed=FALSE,
normalized=TRUE)
```

```
V(g)$closeness <- closeness(g)
V(g)$eigen <- eigen_centrality(g)$vector
# Proteins with the highest degree are identified as
potential hub proteins
top_hubs <- V(g)$name[order(V(g)$degree, decreasing
= TRUE)][seq_len(min(20, igraph::vcount(g)))]
```

8. Detect "communities" to identify densely connected sub-networks (modules) by simplifying the graph through duplicate and self-loop edges removal and subjecting to community detection using the Louvain algorithm.

```
g <- igraph::simplify(g, remove.multiple=TRUE,
remove.loops=TRUE)
com <- cluster_louvain(g, weights = E(g)$weight)
V(g)$module <- membership(com)
```

9. Visualize the network (*see* **Notes 35 and 36**). This visualization highlights hub proteins by degree (node size) and the direction/magnitude of their differential abundance (node-color) (see Fig. 11A and visualization code in **Note 37**).
10. Visualize network modules. This visualization emphasizes structural organization rather than differential abundance. In this case, nodes are colored according to Louvain community membership, allowing visual identification of modules within the network (see Fig. 11B and visualization code in **Note 38**).
11. Create the layout, add nodes, and set the degree:

```
set.seed(27)
# Get layout coordinates
lay <- create_layout(g, layout = "fr")
deg <- igraph::degree(g)
nodes <- lay %>%
as_tibble() %>%
mutate(
label = as.character(label)
degree = unname(deg[name]) # name comes from
layout, matches vertex name)
```

12. Map selected enriched GO terms onto network nodes:

```
terms_keep <- c("GO:0006954", "GO:0006898",
"GO:0006629") # select the desired terms
go_membership <- enrich_data
filter(ID %in% terms_keep) %>%
dplyr::select(ID, geneID) %>%
mutate(geneID = str_split(geneID, "/")) %>%
tidyr::unnest(geneID) %>%
transmute(term = ID, label = geneID) %>%
```

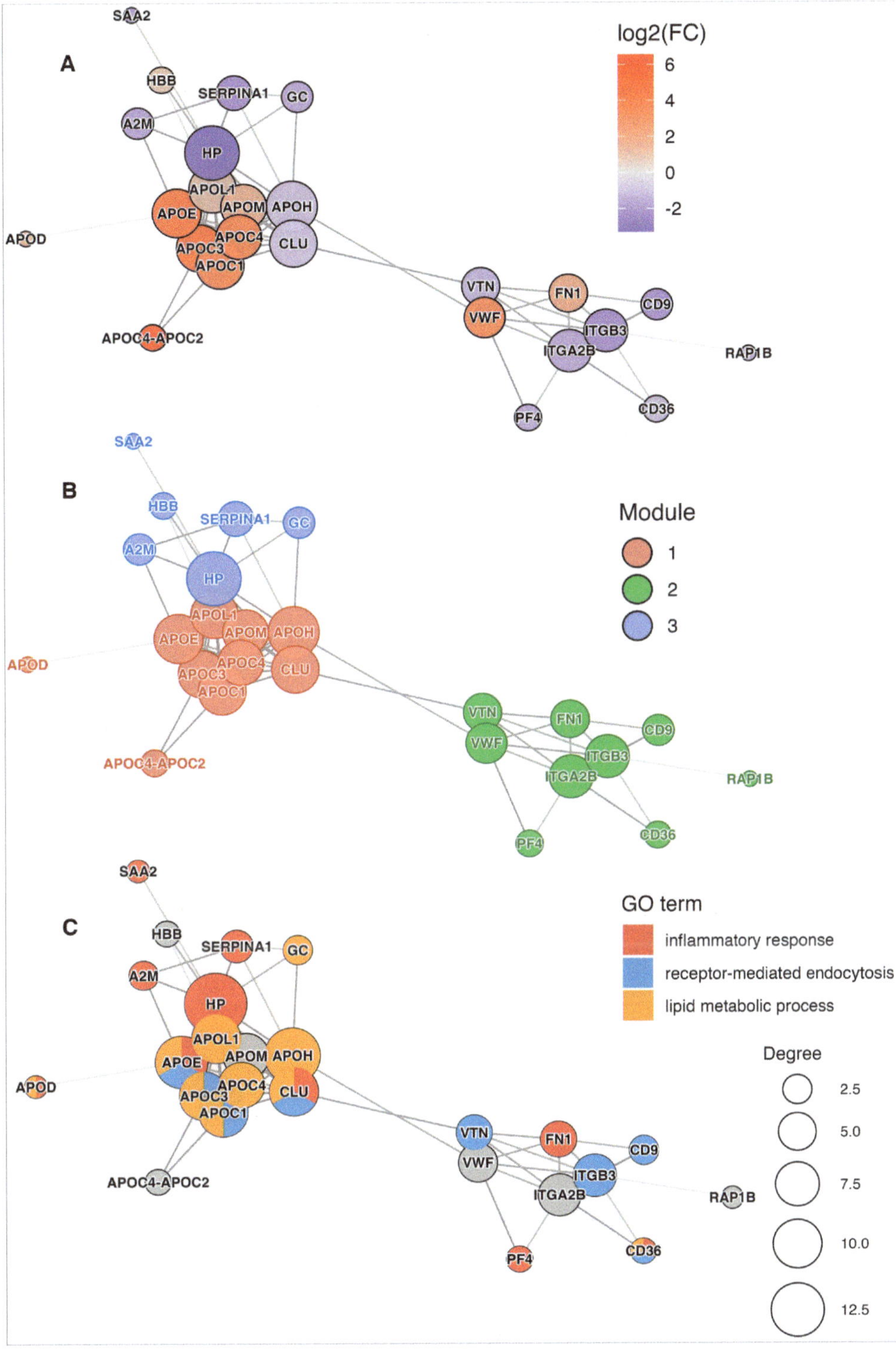


**Fig. 11** PPI networks. Protein–protein interaction (PPI) network generated using STRING for differentially expressed proteins. Nodes represent proteins, and edges represent known or predicted interactions. Node size corresponds to degree (number of interactions). (**A**) STRING PPI network colored by log2 fold change

```
distinct()
go_lut <- enrich_data %>%
dplyr::filter(ID %in% terms_keep) %>%
dplyr::select(ID, Description) %>%
dplyr::distinct()
go_labels <- setNames(go_lut$Description,
go_lut$ID)
go_labels <- c(go_labels, Other = "Other / not in
selected GO terms")
fill_cols <- c("GO:0006954" = "#F15854FF",
"GO:0006898" = "#5DA5DAFF", "GO:0006629"=
"#FAA43AFF", "Other" = "grey90")
```

13. Visualize the GO term membership as a pie-chart node to simultaneously represent multiple functional annotations per protein. For this, use the "scatterpie" package [24]:

```
pie_df <- nodes %>%
dplyr::select(name, x, y, label, degree) %>%
left_join(go_membership, by = "label") %>%
mutate(in_term = 1) %>%
tidyr::pivot_wider(
names_from = term
values_from = in_term
values_fill = 0) %>%
rowwise() %>%
mutate(total = sum(c_across(all_of(terms_keep))))
%>%
ungroup() %>%
mutate(across(all_of(terms_keep), ~ ifelse(total >
0,.x / total, NA_real_))) %>%
mutate(
Other = ifelse(is.na(total) | total == 0, 1, 0)
'GO:0006954' = ifelse(is.na('GO:0006954'), 0,
'GO:0006954')
'GO:0006898' = ifelse(is.na('GO:0006898'), 0,
'GO:0006898')
'GO:0006629' = ifelse(is.na('GO:0006629'), 0,
'GO:0006629')) %>%
mutate(r = scales::rescale(degree, to = c(0.2,
0.55)))
```

14. Visualize the GO on the network. For each node, pie segments represent selected GO terms, and node radius is scaled

**Fig. 11** (continued) (red = upregulated; blue = downregulated in HC compared to PTB). (**B**) STRING PPI network colored by detected modules identified through network clustering. (**C**) STRING PPI network colored by GO term association. Nodes are colored based on association with inflammatory response (red), receptor-mediated endocytosis (blue), lipid metabolic process (orange), or other/unselected GO terms (grey) according to "degree." Proteins not associated with selected terms are assigned to an "Other" category. This visualization integrates network topology, quantitative proteomics results and functional enrichment annotations within a single interpretable graphical representation (see Fig. 11C and visualization code in **Note 39**).

15. Interpret the results. The STRING PPI network reveals two main interconnected protein clusters corresponding to distinct functional modules (Fig. 11). A prominent cluster enriched in apolipoproteins (APOE, APOC1, APOC3, APOC4–APOC2, APOM) shows coordinated upregulation, suggesting modulation of lipid transport and lipoprotein metabolism. In contrast, a second cluster comprising extracellular matrix and adhesion-related proteins (ITGB3, ITGA2B, FN1, VTN, CD9) displays predominantly downregulated expression. Central hub proteins such as HP and APOE exhibit high connectivity, indicating potential regulatory importance within the network. Network clustering reveals three primary modules within the interaction network. Module 1 (red) consists largely of apolipoproteins and lipid-associated proteins, reflecting a tightly connected lipid metabolism subnetwork. Module 2 (green) includes integrin subunits and extracellular matrix components, consistent with adhesion and cell–matrix interaction processes. Module 3 (blue) comprises inflammatory and acute-phase proteins such as HP, SERPINA1, and A2M. The clear segregation of these modules suggests that the observed proteomic changes are organized into discrete functional communities rather than uniformly distributed across the network. Some hub proteins, such as APOE and CLU, appear to bridge functional categories, indicating potential crosstalk between lipid metabolism and endocytic processes. The spatial segregation of GO annotations across the network reinforces the presence of biologically distinct but interconnected functional modules, supporting enrichment findings that implicate inflammatory and adhesion-related pathways in the observed proteomic differences.

## 4 Notes

1. Proteomic data analysis is not extremely memory intensive for moderate dataset sizes (thousands of proteins across tens of samples), but sufficient memory and CPU speed will facilitate data processing and algorithm implementation (especially for steps like missing value imputation and enrichment analysis).

2. The quantification table is typically only a few MBs, but if pathway databases or large annotation packages are used, allow a few hundred MBs for installation.
3. While R packages, databases, and other software resources can be used offline if pre-downloaded, an internet connection simplifies the task when performing enrichment or pathway mapping.
4. This depends on the user's data. Ensure that the protein IDs in your dataset are consistent and publicly available (i.e., UniProt accessions or gene symbols). An ID mapping step may be needed if, for example, the employed quantification software outputs protein IDs that need to be converted to gene symbols or Entrez IDs for enrichment analysis or network mapping.
5. The GO knowledgebase is generally used for functional annotation. In R, GO term mappings are obtained via organism-specific annotation packages (e.g., "org.Hs.eg.db" for human). No manual download is required if using these packages, but it is important to have the correct OrgDb for the corresponding organism.
6. For metabolic and signaling pathways, "clusterProfiler" provides functions to query KEGG, while "pathview" will fetch pathway diagrams as needed.
7. STRING is a comprehensive resource of known and predicted protein–protein associations, which will be used for network analysis. For R, the "STRINGdb" package simplifies retrieval of these interactions.
8. Abundance can be represented as spectral counts, MS1 peak intensities, LFQ normalized intensities, iTRAQ/TMT reporter ion intensities, *Z* scores, etc., depending on the experiment. Additional columns can indicate the gene id or the number of peptide sequences used to quantify each protein.
9. If present, other covariates (like batch/pool, experimental details, etc.) can also be included in the metadata. Metadata can be provided as a separate design matrix or be generated during the pipeline. In the latter case, the sample names should be consistent along the dataset (i.e., batch_condition_replicate or any variation of this formulae).
10. It is important to consider the nature of missingness. "Missing Completely at Random" (MCAR): Random instrument or processing effects; these missing values have no relation to true abundance. "Missing at Random" (MAR): Missingness related to other observed variables (e.g., a protein might be missing in a sample because its total protein load was lower).

"Missing Not at Random" (MNAR): Systematic missingness, often left-censored data where missing indicates an abundance below detection limit. In label-free proteomics, MNAR is common: a protein may be consistently missing in a condition where it is truly absent or very low.

11. Common formats are CSV, TSV, TXT, or Excel. These formats can be read using the corresponding functions "readr::read_csv()," "readr::read_tsv()," "readr::read_table()," and "readxl::read_excel()."
12. The metadata file is optional as the md object can be generated programmatically based on the naming of the columns:

```
# Sample columns naming pattern: batch_condition_rep
(edit to match your data     )
sample_cols_pattern <- "^[A-Za-z0-9]+_[A-Za-z0-9]+
_[0-9]+$"
sample_cols <- grep(sample_cols_pattern,
colnames(data), value = TRUE)
md <- tibble(
Column = sample_cols) %>%
tidyr::separate(
Column
into = c("Batch", "Condition", "Replicate")
sep = "_"
remove = FALSE
extra = "merge") %>%
mutate(
Batch = factor(Batch)
Condition = factor(Condition)
Replicate = factor(Replicate))
```

13. For convenience, it is advisable to drop the rest of unused columns.
14. If there are protein groups (multiple identifiers in one row), you may want to simplify to a single representative ID (e.g., leading accession).
15. Optionally remove proteins identified by only a single peptide as the identification confidence may be questionable, leading to false positives. Nevertheless, modern methods like "DEqMS" allow keeping one-peptide proteins by adjusting their variance to maximize coverage.
16. Many quantitative software tools report undetected signals as zero rather than missing values. These zeros must be recoded as NA to prevent mathematical errors during log2 transformation and to ensure downstream normalization and imputation algorithms correctly identify these as missing data.
17. For example, retain only features quantified in at least 70% of samples. Similarly, samples with a missingness over the 30%

can also be excluded. These missing values can be imputed; however, this should be done carefully, as inappropriate handling of missing data can inflate statistical significance and bias downstream analyses.

18. Different normalization families are tailored to different experimental assumptions. Simple sample-shifting methods correct global technical effects under the assumption of limited differential expression, whereas sample-to-reference methods address distributional or intensity-dependent biases. Variance-stabilizing approaches focus on reducing mean–variance dependency to improve statistical inference, while model-based methods explicitly estimate and remove technical effects, enabling robust normalization even in heterogeneous datasets. A representative of each family is tested herein.

19. If substantial differences remain between sample medians or distributions show poor overlap, normalization may be insufficient.

20. PCA visualization code:

```
# Variance explained for axis labels
ve <- (pca_res$sdev^2) / sum(pca_res$sdev^2)
# Plotting code
ggplot(scores, aes(x = PC1, y = PC2, color = Batch,
shape = Condition)) +
geom_point(size = 5) +
labs(
title = "RobNorm"
x = paste0("PC1 (", round(100 * ve(1), 1), "%)")
y = paste0("PC2 (", round(100 * ve(2), 1), "%)")) +
theme_bw() +
theme(
title = element_text(size = 15)
plot.title = element_text(hjust = 0.5)
axis.title = element_text(size = 15)
axis.text = element_text(size = 10)
legend.position = "right")
```

21. For MCAR/MAR-type missingness, an imputation method that predicts missing values from observed data (i.e., using similarities between samples or proteins) can be used. For MNAR, a common approach is to impute a small value simulating the detection limit [4]. In R, "msImpute package" [17] can be used to handle both MAR and MNAR. For each feature, msImpute first estimates the distribution of intensities under an MAR assumption using a low-rank matrix completion approach that exploits global similarity patterns across samples. In parallel, a group-specific MNAR distribution is estimated using a down-shifted Gaussian model that mimics left-censoring due to the detection limit. The final imputed

values are obtained as a weighted barycentric average of the MAR- and MNAR-based estimates, where the weights are determined by an entropy-based metric that assesses whether missing values occur randomly across samples or in a group-specific manner.

22. Code for NA assessment heatmap:

```
# Construct matrix M to capture missing entries
M <- ifelse(is.na(data_norm),1,0)
M <- M[hdp$msImpute_feature,]
# Build the annotation
ann_col <- md %>%
filter(Label %in% colnames(M)) %>%
dplyr::select(Label, Batch, Condition) %>%
distinct() %>%
as.data.frame()
rownames(ann_col) <- ann_col$Label
ann_col$Label <- NULL
# Ensure factors
ann_col$Batch <- factor(ann_col$Batch)
ann_col$Condition <- factor(ann_col$Condition)
# Build annotation colors
cond_lvls <- levels(ann_col$Condition)
batch_lvls <- levels(ann_col$Batch)
ann_colors <- list(
Condition = setNames(brewer.pal(max(3, length(cond_
lvls)), "Set1")[seq_along(cond_lvls)], cond_lvls)
Batch = setNames(brewer.pal(max(3, length(batch_
lvls)), "Set2")[seq_along(batch_lvls)],
batch_lvls))
# Plotting code
ha <- HeatmapAnnotation(
df = ann_col
col = ann_colors
annotation_name_side = "right"
gp = gpar(col = "white", lwd = 0.5))
hm <- Heatmap(M
column_title = "NA pattern, columns ordered by
dropout similarity"
name = "Pressence"
col = c("#FFEFDB", "#8FBC8F")
show_row_names = FALSE
show_column_names = FALSE
cluster_rows = TRUE
cluster_columns = TRUE
show_column_dend = FALSE
show_row_dend = FALSE
top_annotation = ha
row_names_gp = gpar(fontsize = 7)
column_names_gp = gpar(fontsize = 8)
heatmap_legend_param = list(#direction =
"horizontal"
labels = c("missing", "observed")
legend_width = unit(6, "cm"))#
```

```
#rect_gp = gpar(col = NA, lwd = 0.5))
hmd <- draw(hm
heatmap_legend_side = "right"
annotation_legend_side = "right"
merge_legends = TRUE)
```

23. Proteomic data often includes proteins identified only by a few peptides. These proteins generally exhibit higher variance during quantification compared to proteins supported by a larger number of peptides. Standard limma assumes a common prior variance across all proteins, an assumption that may not hold for this data. As a result, proteins quantified by a single peptide may be penalized due to the expected high variance, even when their observed variance is relatively low. DEqMS accounts for the relationship between variance and the number of peptides used for quantification, allowing more accurate variance estimation and the inclusion of more proteins without increasing false positives.
24. Code for Volcano plot:

```
EnhancedVolcano(res_deqms, # DEA results
x = "logFC", # FC column
y = "sca.adj.pval", # Adjusted P column
lab = res_deqms$gene, # Gene symbols for
visualization
labSize = 5
pCutoff = 5e-2, # P cut-off
FCcutoff = 1, # FC cut-off
title = "HC vs. PTB"
subtitle = bquote(italic("DEqMS"))) +
scale_x_continuous(limits = c(-8, 8), breaks =
seq(-8, 8, by = 2)) +
scale_y_continuous(limits = c(0, 10), breaks =
seq(0, 10, by = 2))
```

25. Differentially expressed proteins heatmap:

```
# Build matrix
top_proteins_deqms <- rownames(res_deqms[abs(res_
deqms$logFC) > 1,])[ order(res_deqms$sca.adj.
pval[abs(res_deqms$logFC) > 1])][1:20]
cols_keep <- md_qc$Label[md_qc$Condition %in%
c("HC", "PTB")]
mat_top_deqms <- data[top_proteins_deqms, cols_
keep, drop = FALSE]
rownames(mat_top_deqms) <- res_
deqms$gene[match(rownames(mat_top_deqms),
rownames(res_deqms))]
# Build annotation
ann_col <- md %>%
filter(Label %in% colnames(mat_top_deqms)) %>%
```

```
dplyr::select(Label, Batch, Condition) %>%
as.data
rownames(ann_col) <- ann_col$Label
ann_col$Label <- NULL
ann_col$Batch <- factor(ann_col$Batch)
ann_col$Condition <- factor(ann_col$Condition)
# Build annotation colors
cond_lvls <- levels(ann_col$Condition)
batch_lvls <- levels(ann_col$Batch)
ann_colors <- list(
Condition = setNames(brewer.pal(max(3, length(cond_
lvls)), "Set1")[seq_along(cond_lvls)], cond_lvls)
Batch = setNames(brewer.pal(max(3, length(batch_
lvls)), "Set2")[seq_along(batch_lvls)],
batch_lvls))
# Plotting code
top_ha <- HeatmapAnnotation(
df = ann_col
col = ann_colors
annotation_name_side = "right"
gp = gpar(col = "white", lwd = 0.5))
ht <- Heatmap(
mat_top_deqms
name = "Intensity"
top_annotation = top_ha
cluster_rows = TRUE
cluster_columns = TRUE
clustering_distance_rows = "euclidean"
clustering_distance_columns = "euclidean"
clustering_method_rows = "complete"
clustering_method_columns = "complete"
show_row_names = TRUE
show_column_names = TRUE
row_names_gp = gpar(fontsize = 10)
column_names_gp = gpar(fontsize = 11)
col = viridis(100)
rect_gp = gpar(col = "white", lwd = 0.5))
htt<-draw(
ht
heatmap_legend_side = "right"
annotation_legend_side = "right"
merge_legends = TRUE)
```

26. The background is formed by all proteins that were "measurable" in the experiment, avoiding the use of the entire genome unless your proteomic results covered essentially all expressed genes. Using the measured proteome as background avoids bias from subsets of proteins being underrepresented in the experiment.
27. Ensure the identifiers match what the GO analysis function expects. For instance, the "clusterProfiler" package takes gene symbols, Entrez gene IDs or UniProt IDs among others. Use

"keytypes(org.Hs.eg.db)" to see all available options. Provide the correct keyType (SYMBOL, ENTREZID, UNIPROTID, etc.) and OrgDb ["org.Hs.eg.db" for human (*Homo sapiens*), "org.Mm.eg.db" for mouse (*Mus musculus*), etc.] for mapping.

28. In the example used in this chapter, no significant terms were retrieved under this threshold, so for the sake of the chapter a raw *P* threshold <0.05 was instead used, by running twice the enrichment and setting a "pvalueCutoff = 0.35" and "qvalueCutoff = 0.33" accordingly.

29. Often, GO enrichment yields many overlapping terms (due to the hierarchical nature of GO), and it can be useful to simplify or cluster them to make the results more manageable.

30. GSEA can be performed for various gene sets in parallel. The user can utilize GO terms, KEGG pathways, or custom gene sets (e.g., MSigDB's Hallmark sets), using "clusterProfiler," dedicated functions like "gseGO" for GO terms and "gseKEGG" for KEGG pathways or alternatively, "clusterProfiler's generic GSEA()" function as it can take any set of gene sets via a "TERM2GENE" input.

31. This code will produce the classic GSEA plot for a set of pathways. This is the running enrichment score on top vs the ranked gene list in the bottom sorted by magnitude of the selected metric. In the middle, ticks show where genes of the corresponding set appear in the list. The GSEA tabular results can be accessed at "gsea_go@result."

32. As pathways of interest, the top enriched KEGG pathways from either ORA or GSEA can be used.

33. STRING database integrates experimentally validated and predicted functional associations across multiple organisms. Each interaction is assigned a combined confidence score (ranging from 0 to 1000) that reflects the strength and reliability of the evidence supporting the interaction, based on multiple sources such as experimental data, curated databases, co-expression, and computational predictions. To increase confidence in the resulting network, only interactions with a combined confidence score ≥700 are retained.

34. Structural properties of the network can serve to identify potential hub proteins, which may represent central regulators within the interactome. For this, several graph-theoretical centrality measures can be computed, such as "Degree": number of direct interactions per protein; "Betweenness centrality": frequency with which a node lies on shortest paths; "Closeness centrality": inverse average shortest-path distance; "Eigenvector centrality": influence of a node within the network.

35. Several example visualizations can be generated directly in R using "ggraph" [25] to illustrate the integration of network topology with quantitative data. However, Cytoscape [26] is also a powerful tool for interactive network visualization, exploration, and analysis. Networks can be exported to Cytoscape using the "RCy3" [27] interface, enabling access to Cytoscape's advanced layout algorithms, extensive styling options, and additional network analysis plugins.
36. In this example, a force-directed layout (Fruchterman–Reingold) is used for visualization. This vertex layout is determined by the forces pulling nodes together, which means that the nodes connected are closer as the edge weight grows.
37. STRING PPI network colored by log2 fold change:

```
set.seed(27)
nw <-
ggraph(g, layout = "fr") +
geom_edge_link(aes(alpha = weight), colour =
"gray", show.legend = FALSE) +
geom_node_point(
aes(fill = logFC, size = degree)
shape = 21, colour = "black", stroke = 0.6, alpha =
0.95) +
scale_size_continuous(range = c(4, 15), name =
"Degree") +
scale_fill_gradient2(
low = "blue", mid = "grey90", high = "red"
midpoint = 0, name = "log2(FC)") +
guides(
color = "none"
fill = guide
size = guide_legend(order = 2)) +
theme_void() +
theme(plot.margin = margin(10, 10, 10, 10)
legend.position = "right"
legend.box = "vertical") +
ggtitle("STRING PPI colored by log2(FC)") +
geom_shadowtext(
#data
aes(x = x, y = y, label = label)
colour = "black"
bg.colour = "white"
bg.r = 0.1
size = 3, fontface = "bold"
inherit.aes = FALSE) + coord_fixed()
nw
```

38. STRING PPI network colored by detected modules

```
set.seed(27)
mods <- V(g)$module
```

```
mod_ids <- sort(unique(mods))
module_pal <- setNames(hue_pal(l = 65, c = 100)
(length(mod_ids)), mod_ids)
module_pal_fill <- lighten(module_pal, amount =
0.30)
module_pal_border <- darken( module_pal, amount =
0.30)
nw_mod <-
ggraph(g, layout = "fr") +
geom_edge_link(aes(alpha = weight), colour =
"grey75", show.legend = FALSE) +
geom_node_point(aes(size = degree, fill =
factor(module), color = factor(module))
shape = 21, stroke = 0.6, alpha = 0.95) +
scale_size(range = c(4, 15), name = "Degree") +
scale_fill_manual(values = module_pal_fill, name =
"Module") +
scale_color_manual(values = module_pal_border,
guide = "none") +
guides
color = "none"
fill = guide_legend(order = 1, override.aes =
list(size = 6))
size = guide_legend(order = 2)) +
theme_void() +
theme(plot.margin = margin(10, 10, 10, 10)
legend.position = "right"
legend.box = "vertical") +
ggtitle("STRING PPI colored by module") +
geom_shadowtext(
#data
aes(x = x, y = y, label = label, colour =
factor(module))
bg.colour = "white"
bg.r = 0.1
size = 3, fontface = "bold"
inherit.aes = FALSE) + coord_fixed()
nw_mod
```

39. STRING PPI network colored by GO term association

```
set.seed(27)
nw_go <- ggraph(lay) +
geom_edge_link(aes(alpha = weight), colour =
"grey75", show.legend = FALSE) +
geom_node_point(
data = pie_df
aes(x = x, y = y, size = degree)
alpha = 0) +
scale_size_continuous(
name = "Degree"
range = c(3, 12),) +
ggforce::geom_circle(
```

```
data = pie_df
aes(x0 = x, y0 = y, r = r)
inherit.aes = FALSE
colour = "black"
fill = "grey80"
linewidth = 0.6) +
scatterpie::geom_scatterpie(
data = pie_df
aes(x = x, y = y, r = r)
cols = c(terms_keep, "Other")
colour = NA, alpha = 0.95) +
scale_fill_manual(
values = fill_cols
name = "GO term"
labels = go_labels[c(terms_keep, "Other")]) +
theme_void() +
ggtitle("STRING PPI colored by GO terms") +
geom_shadowtext(
data = pie_df
aes(x = x, y = y, label = label)
colour = "black", bg.colour = "white", bg.r = 0.12
size = 3, fontface = "bold"
inherit.aes = FALSE) +
guides
fill = guide_legend(order = 1)
size = guide_legend(
order = 2
override.aes = list(alpha = 1, shape = 21, fill =
NA, colour = "black", stroke = 0.4))) +
theme(legend.position = "right", legend.box =
"vertical") + coord_fixed()
nw_go
```

## Acknowledgments

This study was supported by a grant from the Obra Social Fundación "la Caixa" (LCF/PR/HR23/52430018). We are especially grateful to the research group led by Dr. Martín-Cófreces for providing access to the computational infrastructure funded by this project, including the ASRock W790 WS R2.0 workstation [Intel Xeon W5-3525 16-core processor (4.8 GHz), 64 GB RAM, NVIDIA RTX A4000 GPU, and Ubuntu 64-bit operating system], as well as real experimental data, which was essential for the development and execution of this work. We also thank the Methodological Support Unit of IIS-Princesa (Madrid, Spain) for their valuable assistance in the development of the software.

## References


1. Shi Z, Wen B, Gao Q, Zhang B (2021) Feature selection methods for protein biomarker discovery from proteomics or multiomics data. Mol Cell Proteomics 20:100083. https://doi.org/10.1016/j.mcpro.2021.100083. PubMed PMID: 33887487; PubMed Central PMCID: PMC8165452.
2. Carvalho LB, Teigas-Campos PAD, Jorge S, Protti M, Mercolini L, Dhir R et al (2024) Normalization methods in mass spectrometry-based analytical proteomics: a case study based on renal cell carcinoma datasets. Talanta 266:124953. https://doi.org/10.1016/j.talanta.2023.124953
3. Wei R, Wang J, Su M, Jia E, Chen S, Chen T et al (2018) Missing value imputation approach for mass spectrometry-based metabolomics data. Sci Rep 8(1):663. https://doi.org/10.1038/s41598-017-19120-0
4. Shen M, Chang YT, Wu CT, Parker SJ, Saylor G, Wang Y et al (2022) Comparative assessment and novel strategy on methods for imputing proteomics data. Sci Rep 12(1):1067. https://doi.org/10.1038/s41598-022-04938-0
5. Locard-Paulet M, Doncheva NT, Morris JH, Jensen LJ (2024) Functional analysis of MS-based proteomics data: from protein groups to networks. Mol Cell Proteomics 23(12):100871. https://doi.org/10.1016/j.mcpro.2024.100871. PubMed PMID: 39486590; PubMed Central PMCID: PMC11667155.
6. Unidad de Metodología Bioestadística y Bioinformática IIS Princesa (2026) UMBB-IIS-Princesa/Quantitative-Proteomics- Pipeline: Initial release—Quantitative Proteomics Pipeline [Internet]. Zenodo. https://doi.org/10.5281/zenodo.18849537
7. The UniProt Consortium (2025) UniProt: the Universal Protein Knowledgebase in 2025. Nucleic Acids Res 53(D1):D609–D617. https://doi.org/10.1093/nar/gkae1010
8. The Gene Ontology Consortium (2026) The Gene Ontology knowledgebase in 2026. Nucleic Acids Res 54(D1):D1779–D1792. https://doi.org/10.1093/nar/gkaf1292
9. Kanehisa M, Goto S (2000) KEGG: kyoto encyclopedia of genes and genomes. Nucleic Acids Res 28(1):27–30. https://doi.org/10.1093/nar/28.1.27. PubMed PMID: 10592173; PubMed Central PMCID: PMC102409.
10. Szklarczyk D, Kirsch R, Koutrouli M, Nastou K, Mehryary F, Hachilif R et al (2023) The STRING database in 2023: protein–protein association networks and functional enrichment analyses for any sequenced genome of interest. Nucleic Acids Res 51(D1):D638–D646. https://doi.org/10.1093/nar/gkac1000
11. Barkovits K, Pacharra S, Pfeiffer K, Steinbach S, Eisenacher M, Marcus K et al (2020) Reproducibility, specificity and accuracy of relative quantification using spectral library-based data-independent acquisition*. Mol Cell Proteomics 19(1):181–197. https://doi.org/10.1074/mcp.RA119.001714
12. Thompson A, Schäfer J, Kuhn K, Kienle S, Schwarz J, Schmidt G et al (2003) Tandem mass tags: a novel quantification strategy for comparative analysis of complex protein mixtures by MS/MS. Anal Chem 75(8):1895–1904. https://doi.org/10.1021/ac0262560
13. Jiang Y, Rex DAB, Schuster D, Neely BA, Rosano GL, Volkmar N et al (2024) Comprehensive overview of bottom-up proteomics using mass spectrometry. ACS Meas Sci Au 4(4):338–417. https://doi.org/10.1021/acsmeasuresciau.3c00068
14. Biadglegne F, Schmidt JR, Engel KM, Lehmann J, Lehmann RT, Reinert A et al (2022) Mycobacterium tuberculosis affects protein and lipid content of circulating exosomes in infected patients depending on tuberculosis disease state. Biomedicines 10(4):783. https://doi.org/10.3390/biomedicines10040783
15. Perez-Riverol Y, Bandla C, Kundu DJ, Kamatchinathan S, Bai J, Hewapathirana S et al (2025) The PRIDE database at 20 years: 2025 update. Nucleic Acids Res 53(D1):D543–D553. https://doi.org/10.1093/nar/gkae1011
16. Arend L, Adamowicz K, Schmidt JR, Burankova Y, Zolotareva O, Tsoy O et al (2025) Systematic evaluation of normalization approaches in tandem mass tag and label-free protein quantification data using PRONE. Bioinformatics 26(3):bbaf201. http://biorxiv.org/lookup/doi/10.1101/2025.01.27.634993 https://doi.org/10.1101/2025.01.27.634993
17. Hediyeh-Zadeh S, Webb AI, Davis MJ (2023) MsImpute: estimation of missing peptide intensity data in label-free quantitative mass spectrometry. Mol Cell Proteomics 22(8):100558. https://doi.org/10.1016/j.mcpro.2023.100558
18. Ritchie ME, Phipson B, Wu D, Hu Y, Law CW, Shi W et al (2015) limma powers differential expression analyses for RNA-sequencing and microarray studies. Nucleic Acids Res 43(7):e47. https://doi.org/10.1093/nar/gkv007
19. Zhu Y, Orre LM, Zhou Tran Y, Mermelekas G, Johansson HJ, Malyutina A et al (2020)

DEqMS: a method for accurate variance estimation in differential protein expression analysis. MolCell Proteomics 19(6):1047–1057. https://doi.org/10.1074/mcp.TIR119.001646
20. Xu S, Hu E, Cai Y, Xie Z, Luo X, Zhan L et al (2024) Using clusterProfiler to characterize multiomics data. Nat Protoc 19(11):3292–3320. https://doi.org/10.1038/s41596-024-01020-z
21. Sayols S (2023) rrvgo: a bioconductor package for interpreting lists of Gene Ontology terms. microPublication Biol https://doi.org/10.17912/micropub.biology.000811
22. Luo W, Brouwer C (2013) Pathview: an R/Bioconductor package for pathway-based data integration and visualization. Bioinformatics 29(14):1830–1831. https://doi.org/10.1093/bioinformatics/btt285
23. Antonov M, Csárdi G, Horvát S, Müller K, Nepusz T, Noom D et al (2023) igraph enables fast and robust network analysis across programming languages [Internet]. arXiv.. https://doi.org/10.48550/arXiv.2311.10260
24. Yu G, Xu S (2025) scatterpie: Scatter Pie Plot [Internet] . https://cran.r-project.org/web/packages/scatterpie/index.html
25. Pedersen TL, RStudio (2025) ggraph: an implementation of grammar of graphics for graphs and networks [Internet] . https://cran.r-project.org/web/packages/ggraph/index.html
26. Shannon P, Markiel A, Ozier O, Baliga NS, Wang JT, Ramage D et al (2003) Cytoscape: a software environment for integrated models of biomolecular interaction networks. Genome Res 13(11):2498–2504. https://doi.org/10.1101/gr.1239303 PubMed PMID: 14597658; PubMed Central PMCID: PMC403769.
27. Gustavsen JA, Pai S, Isserlin R, Demchak B, Pico AR (2019) RCy3: network biology using Cytoscape from within R [Internet]. F1000Research. https://doi.org/10.12688/f1000research.20887.3

# Index


## A

Actin. *See* F-actin, *See* G-actin
Actin-binding proteins (ABPs) ........... 67–68, 72, 77–78
Aggregation .................................. 18, 49–50, 333, 337
Alignment ........ 51, 122, 133, 142–143, 223, 225–229, 237–238, 266, 269, 272–274, 287, 291–292, 295–296, 359
Ampicillin. *See* Selectable marker
Analysis ............................................ 2, 9–11, 24, 32, 50, 56, 61, 64, 68–69, 71–73, 79–81, 84, 91, 102, 113–115, 123–124, 127, 130, 133–137, 139–140, 149, 154, 156, 158, 160, 162–163, 166, 168–170, 174, 183, 186, 199–200, 206, 213–214, 216–217, 219, 235–237, 239–240, 245–246, 248–249, 252, 257–258, 260–261, 264, 271, 274, 277, 284, 292, 294–295, 310, 316, 319, 332–333, 338–339, 341–344, 350, 353–356, 360–361, 366–371, 374–376, 379, 383–384, 389, 391
Antifade .................................................. 229, 231–232
Arginylation .................................................. 323–329
Avidin ........................................................ 34, 42, 44
Axoneme ........................ 111–113, 115, 123–124, 243

## B

Bacmid .......................... 41, 43, 47–48, 89, 94–95, 108
Bacteria. *See* Protein purification
Baculovirus ...................... 18, 40–41, 43, 48–49, 89, 95
Biosafety level 2 laboratory (BSL2) ................. 131, 153
Biotin ............................ 21–22, 31–32, 34, 42–44, 46
Brightfield microscopy (BF) ...................................... 20
Buffy coat ........................... 56, 58, 131–132, 141, 307

## C

Caco-2. *See* Cells
Calmodulin ............................................. 91, 100, 107
Capillari micropipettes ..................... 112, 117, 120–121
Carbenicillin ....................................................... 19–20
CD3 ..... 57–58, 62, 131–132, 244–246, 249, 251, 256, 258–259
CD19 cells. *See* Cells
CD4 T lymphocytes. *See* Cells
CD8 T lymphocytes. *See* Cells
CD28 T lymphocytes. *See* Cells
Cells
  Caco-2 ..................................................... 149, 158
  CD19 cells ........................................................ 225
  CD4 T lymphocytes .... 55, 130, 132, 134–137, 139, 141–142, 307, 313
  CD28 T lymphocytes ............................. 55, 57–58, 62, 130–131, 251, 256
  *Chlamydomonas* cell lines ........... 112–113, 116, 126
  Expi293 ......................................................... 24, 33
  HEK293 .................... 18, 34, 59, 63, 150, 152, 167
  Jurkat E6-1 ........................................... 59, 63, 255
  MDCKII ........... 149–150, 155, 158–159, 167–168
  MEC1 .............................. 225, 230, 232–234, 236
  RPE1 .........149–150, 153–155, 160–161, 165–167
  Sf9 ............... 40, 43–44, 47–50, 88–89, 91, 93, 95, 106–107
Centrosome ........................................... 55–56, 59–64, 134, 143, 148–149, 151, 155–156, 160–166, 168–169, 248–249, 255, 259–261, 302, 313
Chamber ........ 23, 46, 80, 82, 127, 132–133, 141–142, 158–159, 165, 168, 189–192, 201, 226, 231, 259, 264, 306, 311, 317
Chimera ................................................................ 297
*Chlamydomonas*. *See* Cells
Chloramphenicol, selectable marker ..................... 19–20
Chromatography
  affinity ........... 2, 4–5, 8–10, 12, 14, 18, 40, 84, 199
  FPLC ........................................... 2, 14–15, 90, 97
  Ni-NTA column .................................... 96–97, 198
  size exclusion ........................... 2, 4–5, 9–10, 12, 14
  Strep-Tactin XT column ..................... 19, 21, 31, 42
Classification .................. 266, 270–273, 286, 289–291, 295–297
Closed conformation ............. 264–265, 267–268, 271, 275–276, 278, 296, 302
Computational ....................................... 292, 350, 390
Confocal. *See* Microscopy
Contraction ..................................... 77, 142, 174, 205
Coverslip ........... 42, 46, 70–73, 90, 99–101, 113–114, 117, 119–127, 130, 142–143, 150–151, 154, 156–162, 167–169, 190, 192, 207, 213, 219,



Noa B. Martín-Cófreces et al. (eds.), *The Cytoskeleton*, Methods in Molecular Biology, vol. 3070,
https://doi.org/10.1007/978-1-0716-5515-3,

225–226, 230–232, 237–238, 250, 254–255, 310–312, 317–318
Cryo-electron microscopy (cryo-EM). *See* Microscopy, *See* Microscopy
Cryo-electron tomography (cryo-ET). *See* Microscopy
Cyst.........................................149, 155, 158–160, 168
Cytoskeleton.........................................39, 48, 78, 84, 107, 129–131, 134, 140–143, 148, 158, 168, 173–176, 181, 189, 205, 209, 225, 249, 302, 312, 320, 332, 377–378
Cytosolic chaperonin containing TCP-1 (CCT). *See* TRiC

## D

Data.......46, 72–73, 99, 102, 107, 116, 123, 125, 134, 136–137, 139, 162, 165, 168, 215–217, 235, 257, 270–271, 277, 284, 289, 294, 309, 339, 341–342, 344, 349–350, 353–364, 366–367, 376, 378–379, 383–386, 388, 390–391
Database.................................174, 184, 277, 294, 339, 350, 353, 374, 378, 384, 390
Dialysis.......................................2, 7–8, 13–14, 89, 95
Differential expression.............366, 368–369, 375–376, 379, 386
DNase I.........................................21, 30, 79, 81–82
Doxycycline monohydrate.........................................21
Drying chamber.........................................23, 226
Dwell time.........................................195, 235
Dynamics.............................18, 39, 48, 50, 56, 67–68, 77–78, 115–116, 123, 129–130, 134, 137–141, 143–144, 147–148, 164–165, 170, 173–175, 181, 183, 225, 238, 243, 245, 248, 284, 302, 312, 319, 349, 378
Dynein...111, 124, 206, 208, 210, 215, 217, 243, 246, 248, 250, 255, 260

## E

Electron microscopy. *See* Microscopy
Electrophoresis
  gradient SDS-PAGE.........................................23, 32
  native PAGE.............................266, 278, 306, 310
  SDS-PAGE.........................................11, 80, 82, 84
Electroporation.............................140, 250, 257, 303, 306–307, 313–314
End binding 1 (EB1).............................144, 151, 156, 161–166, 169–170
Enrichment.......33, 59, 61, 63–64, 238, 255, 257, 344, 369–372, 374–376, 378, 383–384, 390
Essential light chain (ELC). *See* Myosin light chain (MLC)
Expi293. *See* Cells
Ethane.........................................43, 46, 51, 269, 286
Expression profiling.........................................362

## F

FIB-SEM. *See* Microscopy
Filamentous actin (F-actin).............77–78, 87, 98, 109, 168, 200
Folding.......................1–2, 18, 78, 264, 284, 301–302
Freezing.........................................20, 26, 43, 51, 108

## G

G418. *See* Selection marker
Gene targeting.........................................249
Gentamycin. *See* Selectable marker
Globular actin (G-actin)..............................78, 90, 107
Glutathione-s-transferase (GST). *See* Selectable marker
Gluthatione.........................................89, 95
Gold nanobeads.........................................224–228
Golgi apparatus (GA).............148–149, 151, 155–156, 158, 160–161, 163–166, 168–170
Green fluorescent protein (GFP). *See* Selectable marker
Grid.................2, 11, 32–33, 43, 46, 48, 51, 194, 203, 268–269, 279, 285–287, 294, 297, 351
Guides.........................................149, 230

## H

HEK293 *See* Cells

## I

Image J...46, 69, 84, 91, 102, 123, 132, 134, 137, 139, 154, 162, 164, 166, 170, 210, 214, 216, 235, 249, 254–255, 260, 318
Imidazole....................................79, 82, 90, 106, 186
Immunoblot.........................................81–82, 84, 256
Immunofluorescence......................................... 69–70, 84, 148, 151, 155, 174, 178, 189, 219, 245–246, 248–249, 305, 310, 320
Immunological synapse (IS).............55, 129–130, 135, 142–144, 244–245, 259–260, 313
Infection.......................................44, 47–49, 153–154
Insect cells. *See* Protein purification
Interactome.........................................378, 390
Intraflagellar transport (IFT)...........111–118, 121–124, 126–127, 243, 245, 249–250, 256–257
iSanXoT.........................332, 334, 341–342, 344–345
*In vitro* translation.........................................83–84

## J

Jasplakinolide.........................................68, 200
Jurkat E6-1. *See* Cells

## K

Kanamycin. *See* Selectable marker
Kinesin.........................................111, 243, 302

## L

*Lacrymaria* ..................... 174, 181–182, 184, 189, 198
Latrunculin ..................... 68–69, 71, 178
Lattice ..................... 40, 123
Lentivirus ..................... 151–152, 154
Liquid nitrogen ..................... 27, 30, 43–46, 57, 61, 64, 93, 98, 154, 199, 256, 269, 287
Luria Broth Base ..................... 19

## M

Mammalian cells. *See* Protein purification ..................... 206
Mannose-6-phosphate receptor (MPR) ..................... 246–249, 254–255, 260–261
Mass-spectrometry (MS). 2, 56, 61, 174, 176, 181–184, 199, 332, 339, 343, 349–350, 353–354, 360
MDCKII. *See* Cells
MEC1. *See* Cells
Microinjection ..................... 114, 118–121
Micromanipulator ..................... 118–122, 125
Microscopy
  brightfield microscopy ..................... 20, 107, 134
  confocal laser scanning microscopy ..................... 159
  cryo-electron microscopy ..................... 2, 40, 43, 46, 48, 50–51, 61, 87–88, 263–266, 268, 271, 275–276, 283–287, 289, 293
  cryo-electron tomography ..................... 62
  epifluorescence microscopy ..................... 74, 306, 312, 318
  focused ion beam scanning electron microscopy ..................... 174–175, 180, 195
  scanning electron microscopy ..................... 175
  spinning disk confocal microscope ..................... 154, 165–166, 249
  stimulation emission depletion microscopy ..................... 193, 223–239, 320
  total internal reflection microscopy ..................... 88, 90–91, 99–100, 113, 119, 130, 133–134, 137–138, 141–143
  transmission electron microscopy ..................... 11, 32–33, 174–175, 180, 193, 195, 265, 268
Microtubule (MT) ..................... 39–40, 42–43, 45–46, 48, 50, 56, 61, 111–114, 116–118, 120, 122–123, 125–126, 131, 133, 134, 137, 139, 143–144, 147–149, 154–156, 160–166, 168–169, 173–175, 205–206, 224, 243, 248, 302, 313, 317
Microtubule-associated protein (MAP) ..................... 123, 144
Microtubule-organizing center (MTOC) ..................... 39–40, 55, 148
Monomeric cherry fluorescent protein (mCherry). *See* Selectable marker
Motility ..................... 40, 67, 77, 88, 90–91, 99–100, 111–115, 121, 123–124, 147, 173, 175
Motor proteins ..................... 205–206, 208
Mounting media ..................... 151, 210, 213, 218, 229
Myosin ..... 67, 87–88, 91, 96, 100–101, 106, 108–109, 174, 184, 186–188, 198–199, 228, 230–234
Myosin light chain (MLC). *See* Regulatory light chain (RLC), *See* Essential light chain (ELC)
Myosin light chain kinase (MLCK) 69, 72, 91, 100, 107

## N

N-ethylmaleimide (NEM) tubulin ..................... 117–118
Neo. *See* Selectable marker
Neuromuscular junction (NMJ) ..................... 205–206, 208–210, 213, 217, 219
Nitrosylation ..................... 78
Nocodazole ..................... 56, 151, 161
Normocin. *See* Selectable marker
Nucleation ..................... 39–40, 42, 45, 48, 50, 148–149, 156, 160, 162–166, 169
Nucleofection ..................... 132–133, 140–141, 251, 304, 307, 314
Nucleoskeleton ..................... 69
Nucleus ..................... 67–69, 72–73, 143, 151, 155–156, 163, 302, 305, 318
Nyquist ..................... 213, 219, 235, 239, 273, 292

## O

Open conformation ..................... 272, 274–275, 284–285, 292, 296
Overexpression ..................... 68, 151

## P

Particle ..................... 50–51, 125, 136–137, 144, 152–153, 266, 269–274, 277, 279, 286–292, 294–297, 313
Pathway ..................... 129–130, 247, 249, 259, 350, 353, 369–371, 374–378, 383–384, 390
Pericentriolar matrix (PCM) ..................... 55, 148, 161–162
Phenotype ..................... 302, 349
PhLP1-TST ..................... 20–21, 24–29, 31, 33–34
Photobleaching ..................... 134, 143, 165, 224, 231–232, 237, 239, 317, 319
Photostable ..................... 224, 237
Pinhole ..................... 226, 228, 239, 320
Pixel ..................... 47, 51, 72, 102, 125, 136–137, 139, 144–145, 195, 213–214, 219, 226, 235, 239, 269–270, 273, 278–279, 287–288, 290, 295, 318–319
Plasmid ..................... 19–20, 24–28, 33, 69, 79–80, 84, 93–94, 106, 108, 130–131, 141, 151–152, 184, 187, 257
Polyethylenimine (PEI) ..................... 21, 27, 41, 43, 47–48, 106

Polymerase chain reaction (PCR) ....................... 20, 26, 42, 45–46, 94–95, 98, 100, 108, 256, 306, 315
Post-processing ............................................ 124, 139, 145, 229, 274, 318
Post-synaptic membrane ....................................... 206, 210, 214–216
Post-translational modification (PTM) ..................... 56, 63, 78, 112, 123, 302, 331
Profilin-1 (PFN1) .................................. 78–79, 82–84
Protein production
  *E. coli* ......................................................... 78, 81, 88–89, 94, 107, 184, 187
  *in vitro* ................................................ 42, 81, 112
  mammalian cells (CD4 T cells, Jurkat cells) ........................................................... 206
  *S.pombe* .............................................. 88, 105, 107
  *Spodoptera frugiperda*: insect cells (Sf9) .............................................................. 88
Protein purification
  bacteria ....................................................... 88, 94
  insect cells ........................................ 18, 40–41, 44
  mammalian cells ........................................ 206, 302
  tissues ............................................. 2, 13, 265, 285
  yeast ....................................................... 17–18, 78
Proteome ........................ 331–332, 338, 341, 349, 389
Proteomics ................................................. 55–56, 61, 63–64, 198, 332–333, 341, 349–350, 353–354, 360, 366, 369–370, 374, 378–379, 383, 385, 388–389
Puromycin. *See* Selectable marker

## Q

Quantification ............................................... 2, 80, 84, 91, 102, 113, 162, 174, 212–213, 216, 248–249, 260, 309, 319, 338, 341, 343–344, 349, 353–357, 362, 368, 384, 388
Quantitative ............................................... 40, 84, 130, 149, 162–163, 169, 319, 332, 339, 349–350, 353–354, 377, 379, 383, 385, 391

## R

Recombinant .................................. 41, 56, 78, 80–84, 131, 133, 142, 152, 175, 184–188, 198, 249, 251, 308
Region of interest (ROI) .......................................... 72, 123, 134–136, 144, 162–164, 170, 214–215
Regulatory light chain (RLC). *See* Myosin light chain
R environment ....................................................... 350
Retrograde flow ............................. 131, 133–134, 136
Retrograde traffic .................................... 247–248, 254
RPE1. *See* Cells

## S

Scanning electron microscopy (SEM). *See* Microscopy
Scipion ................................................ 267, 269–274, 278–279, 285, 287, 289–290, 297
Segmentation ......................................................... 169
Selection marker tag
  ampicillin ...................... 89, 94, 177, 184–185, 196
  doxycycline monohydrate .................................. 21
  G418 ............................................................... 257
  gentamycin ..................................................... 89, 94
  GFP ........................................ 127, 131, 133–134, 137, 139, 141, 143–144, 208–209, 211, 249, 252, 256
  GST ................................................. 95, 174–175, 177, 187–188, 196, 199–200
  6xHIS ................................................... 79, 82, 84
  kanamycin ..................................................... 89, 94
  mCherry ............................................... 79, 81–85, 131, 133–137, 141, 143, 208
  Neo .................................................................... 83
  normocin .................................................... 21, 28
  puromycin ....................................... 21, 27–29, 33
  tetracyclin .................................................... 89, 94
  X-gal ........................................................... 89, 94
Sf9. *See* Cells
Sleeping Beauty transposon system ..................... 18, 26
siRNA ................................................ 169, 249, 251, 302–303, 307, 313
small interfering RNA. *See* siRNA
Statistics ........................................................ 114–115, 140, 240, 354, 367, 370, 374–375
STED. *See* Microscopy
Streptavidin .......................................................... 314
Subcellular ............................................. 32, 167, 196, 331–334, 338, 340–342, 344, 370
Sucrose gradient ...................................... 4, 9, 12–14, 42, 45, 49, 59–60
Super-resolution microscopy ................................. 130, 142, 224, 233

## T

T cell receptor (TCR) ................ 55, 58, 142, 244–246, 252, 256, 258–259
Taxol .................................................................. 118
TEM. *See* Microscopy
Tetracyclin. *See* Selectable marker
TIFF ..................................................... 162, 164, 166
Tissues. *See* Protein purification
Total internal reflection microscopy (TIRFm). *See* Microscopy
Trans-Golgi (TGN) ............................................... 163, 247–248, 260
Transcription ............................. 78, 83, 302, 309, 315
Transfection .............................. 21, 27, 33, 41, 48, 73, 89, 95, 131–133, 140–141, 143, 152–153, 249, 251–252, 256–257, 303, 307–308, 313

Translation ............................78–81, 83, 151, 302, 331
Transport......................................................40, 55, 77, 111, 147, 173, 175, 243, 246–247, 249, 260, 371–372, 383
Transposon.................................................................33
TRiC (TCP-1 ring complex). *See* CCT
Tubulin .....................................................1–2, 18, 40, 42, 45–46, 48, 50, 57, 61, 64, 117–118, 129–130, 141, 143–144, 151, 155, 159–162, 168–169, 192, 225, 248, 250, 255, 260, 264–265, 302, 305, 313, 317
γ-tubulin ring complex (γTuRC)..........................40–51

## U

Ultracentrifuge ..................................2, 4, 7–8, 42, 45, 56, 59–60, 152, 306, 313

## V

Vectors ................20, 25–26, 33, 83–84, 93, 106, 108, 184, 187, 374, 376
Vitrification ................................................................... 43, 48, 50–51, 265–267, 269, 277, 285–287

## W

Western blot ...............................34, 42, 45, 49, 57, 61, 63–64, 80–81, 84, 174, 177, 186, 309

## X

X-Gal staining. *See* Selectable marker

## Y

Yeast. *See* Protein purification

FSC
www.fsc.org
MIX
Papier | Fördert
gute Waldnutzung
FSC® C083411

Kolibri 360 GmbH
Ferdinand-Jühlke-Straße 7
99095 Erfurt, Deutschland
produktsicherheit@kolibri360.de